WHEN TECHNOLOGY FAILS

When TECHNOLOGY Fails

Significant Technological Disasters, Accidents, and Failures of the Twentieth Century

Edited by NEIL SCHLAGER

Foreword by HENRY PETROSKI

Gale Research Inc. • DETROIT • WASHINGTON, D.C. • LONDON

Editor
NEIL SCHLAGER

Contributing Editor
LEONARD C. BRUNO

Staff Editors
CHRISTINE JERYAN
KYUNG-SUN LIM
KIMBERLEY A. McGRATH
BRIDGET TRAVERS
ROBYN V. YOUNG

Staff Associate Editors
JEFFERY CHAPMAN
MEGGIN M. CONDINO
THOMAS LIGOTTI
JEFFREY MUHR
SEAN R. POLLOCK

Staff Assistant Editors
CHRISTINE M. BICHLER
NANCY G. DZIEDZIC
IAN A. GOODHALL
MARGARET HAERENS

Research Manager
VICTORIA CARIAPPA

Research Supervisor
MAUREEN RICHARDS

Research Associates
BOB LAZICH
MARY BETH McELMEEL
DONNA
MELNYCHENKO
TAMARA NOTT

Research Assistants
KAREN FARRELLY
KELLY HILL
JULIE LEONARD

Production Director
MARY BETH TRIMPER

Production Assistant
SHANNA HEILVEIL

Art Director
CYNTHIA BALDWIN

Cover and Interior Design
MARK C. HOWELL

Cover photos
AP/WIDE WORLD PHOTOS

Cover illustration
NASA-COMMISSION
REPORT, VOL. 2, P. L-40

ISBN 0-8103-8908-8

Printed in the United States of America by Gale Research Inc.
Published simultaneously in the United Kingdom by Gale Research International Limited
(An affiliated company of Gale Research Inc.)

I(T)P™
The trademark ITP is used under license.

Library of Congress Cataloging-in-Publication Data

When technology fails: significant technological disasters, accidents, and failures of the twentieth century / edited by Neil Schlager; foreword by Henry Petroski.
p. cm.
Includes bibliographical references and index.
ISBN 0-8103-8908-8 (alk. paper): $59.95
1. System failures (Engineering)—Case studies. 2. Disasters—Case studies. I. Schlager, Neil, 1966-.
TA169.5.W44 1944
363.1—dc20 93-41874
CIP

10 9 8 7 6 5 4 3 2 1

Contents

Chemical and Environmental Disasters

Preface

Overview

Webster's Dictionary defines technology as "the totality of the means employed to provide objects necessary for human sustenance and comfort." These objects can take the form of small, simple items or large, complex systems, but regardless of the form, failure is always a possibility. As Henry Petroski points out in his eloquent foreword, technological failure can result in something as relatively innocuous as a traffic jam or as devastating as a bridge collapse that claims many lives or a spaceship explosion that stuns an entire nation. The impact of a spectacular failure often transcends its immediate consequences: at the same time the public mourns the victims of the disaster, it may begin to doubt the safety of the technological systems that provide the very foundation of modern society.

It is these large breakdowns of technology that are the focus of this book. *When Technology Fails: Significant Technological Disasters, Accidents, and Failures of the Twentieth Century* includes more than 100 such breakdowns from a variety of fields, ranging from transportation to civil engineering to the environment. While the events chronicled in this book had a variety of negative consequences, they also often led to improved standards and safer, better technological systems. *When Technology Fails* explores not just the tragedy inherent in the events but the valuable lessons learned—albeit the hard way—from them.

Scope

As the subtitle explains, *When Technology Fails* covers significant technological disasters from the twentieth century. What constitutes a "technological" disaster? Using *Webster's Dictionary* as our guide, we included some areas that most would agree fall under the heading "technological" (e.g. buildings and spaceships) and others that might take readers by surprise (e.g. medical disasters). This book does not include purely natural disasters such as hurricanes and earthquakes, although natural phenomena did play a role in some of the events included here, such as the Vaiont landslide disaster in Italy in 1963.

The qualifier "significant" is also problematic for its vagueness. Admittedly, we did not apply the term evenly to every event included in this collection. Some events were chosen because they resulted in large losses of life, while others were selected for the extensive press coverage they received and the impact they had on public opinion. Still others in this book led to no loss of life and received little press coverage, but were included because they offer valuable lessons to readers.

In general, the failures included in *When Technology Fails* were caused by poor design, planning, testing, or construction rather than simple human error. Here again, though, exceptions can be found, since many disasters were caused by a combination of the above-mentioned factors. A primary example is the Exxon *Valdez* oil spill in 1989. While human error—including alcohol use and fatigue—was the immediate cause of the accident, technical elements such as the ship's single- rather than double-hull construction and the failure of oil containment booms both contributed heavily to the magnitude of the disaster.

Finally, events caused by deliberate human actions—such as terrorist bombings—are not included here for the most part. Once again, there are exceptions, particularly in the chemical and environmental area, which includes several disasters (e.g. Agent Orange, Minamata) that many feel were the result of deliberate, cruel actions by governments or corporations.

Organization

The topic entries covered in *When Technology Fails* are arranged chronologically within broad subject categories. These categories, in turn, are presented in the following alphabetical order:

- Aircraft
- Airships
- Automobiles
- Bridges
- Buildings and Other Structures
- Chemical and Environmental Disasters
- Dams
- Medical Disasters
- Nuclear Plants
- Ships
- Spacecraft
- Submarines

Within each topic entry, information is divided into standard sections. These sections generally fall into the following pattern:

- Background

- Details of the Disaster
- Impact
- Where to Learn More

To find specific topics, readers can go to two places: the **descriptive table of contents** at the front of the book and the **subject index** at the back. In addition to listing the events included in the book, the subject index lists key terms, concepts, and people.

Other sections of interest to the reader include a **chronological list of events**, which can be found at the front of the book (immediately preceding the main text), and a **bibliography**, located at the back of the book (immediately preceding the index). This bibliography does not repeat the books and periodicals listed in the "Where to Learn More" sections in each entry. Rather, it includes general works about disasters and technology.

Contributors/ Advisors

The entries in *When Technology Fails* were written by a skilled team of **expert contributors**, including academics, journalists, engineers, and technical writers. The "Contributors" section in the front of the book lists the accomplishments and qualifications of these writers.

In addition, the following individuals assisted in the formulation of *When Technology Fails*:

George M. A. Cumming, Jr.
Librarian
Science Reference Department
Boston Public Library

Larayne J. Dallas
Assistant Engineering Librarian
McKinney Engineering Library
University of Texas at Austin

John Loss, FAIA
Architect, Whitehall, Michigan
Former Professor of Architecture and Executive Director, Architecture and Engineering Performance Information Center
University of Maryland at College Park

The editor would also like to thank **Henry Petroski**, who composed the book's foreword, and contributing editor **Leonard C. Bruno**, who performed a myriad of duties, from providing research to writing many entries in the airship, spaceship, and submarine sections.

Suggestions

Comments, questions, and suggestions are welcome. Please send all such correspondence to:

The Editor
When Technology Fails
Gale Research Inc.
835 Penobscot Bldg.
Detroit, MI 48226

Contributors

Henry Petroski, who wrote the Foreword, is Aleksandar S. Vesic Professor of Civil Engineering at Duke University and author of *To Engineer Is Human, The Pencil,* and *The Evolution of Useful Things.* His book of historical case studies of failure will be published by Cambridge University Press in 1994.

Contributing Editor **Leonard C. Bruno** is Senior Science Specialist in the Science and Technology Division at the Library of Congress. He has written several books on science and technology, including *The Landmarks of Science* (Facts on File, 1989), *On the Move* (Gale Research, 1993), and *The Tradition of Technology* (Library of Congress, 1994).

Byron Acohido is a journalist at the *Seattle Times* specializing in aerospace-related news. He received the 1993 Premier Award of Excellence for Aviation Safety Coverage from the Aviation/Space Writers Association for "Jet Engine Pins: How Big a Risk" and "Flight 811, Terror in the Sky."

Leo Argiris is a structural engineer and associate in the engineering consulting firm of Weiskopf & Pickworth in New York City.

Linda Buchaca is a civil engineer specializing in architectural and structural design. She is currently a consultant in Sunnyvale, California, providing plan review and management of unreinforced masonry hazard mitigation programs.

Liane Clorfene-Casten is a free-lance writer who has written extensively about environmental topics. She is the author of the book *Lies and Contempt: Dioxin and Agent Orange,* as well as articles that have appeared in the *Chicago Sun-Times,* the *Chicago Tribune, Mother Jones, Ms.,* and the *Nation.*

Janice Knestout Colvin is a pilot and aviation writer who has contributed to various publications, including *Air Safety Week*, *Private Pilot*, *IFR Refresher*, and *Career Pilot*. She was formerly the editor of *AOPA Airports USA* and *AOPA Handbook for Pilots*, published by the Aircraft Owners and Pilots Association.

Deborah A. Day is an archivist at Scripps Institution of Oceanography/UCSD.

Robert W. Day is Chief Geotechnical Engineer at American Geotechnical in California and the author of numerous articles on soil and clay properties.

Frank M. D'Itri is Associate Director at the Institute of Water Research and Professor in the Department of Fisheries and Wildlife at Michigan State University. He has written numerous articles about environmental and water-related topics.

Virginia Kent Dorris is a free-lance writer specializing in construction design and development. Between 1989 and 1992, she was an assistant editor for the construction industry trade weekly *Engineering News-Record*. She has also contributed to *Architecture*, *Infrastructure Finance*, and *Civil Engineering*.

David J. Elton is Associate Professor of Civil Engineering at Auburn University in Alabama. He has contributed to *Journal of Geotechnical Engineering*, *Journal of Civil Engineering Systems*, and *Earthquake Spectra*.

James M. Flammang is a free-lance writer specializing in automotive topics, including the history and technology of modern and classic cars. His books include *Understanding Automotive Specifications & Data* (TAB Books, 1986) and *Great Book of Dream Cars* (Publications International, 1990).

David N. Ford is an instructor in the Department of Civil and Environmental Engineering at Massachusetts Institute of Technology. He has extensive experience in designing and managing building expansion and renovation projects.

Norman J. Glover is the managing director of International Consultants & Engineers in New York. He has written numerous technical articles about engineering topics.

Neal J. Gruber is a civil engineer affiliated with Dames & Moore in Cincinnati, Ohio. He has extensive experience in the areas of hydrology, hydraulics, hydropower, coastal engineering, and environmental engineering. His work has recently appeared in *Hydro Review*.

Loretta Hall is a free-lance writer and regular contributor to *Southwest Contractor*. She has written several articles in the areas of construction design and techniques.

Gillian S. Holmes is the founder and principal engineer of Geotechnica, a consulting firm specializing in soil engineering and environmental geotechnologies. She has written extensively about soil and environmental topics.

Lamont Ingalls is a technical writer and researcher. He has done consulting work for the Oak Ridge National Laboratory and Chevron Chemical Corporation, among others.

Greg Janicki is a free-lance writer and syndicated columnist covering developments in the automotive industry. He is the author of *Cars Europe Never Built* and the editor of *Cars Detroit Never Built*, both available from Sterling Publishing.

Julian Larson is a free-lance writer and indexer. He has worked extensively on topics related to ground water and hazardous waste issues.

W. H. Marianos, Jr., is a consulting civil engineer specializing in the design and repair of bridges. He has taught structural engineering at Tulane University and has contributed articles on civil engineering topics to *Journal of Professional Issues in Engineering* and *Concrete International*.

Robert Mark is a journalist, teacher, and pilot who has written for magazines including the *Chicago Tribune Magazine*, *Family Circle*, and *Airline Pilot*. He writes a monthly column for *General Aviation News and Flyer* and is the author of *The Joy of Flying*, available from McGraw-Hill.

Kirk Ian Mettam is a structrual engineer and an associate at Skidmore, Owings, and Merrill in New York City, where he heads the Structural Engineering efforts for SOM's New York office. He has played a key role in the design of such New York City structures as 461 Fifth Avenue, St. Lukes/ Roosevelt Hospitals, and the recently completed Tribeca Bridge.

Patrick Moore is a professor at the University of Arkansas, Little Rock, and a technical writer. His recent publications include "Intimidation and Communication: A Case Study of the *Challenger* Accident" and "When Politeness Is Fatal: Technical Communication and the *Challenger* Accident," both for the *Journal of Business and Technical Communication*.

David E. Newton was formerly Professor of Chemistry at Salem (Mass.) State College and Adjunct Professor of Professional Studies at the University of San Francisco. He is the author of more than 50 books, including *Taking a Stand Against Environmental Pollution* (Watts, 1990),

Environmental Chemistry (J. Weston Walch, 1991), and *Global Warming* (ABC-Clio, 1993).

M. Kevin Parfitt is a professor in the Architectural Engineering Department at Pennsylvania State University. He has written numerous articles on architectural and engineering topics.

Jeffrey R. Raines is a novelist and civil engineer with GeoSyntec Consultants in California. He has been active in the design, analysis, and construction of nuclear power plants as well as in the rehabilitation of port and highway structures.

Lewis G. Regenstein is president of The Interfaith Council for the Protection of Animals and Nature, located in Atlanta, and is the author of *Cleaning Up America the Poisoned* (Acropolis Books, 1993).

L. Carol Ritchie is a journalist who has written extensively about environmental topics.

Rita Robison is a free-lance writer specializing in issues related to civil engineering. Formerly, she worked for *Civil Engineering Magazine*, *Progressive Architecture*, and *Architectural & Engineering News*.

Robert J. Serling is an accomplished author of over 20 books and the recipient of several Best Non-Fiction awards from the Aviation and Space Writers Association. His areas of interest include airline histories and air safety.

William B. Shapbell, Jr., is a launch and flight systems manager at the National Aeronautics and Space Administration's Kennedy Space Center. He has participated in launch operations management of all manned spaceflight programs at NASA since the Gemini program.

Joel Simon is a regular contributor to Copley News Service and has published several articles in their *Car Care* and *AutoWord* columns. He has also worked as a training specialist in the automotive industry with such clients as Isuzu Motors and Ford Motor Company.

L. A. Turner is a free-lance journalist who has contributed to *Airport Support*, an international monthly magazine covering technical aviation topics. Her recent article "Signs of the Times" about air safety was reprinted in *Inter Pilot*.

Sally Van Duyne is a free-lance writer who has written articles on various topics, including psychology, medicine, natural history, and naval engineering.

Richard Weingardt is the president of Richard Weingardt Consultants, a structural engineering firm located in Denver, Colorado, and the Vice President of American Consulting Engineers Council. He lectures internationally and is the author of three books and numerous articles on creativity, leadership, business, and engineering.

Ellen Williams is a consultant in the areas of public policy, health care, medicine, and high technology. She has worked within both the FDA and the pharmaceutical industry, and has also taught science writing at the Massachusetts Institute of Technology.

Erin A. Zeno is the communications manager in the engineering consulting firm of Weiskopf & Pickworth in New York City.

Acknowledgments

Photographs and illustrations appearing in *When Technology Fails* were received from the following sources:

The Bettmann Archive: **pp. 7, 12, 35, 37, 41, 68, 117, 136, 146, 166, 182, 542, 544, 549;** AP/Wide World Photos: **pp. 16, 85, 89, 97, 98, 106, 112, 119, 124, 152, 158, 160, 189, 238, 293, 304, 310, 324, 329, 356, 359, 393, 399, 429, 444, 471, 495, 502, 525, 534, 545, 560, 567, 581, 584, 594, 607, 612, 623, 635;** Illustration by Kevin Woest from *Why Buildings Fall Down,* by Matthys Levy and Mario Salvadori. Illustrations copyright © 1992 by Kevin Woest. Reprinted by permission of W. W. Norton & Company, Inc.: **pp. 31, 206;** Reuters/Bettmann: **pp. 42, 104, 531, 565;** UPI/Bettmann: **pp. 48, 56, 63, 73, 91, 129, 144, 186, 187, 211, 217, 222, 231, 280, 285, 298, 311, 320, 322, 335, 374, 381, 408, 414, 422, 424, 433, 437, 447, 451, 456, 483, 568;** Copyright 1979 Time Inc. Reprinted by permission: **p. 62;** UPI/Bettmann Newsphotos: **pp. 131, 138, 194, 405, 512, 515 554;** Archive Photos/Lambert: **p. 141;** The Gazette: **p. 180;** Boston Globe Photos: **p. 248;** Courtesy of the National Archives Trust Fund Board: **p. 263;** Rex Features Ltd.: **p. 274;** Photograph by Mary Ann Cunningham: **p. 342;** © 1975 Allen & W. Eugene Smith, courtesy of Black Star: **p. 368;** © Peter A. Simon/ Phototake, NYC: **p. 477;** Kyodo Photo Service: **p. 521;** Photograph by V. Cheredintsev/Sovfoto: **p. 588;** NASA: **pp. 596, 613, 615;** Photograph by V. Muaelian/Tass from Sovfoto: **p. 600;** U.S. Navy: **pp. 633, 640.**

Foreword

Technology is supposed to work, and by and large it does. We can drive our automobiles for hundreds of miles on interstate highways without incident and without stopping, passing thousands of automobiles going the other way, also without incident. When our trip is going well, we tend not to marvel at how wonderfully reliable all the speeding machines (and their drivers) are; we tend to enjoy the scenery or listen to some pleasant music. It is only when we encounter something out of the ordinary that we may reflect upon the technology itself.

Once in a while, especially on a hot day, we may see an overheated car on the other side, and this may give us pause. Soon, however, we tend to forget that incident or put it out of our minds, perhaps rationalizing that it was the fault of the driver or of an automobile inferior to ours. On another occasion, we might get caught in stop-and-go traffic and wonder what is going on up ahead. We can look for clues from the traffic on the other side: If it is moving normally, we can guess there is a stalled car or something gone wrong on our side; if the other side is moving even slower than ours, we can guess there is an accident over there. Sometimes both directions are equally at a standstill, and only after hours of waiting and inching ahead do we pass by the scene of a terrible collision. We may wonder if the machines or their operators malfunctioned, but the thought of the carnage causes all of us to drive a bit more slowly and alertly—at least for the next several miles, during which we may reflect upon the chance we take that one of those drivers coming the other way will not lose control and head straight for us. It is not comfortable to drive with such thoughts, and we tend to blot them out, somehow accepting the odds or believing that we can steer our own automobile out of the way of another.

Even when we cannot control it in any sensible way, we tend to take for granted technology of all kinds, as long as it is working. The better and more reliably it works, the more invisible and uninteresting it is to us as we go about our daily lives. There is no front-page news value to the fact that traffic was moving along at the usual clip on the freeway yesterday, or that our water, electric power, and telephone service all remained uninterrupted for the three-hundredth day straight. Such would be a dog-bites-man story, and it does not sell newspapers or cause viewers to tune into the evening news on television.

However, when technology fails, no matter how benignly, we tend to take note, and we may even exaggerate its consequences. It takes just an inexplicable traffic jam encountered on our way to work to make us a bit more somber over coffee that morning, perhaps going on and on about how inadequate the roads are and about how something should be done about them. If we are frequent flyers, even though we may in fact miss an airplane connection only once in dozens of trips, when we talk about it we tend to tell the story as if it happens every other time we fly. But such annoyances and inconveniences can be expected to happen in technological systems that rely not only upon the workings of machines but also upon the vagaries of their human operators and the weather. While such incidents serve as icebreakers for conversation and provide material for stand-up comics on late-night television, they are of little consequence in the whole scheme of things.

On occasion, technology fails in catastrophic ways, and when this happens it is no joking matter. The sudden collapse of a bridge over which hundreds of thousands of cars may travel every day, or the separation of an engine from an airliner during takeoff, can and has resulted in great losses of life and raised fundamental questions about the reliability of technology in which millions of people trust day in and day out. Unlike in the case of the oncoming car, there is little we can do to control our destiny when we find ourselves on a bridge that is collapsing beneath us or when we are a passenger on an airplane that is out of control. It is this sense of helplessness that underlies our demand that some technologies be more reliable than others.

More than technologies that are operated on a personal level, technologies over which ordinary people do not have any direct control and for which the consequences of failure potentially affect great numbers of lives tend to be held to a much higher standard of reliability. Not only is it a remarkable catastrophe when something of some magnitude fails, it also is a man-bites-dog story—precisely because it is such an unusual and unexpected occurrence. The potentially great consequences of a technological failure necessarily causes considerable care to be taken in the design, construction, maintenance, and operation of the artifact in the first place—precisely to obviate any failure from occurring.

Those most directly responsible for conceiving, building, and operating the large and complex technological systems upon which we have come to depend have as their foremost objective the safe and predictable functioning of their design. The whole technological enterprise, including the profit motive, is founded upon an understanding of what could go wrong with a given design, and the proper development of any artifact or system incorporates safeguards to insure that operation takes place within prescribed limits of safety. The long and deliberate conceptualization and detailed design process that precedes the realization of our more complex technologies can include seemingly endless "what-if" questions that engineers are expected to answer to everyone's satisfaction.

When technology developed this way fails, it is more than news to the engineers that designed it, for an unexpected failure calls into question the very process by which they design and develop systems of all kinds. Every failure raises fundamental questions about what went wrong: Was there a physical phenomenon that was unknown or unaccounted for in the design? Were there incorrect assumptions or calculations made? Were faulty materials or manufacturing processes employed? Were there forces applied beyond those anticipated? Was the system abused or improperly maintained? Was the system improperly operated?

When technology works, such questions are naturally assumed to have been answered satisfactorily during the process of design, development, and demonstration of the system. Having passed through a rigorous period of critical testing, a technological system confirms the claims of its designers that it will work as intended. Because it is impossible to test fully something so large as a bridge or a jumbo jet airliner until the artifact is actually built, the successful pursuit of advanced technology depends to a large extent upon the validity of the design process and its assumptions. These assumptions become embodied in the artifacts and systems and become fully tested only with their use.

When technology fails, we are presented with an undesired but incontrovertible test that can provide information heretofore unavailable, and whose study can provide answers to the very questions the failure raises. Because of the often unique insights that can result from a close study of how and why a failure occurred, there is always considerable need to recover and preserve the physical debris of an accident and to interview any witnesses or survivors. In especially tragic failures, such as that of the space shuttle *Challenger*, there is often a highly visible and distinguished committee that conducts the investigation in full public view. It is during such proceedings that the true nature of technology and the technological process are often made clear in stark detail.

The failure case studies presented in this book have the value not only of providing details about the particular incidents described, but also of providing in their collectivity a means of understanding the technological process. Failures generally occur in a climate of technological complacency or hubris: The *Challenger* was launched against the better judgement of some engineers in part because of the robustness that two dozen earlier space shuttle launches had demonstrated. In the wake of a failure, there tends to be renewed conservatism about the technology involved: After *Challenger* exploded, space shuttle flights were suspended for a considerable period so that numerous design and procedural changes could be made.

The success-failure dichotomy exhibited in the case of the *Challenger* is not a new phenomenon. While *When Technology Fails* provides an excellent selection of twentieth-century catastrophes, there have been countless examples throughout the history of technology of catastrophic failures occurring in the wake of prolonged success, and of renewed cautions and new departures occurring in the wake of failures. If we do not become aware of and sensitive to this larger cyclic phenomenon in the technological process, we can expect technology to continue to fail at its present rate.

HENRY PETROSKI

Chronological List of Events

	1883 to present	Sudbury, Ontario smelter poisoning
	1907, 1916	Quebec Bridge failures
30	September 1911	Austin Dam failure
14	April 1912	*Titanic* ship disaster
	1912-1955	Toyama, Japan chemical disaster
15	January 1919	Boston Molasses tank rupture
24	August 1921	*ZR-2* (*R-38*) airship crash
28	January 1922	Knickerbocker Theatre roof collapse
21	February 1922	*Roma* airship crash
3	September 1925	*Shenandoah* airship crash
12	March 1928	St. Francis Dam failure
5	October 1930	*R-101* airship crash
3	April 1933	*Akron* airship crash
19	June 1936	Bronx apartment house collapse
6	May 1937	*Hindenburg* airship disaster
23	May 1939	*Squalus* submarine accident
1	June 1939	*Thetis* submarine accident
	1939 to present	DDT pesticide disaster
7	November 1940	Tacoma Narrows Bridge failure
	mid-1940s	Liberty ship breakups

		1942-1953	Love Canal, New York hazardous waste contamination
		1944-1988	Hanford nuclear reservation waste disaster
11	**July**	**1946**	TWA Constellation aircraft crash
30	**May**	**1947**	Eastern Airlines DC-4 aircraft crash
24	**October**	**1947**	United Airlines DC-6 aircraft crash
		1940s-1960s	In-flight aircraft propeller reversals
		1940s-1970s	DES drug disaster
		1954	Comet aircraft crashes
		1955 to present	Minamata, Japan mercury poisoning
30	**June**	**1956**	TWA-United aircraft collision, Grand Canyon
25	**July**	**1956**	*Andrea Doria-Stockholm* ship collision
29	**September**	**1957**	Kyshtym nuclear plant accident
10	**October**	**1957**	Windscale nuclear plant accident
22	**May**	**1958**	Nike missile explosions
17	**June**	**1958**	Vancouver Second Narrows Bridge failure
2	**December**	**1959**	Malpasset Dam failure
		1959-1960	Lockheed Electra aircraft crashes
		1959-1963	Chevrolet Corvair automobile design controversy
		1950s-1960s	Thalidomide drug disaster
24	**October**	**1960**	Soviet rocket fire
16	**December**	**1960**	TWA-United aircraft collision, Staten Island
3	**January**	**1961**	SL-1 nuclear plant accident
15	**January**	**1961**	Texas Tower radar station sinking
		1961-1971	Agent Orange herbicide disaster
14	**December**	**1962**	King's Bridge failure
10	**April**	**1963**	*Thresher* submarine accident
9	**October**	**1963**	Vaiont Dam failure
14	**December**	**1963**	Baldwin Hills Dam failure
1	**November**	**1965**	Ferrybridge cooling towers collapse
27	**January**	**1967**	*Apollo 1* spacecraft accident

24	**April**	**1967**	*Soyuz 1* spacecraft disaster
15	**December**	**1967**	Point Pleasant Bridge failure
16	**May**	**1968**	Ronan Point apartment building collapse
27	**May**	**1968**	*Scorpion* submarine accident
		1960s-1990s	Silicone-gel breast implant disaster
13	**April**	**1970**	*Apollo 13* spacecraft accident
15	**October**	**1970**	West Gate Bridge failure
30	**June**	**1971**	*Soyuz 11* spacecraft accident
		1971	Times Beach, Missouri dioxin contamination
		1971-1972	Methylmercury seed dressing poisoning
		1971-1976	Ford Pinto automobile gas tank controversy
		1972-1978	Firestone 500 tire recall
2	**March**	**1973**	Skyline Plaza building collapse
14	**May**	**1973**	*Skylab* meteoroid shield failure
3	**March**	**1974**	Turkish Airlines DC-10 aircraft crash
1	**June**	**1974**	Flixborough, England chemical plant explosion
22	**March**	**1975**	Browns Ferry nuclear plant accident
5	**June**	**1976**	Teton Dam failure
10	**July**	**1976**	Seveso, Italy dioxin leak
18	**January**	**1978**	Hartford Coliseum roof collapse
16	**March**	**1978**	Amoco *Cadiz* oil tanker accident
27	**April**	**1978**	Willow Island cooling tower collapse
		1978-1986	Audi 5000 automobile sudden acceleration controversy
28	**March**	**1979**	Three Mile Island nuclear plant accident
25	**May**	**1979**	American Airlines DC-10 aircraft crash
4	**June**	**1979**	Kemper Arena roof collapse
17	**September**	**1979**	Air Canada DC-9 aircraft accident
		1970s-1980s	Dalkon Shield medical disaster
27	**March**	**1980**	*Alexander Kielland* oil-drilling rig sinking
9	**May**	**1980**	Sunshine Skyway bridge failure
21	**November**	**1980**	MGM Grand Hotel fire

8	**March**	**1981**	Tsuruga nuclear plant accident
27	**March**	**1981**	Harbour Cay condominium collapse
17	**July**	**1981**	Hyatt Regency Hotel walkways collapse
25	**January**	**1982**	Ginna nuclear plant accident
15	**February**	**1982**	*Ocean Ranger* oil-drilling rig sinking
15	**April**	**1982**	East Chicago ramp collapse
	August	**1982**	Zilwaukee Bridge accident
28	**June**	**1983**	Mianus River Bridge failure
3	**December**	**1984**	Bhopal, India chemical vapor leak
19	**July**	**1985**	Stava Dam failure
11	**August**	**1985**	Institute, West Virginia chemical vapor leak
12	**August**	**1985**	Japan Airlines Boeing 747 aircraft crash
6	**September**	**1985**	Midwest Express DC-9 aircraft accident
28	**January**	**1986**	*Challenger* spacecraft explosion
26	**April**	**1986**	Chernobyl nuclear plant accident
5	**April**	**1987**	Schoharie Creek Bridge failure
23	**April**	**1987**	L'Ambiance Plaza apartment building collapse
24	**February**	**1989**	United Airlines Boeing 747 aircraft accident
24	**March**	**1989**	Exxon *Valdez* oil tanker accident
19	**July**	**1989**	United Airlines DC-10 aircraft crash
26	**May**	**1991**	Lauda Air Boeing 767-300 aircraft crash
4	**October**	**1992**	El Al Boeing 747-200 aircraft crash

WHEN TECHNOLOGY FAILS

Aircraft

- TWA Constellation (1946)
- Eastern Airlines DC-4 (1947)
- United Airlines DC-6 (1947)
- In-flight propeller reversals (1940s-1960s)
- Comet (1954)
- TWA-United Grand Canyon collision (1956)
- Lockheed Electra (1959-1960)
- TWA-United Staten Island collision (1960)
- Turkish Airlines DC-10 (1974)
- American Airlines DC-10 (1979)
- Air Canada DC-9 (1979)
- Japan Airlines Boeing 747 (1985)
- Midwest Express DC-9 (1985)
- United Airlines Boeing 747 (1989)
- United Airlines DC-10 (1989)
- Lauda Air Boeing 767-300 (1991)
- El Al Boeing 747-200 (1992)

TWA Constellation crash

Reading, Pennsylvania

(1946)

A tiny electrical stud caused a fatal accident, resulting in the grounding of all Constellations.

Background

On July 11, 1946, a new TWA Lockheed Constellation carrying five crew members on a routine training flight caught fire and crashed while trying to make an emergency landing at the Reading, Pennsylvania, airport. The only survivor was Capt. Dick Brown, a TWA instructor who was giving Capt. Art Nilsen an instrument check ride.

It was the first fatal accident involving the Lockheed Constellation, a transport in whose design none other than Howard Hughes had played a major role. Hughes, who had acquired control of TWA before World War II, regarded the Connie as the key to achieving postwar competitive supremacy. Other carriers had ordered the big Lockheed, but TWA held first delivery rights. Hughes had even decided to change the airline's name from Transcontinental & Western Air, its official corporate designation since 1930, to Trans World Airlines, reflecting its desire to be one of the leading global carriers. The Constellation was to be the chief instrument for such expansion.

Lockheed actually had started developing the four-engine, triple-tailed craft before Pearl Harbor, but TWA couldn't begin accepting deliveries until the war ended. As the first three hundred-mile an hour, fully pressurized transport, the Constellation went into TWA's domestic and international service in February 1946, but the Reading accident cut short its reign as ruler of the world's airways.

Details of the Crash

During the flight, Nilsen had just begun to practice instrument approach procedures when a strong odor, resembling that of burning insulation, infiltrated the cockpit. Brown ordered the flight engineer to check the cabin for the source of the smell. Seconds later, he returned shouting, "The whole cabin's on fire!"

Nilsen and the flight engineer grabbed fire extinguishers, but by now the smoke and heat were so intense that it was impossible to enter the cabin. Another crew member, a student flight engineer, opened the cockpit floor hatch thinking it would draw off smoke. This had the opposite effect, for unknown to the crew, the fire had started in a forward baggage compartment. When the hatch was opened, clouds of thick smoke began entering the cockpit. The instrument panel and windshield were totally obscured, even as Brown frantically tried to get back to the Reading airport. In desperation, he opened a side window, stuck out his head, and tried to line up the aircraft for a landing on a runway he still couldn't see.

Heat and smoke worsened to such an extent that neither Brown nor Nilsen could control the airplane. Brown gave up trying to see outside the side window, pulled in his head, and as a last resort attempted a blind landing that depended entirely on sheer instinct. He might have made it, but as the stricken "Connie" groped toward the airport, it struck two tall power lines and crashed.

What the CAB found produced startling consequences—embarrassing evidence of faulty design, the grounding of an entire airline fleet for only the second time in the history of U.S. commercial aviation, and the eventual firing of a man who was not only a respected and popular airline president, but one of the industry's greatest pioneers.

The eyewitness account from Brown, whose injuries were so severe that an arm had to be amputated, led Civil Aeronautics Board (CAB) investigators to immediately suspect some kind of electrical fire. What they came up with produced some startling consequences—embarrassing evidence of faulty design, the grounding of an entire airline fleet for only the second time in the history of United States commercial aviation, and the eventual firing of a man who was not only a respected and popular airline president, but one of the industry's greatest pioneers.

Just as a microscopic virus can bring death to the human body, the Constellation's nemesis proved to be a small, supposedly insignificant electrical component. It was called a through-stud, a tiny metal conductor that allowed electrical power to be brought through the fuselage of a pressurized aircraft without destroying the pressurization seal.

The through-studs installed on the Constellation, however, were fatally flawed—in fact, the CAB's final report on the Reading accident bluntly labeled their design "crudely deficient" in more than one respect. First, Lockheed had placed the studs so close to the fuselage skin that electrical "arcing" resulted—in simplest terms, the metal skin could attract an electrical charge from the power being transmitted through the studs. And when arcing resulted, both the stud and skin were burned.

This was only the first step toward catastrophe. The next design error was the unwise placement of the studs too close to the high-pressure hydraulic lines. These produced the necessary boost aids for the landing gear and controls—like power steering and brakes on an automobile. Those lines were filled with flammable fluid. If arcing produced enough heat to burn a stud in two, it was only too easy for one end to fall on a hydraulic line and ignite not only a major fire, but one that was uncontrollable unless discovered immediately.

But the main flaw lay in the composition of the studs themselves. The investigators at first couldn't understand why such a simple component was so vulnerable to serious arcing. The answer was equally simple and devastating in its simplicity. The entire stud assembly was composed of a conglomeration of varying materials that invited disaster.

Lugs were made of copper and aluminum. Nuts contained both brass and steel. The bolts were brass. And all of these metals had one thing in common: each had a different rate of expansion under heat. This is what caused the studs to become loose, move toward the fuselage skin, and produce arcing and pitting.

Civil Aeronautics Board technicians inspected every Constellation in service and were shocked to find evidence of arcing in nearly half of them. They also cited a number of additional design deficiencies, such as inadequate fire and smoke detection systems in the forward baggage compartment—this is where the fire had started on the Reading training flight. Nor was there any fire-extinguishing equipment in such remote fuselage areas as baggage holds.

The CAB "probable cause" findings also criticized the inspection procedures of the three airlines then operating Constellations—TWA, Pan American, and American Overseas. More stringent inspections, the CAB charged, might have prevented the tragedy at Reading, for the investigation disclosed there had been previous maintenance reports of leaking hydraulic lines due to faulty studs. The board said these incidents should have alerted the airlines to a serious potential hazard. Merely noticing the worn studs and hydraulic leaks wouldn't have rung any alarm bells, however, because no one realized they were potentially lethal.

The CAB noted the almost total absence of research on the direction of smoke flow from a fuselage fire, and this was an indirect criticism of the Civil Aeronautics Administration (CAA), the board's sister agency, which had certificated the Constellation as completely safe for passenger service. The CAB, however, failed to challenge the overall certification process itself, which was flawed to a far greater extent than the Reading accident demonstrated—as subsequent events involving another new postwar airliner were to reveal.

Impact

Within twenty-four hours of the TWA crash, the CAA—acting on the CAB's recommendation—ordered the grounding of all Constellations until the cause of the in-flight fire could be determined. Once the deficiencies in the electrical system were uncovered, the grounding order stayed in effect until Lockheed had modified every delivered airplane.

TWA, operating the largest Connie fleet, was the hardest hit. Its Constellations stopped flying July 12 and didn't resume service until September 20, a sixty-nine-day period in which the airline's transcontinental and overseas

Tiny electrical studs allowed electrical power to be brought through the fuselage of a Lockheed Constellation without destroying the pressurization seal. However, the studs were placed too close to the metal fuselage, causing electrical arcing. The results, as seen here, were tragic.

schedules were demolished. All it had for its hard-won postwar transatlantic routes was a handful of lumbering, obsolete, unpressurized DC-4s. It was forced to operate transcontinental service with even older and slower DC-3s, and the domestic advantage the Connies had given TWA over American and United evaporated overnight.

The grounding order itself was almost unprecedented. Only once before had a federal air agency taken such drastic action. In 1938, the then Bureau of Air Commerce had grounded all Lockheed L-14s, a twin-engine, twin-tail transport known variously as the Zephyr or Super-Electra, after structural failure of the fins and rudders caused a fatal crash blamed on flutter. Northwest, the principal L-14 operator, was the carrier most seriously affected.

In addition to the devastating effects of the Constellation grounding, TWA was beset by other problems after the Reading crash. Only a month after the grounding order was lifted, TWA's eight hundred pilots staged an ill-timed strike that shut down the airline for another twenty-five days. Also, a bitter power struggle ensued among TWA executives, eventually resulting in the ousting of president Jack Frye. Frye had headed TWA since 1934 and was a dominant, pioneering force in the airline industry. But he couldn't withstand the unfortunate events of 1946—all started by a poorly designed electrical conduit.

ROBERT J. SERLING

Where to Learn More

"Fire in the Air." *Business Week.* July 20, 1946, pp. 17-18.

Graham, Frederick. "Plane Crash Laid to Electric Fire." *New York Times.* August 1, 1946, p. 15.

Graham, Frederick. "Agree Single Part Caused Air Crash." *New York Times.* August 2, 1946, p. 11.

"Assure Air Return for Constellation." *New York Times.* August 3, 1946, p. 17.

"Changes Ordered on Constellation." *New York Times.* July 25, 1946, p. 23.

"Constellation Put on Ground by CAA." *New York Times.* July 12, 1946, p. 1.

"Goliath Aground: Constellation Crash-landed in Connecticut." *Newsweek.* July 1, 1946, p. 19.

"Ill-starred Constellations." *Newsweek.* July 22, 1946, p. 66+.

Eastern Airlines DC-4 crash

Port Deposit, Maryland

(1947)

One of the few "for reasons unknown" crashes in the history of U.S. air accident investigation still has experts puzzled after nearly fifty years.

Background

The clear, sunny Maryland weather on May 30, 1947, was absolutely perfect for the Memorial Day holiday—and for flying. It was 6:45 P.M., with about an hour of daylight left, and approaching the small town of Bainbridge just north of Baltimore at an altitude of four thousand feet was Eastern Flight 605, en route from Newark to Miami. Aboard the four-engine DC-4 were forty-nine passengers including an infant, two pilots, and two flight attendants—all of whom were about to die in one of the most bizarre accidents in aviation history, a crash that remains unexplained even though it was witnessed by two experienced government pilots.

About three miles behind the Eastern plane and flying slightly higher was a DC-3 carrying a group of Civil Aeronautics Board (CAB) officials. Ironically, they were returning from the scene of an airline crash that had occurred the previous day at New York's LaGuardia Airport. The two CAB pilots, one of them a trained air accident investigator himself, were idly watching the plane ahead of them when for no discernable reason it suddenly went into a dive.

The dive became steeper, and the horrified CAB pilots realized there was no apparent attempt being made to pull out of it. The dive continued until the DC-4's nose was pointing straight toward the ground. Not believing their own eyes, the DC-3 crew watched as the big airliner, by now almost on its back, crashed into a wooded area and exploded. The angle of impact strongly suggested that the DC-4 had gone through the first half of an outside loop, in which the plane dives, flies in an inverted position, climbs, and then resumes normal forward flight; not for several months was any significance attached to this implausible maneuver. At the time, the fifty-three fatalities represented the worst crash in U.S. aviation history.

A number of eyewitnesses had seen the mysterious dive from the ground, and the subsequent CAB investigation centered on the testimony of eighteen people who said they had seen something fall from the plane during the dive. Sixteen of them said the falling part or parts had come from the tail, a strong indication that at least one section of a horizontal elevator might have failed. A search party did find ten pieces of the tail assembly a quarter of a mile from the main wreckage, but it was impossible to determine whether structural failure had caused the dive or, whether just as likely, it had resulted from the stresses imposed by the abnormal maneuver—that half-completed outside loop.

Four months later, while investigators still were trying to solve the riddle of Flight 605, another DC-4 went into the same start of an outside loop, and this time there was no mystery about the cause.

Details of the Crash

An American Airlines DC-4 had left Dallas bound for Los Angeles. The flight was commanded by Captain C. R. Sisto, with another four-striper—Mel Logan—serving as copilot. Riding in the cockpit jump seat was Captain John Beck, a DC-3 pilot familiarizing himself with DC-4 equipment. The aircraft was cruising at eight thousand feet and approaching El Paso when Sisto asked Beck if he'd like to fly the airplane.

They exchanged seats, Beck taking over the controls. He was maintaining level flight when the DC-4 suddenly started to climb. Beck rolled the elevator trim forward to bring the nose down, but the climb continued. Beck kept rolling the trim tab forward to no avail, then started to turn the tab back to normal when the plane pitched violently forward and went into a steep dive.

At first, all three pilots said they thought the automatic pilot had malfunctioned—an explanation both Logan and Beck sincerely believed. But one week after the near-disaster, Sisto confessed the truth. Without the knowledge of the other pilots, he had engaged the "gust lock," curious to see what would happen if this device were activated in flight.

Down it roared, first vertically and then executing the first half of an outside loop until the DC-4 was on its back. Beck and Sisto, who did not have their seat belts fastened, were thrown to the top of the cockpit where their heads accidently hit three of the four propeller feathering switches. Feathering changes the propellers' angle and prevents them from windmilling or continuing to turn when an engine is shut off. Props one, two, and four automatically feathered and the plane immediately lost most of its power. The crew did not know it at the time, but the accidental feathering and loss of power had saved the lives of all fifty-four people on the aircraft.

With the power cut, Logan—the only one in the cockpit whose seat belt was fastened—managed to roll the DC-4 out of its inverted position. When he finally restored level flight, the airplane was only three hundred fifty feet from the ground. Logan landed safely at El Paso.

At first, all three pilots said they thought the automatic pilot had malfunctioned—an explanation both Logan and Beck sincerely believed. But one week after the near-disaster, Sisto confessed the truth. Without the knowledge of the other pilots, he had engaged the "gust lock," curious to see what would happen if this device were activated in flight.

The gust lock was a cockpit lever that locked a DC-4's rudder and elevators in place when it was parked on the ground, protecting these external surfaces from sudden wind gusts. On the DC-3 and other older airplanes, mechanics attached wooden chocks to the elevators and rudder for such gust protection, but this was not possible with the high tail of an aircraft with a tricycle landing gear.

Actually, Beck had suspected the gust lock was engaged. When the nose began to climb without warning, and rolling the trim tab had no effect, Beck asked Sisto, "Is the autopilot on?" Sisto said no, and Beck immediately thought of the gust lock. He was about to neutralize it when Sisto beat him to it and released the lever. Because the trim tab was still in a forward position, when the gust lock was disengaged Beck's controls remained literally locked in a dive. Except for the accidental prop-feathering, and the fact that the American DC-4 was flying four thousand feet higher than Eastern 605, with more altitude in which to recover, the CAB would have been faced with two unexplainable crashes.

The apparent reason Sisto had gotten curious provided one more bizarre element. The May 29 LaGuardia accident from which the CAB officials were returning also had involved the gust lock on a United DC-4 whose captain had tried to take off with the lock still engaged. Obviously, one of the pilots had somehow skipped the mandatory unlocking procedure on the pretakeoff checklist. The aircraft failed to gain altitude and crashed, killing forty-three of the forty-eight aboard. Although Sisto never made a direct admission of motive, the CAB knew there had been some speculation among pilots about the effects of in-flight gust lock engagement. And Sisto received more than a slap on the wrist—he was fired and his license revoked, while Logan was commended for superb airmanship.

Now the probe into the Eastern crash shifted to its crew, the captain in particular. Bill Coney had been a veteran test pilot before joining the airline, and the CAB found it hard to believe that an airman with such experience would have been foolhardy enough to let curiosity override common sense. In its final and sadly inconclusive report on the accident, the CAB noted the possibility of deliberate gust lock engagement but added, "He (Coney) would have been fully aware of the possible results of such experimentation."

That same report admitted the board's inability to establish a definite explanation for the fatal dive. "The probable cause of this accident," it stated, "was a sudden loss of control, for reasons unknown, resulting in a dive to the ground."

Impact

Several months had passed after the unsolved verdict was issued when another DC-4 went into the same half-outside loop near Cleveland, Ohio. The captain regained control the same way Logan had—by cutting power, returning to level flight, and landing safely. This time there was no question that anyone had been experimenting with the gust lock. After the near-fatal

A sailor examines the tail section the Eastern Flight 605. While still open to speculation, official reports indicate the DC-4's controls were frozen as the gust lock was initiated.

American incident, all pilots had been sternly warned of the consequences. CAB technicians swarmed over the DC-4 looking for clues, and on one elevator found a tiny piece of bonding tape that had worked loose.

Subsequent tests showed that if the slipstream caught the loose tape in a certain position, it created a ballooning effect that forced the plane into a dive. The board then began receiving reports of similar loose bonding tape on the elevators of Air Force C-54s, the DC-4's military version. All carriers were ordered to change the tape's location on their DC-4s so that even a loose tape could not catch the wind and balloon out. No further violent dives were ever reported.

Meanwhile, elated CAB Bureau of Safety investigators, confident they finally had the solution to the Maryland crash, located the wreckage in a warehouse—fully expecting to find evidence of loose bonding tape. They found the tape, but it was sealed tightly, in normal position. No further report on the accident was issued, and the Memorial Day tragedy in the clear skies over Maryland remains one of the nation's few unsolved crashes.

Did Captain Coney try experimenting with the gust lock, as Sisto did later? Coney's own contemporaries at Eastern were divided on his possible guilt or innocence. Many agreed with the CAB, refusing to believe that a pilot with his background could possibly have been that foolish. Yet others considered him just the type of inquisitive airman who, precisely because of his test pilot experience, simply might have become curious.

Even if it was a case of human error, it could be said that technology failed, too. Some pilots felt that there should have been a way to prevent gust lock engagement in flight, such as an automatic disconnect once the wheels left the ground. Yet it also could be argued that no one at Douglas could have anticipated such risky experimentation by any responsible airman.

The crash of Flight 605 gradually was forgotten as pressurized DC-6s, Constellations, and Boeing Stratocruisers began relegating the unpressurized DC-4 to eventual oblivion. The military C-54 won far more fame as the backbone of the Berlin airlift than its civilian counterpart ever achieved, and that is how this airplane will mostly be remembered.

ROBERT J. SERLING

Where to Learn More

"Eastern Crash Termed Mystery." *Aviation Weekly*. August 25, 1947, p. 12.

"CAB Lays Blame for United's DC-4 Crash May 29." *Business Week*. September 27, 1947, p. 21.

Gustafson, Philip. "Something Fell Off the Plane." *New Republic*. July 7, 1947, pp. 22-25.

Hotz, R. "Pilot Fiddling with Gust Lock Sends DC-4 Into Outside Loop." *Aviation Weekly*. Ocotber 27, 1947, pp. 40-41.

United Airlines DC-6 crash

Bryce Canyon, Utah

(1947)

An in-flight fire caused the crash of a DC-6 at Bryce Canyon, Utah.

Background

The DC-6 was the Douglas Airplane Company's answer to Lockheed's Constellation—and it was an effective answer. Douglas's first postwar commercial transport was larger and faster than its chief competitor. Introduced into airline service about a year after the Connie by American and United, the launch customers, the DC-6 actually was a stretched DC-4 with full pressurization and more powerful engines.

The DC-6 was a full hour faster than the Constellation on a transcontinental flight. It was so superior to anything else flying that one airline (Capital) painted squares around the circular windows of its DC-4s, so less experienced passengers might think they were boarding a new DC-6. The DC-6 was good enough, in fact, to push Lockheed into designing an improved Constellation—the Burbank firm had already suffered a humiliating setback with the 1946 grounding of the original model Connie after an electrical fire caused a fatal crash. Douglas was confident it had a winner, but only six months after the DC-6's introduction, the world's most respected and successful airframe manufacturer learned that its plane was also vulnerable to design errors of deadly proportions.

Details of the Crash

On October 24, 1947, a United DC-6 en route from Los Angeles to Chicago and cruising routinely at nineteen thousand feet, suddenly reported that a fire had started in a baggage compartment and was raging out of control. The plane was only one minute away from a radio checkpoint at Bryce Canyon, Utah, and the air traffic controllers monitoring the flight held their breaths as the United crew radioed a terse description of its desperate attempt to reach safety.

Two minutes after the initial Mayday, the flight messaged: "The tail is going out . . . we may get down and we may not."

Another tense two minutes passed before the pilot reported they were trying to find a place to land. Only one minute later, Flight 608 sent its last transmission: "We may make it . . . approaching a strip."

But it wasn't to be. The DC-6 disintegrated in flames before it could reach the strip and all aboard—five crew members and forty-seven passengers—perished. Civil Aeronautics Board (CAB) investigators at first thought they had an easy case—they knew where the in-flight fire had started, all they needed to know was how.

The two chief areas of suspicion were seepage of flammable fluid from the hydraulic lines, or the accidental igniting of the emergency magnesium flares all DC-6s carried. As a precautionary measure, DC-6 operators were advised to check hydraulic lines for leaks and remove the flares until further notice. Douglas itself suspected a short circuit in the heater's electrical system and began redesigning the circuitry.

Civil Aeronautics Board Bureau of Safety experts were still examining Flight 608's wreckage when, only seventeen days after the Bryce Canyon crash, another DC-6 caught fire in midair and barely made it to safety. It was an American Airlines flight from San Francisco to Chicago, commanded by Capt. Evan Chatfield. Two hours after takeoff, Chatfield began the routine transfer of fuel from one tank to another, a procedure that maintained aircraft stability by keeping fuel tanks levels equal as fuel was consumed.

The crew was still engaged in the fuel transfer process when cabin temperature began to rise. A stewardess tried to adjust the temperature control and when that didn't work, Chatfield turned off the heating system. This had no effect, either, so he switched from automatic heat control to manual, figuring the thermostat must have malfunctioned. The cockpit gauge registering cabin temperature dropped so sharply that Chatfield went back to automatic. The temperature began to rise again and was still climbing when a heating compartment fire warning light flashed.

With knowledge of the Bryce Canyon crash fresh in his mind, Chatfield immediately began a fast emergency descent toward the nearest airport—Gallup, New Mexico. A second fire warning came from a forward baggage compartment, but the crew didn't need that to tell them they were in real trouble—smoke was seeping into the cockpit, becoming so thick that as the burning DC-6 swept toward the Gallup airport, its pilot couldn't see the runway dead-ahead.

The copilot opened the cockpit window on his own side, and leaned out until he could spot the runway. For the next two minutes, he served as Chatfield's eyes until the plane landed and was surrounded immediately by fire trucks. Everyone evacuated the aircraft safely via emergency chutes, and investigators were handed a virtually intact DC-6 to inspect.

Fifty-two people perished when this DC-6 became engulfed in flames originating from the baggage compartment. It was later determined that overflow from an inboard fuel tank was sucked into the heater air intake system.

Nothing out of the ordinary to pinpoint the source of the fire was found until an American Airlines test pilot, a member of the carrier's engineering staff, arrived at Gallup to join the investigating team.

His name was Glen Brink, and for a brief time he was as mystified as anyone else after futilely poking around the crippled plane without finding a single clue. Then, while he was at an informal investigative meeting, a veteran American mechanic came in and asked a CAB official for permission to drain one of the fuel tanks. "We've got some fuel leaking out," he explained.

Brink said nothing, but that night he phoned his boss and urged him to recommend that American ground its DC-6 fleet. "There's something wrong with it, and until we find out what, we shouldn't be flying any DC-6," he added. Within hours, American voluntarily took all its DC-6s out of service and United did likewise. Meanwhile, Brink began questioning the mechanic who had reported the fuel leak.

"Which tank was it?" he asked.

"The one nearest the fuselage," the mechanic told him, and a gnawing suspicion dawned in Brink's mind—a rather bizarre theory on what might have caused both in-flight fires. He flew to Santa Monica, confided his suspicions to Douglas engineers, and asked for a special test flight that would duplicate the circumstances of both Bryce Canyon and Gallup. He was especially interested in the fact that the fire on the American DC-6 had coincided with the fuel transfer procedure, and he suspected the same thing had occurred on UAL 608. Inadvertent fuel overflow, he knew, was rare but not uncommon if the crew wasn't monitoring the fuel gauges closely.

Accompanied by another American pilot and several Douglas engineers, Brink borrowed a brand-new DC-6 to demonstrate whether an overflow from an inboard fuel tank could be sucked into the heater's air intake system, since the scoop for that system was located close to the tank vents. Just in case Brink's theory turned out to be right, precautions were taken by readying large fire extinguishers and by disconnecting the heater by shutting off all electrical circuits. To determine the exact path of overflowing fuel from an inboard tank, they glued cotton onto the front of the air scoop, and painted whitewash around the belly area adjacent to the air intake.

Brink was especially interested in the fact that the fire on the American DC-6 had coincided with the fuel transfer procedure, and he suspected the same thing had occurred on UAL 608. Inadvertent fuel overflow, he knew, was rare but not uncommon if the crew wasn't monitoring the fuel gauges closely.

The test DC-6 climbed to nineteen thousand feet and Brink began transferring fuel, hoping to reenact Bryce Canyon and Gallup by deliberately allowing an overflow as they pumped fuel from an outboard to an inboard tank. When they landed, fuel was dripping from the belly, all the whitewash had disappeared, the cotton was saturated with fuel, and they had to drain almost half a barrel of gasoline from the heating ducts.

The mystery of the DC-6 fires had been solved.

Impact

Unbelievably, the Civil Aeronautics Administration (CAA) filed charges against Brink, threatening to lift his pilot's license for conducting an unauthorized test flight, while the CAB itself claimed that the flight hadn't proved anything and also accused him of making a dangerous test flight without proper precautions.

But when CAB hearings on the Bryce Canyon crash opened, the case against Brink collapsed. Donald Douglas himself testified that he had personally authorized the test flight, that all precautions had been taken, and that his company had fully accepted the results. All DC-6s, in fact, stayed grounded until the location of the heater air intake was changed so that no fuel overflow could enter, a major structural modification that took four months to complete before the grounding order was lifted entirely.

To further vindicate Brink, the hearings revealed that a top CAB Bureau of Safety official, intrigued by Brink's theory, had asked the CAB investigator in charge at Gallup to accompany the American pilot to Santa Monica for the test flight. The investigator told his superior that Brink was on a wild goose chase and refused to go.

The CAA's motive in going after Brink was decidedly suspect, for the agency had ignored its own rules by certificating the DC-6 air intake's location. The CAA's certification requirement specifically stated: "It shall not be possible for fuel to flow between tanks in quantity sufficient to cause an overflow . . . vents and [fuel] drainage shall not terminate at points where the discharge of fuel will constitute a fire hazard."

Those safeguards dated back to the DC-4, the first transport-type aircraft with fuel transfer capability. Both the DC-4 and DC-6 had similar fuel transfer systems, but the locations of the heater air intakes were different. When CAA certification inspectors approved the location of the DC-6 airscoop directly behind a fuel vent, they apparently assumed no pilot would permit an overflow.

Brink's fine detective work had an ironic—and tragic—sequel. Subsequent to the Constellation and DC-6 in-flight fire incidents, the CAA ordered the airlines to install more sensitive smoke and fire detection systems in aircraft baggage compartments, and also to equip those compartments with mandatory carbon dioxide extinguishers. Several carriers futilely protested the early deadline for such installations, arguing that such improved and obviously more delicate systems needed more testing—mostly to prevent false alarms.

In June of 1948, a United DC-6 was cruising placidly over Mount Carmel, Pennsylvania, when one of the extremely sensitive new cargo bin fire warnings went off. The captain discharged the compartment's carbon dioxide bottles and was giving Air Traffic Control a calm account of the incident when his voice became slurred, and then incoherent.

Eyewitnesses on the ground saw the DC-6 go into a slow, spiraling descent that ended when the aircraft plunged into the ground, killing all aboard. The CAB investigation revealed there had been no fire in a baggage compartment; the new detector had given a false alarm. And when the carbon dioxide bottles were discharged, fumes had leaked into the cockpit—both pilots and the flight engineer were probably unconscious when the plane crashed. In so hastily designing and installing the new detection system, no one had taken into account the possibility that carbon dioxide fumes—especially with no fire to quell—might find a way into the cockpit.

In closing one safety loophole, the government had created an equally deadly new loophole.

Robert J. Serling

Where to Learn More

"Fire and Downdraft Factors in DC-6 Crash." *Aviation Weekly.* November 3, 1947, pp. 13-14.

"CAB Issues Report on AA DC-6 Fire." *Aviation Weekly.* January 12, 1948, p. 41.

"Improvise Crash-Study Technique." *Aviation Weekly*. April 5, 1948, pp. 24-25.

"CAB Ends DC-6 Crash Investigation." *Aviation Weekly*. January 3, 1949, p. 34.

"52 Are Killed as Airliner Crashes and Burns in Utah. " *New York Times*. October, 25, 1947, p. 1.

"DC-6 Fires Traced to Gasoline Leaks." *New York Times*. December 5, 1947, p. 19.

In-flight propeller reversals

Commercial U.S. airline aircraft experienced inadvertent propeller-pitch reversals caused by mechanical and electrical malfunctions in the propeller reversing mechanism.

(1940s-1960s)

Background

In the ten- to fifteen-year period following World War II, U.S. airlines experienced a series of puzzling in-flight and on-the-ground propeller reversal incidents and accidents, several of which claimed the lives of passengers and crew members. The crashes and continued incidents of inadvertent prop reversals received widespread media attention and cast doubts on the safety of the airlines, something which the fledgling industry needed to correct quickly in order to restore faith in the public's mind.

The airlines at that time—just before the introduction of the jet engine—routinely used twin- and four-engine propeller-driven aircraft such as the Boeing 377, Martin 202, Douglas DC-6, Convair 240 and 340, and Lockheed Constellation. Commonly installed on these and other large transport aircraft were reversing propellers manufactured by Hamilton Standard, at that time a division of United Aircraft Corporation.

A reversing propeller enables an aircraft to decrease its landing roll by slowing the aircraft quickly upon landing, thereby reducing brake wear and increasing ground maneuverability. The engine itself does not reverse, just the propeller, by changing the angle of the propeller blade so that the air no longer is thrust backward but is pushed forward, much like reversing a ceiling fan to push down or draw up the air.

The mechanism that changed the blade angle was composed of a network of electrical circuits, wires, and hydraulic fluid pressure lines, controlled from the pilot's throttle. When it was working properly, it could not be activated until the landing gear were firmly on the ground after landing, thereby closing an electrical circuit and giving power and pressure to the reversing mechanism. Reversible propellers were particularly popular with pilots of the time, who often had to land on short runways in the increasingly faster and more sophisticated aircraft being produced by the aircraft manufacturers.

The reversal of a propeller while in flight could lead to a crash because of the resulting loss of lift, uneven or abnormal control forces and slowing of speed. However, inadvertent reversal of one propeller while at high altitude would not be dangerous as such, so long as the other propellers and engines were working properly. In fact, many large aircraft were flight-tested for just such a contingency. But in critical phases of flight, such as near the ground at takeoff or landing, or in an unusual attitude such as during a training maneuver, little time or room is left for a pilot to try to unreverse the propeller.

Over forty-five inadvertent reversals were reported by the airlines to the Civil Aeronautics Administration (CAA), the majority of which occurred from 1949 to 1953, but reversals continued to occur sporadically into the early 1960s. Over ninety-five persons lost their lives in fatal crashes attributed to prop reversals. In those incidents that ended in fatal crashes, the aircraft—all with Hamilton Standard reversing propellers—were in the landing or takeoff phase, were performing an rejected landing called a go-around, or were on training flights when the prop reversal occurred. Many more incidents occurred without injury while the aircraft were on the ground.

Details of the Problem

Although there were several fatal accidents in the early 1950s attributed to inadvertent propeller reversal, the one that received the most attention, and which seemed to spur industry and government action for a final solution, occurred in Elizabeth, New Jersey, on February 11, 1952. On that date, National Airlines Flight 101, a Douglas DC-6 with sixty-three persons aboard including one infant and a crew of four, crashed and burned after striking an apartment house within the limits of the city shortly after takeoff from Newark Airport, Newark, New Jersey. Of those on board, twenty-six passengers and three crew members were killed together with four occupants of the apartment house. The other passengers and crew received injuries varying from minor to serious.

The flight had arrived normally at Newark from New York International Airport (now Kennedy International), on its way to Miami, Florida. The aircraft was loaded with mail, baggage, and passengers and proceeded to take off (fuel had been added at a prior stop). The weather at Newark consisted of a cloud layer at twenty-thousand feet, with flight visibility clear. The aircraft took off normally and the climb out also appeared normal until the aircraft was just a few miles from the airport, where it was observed by the control tower personnel to lose altitude suddenly and veer slightly to the right. The controller contacted the pilot, who replied he had lost an engine and was returning to the field. The flight was immediately cleared to land, but no further radio contact was made by the pilot. The controller noted the aircraft continued to the right at a low altitude and disappeared from sight. Shortly after, the tower personnel observed a fire in the vicinity of Elizabeth, New Jersey, where it was later established that the flight had crashed.

A Civil Aeronautics Board (CAB) investigation disclosed that the aircraft first made light contact with a tree followed by heavy impact of the right wing with the roof of the apartment building. The impact sheared off a portion of the right wing and started a fire in the building, which killed the four occupants. The aircraft continued forward, striking other obstructions and disintegrating to a stop.

Information gathered from interviews with the surviving passengers verified the sudden drop after takeoff and the veering to the right. Associated with the drop and change in aircraft heading was unusual engine noise described variously as a wail, whine, roar or sound like propellers going into reverse. The passengers described the aircraft as shuddering, vibrating, and shimmying from side to side. Two passengers noticed that the propellor on the outboard engine, or engine farthest out on the wing, had come to a stop.

Surviving passengers verified the sudden drop after takeoff and the veering to the right and mentioned a strange noise—described variously as a wail, whine, roar or sound like propellers going into reverse. In addition to the aircraft's shuddering, vibrating, and shimmying from side to side, two passengers noticed that the propellor on the outboard engine had come to a stop.

After examination of the propellers at the scene and in a later technical study and evaluation following disassembly, the two propellers on the left wing were found to be at a normal pitch angle; of the two on the right, one propeller was in full reverse pitch and the other had been feathered; that is the blade's pitch had been turned edge into the wind. No evidence was found to indicate failure or malfunctioning of the aircraft structure, the control surfaces, and their related systems. It was later determined that the flight crew had purposely feathered the outboard engine, which is standard procedure when an engine fails.

The testimony of the passengers who survived the crash was carefully analyzed and described circumstances—such as a sudden increase in noise level of short duration and abrupt settling and veering—that would be expected to accompany a sudden reversal of a propeller under full power in flight.

The CAB determined that if the governor solenoid valve circuit —the device that controlled the movement of the blade angle—was not isolated from other circuits, it could cause reversal of the propeller if the circuit somehow received power. Should that occur due to a fault in the electrical system, such as coming in contact with another "hot" wire, the voltage flowing to the governor would reverse the propeller without any action from the crew.

Engineers at Hamilton Standard disagreed with the CAB's finding that the Elizabeth crash had been caused by prop reversal. Their analysis of the crash indicated that the propeller had been forced into reverse pitch by the impact. However, the CAB analysts prevailed, and inadvertent prop reversal became the official cause of the crash.

Impact

After the Elizabeth crash, the Civil Aeronautics Administration sent a telegram to all CAA regional offices, which ordered that, to stop the possibility of inadvertent propeller reversals on Douglas DC-6 aircraft, the wiring from the reversing circuit was to be isolated from other wires by placing it in a separate conduit (a tube to protect the wires). They did not recommend deactivation of reversing propellers on any aircraft while the fix was being accomplished. Despite this, however, National Airlines decided to permanently deactivate the reversing mechanism on their DC-6 aircraft.

This change order by the CAA was purely a precautionary step to help prevent inadvertent prop reversals. A debate raged within the industry that there was no definite evidence that "bundling" the propeller circuits with other wiring caused either the National DC-6 crash or other crashes. Douglas engineers had never felt that separate conduiting of the wires was necessary on the DC-6, although Hamilton Standard and CAB safety investigators had recommended it.

However, well before the Elizabeth crash, American Airlines, the largest U.S. operator of post-World War II twin and four-engine aircraft, had decided not to wait for government direction and was the first to install its own solution to the problem. When the CAA's order came three days after the crash, the airline was already well into its own three-phase active engineering program.

The first phase made immediate fixes, including isolating the solenoid circuit from other circuits and replacing hardware susceptible to vibration within throttle controls. The second phase added backup hardware and electrical devices to give secondary protection against inadvertent reversal. The third phase added longer-term fixes to the overall design of the reversing system, including backup devices to prevent reversal from occurring prematurely on the ground, which could happen if a hard landing shorted the circuit.

The Air Line Pilots Association also developed their own proposed solution. Officials of both Hamilton Standard and the CAA approved of the basic idea of this modification but asked for further study. The ALPA plan would insert a pressure relief valve, similar to the safety value on a steam engine, into the propeller reversal hydraulic line, which would be pilot-controlled and would bleed off excess pressure that might force the propeller to reverse when and if the controls failed.

Hamilton Standard engineers took the best ideas of the airlines and pilots and derived a new device designed to safeguard hydraulically against short-circuit of electrical relays or against a faulty throttle lock mechanism—two possible causes of the inadvertent prop reversals. The device combined the safety valve feature with additional electrical safeguards. However, the company decided not to make the new device a mandatory fix, but instead made it a provision for its inclusion at an airline's option.

Just when the search for a foolproof propeller reversing system seemed to have been finally concluded, other causes for inadvertent reversals appeared. Later cases were found to have been caused by throttle control wear, improper circuit breaker position, loose control springs or nuts, and pilot error. Modification orders continued to be issued at various times to fix the problems. The CAA noted "It must be recognized that as long as reversing propellers are used there is always some small possibility that reversing may occur in flight. Every effort has been and is still being made to reduce this likelihood to the greatest possible extent." However, "the use of reversing propellers has undoubtedly prevented many more accidents than it has caused. This, and not fancied economic considerations, has been the sole basis for allowing reversing to continue in active airline use."

These incidents didn't end for airlines until the introduction of jet engines in the late 1950s. Within a matter of a few years, the airlines had switched to jet aircraft, and the prop reversal accidents all but ceased. However, if there is a lasting legacy from that early period, it is that much of the research and many of the modifications made were later incorporated into modern piston and jet engine manufacture. Today such incidents are few and far between.

JANICE KNESTOUT COLVIN

Where to Learn More

"CAA Orders Fix on H-S Prop Assembly." *Aviation Week*, April 13, 1953, p. 15.

Civil Aeronautics Board. *Accident Investigation Report*, National Airlines, Inc. DC-6 crash at Elizabeth, N.J. on February 11, 1952, text also reprinted in *Aviation Week*, October 20, 1952, pp. 68-84.

"Detective Story; Crash at Elizabeth, N.J.." *Newsweek*, February 4, 1952, p. 48.

"Last Flight, Crash at Elizabeth, N.J." *Time*, February 4, 1952, p. 12.

"Letter to the Editor," responding to October 27, 1952 editorial. *Aviation Week*, November 24, 1952, p. 102.

McSurely, A. "New Safeguard Against Prop Reversal." *Aviation Week*, November 17, 1952, pp. 76-79.

———. "New Prop Reversal Safeguard." *Aviation Week*, March 9, 1953, p. 18.

Moore, F. L. "ALPA Proposes Prop Reversal System." *Aviation Week*, April 14, 1952, p. 84.

———. "Crashes Spur Action on Prop Reversals." *Aviation Week*, March 10, 1952, pp. 63-64.

"New Approach to Prop Reversal." *Aviation Week*, February 2, 1953, p. 15.

"New Reverse Prop Proposal Circulated." *Aviation Week*, December 29, 1952, p. 7.

"Prop Reversal Failure Blamed in Accident." *Aviation Week*, January 15, 1956, p. 132.

"Prop Reversal; Hamilton Standard Refutes CAB Crash Findings." *Aviation Week*, December 8, 1952, pp. 91-92.

"Prop Reversal Probed in Crash." *Aviation Week*, February 16, 1953, p. 18.

"Propeller Reversal Blamed in Crash." *Aviation Week*, October 17, 1955, p. 16.

"Propeller Reversed in NEA Crash." *Aviation Week*, March 16, 1953, p. 60.

"Repeat Disaster, Elizabeth, N.J." *Life*, February 4, 1952, pp. 26-27.

"Undetected Prop Reversal Blamed in Fatal FAA Constellation Crash," text of CAB accident report. *Aviation Week and Space Technology*, May 20, 1963, pp. 131-143.

Wood, R. H. "The Reversible Prop Controversy." *Aviation Week*, October 27, 1952, p. 98.

BOAC Comet crashes

Mediterranean islands of Elba and Stromboli

(1954)

Two new jetliners exploded in midair; the cause was eventually traced to metal fatigue cracks around a small window in the upper fuselage that developed during pressurization and depressurization cycles.

Background

At 10:50 on the morning of January 10, 1954, a British Overseas Airways Corporation (BOAC) de Havilland Comet en route from Rome to London radioed Rome that it was breaking through the overcast at twenty-seven thousand feet and climbing to an assigned cruising altitude of thirty-six thousand feet. Less than two minutes later, fishermen off the island of Elba in the Mediterranean saw flaming wreckage falling out of the clouds and into the sea.

The last word from the ill-fated aircraft was a routine message to another BOAC flight, reporting on the height of the cloud cover. Whatever destroyed G-ALYP (the aircraft's registration letters, translated into "George Yoke Peter" as its air traffic control call letters) interrupted the report in mid sentence and happened with devastating swiftness. There were twenty-nine passengers and six crew members aboard.

The jet age had been born two years earlier with the takeoff of the same aircraft on a flight to South Africa, a flight that had also inaugurated BOAC's scheduled jet service. The original Comet was only a thirty-six-passenger aircraft (forty on a longer-ranged version with extra fuel tanks), but its speed and ability to fly above virtually all weather revolutionized air travel. Britain was understandably proud of this technological triumph, for in 1952 Russia was two years away from launching its TU-104 (actually a converted bomber), and in America, the Boeing 707 and the Douglas DC-8 were still on the drawing board. At the time of the mysterious Elba tragedy, de Havilland had firm airline orders for fifty Comets, and negotiations for another hundred were well underway.

During the first two years of the enormously popular Comet airline service, there had been two takeoff accidents, one nonfatal but the other resulting in the death of everyone on board. Each was attributed to the pilot's attempt to rotate (lift) the nose before adequate speed had been attained. In a third accident, a BOAC Comet flew into a violent thunderstorm and apparently disintegrated as a result of lightning and turbulence related to the storm. But neither fatal crash had caused any alarm bells to ring; pilot error and storm-caused structural failure were nothing new to aviation. Flaws in the Comet's design were not considered factors in these accidents.

In fact, cognizant that the new jetliner would be operating in an environment of unprecedented speeds and altitudes, chief designer Ronald Bishop and his engineering staff were particularly aware of pressurization stresses on an aircraft structure. Cabin pressurization—a means of compressing thin upper-atmosphere air into heavier, more breathable air as it enters the cabin—dated back to the Boeing Stratoliner of 1938. But there was a vast difference between a piston-engine plane's pressurization/depressurization cycles and those of a jet, which not only climbed and flew twice as fast, but routinely cruised ten thousand feet higher.

On transport aircraft like the Constellation and the DC-6, fuselages—the central compartments of airplanes that house the crew, passengers, and cargo—were designed to safely hold maximum pressure of about four pounds per square inch, sufficient to bring an aircraft's actual altitude of twenty-five thousand feet down to a level of air pressure in the cabin equivalent to that of an altitude of less than five thousand feet. The Comet required eight pounds per square inch to achieve the same breathable air, but British civil aviation authorities insisted that the new jet's cabin structure cope with more than sixteen pounds per square inch. Bishop and his chief structural engineer, Robert Harper, went beyond even that safety margin; the Comet fuselage was designed to hold twenty pounds per square inch.

To avoid excessive weight, the engineers designing the Comet chose a relatively thin gauge for the aluminum skin. It was only twenty-eight one thousandths of an inch thick, not much thicker than a postcard. Another critical and controversial decision involved the windows: the frames were square, the durability of which the Civil Aeronautics Administration questioned.

To avoid excessive weight, the engineers chose a relatively thin gauge for the aluminum skin. It was only twenty-eight one thousandths of an inch thick, not much thicker than a postcard. Another critical and controversial decision involved the windows: the frames were square, the durability of which the U.S. Civil Aeronautics Administration (CAA) questioned. This predecessor agency of the Federal Aviation Administration (FAA), charged with the responsibility of approving the airworthiness of American-operated Comets, suggested that oval-shaped frames would distribute pressurization stresses more equally. Engineers from BOAC responded that the Comet's windows had been tested at up to one hundred pounds per square inch of pressure without any signs of fatigue.

Following the Elba disaster, BOAC voluntarily grounded its entire seven-aircraft Comet fleet for inspection as a precautionary measure while investigators pondered all possibilities. The causal candidates included sabotage; an explosion caused by a ruptured turbine blade penetrating a fuel tank; structural failure from clear air turbulence; an in-flight engine fire that either ignited fuel or weakened the structure to the point of failure; explosion of hydrogen from a leaking battery; or an explosion of fuel vapor in an empty tank.

One BOAC technical expert wondered if metal fatigue might have caused explosive decompression. This seemed impossible to other BOAC officials. The aircraft had logged fewer than four thousand hours of flying time, the equivalent in mileage of an automobile only a few months off the showroom floor, and the de Havilland designers had assured BOAC that the Comets' structures could endure the stress of ten thousand flights' worth of repeated pressurization and depressurization.

What made it so difficult to determine the cause of the crash was the lack of solid clues. Royal Navy salvage crews were groping for wreckage buried five hundred feet deep and scattered over one hundred square miles of Mediterranean sea floor. Meanwhile, a special investigating committee recommended some fifty modifications to the Comet, all preventative measures based on possible but unproven midair explosion theories. Fuel lines were strengthened. Armor-plated shields were installed between the engines and fuel tanks. Improved smoke and fire detectors replaced original equipment. By late March, BOAC's Comet fleet was back in the air even as the salvage crews continued their search.

These measures were taken in vain, however, for on the night of April 8, 1954, a second Comet aircraft explosion occurred, resulting in twenty-one deaths. The aircraft G-ALYY ("George Yoke Yoke"), flying from Rome to Cairo, was climbing to its assigned altitude of thirty-five thousand five hundred feet when radio contact suddenly ended. The following morning, five bodies and two aircraft seats were recovered from the sea near the island of Stromboli, and for the second time in less than four months all Comets were grounded. Prime minister Winston Churchill ordered Britain's most prestigious aeronautical science organization, the Royal Aircraft Establishment (RAE), to take over the investigation of both accidents. The RAE investigation, headed by Sir Arnold Hall, was responsible for attributing the cause for the destruction of the world's first jetliner and, in effect, possibly determining the fate of all future jet travel.

Details of the Disaster

The RAE scientists were aided immensely by the salvage miracle performed by the Royal Navy. There was little hope of recovering Yoke Yoke's wreckage, which had fallen into water three thousand five hundred feet deep. But less than a month after Yoke Yoke's disappearance, the navy had recovered about two thirds of Yoke Peter's wreckage. The remains started to arrive at Farnborough, the RAE's main research facility, via ship, at too slow a pace for Arnold Hall. Without waiting for official permission, he commandeered a huge U.S. Air Force cargo plane to speed up wreckage delivery and inspection of every salvaged scrap.

Almost from the start, Hall suspected metal fatigue and so did Dr. Peter Walker, head of RAE's Structural Department. The investigation detoured temporarily down one blind alley after the discovery that one of Yoke Peter's engine turbine blades was missing, a finding that lent credence to the theory that a severed blade had punctured a fuel tank. But this possibility was discarded after the turbine casing was found intact and it was determined that the blade had been torn off by crash impact forces.

The final results of the autopsies performed in Italy on the fifteen bodies recovered from Yoke Peter and the five found near the site of the Yoke Yoke crash suggested to the RAE that all had been victims of "violent movement and explosive decompression." The Italian pathologist further concluded that the traumatic ruptures of the hearts and lungs had occurred before impact into the sea. This report supported the RAE investigators' suspicions that the pressure cabin had failed prior to the explosion.

Hall ordered the construction of a tank large enough to hold a Comet fuselage. One of the grounded jetliners was placed inside with its wings protruding from a hole on each side of the tank. Hall believed the original pressurization tests on the prototype had not been sufficient. Although they had involved repeated pressurization and depressurization cycles to determine the effect on the aircraft's structural life, the tests had not been accompanied by the use of hydraulic jacks to simulate the motion of an aircraft in flight.

The RAE installed such jacks to flex the wings of the submerged Comet in conjunction with the hydrostatic tests. The fuselage was filled with water (instead of air because water would cushion an explosive decompression), and the pressure was raised to slightly over eight pounds per square inch—the Comet's normal pressurization at thirty-five thousand feet. The level of eight pounds per square inch of pressure was maintained for three minutes in unison with the up-and-down motions of the wings, then reduced to zero pounds per square inch, and then increased back up to eight pounds per square inch for another three minutes. Each cycle, corresponding to the pressure variations of a three-hour flight, was repeated around the clock until the airframe had accumulated the equivalent of nine thousand flying hours or three thousand flights. The hydrostatic test schedule had aged the airframe forty times faster than would have been possible in normal airline service.

In late June, just as the nine thousand-hour mark was reached, the fuselage could no longer maintain the pressure and a crack developed. After the water was drained from the tank, a crack eight feet long could be seen along the top of the fuselage, extending from a fracture in one corner of a small escape hatch window and bisecting a cabin window frame. Metallurgic inspection of the hatch window revealed the discoloration and crystallization associated with metal fatigue.

This evidence was corroborated when in mid-August a section of Yoke Peter's fuselage roof was recovered and flown to Farnborough. One look provided confirmation of the hydrostatic tests. In the corner of a navigation window atop the fuselage was a telltale crack that had grown into a wide split. With the appearance of this final clue, the RAE investigators eventually came to the conclusion that the accidents were primarily "caused by structural failure of the pressure cabin, brought about by fatigue."

The solution to the Comet mystery, like the findings of most air disaster investigations, produced more than one culprit and no single cause. The first factor leading to the Comet crashes was a simply inadequate test program that failed to understand the long-range effects of continued pressurization and depressurization on airframe integrity. Compounding this error was the use of square-shaped window frames and the failure to incorporate mechanisms to prevent a fatigue crack from spreading into the aircraft's design. The third fatal factor was the fuselage skin—so thin that one U.S. airline president, on an early Comet demonstration ride, swore he could see the sides of the cabin walls "moving in and out like an accordion."

Impact

Three versions of the Comet were built—Models 1, 2, and 3—but only the MK-1 had been involved in the two Mediterranean crashes. MK-2 Comets, a larger and more powerful jet carrying up to forty-four passengers, were extensively modified after the RAE investigation and used by a number of European airlines as well as the Royal Canadian Air Force. But de Havilland built only one Comet 3; the lingering stigma of the 1954 disasters, plus the advent of new, larger American-built jetliners, proved fatal to the Comet's viability.

Many aviation historians have assumed that the designers of the Boeing 707 and Douglas DC-8 learned valuable lessons from the Comet's mistakes, prompting them to design safeguards against metal fatigue and explosive decompression. In truth, more was learned from the Comet's pair of takeoff accidents than from Elba and Stromboli. The lesson learned from the several Comet crashes was the absolute necessity of flying jet transports "by the book," rather than the need to carefully test a new aircraft's design. Even before those tragedies, both Boeing and Douglas were already designing jet fuselages with thicker skin; triple-strength, rounded windows to distribute pressurization stresses equally around the frames; and metal bracing to provide additional structural strength (similar to reinforcing a wooden barrel with iron staves). The most important insurance against catastrophic decompression, however, were the small metal tabs or "stoppers" strategically placed throughout the fuselage so that if a fatigue crack should develop, its propagation path would be blocked before explosive decompression could occur. All rebuilt Comet 2's and the single MK-3 had such structural reinforcements as well.

Boeing dramatically demonstrated the efficiency of this preventative system before the 707 ever flew. Engineers deliberately weakened a 707 test fuselage with saw cuts up to twenty-two inches long, then pressurized the cabin and dropped five huge, stainless steel blades that slashed into the top of the fuselage. Small puffs of air escaped from the wounds, but there was no catastrophic explosion.

As a contrast to this successful demonstration, however, Boeing had first shown airline officials a Comet-like pressurized cabin without stoppers and dropped two of the steel blades onto the fuselage. A slow-motion film of the

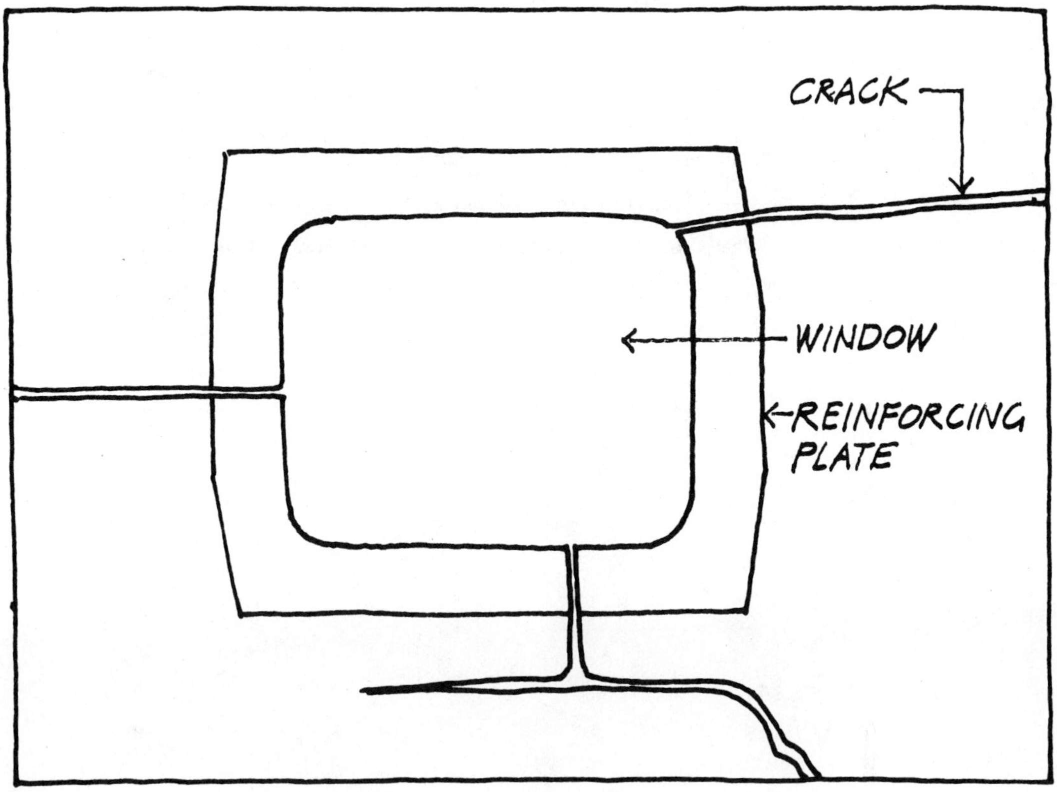

Metal fatique cracks around a small window in the upper fuselage developed during pressurization and depressurization cycles. These cracks resulted in disasterous structural failure of the pressure cabin.

demonstration showed the metal fuselage skin beginning to curl outward at the points of penetration, then curling faster until the entire cabin split open and spit out its contents—seats, dummy bodies, and even the cabin floor. It was a reenactment of the Comet explosions.

ROBERT J. SERLING

Where to Learn More

Anderton, David. "RAE Engineers Solve Comet Mystery." *Aviation Week*, February 7, 1955, pp. 28-42.

———. "How the Comet Mystery Was Solved—Part II: Giant 'Jigsaw Puzzle' Gives Final Clue." *Aviation Week*, February 14, 1955, pp. 26-39.

"Comet Crashes: Dakar Accident." *Aviation Week*, July 27, 1953, p. 16.

"Comet Crash off Elba Attributed by British Inquiry to Metal Stress." *New York Times*, October 20, 1954, p. 14.

"Death of the Comet." *Time*, April 19, 1954, pp. 31-32.

Dempster, Derek D. *The Tale of the Comet*. David McKay, 1958.

Hull, Seabrook. "Fatigue Blamed in Comet Crashes." *Aviation Week*, October 25, 1954, pp. 17-18.

McKitterick, N. "Comet Crash." *Aviation Week*, May 11, 1953, p. 17.

———. "Comet Future?" *Aviation Week*, January 19, 1954, pp. 16-17.

"What's Wrong with the Comet?" *Business Week*, April 17, 1954, pp. 27-28.

"Why the Jets' Skin Ruptured." *Business Week*, November 6, 1954, pp. 121-24.

TWA Super-Constellation and United Airlines DC-7 collision

Grand Canyon, Arizona

(1956)

Two airliners collided in empty airspace—an "impossible" accident that led to the creation of the Federal Aviation Administration.

Background

On June 30, 1956, the Civil Aeronautics Administration's inadequacies—which stemmed largely from its lack of independent status and rather feeble budget requests—were laid bare in a few terrifying seconds over the Grand Canyon when two airliners collided. There were 128 people aboard the two planes, all of whom perished. At the time, it was the highest single-accident death toll in the history of commercial aviation.

Back in the antediluvian days of commercial aviation, the nation's airlines operated their own air traffic control system, which was a relatively simple task. As late as 1932, the entire U.S. airline industry—twenty-four carriers, most of them minuscule—operated fewer than seven hundred daily flights with a fleet of only 450 planes, eighty percent of them single-engine aircraft. (By comparison, some two thousand transports now operate more than fourteen thousand daily flights.)

It was not until 1936 that the Federal government took over responsibility for controlling America's civilian air traffic, under an agency called the Bureau of Air Commerce. Then in 1941, jurisdiction over the nation's civil airways came into the hands of the Civil Aeronautics Administration (CAA), which, like the old Bureau of Air Commerce, was part of the Department of Commerce.

As a relatively small subordinate agency, the CAA was beholden to Commerce for all funding, a budgetary dependency that worried the entire airline industry. With air traffic booming after World War II, more than one air safety expert warned that the money-starved CAA was trying to operate a hopelessly antiquated and potentially dangerous air traffic control system.

"We're tracking three hundred mile-an-hour airplanes with radar designed to track twenty-knot battleships," one official complained.

Details of the Collision

The aircraft involved were a Trans World Airlines triple-tailed Super-G Constellation and a United DC-7. Both flights had left Los Angeles within minutes of each other, heading east and flying "off airways"—using airspace uncontrolled by Air Route Traffic Control (ARTC) centers. Both captains had chosen such routing either because the controlled airways were too clogged or not as direct, and their decision was not unusual given ARTC's handicaps (lack of personnel and proper facilities). Their flight plans, therefore, specified that they would be operating under visual flight rules (VFR), which meant they would assume the responsibility of avoiding other traffic.

The TWA captain had filed a flight plan that put him at a lower altitude than United. But after his Constellation reached its assigned cruising altitude, it flew into clouds and the captain received permission to go one thousand feet higher. This put TWA above the cloud cover, but in doing so, the "Connie" had to climb to twenty-one thousand feet—which happened to be United's assigned cruising altitude.

TWA was told the United DC-7 also was at twenty-one thousand feet, but the undermanned ARTC centers along the route failed to warn United that another flight had been cleared to the same altitude. ARTC, busy with traffic in controlled airspace, could not accept responsibility for flights using uncontrolled airways, and both airliners were on their way to bisecting an angle of death. One controller admitted later he had noticed that the estimated arrival time of both flights over a checkpoint near the Grand Canyon was identical. But the harassed controller, occupied with handling his own traffic, gave the fatal coincidence no further thought.

The two planes, each traveling at approximately three hundred miles per hour, collided at an estimated angle of thirty-five degrees. Searchers later found one of the Constellation's tails with red, white, and blue paint marks, obviously from a DC-7 propeller, which had such colors on its tips. Another search party located a rear cargo door from the TWA plane that also carried red, white, and blue gouges. Civil Aeronautics Board (CAB) investigators deduced that the DC-7 had hit the Connie, although no blame was attached to this finding—the unsuspecting United pilots didn't know TWA had been cleared to their own altitude. Because the collision occurred in wide-open airspace and in relatively clear weather, it was considered to have been an "impossible accident."

But it was far from that—it was an accident just waiting to happen. In fact, the CAB's subsequent investigation steered clear of putting the majority of blame on the two flight crews, and indirectly put it where it belonged: the CAA, Commerce, and a parsimonious Congress, all of whom had allowed aviation technology to outpace an obsolete, overworked air traffic control system.

Two pilots who were in the vicinity of the collision prepare to testify at the crash hearing in Washington, D.C.

Impact

The CAB spent almost ten months investigating the Grand Canyon collision. Its final, understated verdict—"that the pilots did not see each other in time"—akin to saying that the accident happened because it happened.

But the board went far beyond its mundane conclusion. The CAB listed six definite or possible contributing factors that told the real story of the Grand Canyon catastrophe:

1. Limited cockpit visibility.

2. Intervening clouds that may have temporarily obscured vision.

3. Preoccupation with normal cockpit duties.

4. Physiological limits to human vision.

5. Insufficient en route air traffic advisory information due to inadequate facilities and lack of personnel in Air Traffic Control.

6. Preoccupation with unrelated cockpit duties, such as trying to provide passengers with a more scenic view.

The last seemed an unfair charge against the TWA and UAL flight crews alike; the Grand Canyon that day was covered by low-hanging clouds and haze, and it was hardly likely that either flight crew was engaged "in unrelated cockpit duties." (This did not prevent one congressman, who arrived on the scene a few days after the accident, from flatly announcing that "the collision was caused by one or both pilots sightseeing.")

The CAB's other suggested causal factors made far more sense. Airline pilots had frequently and fervently complained about poor cockpit visibility in the transports of that era, and the DC-7 and Constellation were no exceptions to such criticism. "You might as well be trying to see out of a phone booth," one captain remarked. And high-speed airliners flying under visual flight rule conditions were another invitation to disaster. A six hundred miles-per-hour combined closure speed is faster than the velocity of a .45 caliber bullet as it leaves the muzzle.

The board had plenty of evidence on the potential hazards of distracting cockpit workload, inadequate cockpit visibility, and the limitations of the human eyeball, which, in effect, has the same constricted vision as the cockpit itself. There is no way to see directly above, below, or behind from an airplane flight deck, just as you cannot see the top, base, or back of your head. Air Force tests had demonstrated that if a DC-7 pilot spotted a Constellation two miles away coming at him head-on at three hundred miles-per-hour, he had only one-tenth of a second to decide on evasive action. At a distance of one mile, there was no time left.

The Air Force tests disclosed another phenomenon: at high altitudes in clear weather, a pilot actually suffers "empty field myopia," which means he becomes temporarily nearsighted and tends to concentrate on a location less than six feet in front of the cockpit.

But the CAB's fifth suggested contributing factor—the inadequacies of the ATC system—was the most damning of all, and the one that had the most repercussions.

The CAB's final report on the collision merely referred to the system's inadequacies without pinpointing why they existed. It wasn't necessary, however, because the airlines, air safety experts, responsible lawmakers, the White House and media were in unanimous agreement on the real reason for the Grand Canyon tragedy.

The CAA had been too long starved for adequate funding that would have provided the necessary modernization of the air traffic control system. Subservient to its parent Department of Commerce, its budget requests were often too modest, and what money it did ask for usually fell under Commerce's economy axe. Even while Congress was debating solutions, two more midair collisions occurred—a United DC-7 and an Air Force fighter

One of the first on-scene photos, which shows the fuselage of the TWA Constellation. The Connie, apparently, was hit by the United DC-7 after both had been cleared for the same cruising altitude.

over Las Vegas, and a Capital Viscount and National Guard fighter near Brunswick, Maryland. Both military planes were operating under VFR, and the accidents exposed another ARTC weakness—the danger of mixing VFR and controlled traffic, especially when the VFR flights involved swift military planes.

At the time of the Grand Canyon collision, the U.S. airlines were only two years away from inaugurating jet service. They would soon be operating aircraft with closure speeds upward of one thousand miles-per-hour, and they justifiably were asking that something be done about safe separation of such traffic.

In August 1958, Congress passed and President Eisenhower quickly signed into law the Federal Aviation Act, which established an independent Federal Aviation Administration (FAA), and provided it with almost total jurisdiction over civil and military air traffic alike. The new agency was given the necessary regulatory authority and funding to modernize air traffic control and drastically expand controlled airspace. And the FAA administrator reported only to the president of the United States.

Ironically, the chief motivation behind this massive reform—the establishment of a fully independent agency with its own budget and virtually unlimited powers—was to be tossed aside only eight years later. In 1966, Congress, with the enthusiastic blessing of the Johnson administration, placed the FAA under a new Department of Transportation.

Not only did this relegate the air agency to the same subordinate status as the old CAA, but in the opinion of most aviation observers, it also cost the FAA much of its effectiveness. But by then, the Grand Canyon tragedy was just another forgotten chapter in aviation history.

ROBERT J. SERLING

Where to Learn More

"Apparent United-TWA Collision Highlights Traffic-Control Problem." *Aviation Weekly.* July 9, 1956, pp. 39-40.

"CAB Begins Hearings on Mid-Air Collision." *Aviation Weekly.* August 5, 1956, p. 463.

Hotz, R. "We Are All to Blame." *Aviation Weekly.* July 16, 1956, p. 21.

Lewis, C. "Eyewitnesses Describe TWA-UAL Crash." *Aviation Weekly.* August 13, 1956, p. 38.

"CAB Issues Report on Canyon Collision." *Aviation Weekly.* April 22, 1957, pp. 33-34.

"Visual Failure Blamed in Canyon Crash." *Aviation Weekly.* May 13, 1957, p. 113+.

"Grand Canyon Disaster: Averages Catch Up." *Business Weekly.* July 7, 1956, pp. 28-29.

"Perilous Searching Operation: Grand Canyon Crash." *Life.* July 16, 1956, pp. 19-25.

"All 128 on 2 Airlines Found Dead." *New York Times.* July 2, 1956, pp. 1, 14.

"Into the Thunderhead." *Newsweek.* July 9, 1956, p. 25.

"Disaster in the Desert." *Newsweek.* October 8, 1956, p. 51.

"Identify Crash Bodies in Grand Canyon." *Science News.* July 14, 1956, p. 22.

"Painted Desert." *Time.* July 9, 1956, p. 18.

Lockheed Electra crashes

Indiana and Texas

(1959-1960)

An investigation of unprecedented scope into wing failure occurring in a brand-new airliner eventually blamed a relatively obscure physics phenomenon known as whirl mode.

Background

On September 29, 1959, a new Braniff Lockheed Electra, flying in clear, calm weather over Buffalo, Texas, lost a wing and crashed. All thirty-three people on board, including twenty-eight passengers and a crew of five, were killed in the crash. Five-and-a-half months later—March 17, 1960—with stumped investigators about to write off the Braniff accident as "crashed for reasons unknown," a Northwest Electra also lost a wing in flight and crashed near Tell City, Indiana. All sixty-three on board were killed. The Civil Aeronautics Board's Bureau of Safety (now the National Transportation Safety Board) was now faced with two unexplainable fatal crashes of an apparently thoroughly tested new airliner.

The Lockheed Electra L-188 was the first U.S.-designed turboprop (more commonly referred to as "propjet") transport to enter airline service. Four Allison jet engines hitched to huge, four-bladed propellers gave the Electra a cruising speed almost equal to that of pure jets. Pilots loved its enormous reserve power and responsiveness.

At a time when the world's airlines were falling in love with big jets, Lockheed believed there still was a viable demand for a fast, versatile propjet. The venerable Burbank, California, company quickly garnered orders for nearly 180 Electras from sixteen United States and foreign air carriers.

Development costs for the L-188 amounted to some $50 million, with four years of engineering research preceding the Electra's first flight. The very intensity of the testing and certification process made the Braniff and Northwest accidents that much harder to understand.

Civil Aeronautics Board (CAB) investigators at the Braniff site were stumped. The flight engineer's log sheet, recovered from the cockpit wreckage, supplied data only on the aircraft's speed and altitude seven minutes before the unknown menace struck. Structural failure was obvious; the left wing and both its engines were found a mile and a half from the potato patch where the bulk of the wreckage had fallen. From the wreckage distribution—a path of destruction nearly fourteen thousand feet long—it was determined that the left wing had failed first, followed by general disintegration of the entire aircraft. Eyewitnesses reported seeing a giant fireball at around fifteen thousand feet, the flight's assigned cruising altitude.

Clues were sparse and what few there were provided little help until the later Northwest accident at Tell City. One by one, every possible explanation for the Braniff crash was eliminated. They included an unchecked in-flight fire, an out-of-control autopilot, an excessively violent maneuver to avoid another plane or even an accidentally fired missile, and a dive-inducing runaway propeller that startled the pilot to such an extent that he pulled back on the controls too abruptly. Even the remote possibility of a overheated tire or wheel brake somehow rupturing a fuel line was considered. There was no conclusive evidence to support any of these theories, although one peculiar incident nagged at everyone.

Farmers had reported that just before the fireball was observed, "every coon dog for miles around started howling." Other witnesses told of hearing a variety of unusual noises, ranging from thunder to sonic booms, from a "whooshing, screaming noise" to "the clapping of two big boards together." The CAB team went to work recording every known source of intense aeronautical noise—jet aircraft, sonic booms, runaway props, as well as Electras in normal flight and also diving and climbing. The tapes, which included intentionally random noises not associated with airplanes, were played back to witnesses who were asked to select the sound or sounds closest to what they heard. The majority picked two: a runaway propeller whirling at supersonic speeds, and a jet airplane.

Its significance, however, wasn't grasped immediately. In fact, the Board was about to inform Braniff that it could release the remains of the aircraft to its insurance underwriters when the Northwest Electra went down. The two accidents were not identical, since the Northwest plane was flying higher, at eighteen thousand feet in an area of reported clear-air turbulence, and it was the right wing that failed first. But in the two piles of wreckage the CAB found certain similarities that added up to one conclusion: both planes had been victims of destructive flutter. Flutter is the oscillation or vibration of an aircraft part caused by the airstream. An easy example would be what happens when a light object is held out of a moving car's open window.

Details of the Crashes

The Northwest disaster produced the clearest evidence. The right wing structure between the fuselage and the inboard nacelle (engine housing) displayed numerous indications of damage progression during rapid reversals of load forces—in simpler terms, the wing had been subjected to massive, uncontrolled flutter severe enough to cause failure. The area at which wing separation had occurred showed metal damage that could have resulted only

The Braniff Lockheed Electra exploded in mid-air, leaving this tail section the largest piece of debris. Eyewitnesses reported seeing a giant fireball at around fifteen thousand feet, the flight's assigned cruising altitude.

from the wing's folding rearward instead of upward, as would have happened in the case of a high up-gust or positive maneuver. Saw-toothed diagonal fractures in a key wing structural area indicated powerful oscillation loads occurring in a repetitive cycle, consistent with catastrophic flutter.

There were scratches from a vibrating source on the outlet of an air inlet called the sugar scoop. The mounts that held the huge Allison engine in the outboard nacelles had "bottomed" or shaken up to sixty times at a rate of 2.5 times a second for up to twenty seconds. The engine torque housing bore marks of repeated cycling in the area of the bolt attachments. A vane located near the engine exhaust was scarred with elliptical scratches, telltale evidence of violent movements in that portion of the engine casing.

Evidence of destructive flutter was not as clear-cut in the Braniff Electra, yet there were also significant duplications. Flutter damage was greatest in the outboard engine on the failed wing of each aircraft. Both showed severe engine mount oscillation and the marks of repeated, massive vibration in the outboard nacelle structure. Yet even the determination that flutter was the

The Northwest crash in Tell City produced the clearest evidence of severe flutter of the right wing—tragically resulting in wing separation.

culprit in both crashes left a major unanswered question: how could flutter have been allowed to progress to the point of structural failure?

Flutter is caused by the turbulence of air over a fast-moving airplane's surfaces, and is present in every type of aircraft. The trick is to keep it from going unchecked into a continuous, uncontrolled cycle that eventually can produce fatigue even in new metal. An airplane's ability to absorb and control flutter is called damping, and precertification tests had shown the Electra to have excellent flutter-damping characteristics at all speeds. If an external driving force like turbulence "excited" the wing, it was dissipated by a slight speed reduction. Furthermore, the Electra's wing, like that on all airplanes, was designed to flex and recoil, thus absorbing energy from flutter oscillation.

Once the CAB had determined that flutter had been involved in both accidents, it was faced with another riddle: severe clear air turbulence apparently was a factor in the Indiana crash, but there was absolutely no turbulence present in the skies over Texas. Investigators begin searching for a flutter-inducing source other than turbulence.

While the wreckage from Buffalo and Tell City was being shipped to Burbank for even more intensive examination, Lockheed pilots took an instrument-crammed Electra through a series of nearly seventy hair-raising test flights, most of them into areas of known severe turbulence, deliberately courting destructive flutter. Boeing provided a team of flutter experts to help Lockheed interpret the data from the test flights. Douglas loaned its California rival a new flutter-inducing device—vanes mounted on the wings, whose angles were controlled from the cockpit. These missions totaled some fifteen hundred hours of dangerous flying and out of them came the first indication of trouble in the outboard nacelles: severe and extreme turbulence exerted a bending force on the outer wing section, from the outboard nacelles to the tips, that was ten times greater than on the rest of the wing.

Lockheed and the CAB combined this revelation with the flutter damage evidence found in the two wreckages, and began searching for the type of flutter that could have destroyed a sturdy wing. There are more than a hundred kinds of flutter, called modes, and the experts went back to the tape recordings of various sounds, zeroing in on that of an overspeeding propeller. They tried every flutter mode capable of producing such potentially destructive speed and found exactly one in which the propeller tips approached sonic velocity without any increase in revolutions per minute or air speed: whirl mode.

Whirl mode is a form of vibration natural to any piece of rotating machinery including airplane propellers. Until the Electra crashes, however, it was not considered even remotely hazardous. Damping checked all flutter modes in piston airplanes, for if flutter continued, the prop would tear itself off. The baffled investigators went back to the wreckage looking for the reason damping had failed in the L-188. They found the answer in the nacelle or housing structures. In both aircraft, support struts had been bent and severed, the engine mounts had been subjected to abnormal loads in various directions, fractures in the propeller reduction box (which keeps the prop from revolving as fast as the turbines) displayed signs of vicious multidirectional loads, and there were curved scratches on certain engine parts. This was evidence of huge cyclical motions not just of a wobbling propeller but of virtually the entire engine.

Subsequent wind tunnel experiments with an Electra scale model literally reproduced Buffalo and Tell City. After deliberately weakening the struts and braces that would have held the engine package firmly in place, the model was "flown" at speeds equal to or greater than those of the Braniff and Northwest planes. Then at those speeds, technicians created a sudden jolt (such as simulated turbulence) that excited the outboard nacelles or engine housings. Whirl mode started, fed itself on its own increasing energy, and eventually devoured the wing.

A propjet engine package is, in effect, a huge gyroscope with turbines spinning at nearly fourteen thousand revolutions per minute and the prop at more than twelve hundred revolutions per minute. An exciting jolt was like a giant finger reaching out to touch this smoothly whirling mass, causing it to break stride and wobble into a flutter mode that gained new strength from every unending cycle of violent motion until the wing itself surrendered.

Impact

After Tell City, the Federal Aviation Administration (FAA) ordered all U.S.-operated Electras to fly at lower speeds until the CAB could establish the cause of the two accidents. Once the cause was determined, the speed restrictions remained in effect until Lockheed proved the effectiveness of a $25 million modification program. It called for numerous structural changes; approximately fourteen hundred pounds of strategically-placed reinforcements went into the wings as well as the nacelles. In the most daring test flight of all, Lockheed pilots flew an Electra into an area of severe turbulence, and deliberately induced whirl mode with the Douglas vanes to see if the structural improvements would dampen the "death mode." They did, and speed restrictions on modified Electras were lifted.

Although whirl mode running unchecked in a weakened nacelle was definitely the cause, what had weakened the Braniff and Northwest outboard nacelles in the first place was never conclusively established. A previous hard landing was suspected in the case of the Northwest Electra, but the source of earlier damage on the Braniff plane remains unknown.

And another question remained unanswered: was the Electra's Achilles' heel the engine or the wing? Allison claimed that whirl mode should not have fatally affected the structural integrity of the wing unless wing and nacelle strength were marginal to begin with. Lockheed argued that whirl mode could not be transmitted to a wing solely through a weakened nacelle, and was possible only if a failure occurred first within the engine itself. In the outboard engines of both aircraft, investigators found preimpact fatigue breaks in the gear boxes and power sections.

That dispute may never be resolved.

Robert J. Serling

Where to Learn More

"Braniff Electra Crash Investigated by CAB." *Aviation Week and Space Technology.* October 5, 1959, pp. 38-39.

"Lockheed Completes Electra Test; Announces Probable Crash Cause." *Aviation Week and Space Technology.* May 16, 1960, p. 40.

Bulban, Erwin J. "Braniff Electra Disintegrated in Flight, CAB Investigator Says." *Aviation Week and Space Technology.* October 26, 1959, p. 45.

"Electra Accidents." *Engineer.* July 1, 1960, pp. 45-46.

Mooney Richard E. "Parley Will Sift Electra Crashes." *New York Times*. March 22, 1960, p. 74.

"33 Feared Killed in Airliner Crash." *New York Times*. September 30, 1959, p. 1.

"CAB Digs Deeper Into Electra Troubles." *Product Engineer*. August 29, 1960.

United Airlines DC-8 and TWA Constellation collision

New York City

(1960)

A midair collision over Staten Island exposed weaknesses in an air traffic control system supposed to have been completely modernized.

Background

A bleak winter sky hung over the metropolitan New York area in the midmorning of December 16, 1960. A United Airlines DC-8 jetliner, carrying seventy-seven passengers and seven crew members, inbound from Chicago and nearing Staten Island, was about to land at Idlewild Airport. A Trans World Airlines piston-engine Constellation, carrying thirty-nine passengers and five crew members, was also approaching Staten Island, heading for LaGuardia Airport after a routine flight from Columbus, Ohio. The blip representing the "Connie" was moving steadily on a radar scope in LaGuardia approach control, when the controller saw another blip at the same altitude of five thousand feet.

"Unidentified target approaching you . . . six miles . . . jet traffic," he advised the TWA flight.

"Roger, acknowledged."

The controller kept his eyes on the two blips, frowning as he watched their progress. He paged TWA again.

"Identified object . . . three miles . . . two o'clock."

"Roger. Acknowledged."

Before the controller's horrified eyes, the two blips moved unerringly toward each other—and merged.

The time was 10:33 A.M.

One of the DC-8's outboard engines sliced into the Constellation's fuselage like a giant meat cleaver. The stricken smaller plane fell onto a field only a hundred yards from a public school filled with children.

The crippled jet wobbled steadily earthward, staying airborne just long enough to reach the borough of Brooklyn where it yielded to the inevitable. One wing tilted and hit the roof of a four-story apartment building. The big jetliner bounced onto a street, cartwheeled, and smashed into the Methodist Pillar of Fire Church. Aircraft and house of worship exploded into flames, sending burning debris in every direction and setting ten more buildings on fire. Where the DC-8 finally crashed was one block away from two parochial schools occupied by seventeen hundred students and teachers. The only part of the plane left intact was the tail section, the word *United* plainly visible on the fin. Pieces were sprawled over the entire width of a Brooklyn street called Sterling Place.

The initial death toll was one hundred thirty-three, including six bystanders unlucky enough to have been caught in the open when the jet came down, and all but one of the one hundred twenty-eight passengers and crew members aboard the two planes. The one hundred thirty-fourth fatality was an eleven-year-old boy who had been sitting in the rear of the DC-8. Miraculously he was pulled out of the flaming wreckage alive. But with burns over 80 percent of his body, he died the next day.

The entire nation was asking some tough questions. Since a 1956 TWA-United collision over the Grand Canyon in which one hundred twenty-eight people had died, the new Federal Aviation Administration (FAA) had poured nearly half a billion dollars into various air traffic control projects. Only a month before the Staten Island collision occurred, FAA administrator Elwood "Pete" Quesada had proudly told a National Press Club audience in Washington that jets were being monitored by radar "from takeoff to touchdown."

If that was true, what had gone wrong at five thousand feet above New York?

Details of the Crash

Quesada's boast of "takeoff to touchdown" radar monitoring turned out to be empty hyperbole, as testimony and tape recordings of air traffic control communications sadly demonstrated.

Two minutes before the collision, the Idlewild (later to become Kennedy International) air control center had, indeed, been monitoring the United jet on radar. It advised the United flight that if it had to hold at Preston, New Jersey, a radio checkpoint, it was to execute the usual "racetrack" pattern for circling at an assigned altitude until cleared for final approach. United confirmed this instruction.

Pieces of the United DC-8 were sprawled over the entire width of a Brooklyn street. The DC-8 crashed there after colliding with a TWA Constellation.

One minute later, the pilot told Idlewild traffic control it was descending from six thousand to five thousand feet. Traffic control acknowledged and added these key words:

"*Radar service is terminated* [italics ours]. Contact Idlewild approach control. Good day."

United immediately switched radio frequencies and informed Idlewild approach control, "Approaching Preston at five thousand."

Approach control advised United to maintain five thousand feet, reported that no traffic delays were expected, and provided the flight with the current airport weather conditions along with landing instructions. That communication took only sixteen seconds to complete, but the United flight crew could not acknowledge it—the collision already had happened.

To fly in only sixteen seconds from the Preston checkpoint, where United said (or thought) it was, to the point of collision over Staten Island would have required a speed of more than one thousand miles per hour. And this impossible timetable established two facts:

Only 88 seconds before the collision, Idlewild traffic control had advised, "Radar service terminated." At that moment, the DC-8 was already far off course, a glaring discrepancy that should have been noticed and United warned immediately that it was not "approaching Preston," but had already passed it.

1. The DC-8 jet was not where its instruments apparently indicated it was. At the moment United radioed "approaching Preston," the jet actually was *eleven miles beyond Preston* and racing at nearly five hundred miles per hour through the buffer zone that separated Idlewild's traffic from LaGuardia's. Both figuratively and literally, it had run an aerial red light.

2. Radar could not have been monitoring the United flight very closely. Only eighty-eight seconds before the collision, Idlewild traffic control had advised, "Radar service terminated." At that moment, the DC-8 was already far off course, a glaring discrepancy that should have been noticed and United warned immediately that it was not "approaching Preston," but had already passed it.

The subsequent Civil Aeronautics Board (CAB) investigation revealed that the DC-8 crew had failed to inform Idlewild traffic control that one of its two key navigation instruments was inoperative. The board's final report, putting the entire blame on United, listed this as the accident's probable cause. If Idlewild traffic control had known of the malfunction, the CAB said, the DC-8 flight would have been handled differently.

Few more controversial and angrily challenged accident reports were ever handed down, as subsequent events demonstrated.

Impact

The CAB itself and the FAA accepted the board's report blaming United and its two dead pilots. The wrongful death claims from one hundred fifteen individual lawsuits filed after the accident totaled $77 million, and the codefendants included not only United and TWA, but Quesada's own agency.

Quesada insisted that neither the FAA's air traffic control system nor any individual controller was in the least way responsible, and that the CAB's indictment of United was eminently fair. There was irony in his eager acceptance of that finding, for the combative FAA chief had been feuding bitterly with the board. Quesada thought the FAA, not the CAB's Bureau of

Safety, should have sole authority to investigate air crashes—a stand generally regarded as specious, for in some cases the FAA, as a regulatory agency, might end up investigating itself.

The bureau, in turn, resented Quesada's practice of showing up occasionally at the scene of major accidents (as he did after the Staten Island collision, resulting in a shouting match with a CAB official) and providing the media with speculation about causes.

Yet Quesada's own actions belied his denial of any culpability on the part of his controllers or the air traffic control system. Acting on his orders within days after the collision, the FAA drastically reduced speed limits for aircraft entering a terminal area. It began assigning extra controllers at high-density traffic centers, their specific duty being to watch radar scopes for any planes straying from terminal clearances and holding patterns. And the agency also put into effect improved radar hand-off protection, the process under which one air traffic control center hands off traffic to another. All this was at least an indirect admission of involvement, and was not ignored when the lawsuits came pouring in.

Quesada had left office by the time the legal wars ended. After five months of hearings before a federal judge, with the testimony filling some fourteen thousand transcribed pages, the two airlines and the Justice Department—the latter actually representing the FAA—agreed to share responsibility in varying degrees: United accepted sixty percent of the liability, TWA fifteen percent, and the Federal government twenty-five percent.

Quesada's successor, Najeeb Halaby, fought the compromise settlement in vain, although he managed to take his case all the way up to then-attorney general Robert F. Kennedy. He argued that it was a slap in the face to controllers—"I didn't want our men to carry the stigma of implied guilt when they were so sure in their own minds that they weren't guilty," he explained later.

"I thought the compromise ignored the true role of the controller," Halaby wrote in his autobiography *Crosswinds*. "The term 'controller' implies authority to dictate to flights. But a controller doesn't really control; he *cooperates*. A pilot . . . can ignore advisories and warnings, just as the TWA crew did when Idlewild traffic control warned them that radar had spotted approaching traffic."

Halaby's defense of his men was in character; he enjoyed greater popularity among controllers than any FAA administrator before or since his tenure. But long after he left government service, he admitted to a journalist friend that the warnings issued to TWA were futile anyway—in an overcast sky, the prop-driven Connie was helpless to take evasive action against a fast-approaching jet the crew couldn't even see.

It seems fair to conclude that the system, not so much individual controllers, had to share the blame for the Staten Island crash. The Grand Canyon collision four years earlier—ironically involving the same two airlines—had spurred the FAA to spend millions on what amounted to multilane aerial turnpikes, a massive and worthwhile achievement.

A sad lesson of the disaster was that not nearly enough had been done about the one-lane traffic at the end of these airways.

ROBERT J. SERLING

Where to Learn More

"Crash Stirs New York Air Traffic Probe." *Aviation Weekly.* December 26, 1960, pp. 27-29.

"New York Collision Hearings Point to New CAB, FAA Friction." *Aviation Weekly.* January 9, 1961, p. 31.

Hoffman, D. H. "CAB Probe Centers on DC-8 Recorder." *Aviation Weekly.* January 16, 1961, pp. 38-39.

"Civil Defense." *Nation.* December 31, 1960, pp. 513-514.

Ruppenthal, K. M. "Crash of a System: New York's Air Tragedy." *Nation.* December 31, 1960, pp. 516-519.

"Death in the Air: Collision in New York." *Time.* December 26, 1960, pp. 14-15.

Turkish Airlines DC-10 crash

Ermenonville, France

(1974)

All three hundred and forty-six persons aboard a new DC-10 perished when the jet crashed just after takeoff from Paris.

Background

Turkish Airlines (also known as THY, for Turk Hava Yollari) Flight 981 left Orly Airport bound for London shortly after noon, March 3, 1974, the cabin of the wide-body DC-10 jammed wall to wall with 334 passengers and nine flight attendants. The experienced Turkish cockpit crew consisted of two pilots and a flight engineer. They all had less than ten minutes to live.

The start of the takeoff roll was logged at 12:31 P.M. At thirteen seconds past 12:41, the big jet—completely out of control—plunged into a pine-covered area known as Ermenonville Forest. Big trees snapped like matchsticks under the two hundred twenty-five-ton weight of a fully loaded jetliner impacting at nearly five hundred miles an hour.

This was only the second fatal crash of a so-called "jumbo" or wide-body jet transport since the Boeing 747's introduction in 1970, but it was the first to involve total fatalities. The THY accident was, at the time, the worst in the history of commercial aviation, and at this writing its death toll is exceeded only by two other air disasters: the record five hundred and eighty-two fatalities in the 1977 collision between two 747s on a Tenerife runway in the Canary Islands, and the five hundred and twenty who died when a Japan Airlines 747 on a 1985 domestic flight hit a mountain.

Even as the phrase "jumbo jet" came into common use, it was accompanied by widespread concern over the potential for catastrophe—a single crash of a wide-body jetliner could involve more fatalities than the combined death toll in three or even four separate accidents involving transports like the 707 and DC-8. And no one was more conscious of this dread possibility than the three U.S. manufacturers then producing the huge new planes: Boeing, Lockheed, and McDonnell Douglas (MCD).

Safety has been defined as "the art of reducing risk to the least possible chance of occurrence." McDonnell Douglas was convinced it had adhered faithfully to that definition in designing the DC-10, an airplane of a size and with performance parameters first laid down by American Airlines.

American, later joined by United as the DC-10's launch customers, wanted a jetliner sized between the 747 and the 707 or DC-8—and for operating economy, three engines instead of four. Lockheed aimed its similar L-1011 TriStar at the same market, bringing these two aerospace giants into bitter competition. Although both companies began design work almost simultaneously in 1966, the DC-10 went into airline service nearly a year ahead of the TriStar. One of the postscripts to the THY crash was a claim that MCD, in a rush to beat Lockheed to a dominant market share, had cut a few design corners—an accusation MCD bitterly denied.

McDonnell Douglas itself was the product of a 1967 merger between the venerable but ailing Douglas Aircraft Company and the McDonnell Aircraft Corporation, primarily a builder of fighter planes. McDonnell had, in fact, ventured only once into the field of commercial transports with an unsuccessful small business jet. Its merger with Douglas was more of an acquisition by McDonnell—not only of the Douglas company, but its reputation as one of the world's finest manufacturers of airliners.

During the DC-10's first three years of safe, worldwide operating experience, there had been one particularly disturbing incident that virtually scripted the eventual fate of THY Flight 981. In 1972, an insecurely latched aft cargo door on an American Airlines DC-10 blew off over Windsor, Ontario. The explosive decompression blast ruptured most of the hydraulic control lines, opened a seventeen-foot hole in the bottom of the fuselage, and collapsed part of the cabin floor.

The DC-10 was the first jetliner born of this corporate marriage, although it had begun strictly as a Douglas project prior to the merger. Except for one major difference, the DC-10 and L-1011 could have been twins; they were roughly the same size, designed to meet almost identical performance specifications, and both were trimotors. The DC-10's center engine, however, was mounted midway on the tail section, whereas the Lockheed's center engine was buried in the rear fuselage, with air being fed into the compressor via an "S"-shaped duct similar to the center engine arrangement on the Boeing 727 and Britain's Trident. The TriStar was considered more pleasing aesthetically, but MCD engineers argued—with some justification—that sticking that engine on the middle of the tail section created a more efficient air intake system.

Both airplanes won instant pilot praise for their flying qualities. The DC-10, however, was the clear sales winner; by mid-1973, MCD already had delivered one hundred DC-10s and had firm orders for another one hundred and ninety-seven, plus options for thirty-seven more. In the same period, Lockheed trailed with only one hundred and thirty-five firm orders and sixty-five options.

Yet during the DC-10's first three years of safe, worldwide operating experience, there had been one particularly disturbing incident that virtually scripted the eventual fate of THY Flight 981. On June 12, 1972, an insecurely latched aft cargo door on an American Airlines DC-10 blew off over Windsor, Ontario, shortly after the flight departed Detroit. The explosive decompression blast ruptured most of the hydraulic control lines, opened a seventeen-foot hole in the bottom of the fuselage, and collapsed part of the cabin floor.

The pilots were left flying an airplane with no rudder control, no stabilizers, extremely stiff elevators, and only the engine throttles and ailerons—a movable wing part that provides lateral control—providing partial maneuverability. Captain Bryce McCormick brought the crippled plane in safely by manipulating engine power—cutting off the center engine while adding thrust to the other two just before landing—a technique taught in simulator emergency training.

Once the poorly latched cargo door was established as the culprit, corrective action was taken, including beefing up the door structure and modifying the electric locking system. Engineers also installed a small peephole in the door, only an inch in diameter, which enabled baggage handlers to make a visual check of the locking mechanism. Actually, a ground crew member could see only one of the four lockpins located behind the door latches; it was assumed that if this lockpin was positioned properly, the other three were also, and likewise all locking components could be assumed to be correctly hooked and properly aligned. Admittedly, the peephole method wasn't foolproof, but even if a baggage handler failed to make a visual check, an improperly secured door was supposed to activate a warning light on the cockpit instrument panel.

Not all safety experts were convinced McDonnell Douglas had really fixed the door problem. During early pressurization tests on a DC-10 fuselage, a rear cargo door had failed under far less pressurization stress than the American flight was to encounter later. And the cabin floor had buckled in that precertification test, too, offering plenty of advance warning of a potentially catastrophic structural failure that could involve the hydraulic control lines located under the floor. More than one National Transportation Safety Board (NTSB) official had also questioned MCD's decision to use three hydraulic systems on the DC-10; the 747 and L-1011 each had four, this redundancy providing an extra margin of safety even though the Federal Aviation Administration (FAA) had certified three systems as adequate.

McDonnell Douglas believed strengthening the door structure and installing the peephole had solved the cargo door problem.

What happened March 3, 1974, proved otherwise.

Details of the Crash

Although the wreckage of Flight 981 resembled an aluminum junkyard, investigators had plenty of clues to work with—starting with the cockpit voice recorder and recorded cockpit-Air Traffic Control communications. The DC-10 was nearing twelve thousand feet in a steady climb when Orly controllers saw its radar image split in two. The larger blip was that of the airplane; the smaller, horrified controllers realized later, represented a cargo door blown from the fuselage, carrying bodies and part of the cabin floor along with it.

All contact with the flight ended, but the recovered cockpit voice recorder provided indisputable evidence of a violent explosive decompression—including the copilot's startled exclamation, just before the radar image split in two: "The fuselage has burst!"

As the stricken jet fell out of control, the voices of the two pilots vividly described the twelve thousand-foot descent to oblivion.

"Bring it up! Bring her nose up!"

"I can't bring it up! She doesn't respond . . . nothing is left!"

Then, seconds before impact, there were three final, desperate words.

"We've lost it."

It was subsequently determined that the aft cargo door blew off during the required increase in fuselage pressurization as the aircraft climbed to its assigned cruising altitude. The resulting explosion severed every control component—backup as well as primary—and collapsed part of the cabin floor, sucking passengers still strapped in their seats out of the airplane. It was a replay of the Windsor incident, but with a far more tragic ending.

Investigators learned that an Orly baggage handler hadn't made sure the cargo door was locked securely. A warning light should have flashed on the cockpit instrument panel, and at first it was assumed the bulb must have burned out. But later it was determined that the design of the supposedly fail-safe locking mechanism had sent a false "secure and locked" signal to the cockpit.

It would have been only too simple to place the majority of blame on the Orly baggage handler who had closed the door as he had been taught, but hadn't checked the peephole. But this didn't surprise investigators; some baggage handlers working for DC-10 operators in the United States didn't even know why the peephole was there—they assumed it was for the convenience of mechanics.

Not until subsequent litigation hearings was culpability extended far beyond a simple baggage handler's error.

Impact

The hearings spawned by scores of lawsuits uncovered some unpleasant facts, and the most objective conclusion is that there was plenty of blame to go around.

First, there had been a glaring if inadvertent assembly line mistake by McDonnell Douglas. The post-Windsor modifications to the cargo doors included a crucial support plate for additional strength, but when the modifications were implemented on Ship 29—the ill-fated THY DC-10—the

Shortly after its takeoff from Paris Orly airport, this Turkish DC-10 crashed into a nearby forest. An improperly fastened cargo bay door caused the fuselage to explode.

plate somehow had been left off. Yet MCD records showed all modifications had been accomplished and inspected, a discrepancy the company attributed to human error.

There was general agreement that the Federal Aviation Administration was partially at fault. It had certified the original cargo door design as airworthy, despite the precertification test failure. Then, after the near disaster at Windsor, the FAA compounded its rather careless design analysis by playing down the cargo door problem. Instead of issuing a strong airworthiness directive, which would have emphasized the potential for catastrophe by requiring mandatory compliance with door modifications on DC-10s already in operation, the agency merely advised all DC-10 operators to follow a manufacturer's service bulletin—a watered-down warning at best.

Since the Windsor incident and the THY crash, the NTSB had been questioning not only the ability of the DC-10's cabin floor to withstand an explosive decompression, but the vulnerability and possible inadequacy of the three hydraulic systems. On May 25, 1979, while the FAA was still dragging its bureaucratic heels on the safety board's concerns, a Los Angeles-bound

American Airlines DC-10 crashed a few minutes after taking off from Chicago's O'Hare Airport.

The aircraft was in a routine climb when the entire port engine and pylon ripped away, momentum carrying these heavy components over the wing and destroying part of the wing's leading edge. The impact severed the hydraulically operated leading edge slats, creating an asymmetrical flap condition akin to locking the wheels on one side of an automobile. The DC-10 stalled and fell out of control, carrying all two hundred and seventy-five people aboard to their deaths.

Cargo door deficiencies, long since eliminated, played no role in this accident. Major responsibility was placed on American for using an unapproved procedure in changing engines, which had led to a cracked pylon, and on McDonnell Douglas for the original pylon design. But again the wisdom of relying on three control systems instead of four came under scrutiny. The FAA took the drastic step of grounding virtually all U.S.-operated DC-10s pending inspection and redesign of the pylon attachments, an order many felt was sheer overreaction taken mainly to give the agency a "tough guy" image, and to draw attention away from the FAA's alleged incompetence during certification.

Critics pointed out that the pylon design—like that of the aft cargo door, the cabin floor structure, and the use of three hydraulic control systems—all had FAA approval.

All DC-10s flying today have a backup system that provides control capability if complete hydraulic failure occurs. It is, in a way, the legacy of Flight 981.

ROBERT J. SERLING

Where to Learn More

"Cargo Door Focus of Paris Crash Study." *Aviation Week and Space Technology*. March 11, 1974, pp. 198-199.

"Airlines Expand DC-10 Cargo Door Fixes." *Aviation Week and Space Technology*. March 18, 1974, pp. 26-27.

"DC-10 Evidence Points to Cargo Door." *Aviation Week and Space Technology*. April 1, 1974, pp. 26-27.

"Shock Waves from the DC-10 Crash." *Business Weekly*. March 30, 1974, pp. 26-27.

Eddy, Paul, Elaine Potter, and Bruce Page. *Destination Disaster: From the Tri-motor to the DC-10: The Risk of Flying*. New York: Quadrangle, 1976.

Fink, D. E. "Turkish Airlines DC-10-30 Court Case." *Aviation Week and Space Technology*. June 9, 1975, p. 27; June 23, 1975, pp. 26-28; July 14, 1975, p. 30; July 28, 1975, p. 39+.

Godson, John. *The Rise and Fall of the DC-10*. New York: David McKay Co., Inc., 1975.

"Death Comes at Ermenonville: Crash of Turkish Airlines DC-10." *Time*. March 18, 1974, p. 40.

"Great DC-10 Mystery." *Time*. April 8, 1974, p. 71.

Witkin, Richard. "Maker to Improve Safety on New DC-10 Airliners." *New York Times*. November 1, 1974, p 77.

Witkin, Richard. "Cargo Door Clue Studied in Crash of Turkish Plane." *New York Times*. March 9, 1974, p. 58.

American Airlines DC-10 crash

Chicago, Illinois (1979)

Crash after left engine ripped away from aircraft during takeoff killed 273 and prompted the FAA to temporarily rescind the airworthiness certification of the DC-10.

Background

On May 25, 1979, the left engine of a DC-10 aircraft broke away during the takeoff of American Airlines Flight 191 from Chicago's O'Hare International Airport. Although the DC-10 was heavily loaded for its trip to Los Angeles, the aircraft managed to climb 325 feet above the ground before it rolled to the left and began its fatal dive. The aircraft crashed less than a mile from the end the runway, cutting a swath in the ground with its left wing and narrowly missing three large fuel storage tanks. The explosion resulting from the crash, described by eyewitnesses as a fireball similar to a nuclear "mushroom cloud," killed 271 persons onboard the DC-10 and 2 persons on the ground. The crash of Flight 191 was the worst American aircraft accident to date.

The American Airlines crash was not the first major accident of the McDonnell-Douglas DC-10, nor would it be the last. In 1974, a Turkish Airlines DC-10 crashed in the course of takeoff from Paris' Orly Field after a cargo door blew out due to a poorly-designed latch. The resulting gap in the side of the fuselage depressurized the area underneath the floor of the aircraft, causing the floor to collapse and sever the aircraft's control cables. The DC-10 entered a free-fall dive from fourteen thousand feet that left 346 people dead.

Most of the controversy that was to surround the DC-10, however, involved the pylons that attached the aircraft's underwing engines to the wing itself. McDonnell-Douglas issued a service bulletin (a recommendation from the manufacturer about how a specific maintenance procedure should be handled) relating to these pylons in May 1975 and again in February 1978. These two service bulletins called for the replacement of some of the upper and lower bearings on the pylons.

Since a pylon weighed over 1,800 lbs, and the combined engine and pylon weighed 13,477 lbs, the replacement of the bearings could not be accomplished anywhere or at anytime. In the service bulletins, McDonnell-Douglas recommended that aircraft operators change these bearings when an engine needed to be removed or replaced, since this larger procedure would expose and allow easy access to the pylons attached to the wing. Therefore, all of the American commercial airlines operating the DC-10 at the time—American, United, Continental and National—planned the replacement of the pylon bearings around the dates when engines were scheduled to be dismounted at well-equipped maintenance facilities.

American Airlines eventually developed an internal Engineering Change Order (ECO) to direct their maintenance technicians in the bearings replacement procedure. In the formulation of this ECO, American's engineering and maintenance personnel assessed the feasibility of using a large forklift to remove the entire engine/pylon package, thereby reducing the man-hours and costs involved. With the aid of the forklift, American believed it could save over two hundred man-hours per engine, as well as reduce the number of hydraulic and electric line connections and disconnections necessary. This strategy seemed feasible in principle since American's competitor, United Airlines, removed the engine/pylon package from their DC-10s with an overhead crane.

Before this time-saving method was adopted, American Airlines contacted McDonnell-Douglas to discuss the procedure. The engineering staff at McDonnell-Douglas expressed reservations about the proposed method: "Douglas would not encourage this procedure due to the element of risk involved in remating the combined engine and pylon assembly to the wing." However, since American itself owned the aircraft, McDonnell-Douglas did supply recommendations. McDonnell-Douglas indicated that during the lowering and raising of the engine/pylon package, the forklift should lift the engine assembly precisely at its center of gravity (CG) to reduce the risk of damage while the engine was being supported by the forklift. To this effect, American Airline's engineering staff mathematically calculated the CG of the engine/pylon package. In practice, however, the forklift was positioned under the engine assembly on the basis of a visual estimate.

While the engine was being lowered and raised, it was not unusual for the assembly to shift slightly. An American Airlines maintenance record indicated that the forklift lowered slightly under load (when holding up the engine assembly), so that any portion of the engine that remained attached to the aircraft was subjected to great stress. Moreover, because of the close fit of the pylon to the wing and the minimal clearance between structural elements, American's maintenance personnel had to be extremely careful when performing the dismounting procedure. A minor mistake or movement by the forklift operator could easily damage the rear attachment bulkhead and flanges. In the case of the DC-10 aircraft used for Flight 191, this is precisely what happened.

Details of the Crash

On May 25, 1979, at a few minutes past three o'clock in the afternoon, Flight 191 began its takeoff roll on runway 32 right of Chicago's O'Hare International Airport. The position of the engines on the DC-10 makes them impossible to see from the cockpit, and the pilots did not realize that a problem existed with the left (number one) engine until just before liftoff from the runway. The number one engine had broken away from the aircraft, trailing behind the pylon designed to connect it to the wing. As the number one engine ripped from the aircraft, three feet of the leading edge of the left wing were also torn away.

By the time the pilots became alerted to the problem, the stretch of runway that remained was insufficient to brake the aircraft. Therefore, as was generally indicated under these circumstances, the pilots completed the takeoff. Under the load conditions of the flight, the aircraft normally would have been operable using the two remaining engines, although its performance would be substantially reduced. The procedures the pilots relied on, however, were based on a power failure in an engine, and not on the engine physically breaking away from the aircraft.

When the number one engine separated from the aircraft, it tore loose all associated hydraulic and electrical lines. The slats (devices to increase lift) on the wing's leading edge were hydraulically operated, and when the hydraulic pressure to lower the slats for takeoff was lost, the slats retracted back into the wing. The backup system for holding these slats down failed because it too relied on hydraulic lines that had been torn loose with the breaking away of the engine. The result of the slats retracting back into the left wing during takeoff was that the wing would stall or stop "flying" at a much higher airspeed than normal. With an engine out, the aircraft was incapable of maintaining speed during a climb, and the aircraft began to decelerate and approach the stall speed of the left wing.

Under ordinary circumstances, a crew is able to react to signals of an approaching stall. In this case, however, the stall warning system as well as the slat position warning system was inoperative because the electrical system that powered them were dependent on the generator from the number one engine. No backup system was designed into the DC-10 to cope with this situation, and the pilots were never warned that the wing was about to stall. If they had known precisely what was happening to their aircraft, they could have reacted effectively. In post-crash tests conducted in a DC-10 simulator, a device capable of recreating the last moments of Flight 191, cockpit crews who knew their stall warning device was inoperative were able to take steps that would lower the nose of the DC-10, keep the aircraft flying, and prepare for a landing.

As the crew of Flight 191 tried to keep the DC-10 climbing, its airspeed decreased and reached the 159 knot stalling speed of the wing, at which time the aircraft began a slow roll to the left that the pilots were unable to control. When the aircraft rotated past the point where the wings were perpendicular to the ground, the nose of the DC-10 dropped and the aircraft began its final descent.

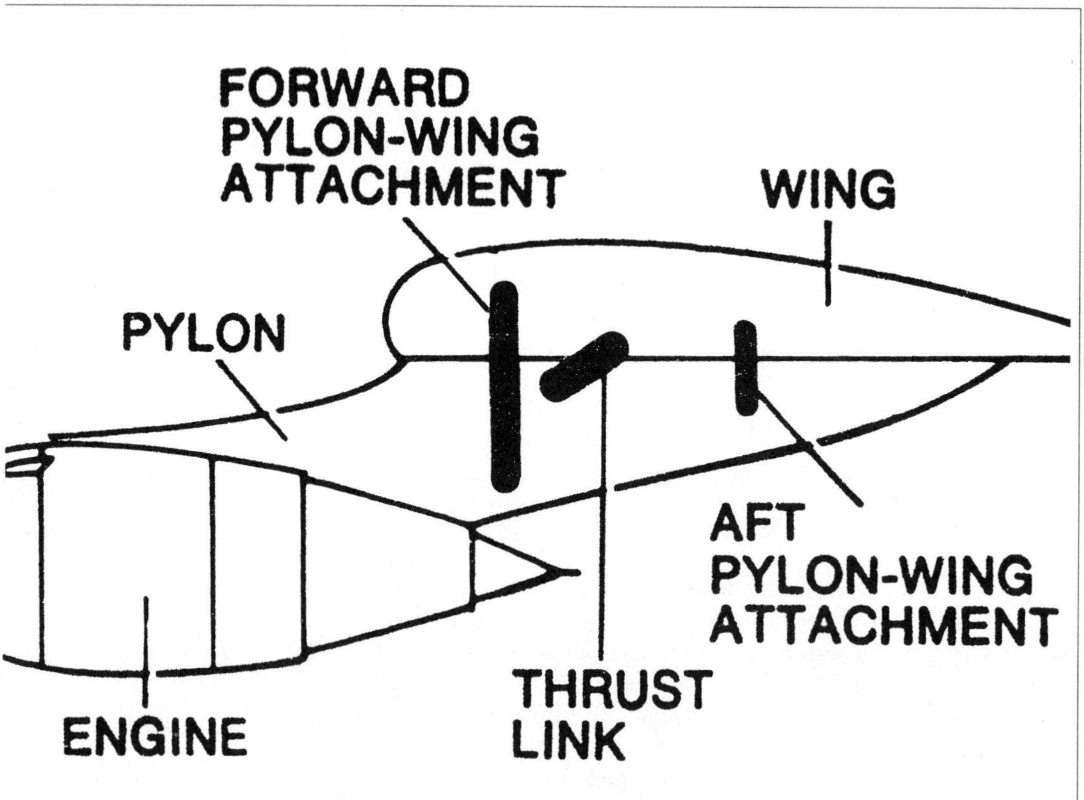

This diagram shows the initial takeoff position of the left-wing pylon-engine assembly.

Impact

The report prepared by the National Transportation Safety Board (NTSB) about Flight 191 exonerated the pilots of any responsibility for the accident, given the various system failures that occurred. The NTSB report blamed the crash on the maintenance procedures devised by American Airlines and on the design of the DC-10 engine pylons.

The NTSB determined that method of engine/pylon removal developed by American Airlines, combined with the stresses exerted by movements of the forklift, caused overstress cracks in the flanges in the rear of the engine pylons. Subsequent tests performed by both American Airlines and McDonnell-Douglas confirmed that a movement of the forklift as slight as six-tenths of an inch could damage the engine attachment flanges and bulkheads. On December 19, 1978 and on February 22, 1979, Continental Airlines DC-10s undergoing maintenance on engine pylons had been damaged in exactly the same way. Continental Airlines did not report this damage to the Federal Aviation Administration (FAA), however, nor were they required to do so.

The NTSB was very pointed in its comments on the design of the DC-10 engine pylons. The NTSB cited the aircraft manufacturer for an engine pylon design that left attachment devices susceptible to damage during mainte-

Firemen examine the smouldering crash site of Flight 191. The DC-10 plunged to the ground after its left engine fell off; the engine's rear attachment bulkhead and flanges had been damaged during maintenance procedures.

nance procedures that were inherently difficult to perform. The NTSB report maintained that "McDonnell-Douglas should have foreseen that pylons would be removed, and therefore the parts of the rear bulkhead [the flanges] should have been designed to eliminate, or at least minimize, vulnerability to damage during maintenance."

The NTSB was also concerned with the design of the flight control, hydraulic and electrical systems, the loss of which kept the cockpit crew on Flight 191 from determining the nature or extent of the problem. Although the original certification of the DC-10 did require the aircraft to be operable despite the failure of certain lift devices, the regulation applied only to the wings' trailing edge flaps, and not to the leading edge slats that retracted during the Flight 191 takeoff. Additional NTSB investigation confirmed that, with an engine and a slat failure, the ability of the DC-10 to climb safely was jeopardized. Finally, the NTSB acknowledged that although the certification of the DC-10 complied with all current regulations, the regulations themselves may have been inadequate and have contributed to the Flight 191 crash.

As a part of the NTSB's final report, it recommended that McDonnell-Douglas provide to all airlines flying DC-10s some emergency instructions about the improper maintenance procedures that caused the American Airlines accident. On June 6, 1979, the FAA took the NTSB's recommendations one step further and revoked the airworthiness certificate of all DC-10 models, effectively grounding the American commercial aircraft fleet until repairs of the pylons were completed and maintenance procedures were corrected. On June 26, 1979, the FAA broadened the restrictions on the DC-10 through a Special Federal Aviation Regulation that "prohibited the operation of any Model DC-10 aircraft within the airspace of the United States." Foreign airlines, therefore, were also prohibited from using the aircraft to land in or depart from this country. On July 13, 1979, after the FAA confirmed that the airlines had performed the necessary repairs and updated the maintenance procedures, the agency pronounced the McDonnell-Douglas DC-10 airworthy.

ROBERT MARK

Where to Learn More

North, David M. "Crash to Boost FAA Scrutiny." *Aviation Week and Space Technology,* June 4, 1979, pp. 12-15.

"Worst U.S. Air Crash." *Time,* June 4, 1979, pp. 12-13.

"Disaster on a Jumbo Jet: Crash of an American Airlines DC-10 at Chicago's O'Hare International Airport." *U.S. News and World Report,* June 4, 1979, p. 14.

Williams, Dennis A. et al. "It Just Disintegrated." *Newsweek,* June 4, 1979, p. 24+.

Witkin, Richard. "U.S. Halts DC-10's Indefinitely, Demanding New Safety Tests: Travel Is Disrupted for 60,000." *New York Times,* June 7, 1979, p. 1.

Air Canada DC-9 accident

(1979)

The rear portion of a DC-9 fuselage ripped open in flight owing to fatigue cracks that had gone undetected in a recent inspection.

Background

On September 17, 1979, about fourteen minutes after takeoff from Boston's Logan International Airport, an Air Canada DC-9 passenger jet bound for Yarmouth, Nova Scotia, lost its rear tail cone. Compressed air inside the cabin rushed explosively out a gaping hole at the rear of the aircraft, knocking a flight attendant off her feet and tossing out a fully loaded drink cart and lavatory water supply tank.

The accident, and subsequent scrutiny of fatigued tail cones on other DC-9 jets, showed how efforts to monitor aging jets and keep them safe by means of intensive inspection of parts known to be susceptible to failure can be undermined by human error.

Early versions of the DC-9, a short-hop jet designed and manufactured by Douglas Aircraft Company, included an auxiliary emergency exit door in the aft pressure bulkhead, or rearmost interior wall of the fuselage. In a ground emergency, the tail cone (the tapered rear section of the jet) was designed to be jettisoned to allow an escape route through the auxiliary door in the aft pressure bulkhead. Later, Douglas changed the design, replacing the escape door with a stairway that tucks into the cabin for flight and could be lowered onto the ground for ingress or egress.

In January 1976, it came to the attention of Douglas Aircraft officials that older-model DC-9s, in particular aircraft with fifteen thousand or more flights, were susceptible to cracking in the aft pressure bulkhead. Because the DC-9 is a smaller jet designed for short trips, it often makes two or more flights each day. Each time the airplane rises from the ground to twenty thousand or thirty thousand feet cruising altitude, the cabin is pressurized by compressed air pumped into the cabin from the jet engines. The result is that at seven or eight miles above the surface of the earth, passengers feel about the same atmospheric pressure they would experience standing on the ground in Denver, Colorado, the Mile High City. With each pressurization cycle, the aircraft fuselage actually expands slightly, like a balloon, then contracts as the aircraft descends and the cabin depressurizes. The effect is akin to bending one of the loops on a paper clip back and forth. Fatigue stress accumulates at the molecular level of the metal, until the metal eventually cracks.

Douglas notified all DC-9 operators and recommended visual inspections for cracks and repair of any cracks found in the aft pressure bulkhead of older-model DC-9s. Concern was focused on the jets with the old-style escape door. About a month later, in February 1976, the company issued a letter that gave airline maintenance crews additional details, including diagrams outlining suggested maintenance steps. Concern about the cracks continued to mount. In May 1976, Douglas issued a higher-level alert, known as a service bulletin, advising DC-9 operators that the cracking could be attributed to fatigue caused by pressurization cycles. The service bulletin also set forth extensive corrective measures.

Operators were advised to inspect the bulkhead for cracks periodically. If no cracks were found, the airline could elect to perform preventative modifications, namely, beefing up the problem area by installing reinforcing doublers and fillers on the emergency exit door jam. But the airlines were given a less costly alternative as well. If no evidence of cracking was found, and the airline chose not to reinforce the problem area, the company recommended ongoing, periodic visual or X-ray inspections of the area until such time as cracks developed in the door jam. The proper corrective steps to refurbish the jam were outlined in the bulletin.

On May 5, 1979, Air Canada DC-9, tail number CF-TLU, underwent a routine inspection and refurbishment known as a C-check. The aircraft at that point had made twenty-five thousand eight hundred seventy-nine flights, well above the fifteen thousand-flight benchmark for which there was heightened concern about aft bulkhead cracks. X-rays were taken of the aft bulkhead door jam, as prescribed by the manufacturer's service bulletin. No cracks were detected. The airline elected not to take the opportunity to reinforce the door jam area.

Details of the Accident

At 12:12 P.M., September 17, 1979, Air Canada Flight 680 departed routinely from Logan International. The DC-9 jet had made more than nine hundred thirty flights since the X ray of its aft bulkhead proved negative five months earlier. On board were two pilots, three flight attendants and forty passengers heading for Yarmouth, Nova Scotia.

At 12:26 P.M., with the first officer flying the aircraft, the flight was about to level off at twenty-five thousand feet. At that point, there was an extremely loud bang followed by complete loss of cabin pressurization. The first officer placed the aircraft on autopilot as he and the captain donned their oxygen masks. The captain then assumed control of the aircraft and started an emergency descent.

The pilots radioed Boston tower that they were in an emergency descent at twenty-three thousand feet and dropping. Less than thirty seconds later the pilots advised air traffic control that they were at twenty thousand feet, requesting clearance to level off at nine thousand feet. The door to the cockpit was missing, apparently having torn loose during the rapid decompression. Looking behind them through the cabin, the pilot and first officer saw blue sky where the rear of the cabin should have been. The captain noticed that the passenger's oxygen masks had deployed, so he slowed the rate of descent. The purser notified the captain that everyone was all right except for a flight attendant who had been knocked briefly unconscious. When the explosive decompression occurred, the flight attendant was picking up trays in the rear of the cabin. Her leg was caught in a seat row as she fell to the floor, letting the trays go. She was unconscious for about fifteen seconds and suffered minor leg, head, and hand injuries.

The flight turned back for Boston missing the tail cone, the aft pressure bulkhead access door, a fully loaded drink cart, and the lavatory water supply tank. As the pilots descended into the Boston area, they experienced problems with the throttle controlling the right engine. However, cockpit readings showed normal engine power and normal operation of the hydraulic system. The aircraft was purposely kept high on approach for landing. The wing flaps and landing gears were used to reduce speed and altitude. This was done because of limited control of the throttle setting for the right engine. The landing was normal, except that an engine braking device, called a reverser, failed to activate after touch down.

Impact

This accident heightened concern about lax oversight of known safety problems related to upkeep of aging jets. Just four months earlier, more than two hundred seventy people were killed in an American Airlines DC-10 jet crash in Chicago. That accident, one of the worst aviation disasters in U.S. history, occurred after one engine tore off from the wing on takeoff. Investigators blamed a maintenance shortcut that damaged the DC-10's engine mount.

On examining the damage on the DC-9, investigators found numerous fatigue cracks in the area around the aft bulkhead door. Some of the cracks clearly were caused during the explosive decompression, but many others showed signs of existing well before the accident. Metallurgical analysis showed some of the cracks had been growing for at least three thousand flights. On reexamining the X-rays taken during the jet's last inspection the previous May, investigators confirmed their suspicions that the cracks had been in the making for some time. The X-rays clearly showed a substantial crack near the lock bolt fastener hole in the jam below the door.

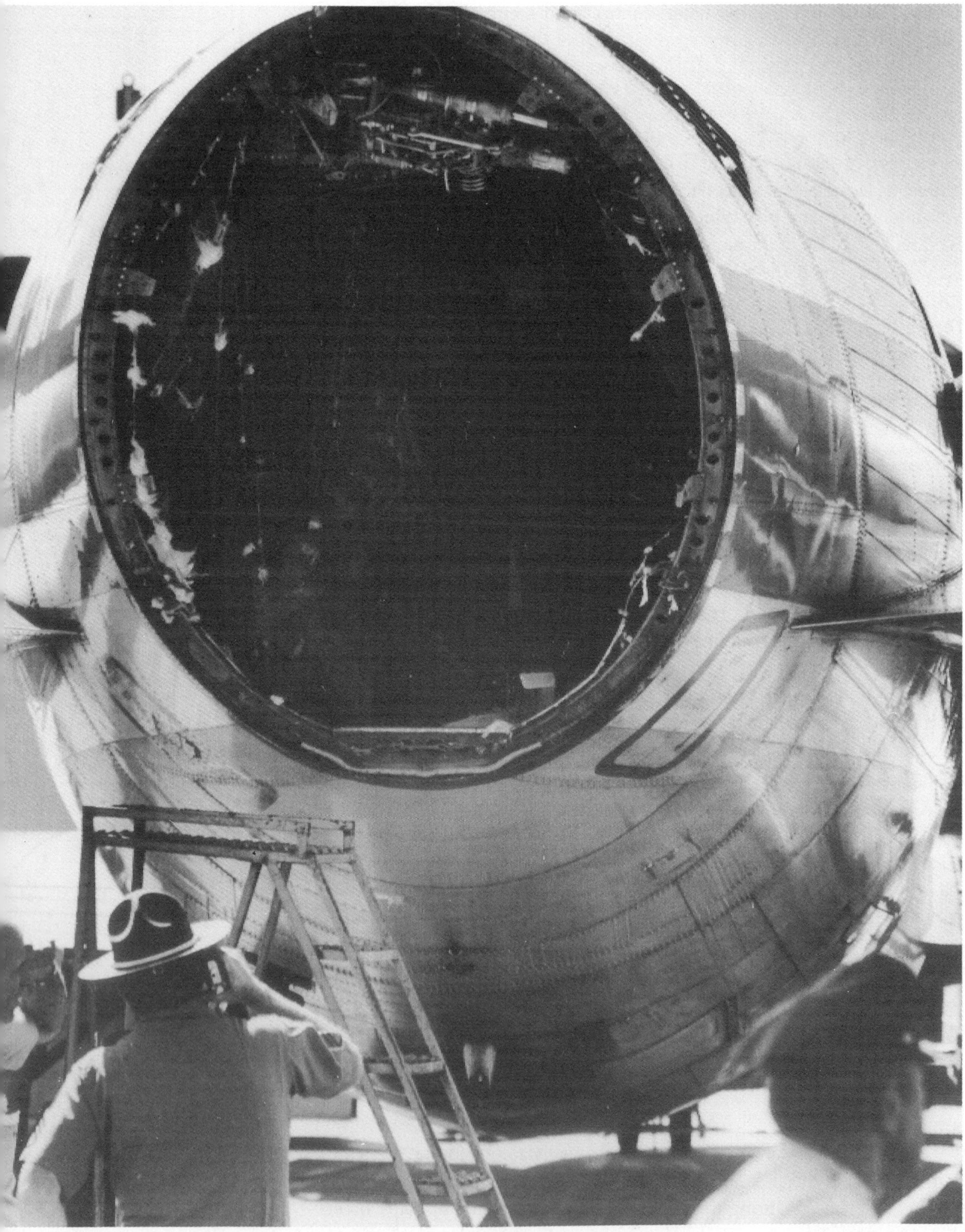

The rear portion of this DC-9 ripped away during a flight in 1979. Investigations revealed fatique cracks in the aft bulkhead door area.

Air Canada officials conceded that the crack had been overlooked during the inspection. Airline officials rationalized that the X-ray plate was difficult to read and that the crack turned up on the postaccident reexamination only because investigators knew precisely were to look and what to look for. Investigators from the National Transportation Safety Board countered that the X-ray plates showed fatigue cracks that were clearly discernible and easily identifiable. The safety board concluded that the airline's inspection and quality control procedures allowed the discrepancy to go undetected.

On examining the damage on the DC-9, investigators found numerous fatigue cracks in the area around the aft bulkhead door. Some of the cracks clearly were caused during the explosive decompression, but many others showed signs of existing well before the accident. Metallurgical analysis showed some of the cracks had been growing for at least three thousand flights.

After the accident, the Federal Aviation Administration, the federal agency principally responsible for air safety, issued a series of emergency orders calling for immediate inspection and mandatory upgrade of the aft bulkhead area on older DC-9s. Air Canada checked its fleet of forty-three older DC-9s for similar cracks. One jet in Regina, Saskatchewan, was found to have cracks in the same area as the jet involved in the accident and was withdrawn from service. The jet was ferried, unpressurized, to the airline's maintenance base in Dorval, Quebec, and refurbished. A second Air Canada jet was found with similar cracks, and was likewise repaired.

Meanwhile, other airlines scrutinized the aft bulkhead area of their older DC-9s as well. Despite the inspection and upgrade bulletins in effect for three years, pervasive problems were uncovered. Cracks of varying lengths were discovered in the aft bulkhead area of thirty-three other DC-9 jets belonging to seven other airlines. In its final report on the accident, the safety board noted that left undetected, the cracks could have resulted in catastrophic accidents, the causes of which would have been extremely difficult to determine.

Safety officials ultimately concluded that the actions taken by the FAA after the Air Canada accident to insure more thorough and timely inspection of the DC-9's aft bulkhead and to promote effective quality control procedures were enough to eliminate the likelihood of a repeat of this type of fatigue failure.

BYRON ACOHIDO

Where to Learn More

"Air Scares: Tail Cone Falls, Flaps Break." *Time*, October 1, 1979, pp. 25-26.

Lenorovitz, Jeffrey. "DC-9 Tail Cone Loss Spurs Aft Bulkhead Inspections." *Aviation Week & Space Technology*, September 24, 1979, p. 32.

National Transportation Safety Board. Aircraft Accident Report no. NTSB/AAR-80/13. Air Canada McDonnell Douglas DC-9-32 (CF-TLU) East of Boston, Massachusetts, September 17, 1979.

"The Trouble with the DC-9." *Newsweek*, October 1, 1979. pp. 24-25.

Japan Airlines Boeing 747 crash

Gumma, Japan

(1985)

Crash of a Boeing 747 passenger aircraft caused by metal fatigue from previous repairs found to be faulty and undetected.

Background

On August 12, 1985, the eve of the three-day celebration of Bon (Feast of Lanterns), Japan Airlines' (JAL) fully booked Flight 123 to Osaka crashed into the side of Mount Osutaka in remote Gumma prefecture only fifty-five minutes after its takeoff from Tokyo's Haneda Airport. Five hundred and twenty people were killed. The outcome of this disaster led to improved worldwide inspection schedules and secured the integrity of the Boeing 747 aircraft design.

The aircraft, registration JA8l19, was a Boeing 747SR (short-range)-100, manufactured and delivered in 1974. The short range version of Boeing's wide-body transport was developed specifically for Japan's internal market. Japan's domestic air route system is characterized by high-density, short-distance routes, and calls for transport that can withstand the rigors of frequent takeoffs and landings for relatively brief flights with large passenger loads.

With the greater number of takeoff/landing cycles, pressurizations, and depressurizations experienced in short range operations, certain parts of an aircraft fatigue more rapidly. Over three thousand pounds of structural material was added to strengthen the 747SR, whose dimensions and external appearance were otherwise similar to the basic 747 version, to withstand these stresses and to assure a fatigue lifetime equivalent to other 747 models. In a twenty-year lifetime of short-range service, a 747SR is expected to experience fifty-two thousand landings. The JA8119 aircraft had flown over twenty-five thousand hours and logged nearly nineteen thousand cycles since delivery.

Records showed that JA8l19 had previously sustained damage that would later prove critical to Flight 123's crash. On June 2, 1978, the aircraft sent up a shower of sparks as it dragged its tail down on the runway during a hard landing at Osaka International Airport. The airframe was substantially damaged and thirty people on board were injured. After provisional repairs were made by JAL at Osaka, the aircraft was ferried to Tokyo for regular repairs to be carried out by a Boeing repair team.

It was normal procedure for repairs to be contracted out to Boeing since the company manufactured the aircraft had a satisfactory repair record. Boeing allegedly replaced fifty-four feet of skin under the lower fuselage in addition to replacing and riveting the entire lower half of the rear pressure bulkhead to the original upper half. The rear pressure bulkhead, located near the lavatories at the back of the passenger cabin, is made of aluminum alloy and is a thin partition shaped like an umbrella canopy that separates the passenger cabin, which is highly pressurized in flight, from the unpressurized tail cone in the rear.

The aircraft was out of operation for two weeks. A JAL spokesperson at the time of the 1978 incident was reported to have said, "As far as we are concerned it [is] as good as new." But flight attendants reported unusual noises in the stern, which would worsen over time. The aircraft had accumulated over sixteen thousand flying hours and more than twelve thousand landings since repairs were made subsequent to the Osaka incident, after which two prescribed maintenance inspections were conducted.

During the repair work a small part of the bulkhead splice, the area where the new and old components of the pressure bulkhead overlapped, should have joined using double rows of rivets; instead, the joining used only single rows, decreasing the strength of the splice to about seventy percent of its original strength and leaving it susceptible to fatigue.

Details of the Crash

The aircraft landed at Tokyo International Airport from Fukuoka less than an hour before its next scheduled flight at 6:00 P.M., a two hundred and fifty-mile one-hour flight to Osaka International Airport, and was inspected in preparation. On this particular flight, the copilot was being trained for the position of captain and was accordingly seated on the left; the captain was on the right.

After a normal takeoff approximately fifteen minutes passed while the aircraft climbed to its cruising altitude of twenty-four thousand feet and made a routing request. Just as it reached the cruising height, an unusual vibration raised the nose of the aircraft and caused a loss of rudder control. A loud boom followed, immediately after which the captain and copilot signaled an emergency to the tower at Tokyo.

The captain requested clearance to descend to and maintain twenty-two thousand foot altitude, and to return to Haneda; and the ground crew responded with route instructions. But suddenly the plane began to "dutch roll," its nose swaying uncontrollably left and right, and to suffer altitude and speed changes known as *phugoid oscillations*, which would continue until just before the crash. At this point, the plane began to rise again.

When ground asked the nature of the emergency, there was no response from the aircraft. Route instructions were repeated but received the response "now uncontrollable" at about 6:30 P.M. In the next two minutes the aircraft changed course northbound, responded "now descending," and reported an altitude of twenty-four thousand feet.

The aircraft continued a rapid descent eastward. The crew reported again that the plane was "uncontrollable" at about 6:45 P.M., flying wildly between altitudes of seven thousand feet and thirteen thousand feet. After acknowledging a transmission that both Haneda and Yokota U.S. Air Force air base approach controls were ready for an emergency landing at 6:55 P.M., the crew made no further communications.

Only seventy-two nautical miles from Nagoya Airport the aircraft requested a return to Haneda despite its position. After veering right and left near Mount Fuji, at approximately 6:40 P.M. the aircraft began its descent from twenty-one thousand feet over Otsuki City to about seventeen thousand feet, actually flying in a complete circle in three minutes.

The aircraft continued a rapid descent eastward. The crew reported again that the plane was "uncontrollable" at about 6:45 P.M., flying wildly between altitudes of seven thousand feet and thirteen thousand feet. After acknowledging a transmission that both Haneda and Yokota U.S. Air Force air base approach controls were ready for an emergency landing at 6:55 P.M., the crew made no further communications.

Eyewitness statements from the four survivors indicated that an abrupt right turn was made before the plane suddenly plunged into a dive, banking to the left behind Mount Mikuni. The aircraft hit the top of a hill with its right wing, cutting several trees, and lost an engine. The aircraft crossed a ravine, losing another engine, and covered 656 yards before striking the top of a 6,233-foot peak and exploding at about 6:56 P.M. It almost totally disintegrated on impact, sections catapulting into ravines on either side. Flight 123 crashed in a remote mountain forest, where even helicopters could not land safely. It wasn't until over fourteen hours after the crash that firemen first reached the area.

After being notified of the accident by Japan's Ministry of Transport, the Aircraft Accident Investigation Commission appointed an investigator-in-charge. Members of Japan's Ministry of Transport, the Federal Aviation Administration (FAA), and teams from the U.S. National Transportation Safety Board (NTSB) and the Boeing company assisted in the investigation of the crash.

The Japanese culture's traditional reverence for the dead placed a priority on locating and recovering the bodies of the 520 who died in the crash, which involved cutting and moving pieces of wreckage that could be critical to the investigation, including the rear pressure bulkhead. The first clue to the cause was revealed when parts of the tail fin and lower rudder were found in Sagami Bay. Almost half the tail fin had detached in flight—it was amazing the aircraft stayed in the air for as long as it.

Examination of bulkhead showed that a single line of rivets had been for part of the repair after the Osaka incident instead of the double line of rivets called

Flight 123 crashed in a remote mountain forest; it wasn't until over fourteen hours after the crash that firemen first reached the area. The accident was caused by metal fatigue from previous faulty repairs.

for in the manual. The faulty repairs allowed abnormal weakening of the bulkhead so that it could not contain the air pressure in the cabin, and the bulkhead blew out in flight. This led to a series of succeeding ruptures, like a chain reaction. The rapid release of pressurized air into the tail section destroyed the vertical tail fin and all four hydraulic systems vital to aircraft control, which is what caused the aircraft to fly wildly out of control.

Impact of the Accident

News of the disaster shocked the Japanese nation. Of the 509 passengers (including 12 infants) and 15 crew members on board the aircraft, 4 passengers, including an off-duty JAL stewardess, who were seated together in the center of row 56, survived with severe injuries. Japanese Airlines president Yasumoto Takagi resigned in a gesture of corporate "guilt," shouldering full responsibility. A JAL maintenance manager at Haneda shamed by the high death toll took his own life in apology for the disaster. The summer of 1985 was one of the darkest times in aviation history—in only two accidents two months apart two 747s (the other aircraft was Air India Flight 182), apparently suffering from some kind of structural damage, had crashed, killing 849 people. The aviation world was deeply alarmed by the events. If the 747 suddenly and unaccountably was proving to be structurally unsafe, answers were needed urgently.

A joint fund was set up by Boeing and JAL on a fifty-fifty basis to compensate victims of the disaster although no recognition of liability was volunteered by either company. The investigation's conclusions allayed any suspicions of a link to the Air India disaster, in which the tail section was found intact. Improper repair, along with JAL's failure to detect the cracks in the bulkhead during regular inspections, proved to have caused the crash. Boeing officials and aviation authorities were relieved, as it dispelled fears that over six hundred 747s in the world have a generic defect and that a repeat of the Japanese disaster was a present danger. Careful inspection programs under way since the crash would find any similar repair defects before accidents occurred.

Actions and countermeasures taken up as a result of this accident included the modification of the 747's empennage (tail assembly) design to protect control surfaces from catastrophic failure in the event of sudden depressurization. The existing vent door designed to accomplish this purpose was found to be inadequate when a large area of the bulkhead suddenly opens. A structural cover for the opening within the empennage was recommended to provide access to the vertical fin. The hydraulic systems design was also modified so all four systems would not be impaired at once in case of pressure buildup. Repair procedures of rear pressure bulkheads to assure they would reinforce Boeing's fail-safe concept were also reevaluated.

The idea behind the initial design of the rear pressure bulkhead was the so-called one-bay fail-safe, which assumed that a crack in the bay could be detected and repaired while remaining within that bay—a single area surrounded by stiffeners and tear straps. In no scenario were cracks presumed to be present simultaneously in several bays, as proved to be the case with JA8119. It was difficult to detect the cracks visually or by air leakage, because repairs were covered with thick sealants and the cracks were small and spread along rivet holes at web overlaps.

Government inspectors were urged to check rear bulkheads in cases of doubt about the validity of repair records. A long-range program to monitor damaged airframe structures was set up. Revisions were made to the 747's rear pressure bulkhead inspection program to provide inspections beyond routine visual inspection to detect the extent of possible multiple-site fatigue cracking. Boeing also called for a switch to more corrosion-resistant bolts in the empennage.

In January 1986 the JAL engineers, still carrying out thorough checks in the aftermath of the disaster, discovered cracked and broken ribs in the frames of the forward sections of some 747 fuselages, later found in other airlines. The FAA issued an emergency directive requiring worldwide inspection of the fleet and Boeing recommended internal visual inspection. Detailed aircraft inspection and prompt repair work overcame the frame-cracking problem and outlined the success of the structural monitoring program.

The JAL crash was recognized as a maintenance error. After a year of difficulties, the lack of evidence suggesting any problem with the 747's basic design, manufacture, or structural integrity confirmed the airworthiness of the aircraft.

Thirty-three (in Japan and the United States combined) damage suits were filed against JAL as a result of the 1985 disaster. All cases were settled out of court, with compensation averaging approximately $850,000 per person.

In addition, in 1992, the Japanese Ministry of Transport set forth a plan removing the 1929 ceiling of compensation for victims on international flights. (That for victims on domestic flights had been unlimited since 1982.) This plan, enacted in January 1993, also facilitates (by eliminating the need for litigation) obtaining compensation for victims of air disasters.

L. A. Turner

Where to Learn More

Aircraft Accident Investigation Commission, Ministry of Transport (Japan). *Aircraft Accident Investigation Report* (JAL JA8119 August 12, 1985), June 19, 1987 [Trans. L. A. Turner].

Allen, Glen. "The Fatal End of JAL Flight 123." *Macleans.* August 26, 1985, p. 42.

Anderson, Harry. "What Went Wrong?" *Newsweek.* August 26, 1985, pp. 14-17.

Haberman, Clyde. "Part of Jet's Tail is Found 80 Miles From Crash Scene." *New York Times.* August 14, 1985, p. 1.

ICAO. *Japanese Civil Aircraft Accident Interim Report,* Report 1985-6 (JAL), August 1985.

"Japan Orders Checks of 747 Tail Sections After JAL Crash." *Aviation Week and Space Technology.* August 19, 1985, p. 30.

"Japanese Cite Faulty Bulkhead Repair as Cause of JAL 747 Crash in 1985." *Aviation Week and Space Technology.* June 29, 1987, p. 34.

O'Lone, Richard G. "747/767 Pressure Bulkhead Tests Precede Suggested Safety Changes." *Aviation Week and Space Technology.* December 16, 1985, p. 29.

Witkin, Richard. "Crash Mystery: What Damaged Tail?" *New York Times.* August 14, 1985, p. 8.

Witkin, Richard. "Clues are Found in Japan Air Crash." *New York Times.* September 6, 1985, p. 7.

Midwest Express Airlines DC-9 crash

Milwaukee, Wisconsin

(1985)

An inexperienced and improperly trained crew became disoriented after an engine failed during takeoff and flew the otherwise controllable airplane into the ground.

Background

On September 6, 1985, less than a minute after lifting off the runway at General Billy Mitchell Field, Milwaukee, Wisconsin, a Midwest Express DC-9 jet experienced the shutdown of its right engine. As the airplane began to twist, or yaw, to the right, the captain reacted instinctively, for a few seconds making the correct adjustments to successfully complete the takeoff. But then, apparently disoriented, the captain reversed the maneuver and flipped the aircraft into an earthward dive, killing all thirty-one people on board.

This air disaster led to increased regulatory pressure for a timely upgrading of certain internal parts of the widely used Pratt & Whitney JT8D series engine, long known to be susceptible to weakening due to corrosion and cracking. It also put a spotlight on substandard surveillance of pilot training programs and independent jet engine repair facilities on the part of the Federal Aviation Administration, the federal watchdog agency in charge of assuring safe skies.

A jet engine works by rapidly spinning a series of fan blades, which compresses large amounts of air and pushes it out the rear of the engine to create thrust. As part of its design, the JT8D used three-inch-long metal alloy sleeves, called spacers, that slipped over the engine's rotating central shaft. The spacers served to keep the compressor blades properly distanced from each other. In the early 1970s, Pratt & Whitney discovered corrosion in and around the spacers had resulted in eventual cracking and failure of the spacers. From the time the JT8D was introduced in the mid-1960s to the mid-1980s, forty-five engine failures related to spacer failure had been reported. In most of those cases, the result was excessive vibration or shutdown of the engine in flight. But in seven incidents, loose spacer parts were ejected through the engine casing. In five cases, the parts actually penetrated the fuselage. Up until the Midwest Express crash, however, no injuries had resulted from problems with the JT8D spacers.

Pratt & Whitney worked steadily over the years to solve the problem. It ran elaborate experiments using spacers with various size cuts in them to determine how many flights it would take before a spacer with a given size crack would break. In 1971 and again in 1981, the engine manufacturer formally advised airlines how to inspect for corrosion and cracks in the spacers and how to make the necessary repairs. In 1983 Pratt & Whitney came up with a new design for the JT8D spacer, making it more corrosion resistant. However, the FAA did not make replacement of the old-style spacer with the new spacer mandatory. The result: both new- and old-style spacers often were mixed on a given engine.

Part of the reason there may have been a lack of urgency regarding upgrading of the JT8D spacers was that twin-engine airplanes like the DC-9, and tri-jets like the Boeing 727, were designed to be flyable if one engine shut down, even during takeoffs and landings, the two most critical stages of flight.

Details of the Crash

At 3:21:26 P.M., September 6, 1985, Midwest Express Flight 105 had just lifted off from Mitchell Field and was climbing at about four hundred fifty feet above the runway when there was a loud clunk and a noticeable decrease in engine sound. The captain asked, "What the — was that?" The first officer did not respond. About three seconds later, the control tower observed smoke and flame emanating from the right engine. Spacers inside the high-pressure section of the right engine had snapped loose and penetrated the engine casing. In an instant, the right engine went from maximum thrust to zero thrust. While the loose spacer parts broke through the engine casing, they did not cause any significant damage to the airplane fuselage, to the system of hydraulic flight controls, or to the left engine.

With the left engine still under full power, the airplane, its nose pointed skyward at about a twelve-degree angle relative to the runway, began to yaw to the right. At 3:21:29 P.M., the captain, addressing the first officer, remarked, "What do we got here, Bill?" Significantly, the first officer again did not respond to the captain. Perhaps the first officer was just as confused as the captain, or he may have been abiding by airline cockpit procedures that directed him to keep conversation to a minimum during emergencies. In any case, the first officer simply notified the control tower, "Midex 105, Roger, we've got an emergency here."

The captain, addressing the first officer, remarked, "What do we got here, Bill?" Significantly, the first officer again did not respond to the captain. Perhaps the first officer was just as confused as the captain, or he may have been abiding by airline cockpit procedures that directed him to keep conversation to a minimum during emergencies. In any case, the first officer simply notified the control tower, "Midex 105, Roger, we've got an emergency here."

As this exchange was unfolding, the captain, with only blue sky in view—and thus no visual points of reference in front of him with which to gauge the relative position of the aircraft—apparently sensed the yawing movement to the right and instinctively made the correct move. He depressed the left rudder pedal, aligning the rudder (the rear, movable part of the vertical tail section) to the left. This maneuver briefly counteracted the yaw to the right, and could have ultimately prevented the crash. However, after a couple of seconds, the captain released the left rudder pedal and let the rudder slip to a neutral position. Then, apparently disoriented, he depressed the right rudder pedal, stalling the airplane and putting it into an earthward dive. Within ten seconds after the failure of the right engine, the aircraft was totally out of control, hurtling toward the ground.

Impact

This accident triggered an extensive investigation that revealed substandard surveillance by FAA officials in charge of monitoring pilot training and certifying cockpit policies at Midwest Express in Milwaukee as well as tracking the proficiency of the Miami-based jet engine repair facility, AeroThrust Corporation, that last inspected the spacers and certified them as being in sound working order.

Before zeroing in on the performance of the pilot and first officer during the crucial seconds after the right engine failed, investigators from the National Transportation Safety Board carefully walked through an elaborate process of elimination. Thorough analysis of the wreckage was compared with information retrieved from the flight data recorder about how the flight controls were operated during the final minutes of Flight 105. More than one hundred witness accounts, many from veteran pilots, were gathered and analyzed. To determine conclusively that the aircraft, indeed, was flyable after the right engine shut down, extensive test flights were undertaken. A series of one-engine power reduction tests were conducted above ten thousand feet, to leave plenty of room to recover. The tests were repeated again at five thousand feet.

The safety board concluded that the test flights proved the DC-9 was docile and easily controllable after a sudden loss of right engine thrust at high speed, and that the airplane did not require unusual pilot skill or strength to maintain continued flight following an engine failure on takeoff. The safety board did acknowledge, however, that at four hundred fifty feet, there was little room for uncertainty or confusion in the cockpit, and that the captain and first officer may not have been adequately trained or sufficiently experienced to recognize that an engine failure had taken place, and that it was imperative for one of them to make the proper maneuvers quickly.

On ascertaining the airplane very likely was still flyable after the engine failure, investigators focused on the experience and training of the pilot and first officer. They discovered that the flight crew was relatively inexperienced. The captain, thirty-one, had been employed by Midwest Express for twelve months and had flown just six hundred hours as a DC-9 first officer before being upgraded to captain on February 7, 1985. By contrast, pilots at large airlines typically put in ten years and upwards of seventy-five hundred flight hours as a DC-9 first officer before earning their captain's wings. The first officer was thirty-seven years old. He was employed by Midwest Express on February 3, 1984 and at the time of the accident had about sixteen hundred flight hours as a DC-9 first officer and eleven hundred hours as a DC-9 captain.

In addition to being relatively inexperienced flying DC-9s, the pilot and first officer had been trained to use some emergency procedures unique to

Midwest Express that contradicted established wisdom. The FAA's principal operators inspector, or POI, in charge of overseeing the Midwest Express pilot training program was a general aviation pilot with no experience flying large commercial jetliners. The POI testified that she was unfamiliar with the DC-9, yet was responsible for approving all of the airline's DC-9 training programs and cockpit procedures, including a controversial "silent cockpit" rule. This policy advised against verbalizing the nature of an emergency, and dictated that pilots should not call out emergency procedures when they begin executing such procedures during an emergency. Midwest's rationale was that excessive chatter could add to the confusion during an emergency, and possibly result in a focus on inaccurate information. However, this policy contradicted longstanding FAA emergency procedures aimed at getting the captain and first officer to verbally coordinate their response to an emergency.

Investigators also found that the flight simulators Midwest Express used to train and license its crew members were somewhat limited. The simulators, for instance, could not imitate the yawing motion, rapid changes in speed, peripheral visual cues, or sounds that occurred as the right engine shut down during takeoff of Flight 105. A veteran DC-9 pilot instructor informed the safety board that engine failure is difficult to recognize quickly, even for seasoned pilots. The instructor testified that one in fifty students (mostly veteran commercial pilots) depresses the wrong rudder pedal when faced with a simulated engine failure on takeoff. Going back to the conversation in the cockpit in the final few minutes of Flight 105, investigators concluded that the captain very likely was confused about the rapidly unfolding events, having never experienced an in-flight engine failure on a DC-9, nor even really coming close to that while training in a simulator. The two questions the captain asked the first officer ("What the—was that?" and "What do we got here, Bill?") seem to underscore his uncertainty in the crucial few seconds where great presence of mind and the rapid execution of the proper maneuvers was required to save the aircraft.

As to why the captain was put in that position in the first place, investigators turned their attention to AeroThrust, the jet engine repair company, and found that the FAA over the years had been lax about requiring adequate record keeping on the part of the repair company. Investigators noted that the FAA seemed satisfied that AeroThrust personnel received mostly on-the-job training for highly complex engine repair duties. It was noted that there were no records of the mechanic who last worked on the spacers involved in the accident having his vision checked annually, as required by FAA rules. Based on research about how quickly cracks grow, investigators determined that a crack of significant length probably existed in the problem spacers when they were last checked by AeroThrust in 1981. The safety board determined that a thorough inspection of the spacer at the time of the 1981 spacer overhaul conducted by AeroThrust should have revealed the cracks and prevented the subsequent failure of the spacer.

The Midwest Express accident ultimately led to new rules requiring timely installation of upgraded spacers in all JT8D engines and improved training for pilots about how to recognize and respond to engine failure during a critical phase of flight.

BYRON ACOHIDO

Where to Learn More

Browne, Malcom W. "U.S. Had Ordered Engines Modified: Agency in 1984 Cited Faults in 11 Midair Emergencies." *New York Times*, August 24, 1985, p. A3.

"FAA Proposed Inspection of JT8D Spacers After Crash Investigation." *Aviation Week & Space Technology*, January 6, 1986, p. 31.

National Transportation Safety Board. Aircraft Accident Report no. NTSB/AAR-87/01, Midwest Express Airlines Inc., DC-9-14 (N100ME) General Billy Mitchell Field, Milwaukee, Wisconsin, September 6, 1985.

"NTSB Orders Engine Tear Down in DC-9 Crash Investigation." *Aviation Week & Space Technology*, September 16, 1985, p. 30.

Salpukas, Agis. "JT8D Engine Gets Scrutiny." *New York Times*, September 20, 1985, pp. D1-D3.

"The Suspect Jet Engines." *MacLean's*, September 16, 1985, p. 51.

United Airlines Boeing 747 explosion

Hawaii

(1989)

Explosive decompression caused by a design that allowed a stray electrical signal to overcome a complex electromechanical system of latches and locks.

Background

On February 24, 1989, about eighteen minutes after takeoff from Honolulu, Hawaii, a United Airlines 747-100 passenger jet bound for Auckland, New Zealand, underwent explosive decompression. The forward cargo door opened, disrupting the even distribution of pressure in the airplane's cabin and cargo hold. The effect was akin to a truck tire blowout. A massive rush of air punched a gaping hole in the fuselage above the door, and nine business-class passengers vanished out the cavity.

The accident triggered a three-year investigation that ultimately revealed how a stray electrical signal could overcome an elaborate system of latches and locks to open a 747's cargo door without being commanded to do so. It also resulted in extensive improvements in the 747's cargo door system.

Designed in the late 1980s, the Boeing 747 jumbo jet, with its distinctive humped fuselage, is one of the most successful jetliners ever built. More than nine hundred 747s are in commercial service, and production is scheduled to continue into the foreseeable future. Each 747 has two nine-foot-square cargo doors located on the starboard (right) side of the airplane's belly, one forward and one aft. Unlike inward-opening "plug doors," which wedge into the passageway as the airplane pressurizes, the cargo doors on a 747 swing out and up, gull-wing fashion.

Until the 747 came along, plug doors were used almost exclusively for both passenger entry and cargo access doors on commercial jetliners. This was considered an extra measure of safety against the possibility of breaching the integrity of the fuselage while the airplane was pressurized. But with the 747, designers wanted to avoid the heavy tracks and wide inside clearances implied by a giant plug door.

So they designed a relatively light door (about eight hundred pounds) that swung out and up and did not obstruct any cargo space. To assure that the door would always hold fast against enormous pressurization forces, an intricate system of electromechanical latches and locks was devised. The system permits a ground worker to lower and shut a door in about fifteen seconds by depressing a toggle switch that sets into motion a series of electrical motors. As a final step the worker locks the door manually by depressing a handle in the middle of the door.

The 747's outward-opening cargo door proved so successful that it became the standard for the ensuing generation of jetliners. Yet soon after the 747 entered commercial service in the early 1970s, a vexing trend emerged. One by one, the door's complicated system of sequential electronic actuators began to fail: locks became battered and bent, and clutches and motors stripped; sensors and switches burned out; latches and locks jammed. Boeing advised airline companies how to rectify each problem.

As the forward cargo door heaved away, it tore a thirteen-by-fifteen-foot opening on the fuselage alongside rows 9, 10, 11, and 12 on the starboard side. In the blink of an eye, business class seats G and H, in rows 8 through 12, vanished along with a chunk of the floor, the starboard aisle, and the eight passengers who were sitting in those seats. Seat 9F, bordering the left edge of the starboard aisle, remained on its mounts, but its occupant, a forty-nine-year-old man, disappeared.

Then on March 10, 1987, the forward cargo door of a Pan Am 747-100 about to depart London for San Francisco was lowered and shut—manually—by a ramp worker using a speed wrench to do the work of the electrical motors. This was a routine procedure prescribed for instances when the door's electrical system malfunctioned. Shortly after takeoff, the pilot found he was unable to pressurize the cabin, so he returned to London, where the worker discovered the forward cargo door open 1.5 inches. The door was reshut and relocked and the plane sent on its way.

A closer inspection in San Francisco revealed severe damage to eight boomerang-shaped door locks designed to secure eight door latches in a closed position during flight. The C-shaped latches work by engaging and then rotating to a closed position around latch pins fastened to the door sill. The locks then swing into place over the open end of the latches, blocking them from reverse-rotating to the open position.

Lab tests conducted by Boeing after this incident revealed that a switch designed to cut all electrical power from the door once the locks are set could jam. This left the door's mechanisms vulnerable to stray electrical signals that could inadvertently actuate the latch motor. The motor could then "backdrive" the latches to an open position, bending the locks out of the way in the process. Tests showed that the locks, made of aluminum, were too weak to prevent the latches from rotating open should something activate the latch motor.

Testing showed that a stray electrical signal could almost instantly backwind the latches past weak locks to a fully open position. An inspection of Pan Am's 747 fleet revealed numerous damaged locks, suggesting a chronic pattern of backwinding of latches against weak locks.

Rather than searching for the possible source of stray electricity activating the latch motor, officials blamed the Pan Am incident on the ground worker who manually operated the door. Though the worker denied doing this, he was said to have used his speed wrench to backwind the latches after setting the locks in place. Officials theorized that this was a shortcut procedure devised by ground workers to confirm that the latches were snugly secured against the locks. It is noteworthy that the mechanic would have had to make ninety-five full turns of his speed wrench to fully open the latches, something the latch motor could do in 1.5 seconds.

About one month after the Pan Am incident, Boeing alerted airlines worldwide that opening of a cargo door in flight "could result in rapid decompression ... resulting in collapse of the passenger cabin floor and the possible damage to airplane electrical and hydraulic systems." That warning was accompanied by a call for airlines to reinforce the weak aluminum locks with special steel braces. Pan Am immediately fabricated the braces and installed them on its 747 fleet. Meanwhile, the Federal Aviation Administration (FAA) assigned a low priority to the matter, giving airlines up to two years to make the upgrade.

Details of the Explosion

On the morning of February 24, 1989, all was routine during preparations for United Airlines Flight 811. Two months earlier, in December 1988, maintenance logs showed the aircraft's forward cargo door's electrical system had malfunctioned repeatedly, necessitating manual operation in fourteen separate instances. But since then, there had been no "write-up" on the door. Flight 811's ground crew used the toggle switch to routinely lower and latch the door, and a ground worker noticed nothing unusual when he depressed the lock handle to swing the locks into place over the closed latches. The locks were the weak aluminum type. United had not yet installed the steel braces.

The flight, carrying 337 passengers, 15 cabin attendants, a pilot, a copilot, and a flight engineer, was cleared for takeoff at 1:52 A.M. As the airplane was climbing past twenty-two thousand feet at more than five hundred miles per hour, pressurization, accomplished by continually pumping compressed air into the cabin and cargo hold, proceeded as usual. Although the aircraft was flying higher than the summit of Mount McKinley, passengers felt no different than if they were crossing a four thousand-foot mountain pass in an automobile.

At 2:09 A.M., all on board were stunned by a loud thump. Captain David Cronin remarked, "What the hell was that?" The thump was followed 1.8 seconds later by a tremendous boom, the sound of eleven million pounds of force—to this point distributed evenly on each square inch of the aircraft's interior—gushing explosively toward a breach in the fuselage. As the forward cargo door heaved away, it tore a thirteen-by-fifteen-foot opening on the fuselage alongside rows 9, 10, 11, and 12 on the starboard side.

In the blink of an eye, business class seats G and H, in rows 8 through 12, vanished along with a chunk of the floor, the starboard aisle, and the eight passengers who were sitting in those seats. Seat 9F, bordering the left edge of the starboard aisle, remained on its mounts, but its occupant, a forty-nine-year-old man, disappeared.

At the moment of explosive decompression, passengers and cabin attendants reported experiencing a hurricane force blast of cold air. Cabin attendant Curt Christensen described it this way: "Immediately the air filled with a hazy smoke and flying debris. I felt a cold wind and sucking, if you will, but it wasn't a sucking. It was like being in the middle of a huge cannon blast, a blast of cold air. There was gray, swirling smoke and debris flying everywhere."

For the next twenty minutes, passengers and cabin attendants, some severely injured, worried that the jetliner would splash down in the Pacific Ocean. They watched as the two starboard engines flamed out. However, Captain Cronin heroically nursed the crippled aircraft to a safe landing in Honolulu.

Impact

This stirring disaster prompted the FAA to order steel braces installed on locks of all 747s within thirty days. Over the next several months, Boeing developed a number of improvements to the door's warning system. As for what caused the accident, officials were stumped. The most telling evidence—the door itself—lay at the bottom of the Pacific Ocean, one hundred miles south of Honolulu.

Even so, after an intensive investigation, but before the door was retrieved from the ocean floor, the National Transportation Safety Board ruled on April 16, 1990, that the weak locks probably were severely damaged during the fourteen instances of manual operation two months before the accident. The safety board concluded that the ground crew somehow did not fully close the latches prior to takeoff, even though they reported no problems using the toggle switch to lower and shut the door, and nothing unusual about locking the door.

Then on September 27, 1990, the U. S. Navy deep-submergence vehicle *Sea Cliff* plucked the lower two-thirds of Flight 811's cargo door from the ocean floor. The eight locks were found in the locked position, but bent out of the way by the eight latches, which had somehow reverse-rotated to the open position. All the locks were found in relatively good shape, which contradicted the safety board's official explanation, based on severely battered locks. A few days later, *Sea Cliff* recovered the upper half of the door. Both parts were then shipped to a Boeing laboratory in Seattle, where they underwent thorough analysis under safety board supervision.

While the door was being examined, a related incident took place on June 21, 1991, at Kennedy International Airport on a newer model United Airlines 747-200 being prepared for a flight to Tokyo. The aft cargo door had been closed and latched, but not yet locked—and no one was touching the toggle switch—when a stray electrical signal rotated the door latches open, then lifted the door up. The powerful door-lift motor continued to run, trying to raise the door past its maximum open position, until a mechanic shut it off by

This photograph shows the starboard side opening on the fuselage. At the moment of explosive decompression, passengers and cabin attendants reported experiencing a hurricane force blast of cold air.

popping a circuit breaker. The stray signal was subsequently traced to a cracked wire-bundle conduit that was located near the door hinges and exposed to wear and tear each time the door was opened or closed.

After that incident, as an extra precaution, United ordered its ground crews to open a pair of circuit breakers just before pushing any 747 back from the gate. This removed all potential power from the cargo door area, above and beyond other power-cutting safety features.

Based on this real-life example of stray signals at work in the door and detailed analysis of Flight 811's cargo door, on March 18, 1992 the safety board took the unusual step of reversing its earlier ruling. The board attributed the opening of the door to a "faulty switch or wiring in the door control system which permitted electrical actuation of the door latches toward the unlatched position after initial door closure and before takeoff."

Although the wire bundle that may have contained conclusive evidence of such a fault was never found, the board weighed other evidence uncovered by United. After the Kennedy door-opening incident, the airline found twenty-one different simple short circuits that could lead to uncommanded actuation of the latch motor. Boeing maintained that steel reinforced locks now used on all 747s would prevent the latches from moving, even if a stray signal did come into play.

Even so, the board called the design of the cargo door locking mechanisms "deficient" and said that a lack of timely corrective action by Boeing and the FAA after the 1987 Pan Am incident contributed to the tragedy of Flight 811.

BYRON ACOHIDO

Where to Learn More

Acohido, Byron. "Flight 811: Terror in the Sky" *Pacific Magazine* (*Seattle Times/Seattle Post Intelligencer* Sunday supplement), January 5, 1992, pp. 10-19.

"Airlines Inspect Cargo Doors in Wake of United Accident." *Aviation Week & Space Technology*, March 6, 1989, p. 22.

"Close Look at Cargo Door Workings Fails to Reveal Cause of Accident." *Aviation Week & Space Technology*, May 22, 1989, p. 86.

"Board Suggests Changing Cargo Door Latch Design." *Aviation Week & Space Technology*, August 28, 1989, p. 29.

Hackett, George. "Flight 811: A Nightmare in the Sky." *Newsweek*, March 6, 1989, p. 26.

Henderson, Breck W. "Investigators Believe Cargo Door Failed, Ripping 747 Open." *Aviation Week & Space Technology*, March 6, 1989, p. 18.

National Transportation Safety Board. *Explosive Decompression—Loss of Cargo Door in Flight*. Aircraft accident report no. NTSB/AAR-92/02, March 18, 1992, pp. 1-113.

United Airlines DC-10 crash

Sioux City, Iowa

(1989)

The crash of a DC-10 aircraft due to an engine explosion and the subsequent loss of hydraulic power led to a massive National Transportation Safety Board investigation.

Background

A regularly scheduled United Airlines DC-10 flight departed from Denver's Stapleton Airport at about 2:00 P.M. enroute to Philadelphia, with a stop scheduled at Chicago's O'Hare International Airport. After an hour of flight, the number 2 (tail section) engine exploded, cutting off the aircraft's hydraulic power. The crew, forced to attempt an emergency landing, almost miraculously was able to bring the aircraft in line with a runway at the Sioux City, Iowa airport. Nonetheless, the crew had no way to slow the airliner's speed of over two hundred forty m.p.h. until immediately prior to touchdown. At this point, the captain's attempt to use the throttles to decelerate sent the aircraft into an uncontrolled spin. In landing, the DC-10 rolled over to the right and blazed into flames. An equipment failure sabotaged the efforts of the Sioux City airport's emergency fire-fighting crew, and one hundred eleven of the two hundred ninety-six people on board the flight were killed.

Details of the Crash

About an hour into United Airlines Flight 232, the cockpit crew heard an explosion in the rear of the aircraft that was followed by a shuddering of the airframe. When the pilots checked their instruments, they learned that the number 2 engine, the engine mounted in the center of the vertical tail section, had failed. As the captain began the emergency shutdown procedure, the flight engineer noticed that the hydraulic system's quantity level as well as pressure had dropped to zero. Hydraulic power is necessary to move the aircraft's elevators, rudders and ailerons on the wings and tail. The co-pilot, who had been flying the aircraft at the time of the explosion, indicated that he could no longer control the aircraft. The cockpit crew attempted to use an emergency backup power source for the hydraulic system, but to no avail.

When the captain took command of the controls, he found the only form of control he had over the four hundred thirty thousand pound aircraft was to use, alternately, the throttles that controlled the two remaining engines—number 1 (on the left wing) and number 3 (on the right wing). When he pulled the number 1 throttle back, the remaining engine on the right wing tended to pull the aircraft around to the left; the opposite reaction occurred with use of the left engine throttle. At this point, the captain reported to air-traffic control that the aircraft was almost out of control and needed to be landed immediately. The nearest airport was in Sioux City, Iowa, where the emergency crew was put on alert.

The captain then instructed the lead flight attendant to prepare the cabin for an emergency landing. He also told her that the chances for a successful landing were not good, since he had very little control over the aircraft's flight path. Another United pilot, who by chance was travelling on the flight, came forward to the cockpit to help. After half an hour of zig-zag turns across Iowa, the crew managed to align the aircraft to land on a short runway at Sioux City; they had too little control over the DC-10 to maneuver it toward a longer runway. The cockpit crew used an emergency system to pump down the landing gear. They did not have use of any hydraulic landing flaps to slow the airliner, forcing them to maintain an aircraft speed of over two hundred forty m.p.h. (the normal landing speed for a DC-10 is about one hundred sixty m.p.h.) until just before touchdown.

A quarter mile from the runway, the captain pulled back the throttles in a last-ditch effort to slow down the aircraft. When he did so, the aircraft began a roll to the right that could not be stopped. As the aircraft touched down, the right wing-tip struck the ground, and the aircraft skidded, turning over to the right as it grounded to a stop. The aircraft was quickly engulfed in flames that killed one hundred eleven of the two hundred ninty-six people on board.

The failure of a critical piece of emergency fire equipment at the Sioux City Airport, which was not prepared to cope with a fire as large as that produced by the Flight 232 crash, contributed to the loss of life after the DC-10 landing. The airport's Kovatch P-18 water-supply vehicle developed a supply hose blockage soon after the crash, which prevented the pumping vehicles from receiving a fresh supply of fire retardant to use against the blaze. Although the fire crew vigorously attacked the fire immediately after the crash, the fire flourished after the supply vehicle failure and burned strongly for another two hours.

As with most aircraft accidents, the crash of United Airlines Flight 232 was the result of a combination of problems, any one of which might have been manageable, but which in conjunction proved almost insurmountable. Flight 232 crashed after the failure of all the aircraft's hydraulic systems left the cockpit crew virtually helpless in its attempts to direct the flight. The hydraulic systems failed because the number two (tail section) engine had not only failed, but exploded.

Two investigators examine the tail piece of the United DC-10 which crashed in Sioux City after an engine explosion resulted in the loss of hydraulic power.

The first stage fan, a device within the opening of the engine that draws in fresh air to help produce the engine's thrust, had failed. The failure of a section of the fan disc unbalanced the remaining portion, causing it to tear itself loose. As the bits of shrapnel-like metal from the fan disc shot through the tail section of the DC-10, they severed the hydraulic lines, which ran through the tail section near the center engine, draining all control fluid from the system.

McDonnell-Douglas had not planned for a total hydraulic failure when they designed the DC-10. Moreover, since the forces required to control an aircraft as large as the DC-10 were beyond manual control by a single individual, the aircraft was not equipped with a manual backup to the flight controls. A large portion of the aircraft tail section was reconstructed after the accident in a hanger at the Sioux City Airport, from pieces of the aircraft gathered during an intensive search across hundreds of square miles of farmland near the airport. The reconstruction proved that pieces of the titanium fan disc had, in fact, punctured the skin of the rudders and horizontal stabilizers in at least seventy-five different places, as well as cutting the hydraulic lines. Although a shroud is installed around the outside of the fan assembly to catch parts that might fly loose from an engine, the shroud could not contain the explosive force of this kind of breakup.

The catastrophic failure of the fan assembly on the number 2 engine occurred because of a fracture to the metal fan disc induced during its production eighteen years earlier by ALCOA. The engine, produced by General Electric Aircraft Engines of Evendale, Ohio in 1972, had flown a total of forty thousand two hundred sixty-six hours. In the life of a turbine aircraft engine, forty thousand hours is old but not unusable, due to the hundreds of maintenance inspections that are performed during an engine's lifetime and in which worn-down components are replaced with serviceable parts. It was in July 1988, during one of these maintenance stops, that the fan disc (serial number MPO 00385) that would eventually fail on Flight 232 was installed on the DC-10. Although routine writeups on this DC-10 aircraft were submitted to the United Maintenance department between July 1988 and July 1989, none indicated any problems with this particular component of the engine.

Impact

The crack in the fan disc initiated a great deal of controversy after the Flight 232 accident, not only in that a defective disc had been shipped out from the ALCOA factory, but in that, over the course of nearly eighteen years of maintenance inspections, the crack was never detected. During those inspections, the disc was repeatedly tested through the use of a fluorescent penetrant (FPI), an industry-accepted method for checking nonmagnetic material, such as the titanium fan disc, for cracks too small to be seen by the unaided human eye. A penetrant fluid is applied to and allowed to flow over the part. The fluid will seep into any cracks in the material and leave a trace later detectable under ultraviolet light. Although metallurgists believe the

The crash during an emergency landing attempt of United Flight 232 resulted in the deaths of 111 people on board.

crack—which was almost half an inch in length—should have been detected, it was not.

During the post-crash investigation, it also emerged that ALCOA had manufactured two discs with the same serial number, MPO 00385. One was discarded due to a defect, and another given the same part number. Experts initially believed that the two discs had become switched, with the defective part shipped out. This theory proved to be erroneous, but only after much investigative work, because GE's records for the manufacture of this part were so poorly maintained. It was ultimately the conclusion of metallurgists that the final machining (known as shot-peening) of the fan disc at ALCOA was responsible for the crack that caused to the accident.

The Federal Aviation Administration (FAA) specifies that the owner of an aircraft—in this case United Airlines—is responsible for the proper maintenance of that craft. Although the aircraft and its engines were indeed maintained in accordance with FAA and GE procedures, United's maintenance department failed to detect the crack on the fan disc during any of its routine checks of the engine parts, the latest of which was performed about a year before the crash.

The fan disc is round and has a large shaft hole in its center. When the fan disc is removed from the engine for the required FAA maintenance, this hole becomes convenient to string the part up for inspection. It was inside this hole, where the cable to hang the part passed through, that the crack was located. Most people familiar with the GE engine believed this spot to be a next to impossible location for a crack to develop, which is why the National Transportation Safety Board (NTSB) concluded that the United Airlines mechanic gave the part only a cursory inspection.

The NTSB determined that the cause of the Flight 232 accident was the inadequate inspection of the fan disc and the fan disc's subsequent breakup in flight. However, one dissenting NTSB member believed the cause to be threefold: the poor manufacturing of the disc at ALCOA; the failure by United Airlines inspectors to detect the crack; and the failure by McDonnell-Douglas to recognize the vulnerability of the DC-10 airframe to a complete hydraulic malfunction after an explosion of the number 2 engine. The FAA determined, however, that McDonnell-Douglas had indeed complied with all certification regulations in effect when the DC-10 was built.

The Flight 232 accident did initiate the formation of a Systems Review Task Force (SRTF) to explore design solutions that would prevent this kind of accident from recurring. Additionally, it was held that "where applicable, the concepts developed by the SRTF should be considered for retrofit of current fleet aircraft." The SRTF includes representatives from Boeing, McDonnell-Douglas, Airbus, Lockheed, General Electric, Pratt & Whitney, and Rolls-Royce.

The NTSB also recommended that the FAA search for new inspection methods to automate procedures and eliminate human inspection error (possible by using a second, back-up inspector). They requested that the FAA to issue an Airworthiness Directive removing from service any aircraft equipped with engine discs manufactured by ALCOA that were similar to the failed Flight 232 disk and made during the same time period. The NTSB also asked the FAA to require that all occupants, even those under two years old, be secured in their seats with an approved restraint device during landing and takeoff. In the initial moments of the Flight 232 crash, the mother of one of the four infants onboard the aircraft could not hold the child in her lap, and he eventually died of post-crash asphyxiation.

While one hundred eleven people onboard Flight 232 died in the crash, one hundred eighty-five survived. Those individuals owe their life to the cockpit crew, who used their combined experience to devise a method of flying an airplane whose normal flight controls no longer had any effect. The captain, a thirty-three year veteran of United Airlines, relied on a sometimes controversial plan called Cockpit Resource Management (CRM)—in which all the pilots truly work as a team to solve a problem—to save those one hundred eighty-five lives.

The McDonnell-Douglas Corporation, the manufacturer of the DC-10, had assumed that a triple hydraulic failure such as occurred in Flight 232 was impossible. After the accident, they spent hundreds of hours in a DC-10 simulator trying to develop a procedure to help a crew cope with this particular systems failure. McDonnell-Douglas discovered, as had the pilots of Flight 232, that control of a DC-10 with no hydraulic systems becomes "very limited." The McDonnell-Douglas simulations underscored how much the Flight 232 crew achieved in maneuvering the DC-10 back to earth.

ROBERT MARK

Where to Learn More

Fotos, Christopher P. "Sioux City Hearings Focus on History of Failed Disk." *Aviation Week & Space Technology*, November 6, 1989, p. 56.

Hughes, David. "DC-10 Accident Leads to Basic Review of Jet Engine Rotating Parts Treatment." *Aviation Week & Space Technology*, May 20, 1991, p. 35.

——— and Michael Dornheim. "United DC-10 Crashes in Sioux City, Iowa." *Aviation Week & Space Technology*, July 24 1989, p. 96.

Ott, James. "Investigators Find Reconstructed Tail of DC-10 Riddled with Damage." *Aviation Week & Space Technology*, July 31, 1989, p. 30.

———. "Probe Focuses on Failure of Fan Disk in DC-10 Crash." *Aviation Week & Space Technology*, July 31, 1989, p. 30.

Springen, Karen. "The 'Fear of Death' Lawsuits; Taking Legal Action after the Flight 232 Crash." *Newsweek*, August 7, 1989, p. 27.

Lauda Air Boeing 767-300 crash

Thailand

(1991)

An inadvertent thrust reverser deployment flipped a jetliner into a crash dive that killed all two hundred twenty-three people on board.

Background

On May 26, 1991, an eighteenth-month-old Boeing 767-300ER (extended range) jetliner operated by Lauda Air was climbing out of Bangkok bound for Vienna when an engine-braking device inadvertently deployed, throwing the left engine into reverse thrust. The airplane instantly flipped into a supersonic crash dive and began breaking up even before it slammed into the jungle floor, killing all two hundred thirteen passengers and ten crew members on board.

This air disaster raised questions about the reliability of electronically controlled thrust reversers put into wide use on all Boeing models starting in the late 1980s. It also showed that aircraft designers had inaccurately predicted how a modern jetliner like the 767, powered by just two large-diameter jet engines, one on each wing, would behave if a reverser deployed in flight.

Reversers slow a jetliner by diverting engine thrust forward. On most modern jetliners they are designed to be used only on the ground as a supplement to the wheel brakes. Air travelers hear the reversers at work moments after a jetliner's wheels touch down on landing; the increased roar of the engines is the sound of thrust being reversed to slow the aircraft down. When Boeing designed the 767 twin jet in the late 1970s, it followed a principle for the reversers used successfully on its older 747 model. The rear portion of the engine casing slid aft, swinging panels into position to block the engine airflow. Thrust was diverted out the opening in the casing at a forward angle.

Power to operate the reverser was supplied by the engine's hydraulic system and was controlled by a series of valves, including a directional control valve, or DCV, designed to sit in two positions. In one position, the DCV directed pressurized fluid to open the reverser. In the other position it directed fluid to close the reverser. On 767s built through the mid-1980s, operation of the reversers was controlled by a mechanical cable linked to a lever in the cockpit. During flight, mechanical locks kept the engine casing secured in the closed position. Another safety feature was found on the reverser actuation lever in the cockpit. The lever was designed to remain in the closed position whenever the throttle was open, powering the engines. The lever could not move to the deploy position unless the throttle was first moved to the idle setting.

In the late 1980s, more sophisticated electronic engine controls became available and Boeing made several major refinements to this system. Boeing began installing the upgraded reverser system on all of its models, including 747s, 737s, and 757s. The main change involved electronic controls replacing the mechanical link to the cockpit actuation lever. Now when the pilot moved the reverser lever in the cockpit, he was not physically actuating the system, but setting into motion electronic signals to perform that function.

Also, an elaborate safety feature, called the auto-restow, was added to the system. Sensors were installed on the casing to gauge if the casing, despite the locks, somehow began to slip open in flight, thus starting the reverser deployment sequence. At the first sign of the casing moving in flight, the auto-restow sensor was designed to signal the electronic controls to send pressurized fluid into the reverser. Since the DCV should always be in the closed position in flight, the DCV would direct the fluid to drive the casing shut, thus automatically restowing the reverser.

Back in the cockpit, an amber "REV ISLN" light was designed to illuminate when the auto-restow activated during flight, and extinguish when the restow process was completed. Pilots were told in the cockpit handbook to expect normal landing if the REV ISLN light came on during flight. While pilots reported seeing the REV ISLN light flickering from time to time, a more common sight was the separate "REV" light, designed to keep pilots advised of various stages of operating the reverser upon landing. The REV light was designed to be blank when the reversers were locked, amber when they were partially extended, and green when the reversers were fully deployed.

In certifying the electronically controlled 767 reverser system and this auto-restow feature with the Federal Aviation Administration, Boeing asserted that there were no known ways the reverser could open in flight. Even so, the FAA asked Boeing to show in a test flight what would happen if a reverser deployed in flight. For this test, Boeing took a 767 up to a relatively low altitude of ten thousand feet. At a moderate speed of about two hundred fifty knots, test pilots were instructed to deliberately slow one engine to idle before deploying the reverser on the idled engine. When they did this, the airplane buffeted and began to swerve, or yaw, in the direction of the reversed engine, but the pilots were able to maintain control. The data from this test was programmed into 767 simulators, and the FAA approved the 767's electronically controlled reverser and auto-restow system without requiring Boeing to calculate or test to see what would happen if a reverser somehow deployed at high altitude and high speed.

Details of the Crash

Shortly after 11:00 P.M. the night of May 26, 1991, a packed Lauda Air Boeing 767-300ER jetliner took off from Bangkok airport with two hundred thirteen passengers and eight cabin attendants on board. At the controls were Captain Thomas Welch, an American, and co-pilot Josef Thurner, of Austria.

At 11:21 P.M., as the jet was climbing under near full thrust, the cockpit voice recorder recorded Welch mentioning to Thurner that the REV ISLN light kept flickering on and off. Thurner asked if he should alert ground staff. Welch told him not to bother, noting that the REV ISLN light is "an advisory thing" and that it was probably flickering because of "moisture in there or something." At 11:26 P.M. Thurner told Welch to add rudder trim to the left to compensate for the airplane beginning to yaw to the left.

At 11:30:37 P.M., the airplane was at twenty-five thousand feet and climbing at a speed of three hundred twenty knots. Thurner spotted the REV light illuminate amber, then green, something that is supposed to happen only on the ground, and says, "Ah, reverser's deployed." At 11:30:39, there was the sound of a snap and at 11:30:41, Welch exclaimed, "Jesus Christ!" At 11:30:44, a series of warning tones began.

The thrust reverser on the left engine had deployed. Because the large-diameter engine was mounted close to the wing and jutted out well in front of the leading edge, the sudden blast of reverse thrust interrupted the air flow over the wing, disrupting lift, or the steady flow of air over the curved top of the wing that kept the airplane aloft. This sudden loss of lift on the left wing caused the airplane to twist, or roll, counterclockwise looking from the tail to the nose.

As the airplane rolled with the left engine, the right engine's high forward thrust caused the airplane to yaw to the left as well. This combination of roll and yaw snapped the right wing forward into the air stream, which produced even more lift on the right wing. With no lift on the left wing and an abundance of lift on the right wing, the airplane's roll was accentuated, and it began a twisting dive to earth. Careening at ten miles per minute, the jet fell eight thousand feet in less than thirty seconds and began to break up at speed and forces it was not designed to withstand. This fury was captured on the final hundredths of seconds on the cockpit voice recorder.

At 11:30:53, Welch said, "Wait a minute." At 11:30:58, he said, "Damn it." At 11:31:05 there was the sound of a bang. At 11:31:06, the recording ended.

Impact

This dramatic air disaster revealed the auto-restow to be an overly complex safety device that probably caused the accident. It also showed that Boeing and the FAA had inadequately examined what would happen if a thrust reverser deployed at a high altitude under high thrust on a modern two-engine jetliner.

The flight recorder of the Lauda Air 767 was analyzed in Washington after it was released by Thai officials. As the plane careened to the ground, the recorder captured the fury in the cockpit.

However, it took several months for authorities to reach consensus that the auto-restow was to blame. In the weeks following the accident, speculation on the cause ranged from a terrorist bomb, to catastrophic engine failure, to spontaneous combustion of a crate of lithium wristwatches in the cargo hold. Adding to the confusion was the fact that the airplane crashed into a remote jungle hillside one hundred thirty miles from Bangkok, whereupon villagers quickly looted the wreckage, including the bright metallic directional control valve from the left engine. The engine itself was recovered with the reverser in a fully deployed position a few days after the accident.

In the early part of the investigation, there was a widespread belief that, even though hard evidence showed the reverser had, indeed, deployed, the pilots should have been able to handle the accompanying roll and yaw, as Boeing's test pilots had done in their low-altitude, low-speed test flight. Airline owner Niki Lauda, a seasoned pilot, himself accepted that line of reasoning after flying through a reverser deployment in a 767 simulator programmed with flight parameters derived from the flight test data. A week after the accident, Lauda said he thought that a bomb or fire caused the crash.

A Boeing investigator examines the debris of the Lauda Air 767. An inadvertent thrust reverser deployment flipped a jetliner into a crash dive that killed all two hundred twenty-three people on board.

Eventually, authorities homed in on the auto-restow system, surmising that something must have shifted the DCV to the wrong position in flight. In that scenario, the engine casing could have been slipping, or an out-of-adjustment sensor could have been signaling the auto-restow to activate. In either case, the auto-restow would activate and illuminate the REV ISLN light. Perhaps the DCV was in the proper position the first couple of times this happened, directing the hydraulic fluid to restow the reverser. This would explain Captain Welch's early nonchalant comments. Then something, perhaps a stray electrical signal or quirky vibration, might have moved the DCV to the open position, so that when the auto-restow activated again the reverser deployed.

In mid-August 1991, Boeing lab technicians were probing this theory when one of them came up with the idea to place a worn seal on the DCV to see what would happen on activation of the auto-restow. The technicians were surprised when bits of the seal contaminated the hydraulic lines and the DCV deployed the reverser. The worn-seal tests prompted the FAA to order a ban on use of reversers on 767s that used the same electronically controlled restow valve as the Lauda airplane. The agency noted that out-of-adjustment sensors, which could have been unnecessarily activating the auto-restow system in flight, had been discovered on nine out of ten airplanes checked to that point. The inspections also turned up chafed wires and short circuits in the wiring bundles integral to operation of the DCV.

Still, the FAA was not yet willing to completely blame the auto-restow, because Boeing continued to maintain that the aircraft should have been flyable, even with the left reverser deployed. However, under pressure from the National Transportation Safety Board, Boeing agreed in early September 1991 to throw out the limited flight test data and run fresh aerodynamic calculations, based on new high-speed wind-tunnel tests, to determine what might have happened to the Lauda jet after the reverser deployed. Once again, officials were surprised.

The fresh data were fed into a 767 simulator in Seattle and airline owner Niki Lauda brought in to fly it. Lauda told news reporters after the flight that, upon deploying the left reverser, the simulator twisted and flipped completely over to its left in a couple of seconds. A few seconds more and it was careening, nose aimed at the darkened jungle below, at past the speed of sound, more than six hundred miles per hour. Designed to withstand eighty-six percent the speed of sound, the simulator plummeted three miles—from twenty-four thousand seven hundred feet to about eight thousand feet—in less than half a minute and then began breaking apart.

"You're suddenly upside down in a dive," Lauda said. "There's nothing you can do. It turns over, then it goes nose down. You go right into overspeed and then you lose all of your controls. The airplane broke up at exactly seventeen seconds after it went out of control."

The Lauda Air crash served as a reminder to aircraft designers that the increasing reliance on electronic controls and computers to operate modern aircraft have dangerous limitations that can be very difficult to anticipate. Ultimately, the accident prompted the FAA to order extra locks for reversers on two thousand late model 767s, 737s, and 757s—all Boeing's twin jet models—as well as upgraded DCVs for some 757s and substantially modified wiring and DCVs for about fifty-seven 767s with the same type of engine as the Lauda airplane.

BYRON ACOHIDO

Where to Learn More

Acohido, Byron. "Agencies Clash on Jet-Safety Rule: Crash in Thailand Spurs Call for Emergency Action." *Seattle Times*, July 23, 1991, p. C1.

———. "Air Disasters: Critics Question FAA's Response." *Seattle Times*, September 1, 1991, p. E3.

———. "FAA: Don't Use 767 Thrust Reversers—Boeing Expands Call for Inspections for Other Jets." *Seattle Times*, August 16, 1991, p. A1.

———. "FAA Scrutinizes Reversers: Agency Orders Changes in Faulty Valve System on Some 757s." *Seattle Times*, September 12, 1991, p. F1.

———. "Flaw in the Fleet: Boeing Executive Says Precautions Adequate for Jets with Suspect Reverser." *Seattle Times*, September 15, 1991, p. E1.

———. "No Worst-case Test for 767 Part: FAA Accepted Boeing Assurances on Thrust Reversers." *Seattle Times*, June 14, 1991, p. A1.

———. "Safety Board Wants 767 Part Banned: Agencies Fight over Safety of Reverser." *Seattle Times*, August 8, 1991, p. A1.

Daly, Kieran. "FAA Narrows Thrust-Reverser Directive." *Flight International*, September 4 - 10, 1991, p. 10.

Lane, Polly and Byron Acohido. "Boeing Tells 757 Owners to Replace Part—Faulty Thrust Reverser Valve Blamed in 767 Accident That Killed 223." *Seattle Times*, September 9, 1991, p. A1.

———. "Interpreting Last Words of Doomed Crew: Boeing Part Occupied Final Minutes as 767's Pilots Struggled with Riddle." *Seattle Times*, June 6, 1991, p. A1.

"New Boeing Lock Targets Unwanted Reverser Activity," *Aviation Week & Space Technology*, June 1, 1992, p. 32.

Norris, Guy. "FAA Lauda 767 Reply Inadequate, Says NTSB." *Flight International*, August 14 - 20, 1991, p. 4.

———. "FAA Queries Thrust-Reverser Designs." *Flight International*, August 28 - September 3, 1991, p. 4.

Norris, Guy and Kieran Daly. "FAA Orders Review of All Thrust-Reverser Designs." *Flight International*, September 18-24, 1991, p. 4-5.

Ott, James. "Lauda Crash Probers Focus on Midair Thrust Reversal." *Aviation Week & Space Technology*, June 10, 1991, pp. 28-29.

Proctor, Paul. "Sabotage Feared in Destruction of Lauda Air 767-300ER over Thailand." *Aviation Week & Space Technology*, June 3, 1991, pp. 29-30.

Richards, Bill. "FAA Calls for Shutdown of 767 Thrust Reversers." *Seattle Post-Intelligencer*, August 17, 1991, p. A1.

El Al Boeing 747-200 crash

Amsterdam, The Netherlands

(1992)

Separation of both engines from the right wing caused a 747 to plunge fatally into apartment complex, drawing public attention to safety hazard raised by corroded Boeing engine mounting bolts.

Background

On October 4, 1992, after a year-old Boeing 747-200F (freighter) operated by El Al Israel Airlines had taken off from Amsterdam's Schipol Airport at 6:22 P.M., the inboard engine came loose from the right wing. Veering sideways, it knocked off the outboard right engine in turn. The jetliner lurched to the right and, constantly losing altitude, made two wide clockwise circles over the city. The airplane finally smashed into a crowded, mostly low-income ten-story apartment complex, where the residents were just sitting down to dinner. The six El Al crew members and more than fifty people on ground at the crash site were killed.

The El Al Boeing 747 catastrophe brought to public light a hazard that air safety officials had been trying to solve since 1978: the potential for corroded or cracked engine mounting safety bolts—called fuse pins—to fail during flight. The disaster also revealed that aircraft designers had miscalculated how an engine under high thrust would behave if it came loose in flight. Instead of safely falling down and away from the aircraft, as designed, the El Al jet's inboard right engine veered forcefully to the side and into the outboard right engine.

The Boeing 747's fuse pins were hollow cylinders of cadmium-plated steel four inches long and 2.25 inches in diameter. These fuse pins were designed to be sturdy enough to endure enormous stress from multiple directions throughout the flight's trajectory—as the airplane takes off, climbs, banks, bounces, descends, and lands. Four fuse pins connect the engine strut to the wing. The pins slip through lug fasteners on the strut and then through parallel lugs on the wing.

Like automobile electrical fuses, these Boeing fuse pins are designed to be fragile enough to break under certain circumstances. For example, designers intend the fuse pins to snap if an engine starts coming apart in flight, so that the damaged engine might sheer cleanly away without damaging the wing. Similarly, if a 747 were forced to belly land, the design specifications call for the engines, upon striking the runway, to sheer away from the wings. This would greatly reduce the chance of the hot engines igniting any fuel that spilled from storage tanks in the wings that had ruptured.

From 1969 through 1980, Boeing mounted its 747 engines with fuse pins that machinists honed to just the proper strength by boring an intricate hourglass shape across the interior surface of the pin. But this original technique turned out to be flawed. Officials discovered that the machining process sometimes left tiny nicks where corrosion could develop. Under stress, the corroded nicks developed into cracks that made the pins dangerously weak. Therefore, in 1979, the Federal Aviation Administration (FAA) ordered airlines to periodically inspect these "old-style" fuse pins for corrosion and cracks.

Boeing attempted to correct the problem by designing a "new-style" pin that featured a simplified hourglass core supported by inserts wedged permanently into the two open ends of the fuse pins. Factory installation of the new-style pins on new jetliners began in 1980. In 1982, the FAA advised airlines operating older 747s that regular inspections of fuse pins would no longer be required if the old-style fuse pins were replaced with the newer version.

However, reports soon reached Boeing that the new-style pins were also corroding and cracking. In 1985 or 1986, an alert ground inspector averted a possible tragedy when he spotted an engine drooping on a 747 passenger jet just minutes before it was to take off. The flight was aborted, and inspectors discovered that the droop was caused by a cracked new-style fuse pin. Over the next few years, several more cases of weakened new-style pins came to light. In May 1991, therefore, the FAA ordered airlines operating 747s equipped with the new-style pins to perform a one-time check to determine whether the anti-corrosive primer on the inside core of the pins was intact. If corrosion or cracks were discovered, the airlines were instructed to install fresh pins.

Nonetheless, the FAA made no provisions to formally tally or assess the cases of corroded or cracked pins uncovered by their call for inspection. Moreover, some airlines balked at the requirement, which meant that they had to squeeze a time-consuming task into a rigorously tight maintenance routine. Mechanics were not pleased with the requirement either. The inspection procedure called for the tightly wedged inserts to be yanked loose with a special tool to check for corrosion or cracks inside the pin. As it turned out, the inserts were the reason why the new-style pins proved susceptible to corrosion. Upon being wedged into place in the ends of the pins, the inserts scraped away small amounts of the anti-corrosive primer coating the internal bore. Moisture, probably from intense condensation, then seeped into the bore and corroded the scraped area.

On December 29, 1991, eight months after the FAA authorities determined that a one-time examination of the new-style pins and periodic inspections of old-style pins would ensure the safety of the 747 fleet, a twelve-year-old

China Airlines 747-200 jet freighter crashed near Taipei, Taiwan, killing all five crew members. The 747's original, old-style fuse pins had recently been replaced with the new-style pins. Even so, a few minutes after take-off from Taipei, the airplane's right inboard engine ripped loose and slammed into the right outboard engine, knocking it, too, loose. Both right-side engines fell into the sea, and the airplane, stuck in a right hand turn, smashed into a cliff.

For several months, very little information was made public about this Boeing 747 crash in Taiwan. Then, on September 11, 1992, an inspector checking an Argentina Aerolineas 747-200 passenger jet prior to take off noticed that the aircraft's right inboard engine was sagging. As was the case with the 1985 or 1986 near-disaster, mechanics discovered the droop was caused by a cracked new-style pin. A week later, Boeing convened a meeting in Seattle, Washington of several major 747 operators to discuss the fuse pin problem. An advisory was worked out that requested that airlines once again perform a one-time inspection of the new-style fuse pins. October 8, 1992 was selected as the date the advisory would be issued.

Details of the Crash

At 6:22 P.M., October 4, 1992, the El Al Boeing 747-200 jet freighter, laden with one hundred fourteen tons of commercial cargo, took off from Amsterdam. Heading southeast over the Dutch city, the airplane was bound for Tel Aviv, with Captain Isaac Fuchs at the controls. At around 6:25, the jet reached four thousand feet and continued climbing under high thrust.

A minute later, Captain Fuchs reported to the control tower that a fire warning light for the right inboard or No. 3 engine had come on. An engine separating from the wing could trigger this warning light, and evidence from the wreckage later confirmed that a fuse pin had failed on the No. 3 engine, setting the accident into motion. At 6:28, Fuchs reported that the right outboard or No. 4 engine was out. It was later determined that the No. 4 engine was operating properly until it was knocked off the wing by the loosened No. 3 engine.

Fuchs had no way of knowing from these control panel indications that the jet had completely lost its two right-side engines. By 6:34, after flying a wide right hand circle over Amsterdam, Fuchs reported a "problem with the flaps," the movable control surfaces on the trailing edge of the wings. Only after this point did Fuchs acknowledge that "We have a control problem." A minute later, at 6:35, Fuchs radioed: "going down, 8162 going down, going down." The jetliner slammed into the ten-story apartment building at 6:35:53 P.M., sundering the building in two. Along with the six-member El Al crew, more than fifty people at the apartment building site were casualties of the crash.

The El Al jetliner was one of about two hundred older model 747s that had continued to use the old-style pins. Airline records showed that El Al mechanics inspected the pins four months before the accident, per regulations, and reported no problems. In January 1993, Israeli authorities, citing

After losing its two right engines, the El Al Boeing 747 slammed into a ten-story Amsterdam apartment building. The plane's crew and more than 50 people on the ground were killed.

evidence uncovered in the wreckage, blamed the broken fuse pin on the inboard right engine for causing the engine to tear off and veer into the outboard engine. El Al officials indicated there were no signs of sabotage and that the crew did everything possible to maintain control of the damaged airplane. An El Al director reported that he would demand payments from Boeing for the victims' families and, given the damage to its reputation, to the airline itself.

Impact

Coming only eleven months after the crash of the China Airlines 747 freighter in Taiwan, the El Al Boeing 747 catastrophe indicated that airline authorities probably should have acted more quickly and definitely to address the long-recognized danger posed by both versions of the Boeing 747 fuse pins. After the Amsterdam accident, the FAA and Boeing did take more aggressive steps to identify and replace weak fuse pins in the 747 fleet of more than nine hundred jets. However, it remains to be seen whether the design of the 747 engine mount systems will continue to pose an air safety hazard.

The day after the crash, Boeing issued the advisory written in connection with the crash in Taiwan, which called for a one time re-inspection of new-style pins. Boeing exempted the newer model 747-400 from the advisory, however, and gave no recommendations regarding 747s still equipped with old-style pins. A few days later, the FAA widened the call for inspection by including the 747-400 model and requiring more frequent inspections of 747s still using the old-style pins. Inspections of the first three hundred jets turned up four hundred ninety-nine corroded pins and fourteen cracked pins, a one in five defect rate. Airlines were permitted to refurbish lightly corroded pins, but were required to replace heavily corroded and cracked pins.

During this period, Boeing has continued its development, initiated several months before the El Al crash, of a corrosion-proof fuse pin. This has proven a difficult design goal, and the effort was ongoing as of April 1993. Like other exterior parts, the fuse pins are exposed to intensive condensation as a jet rapidly ascends to and descends from the sub-Arctic levels of cold found in jetliner cruising lanes seven miles above the Earth. Other potential sources of corrosive moisture include de-icing agents and various types of liquid cleaners and polishers routinely sprayed on the wings near the engine mounts.

Boeing uses an engine mount system similar to the 747's system, including fuse pins, for all its models. As of late 1992, Boeing was working to address widespread problems with fatigue cracking of model 757 fuse pins. The design philosophy of the Airbus airplane manufacturer, in contrast, dictates that engines remain mounted to the wings under all conditions, including massive engine failure and emergency belly landings. Thus, no breakaway fuse pines are used on the engine mounts of Airbus models.

Boeing officials declined to publicly discuss the separate phenomenon whereby a loosened engine, contrary to design expectations, veers into the paired engine mounted on that wing. Aeronautical experts theorize that substantial gyroscopic forces could cause a suddenly loose engine to rocket forward in an arc either to the right or the left, depending on which direction the engine fan blades were rotating. This new information about the trajectory of loose engines has raised a separate concern: the increasing use of jetliners with only one powerful, wide-diameter engine under each of the two wings. These models include the Airbus A321, A300, A310, A330 and A340 and the Boeing 737, 757, 767 and 777. A loose engine on a such an

A large piece of the 747's wreckage is guarded by police in Amsterdam. Investigations showed that corroded engine mount systems allowed the two right engines to drop from the plane, causing the catastrophe.

airplane might someday strike and severely damage the fuselage—the central portion—of a passenger-carrying airplane.

BYRON ACOHIDO

Where to Learn More

Acohido, Byron. *Seattle Times*, 1992: October 9, October 17, October 22, November 3, November 4, November 14, (2) December 27, December 28, December 31. 1993: January 9, January 16.

"Fuse Pins Suspect in El Al Crash." *Aviation Week & Space Technology*, October 12, 1992, pp. 30-33.

Egozi, Arie. "El Al Crash Centered on 'Engine-pylon Fittings.'" *Flight International*, October 28-November 3, 1992, p. 5.

Federal Aviation Administration. Emergency Airworthiness Directive No. T92-24-51. November 13, 1992.

Hornblower, Margot. "Death from the Sky." *Time*, October 19, 1992, pp. 50-52.

Lane, Polly. *Seattle Times*, 1992: October 13, October 15, (2) October 30, November 13. 1993: January 11.

Mecham, Michael. "El Al Strut Shows Possible Fatigue." *Aviation Week & Space Technology*, October 26, 1992, p. 30.

National Transportation Safety Board. Safety Recommendation Nos. A-92-114 through 117. November 3, 1992.

Nelan, Bruce W. "Are 747s Safe To Fly?" *Time*, October 19, 1992, p. 52.

Sharn, Lori. "Faulty Pins Found on U.S. 747s." *USA Today*, October 20, 1992, p. A6.

West, Karen. "Boeing Suggests Checkups: Airlines Advised To Inspect Steel Pins." *Seattle Post-Intelligencer*, October 6, 1992, p. A6.

Airships

- *ZR-2* (*R-38*) (1921)
- *Roma* (1922)
- *Shenandoah* (1925)
- *R-101* (1930)
- *Akron* (1933)
- *Hindenburg* (1937)

ZR-2 (R-38) disaster

Hull, England

(1921)

Anglo-American airship's break-up during flight due to faulty design marked the beginning of America's ill-fated airship program.

Background

On August 24, 1921, during a test flight of a British-designed airship carrying both British and American crews, a series of high-speed turns was attempted. The structurally unsound hydrogen airship sagged and began to break in two, casting out both men and equipment. The front half of the airship, becoming engulfed in flames, exploded. The rear section of the airship, burning only slowly, drifted down to a sandbar. Of thirty-two British and seventeen American crewmen onboard, only five survived.

During World War I, Britain began the construction of a military airship (a lighter-than-air craft) designed to patrol in a radius of 300 miles for up to three days. When the end of the war and the postwar economic depression led the British government to close down this program, the United States offered Britain more than $2,000,000 for the completed airship. The United States had no rigid airships, and could not buy one from the pioneering German Zeppelin firm because the peace treaty had not been concluded. Britain, strapped for cash, readily agreed to finish building the airship to sell it to its World War ally.

Despite pretensions to originality, the British airship, designed by C. I. R. Campbell, was essentially a clumsily-enlarged zeppelin. A center bay was added with the intention of increasing the airship's lift and raising its flight ceiling to 25,000 feet. With its 699-foot length (57 feet longer than the German original) and 85-foot diameter, the British airship held 2,724,000 cubic feet of gas. This volume of hydrogen gas gave the airship a disposable lift (actual lifting ability) of nearly 46 tons. The airship was powered by 6 Sunbeam Cossack engines, each 350 HP, providing a maximum speed of 71 mph. Finished in late June 1921, the airship was designated as the *R-38* by the British, and as the *ZR-2* by the Americans (the "R"s stood for "rigid"; the American "Z" for "zeppelin").

The *R-38* first flew on the night of June 23, 1921. It immediately experienced control problems; those who flew that night commented that the extra bay made the airship seem to "hump." A second trial flight demonstrated that ship's rudders and fins were overbalanced. Even after modifications, the *R-38* at moderate to high speeds proved difficult to control.

During its third flight, on July 17, the *R-38* nearly came to its end. During a trial speed run at two-thirds power, the airship shot up in the air at a dangerous angle, causing some of its central girders to buckle and give way. The ship was only saved from complete break-up by a stroke of luck. A rigger who was walking back to his bunk in the crew's quarters came across the initial buckling, and was able to warn the command bridge just in time to intercept an order from British Flight Lieutenant A. H. Wann to increase engine power. Wann immediately reversed his order, instructing that the motors be cut to half speed. The *R-38* shivered in slowing down, but held together.

Given the various political and financial considerations, even this near-disaster was largely disregarded. The British proceeded on the assumption that the airship merely needed strengthening, and the enthusiasm of the Americans seemed little dampened. During the next two weeks, the central girders were repaired and reinforced, making them stronger than the rest of the ship's girders.

Following these repairs, the *R-38* was readied for one last test flight, to Pulham, before the airship was to be formally turned over to the Americans. Since the United States had not previously owned an airship, the British also were using this flight to provide training to an American crew. The Americans were then prepared to take the airship to Lakehurst, New Jersey.

Details of the Crash

Just after dawn on August 23, 1921, the airship was set for its final trial flight. American insignias had already been painted on its tail, and the ship was now called by its American designation, the *ZR-2*. Seventeen Americans, the British crew of 28, three laboratory workers, and C. I. R. Campbell were aboard as the airship lifted off from Howden and flew over the North Sea. The American Lieutenant Richard E. Byrd, who would later achieve fame as a polar explorer, was to board the flight as a navigator. Having missed his train in London, Byrd arrived by taxi at the field just as the "Cast off" command was given.

After a day-long series of tests, the *ZR-2* headed to dock at Pulham. Once fog began enveloping its aerodrome, however, the ship turned back to sea for the night. The following morning, when Pulham proved to be still fogged in, the crew decided to return to Howden.

At 2,500 feet over the city of Hull and the Humber River, the British officers in charge of the airship initiated a turning test. With its engines almost at full power, the *ZR-2* had reached 62 mph when the rudders were put over hard

The wreckage of the ZR-2, *after the airship broke apart and crashed into the Humber River in Hull, England.*

to port, then to starboard, and then back again. It was never determined why this test was considered necessary, or which officer actually ordered it. Whatever the case, this series of sharp right and left turns at high speed and fairly low altitude proved too much for the airship to bear.

David Phipson, standing on the banks of the Humber River, had come to photograph the *ZR-2* in flight. He later testified to noticing a wrinkle in the great silver belly of the airship, "like a frown in a man's forehead." As the wrinkle deepened, the ship suddenly sagged in the middle, with both its nose and tail shooting upward. A huge gap appeared as the airship began to split into two; out of this gap fell men, fuel tanks, equipment, beds, and blankets. As the ship broke apart, fuel lines, bracing wires, and electric cables were stretched and finally ruptured. The combination of spilt fuel and torn electric cables set fire to the spewing hydrogen in the front half of the ship.

When the explosion came, it was heard fifteen miles away at Grimsby. Tremors were felt fifty miles away. The city of Hull escaped having its center turn into an inferno when the burning wreckage from the front half of the *ZR-2* fell into the Humber river. The rear section of the airship burned only slowly, drifting lazily down onto a sandbar covered with a few feet of water. Only four British crewmen and one American survived; the remaining forty-four men, with Campbell among them, were killed.

Impact

An American Court of Inquiry made various investigations, and Britain's Aeronautical Research Committee reviewed its findings. Most of these investigations focused on the weak design of the *R-38*. The British investigation led to the conclusion that "the structure possessed inadequate strength."

Subsequent investigators offered more telling assessments of the structural inadequacies of the *R-38*. The British airship copied a German zeppelin designed to operate at altitudes of over 20,000 feet. Given this purpose, the German zeppelin required an extremely light frame. The German engineers understood that such a lightly framed airship could not withstand abrupt turns at high-speed, particularly at low altitudes; German airshipmen, in fact, were specifically instructed not to attempt such maneuvers.

Furthermore, neither the British nor the Americans involved with the project seemed aware of the significance of aerodynamic loads (the forces that act upon an object as it moves through the air) to airship design. The British committee, to its amazement, learned that Campbell had not taken aerodynamics into account when formulating his design. Specifically, Campbell's design ignored the principle that the aerodynamic loads or stresses become much higher when an object turns during flight. Years later, the novelist Nevil Shute (who as an aeronautical engineer worked on the British *R100*) recalled in his autobiography how shocked he was, upon studying the *R-38*'s plans, to discover that no calculations had been made of the stresses that turning would impose on the airship's structure. This neglect was shocking; as early as 1900, the Wright Brothers knew to study aerodynamics before attempting to build an airplane.

Neither the British nor the Americans involved with the project seemed aware of the significance of aerodynamic loads to airship design. Specifically, C. I. R. Campbell's design ignored the principle that the aerodynamic loads or stresses become much higher when an object turns during flight.

An elementary design error supplemented this neglect of aerodynamics. After the third test flight, during which the airship's center girders failed, Campbell only reinforced the damaged girders. This was a fundamental miscalculation: a ship's structure must remain balanced in strength.

The fatal crash of *R-38* cast a shadow over both the American and the British airship programs. The United States temporarily halted its development of the *ZR-1*, later christened the *Shenandoah*. The British cancelled several projects then underway toward developing a commercial airship. Related work did not resume until 1924, and more than ten additional years would elapse before a British airship crossed the Atlantic.

Although the *R-38* was never actually handed over to the United States, the Americans were billed $884,095 for the airship. The bill for training the American crew came to another $98,069.

LEONARD C. BRUNO

Where to Learn More

Deighton, Len and Arnold Schwartzman. *Airshipwreck*. Holt, Rinehart and Winston, 1978.

d'Orcy, Ladislas. "Lesson of the *ZR-2* Disaster." *Scientific American*, September 17, 1921, p. 200.

"Startling Report on *R-38* Shows Strains in Motion Were Forgotten." *New York Times*, February 23, 1922, p. 1.

Robinson, Douglas H. *Giants in the Sky: A History of the Rigid Airship*. GT Foulis, 1973.

"Loss of *ZR-2*." *Scientific American*, September 3, 1921, p. 158.

Toland, John. *The Great Dirigibles: Their Triumphs and Disasters*. Dover Publications, 1972.

Roma crash

Langley Field, Virginia

(1922)

Faulty engines and use of a flammable gas caused an early twentieth-century airship to crash and explode, resulting in a large loss of life.

Background

On February 21, 1922, the large semi-rigid Italian-built airship called the *Roma* crashed and burned at Langley Field, Virginia, killing thirty-four of its crew of forty-five Americans. This terrible accident cooled American enthusiasm for airships, but more importantly, it assured that the United States would never again use flammable hydrogen in an airship.

The United States entered the post-World War I period having experience only with blimps (nonrigid airships whose shape was maintained by gas pressure). Rigid airships, or those with a rigid internal frame that holds its shape regardless of whether it is inflated with gas, were built only in Europe. While waiting for its own rigid airships to be built in Philadelphia, the United States bought the large, semi-rigid airship *Roma* from Italy in 1920 for $165,000. Unlike a rigid ship, the *Roma* had a metal skeleton only in its nose and keel, and needed gas for its huge silk and cotton bag to keep its shape. Its bag was filled with hydrogen, a highly explosive gas that was used because of its excellent lifting power and low cost compared with the safer but very expensive helium gas.

The *Roma* was the largest semirigid airship ever built. It was four hundred and ten feet long and held nearly 1.2 million cubic feet of gas, but it was far from being a proven vehicle. Designed by the Italian aeronautical engineer Umberto Nobile (1885-1978) and three others, it was shipped in sections across the Atlantic to be assembled at Langley Field, Virginia, where it would enter service with the United States Army. Before this, the *Roma* had successfully completed twenty-two flights in Italy. When assembled in the United States, the silver grey airship was equipped with six four hundred and fifty-horsepower Italian Ansaldo engines which moved the ship along at about fifty miles per hour.

Several problems would emerge as a result of this transition from Italy to America. First, aside from the two months it took to dismantle the *Roma* and pack its sections in huge crates, much time was spent aboard the freighter *Mars* during its transatlantic journey from Genoa to Langley Field. The airship arrived in August and assembly by the Americans began. Upon inflation and inspection on September 2, 1921, the airship's bag was seen to be covered with mildew and its thin, aluminum preservative coating had been mostly eaten away. The worst spots were patched. Despite the deteriorated bag, the first test flight in the United States, held on November 15, 1921, was generally successful, although the Italian-made engines exhibited some problems. On its first long-distance flight, from Langley to Washington, D.C., the engines proved especially troublesome, and the *Roma* arrived at Bolling Field three hours late with three engines dead. As the weather grew colder, it seemed almost impossible to warm up engines that were designed for the more moderate climate of the Mediterranean.

After several incidents of engines either not starting or malfunctioning in midflight, most agreed that the Ansaldo engines should be replaced with American-built Liberty engines. Advocates of the plan argued that the parts were readily available and that American operators and mechanics were more familiar with the American machines than with the Ansaldo engines. At the beginning of 1922, the decision was made to replace the troublesome Italian engines with new Liberty engines. By February 1922, this was accomplished, and the airship's first flight with the new engines was scheduled for February 21, 1922.

Details of the Failure

On the morning of February 21, 1922, a steady, drizzling rain fell, but by noon it had stopped raining and the *Roma* was walked out of its hangar for its first flight with the new, more powerful Liberty engines. As the drag lines were released by the ground crew, the airship soared into the still-grey sky with forty-five men aboard, five of whom were civilians. The ship flew southward toward Hampton Roads Naval Base. Twenty five minutes into the test flight, following a particularly high speed run, the crew began having trouble keeping the airship's nose level. As it passed over the town of Phoebus, a young witness described the appearance of the *Roma*'s nose as "crinkled up." At this point, the ship responded to controls less and less, and its rear end tilted up higher than its nose. The crew, suspecting a critical malfunction, ordered that the engines be turned off, and the machine began to drift slowly, nose-down, toward the ground. According to one survivor, the *Roma*'s elevation was only about 200 feet at this point, too low for a successful parachute jump. One crew member tried but broke his neck on impact and died on the way to the hospital. Others desperately began throwing anything loose overboard, but there was no time to release enough of the 3,950 pounds of water and sandbag ballast the ship carried to make a difference.

To the soldiers watching the airship's gentle descent from the hangar, it appeared that the *Roma*'s crew would be able to maneuver a successful crash landing with very little damage or injury. The airship dropped out of sight

The Roma *before its trial flight.*

behind a hill, however, and seconds later they saw a brilliant flash, followed by billowing smoke. Then a loud, belated roar announced the obvious. The hydrogen had exploded. To its great misfortune, the *Roma* had struck high voltage wires atop telephone poles. Sparks from the snapping wires instantly exploded the hydrogen, which in turn set the ship's gasoline tanks afire.

Only eleven men on the *Roma* survived the crash and subsequent explosion. The fire that followed raged especially hot and long, fueled by highly flammable hydrogen, and observers were unable to approach the ship to rescue its passengers. After the fire was put out, rescue work was especially grisly. The *Roma*'s Captain, Dale Mabray, was found burned, his hands still gripping the wheel. One officer at Langley, Captain Kepner, later recounted, "I have never forgotten that night. Arms, legs, and heads burned away . . . a head shrunk in heat and burned to the size of a fist with some teeth remaining that the dentist record could identify . . . all my friends and boon companions."

A board consisting of three Langley officers was appointed to investigate the crash, and they held their first meeting at 4:30 P.M at the crash site the same day while bodies were still being removed from the wreckage. Upon examining the wreckage and later questioning survivors and fifty witnesses, the board could reach no positive conclusion about the exact cause of the crash. But it did agree with the testimony of all survivors and witnesses that the ship's controls would not respond.

As to what exactly caused the controls to malfunction, the board speculated that the *Roma*'s air scoops were not kept open, and that this allowed the nose to collapse, buckling the metal structure and jamming its controls. Air scoops were openings in the front of the airship that could be manually operated to allow air to enter the gas bag. This would be done to replace the small amounts of hydrogen that always escaped during a flight. Allowing in some fresh air while the ship was moving forward was an easy way to keep the bag fully inflated.

This explanation has never been fully accepted, since it is difficult to understand how one of the two rigid parts of the *Roma*, in this case its metal framework nose, could collapse simply by the loss of gas pressure. The framework did collapse, however, and most now agree that it did so because of the excessive strain put on it by the new, more powerful American engines. These Liberty engines, each two hundred and forty horsepower, provided much more power than the ship was designed for. Something had to give during its high-speed test run. This theory was given more plausibility when later tests of an American-built semirigid airship, the *RS-1*, revealed that its nose caved in at fifty-nine miles per hour. United States General Orvil Anderson conducted these tests and later wrote, "The weaker nose structure of the *Roma* and the lower internal pressure at which it was flown should have triggered a nose collapse at fifty-five to fifty-six miles per hour. It was flying at about that speed when it crashed."

Impact

Following this dramatic and seemingly avoidable disaster, there was an immediate public outcry against the further use of hydrogen in airships. When the press and public realized that, like the crash of an *R-38* six months before, a hydrogen fire was responsible for nearly every life lost, they vigorously protested the continued use of such a dangerous gas, deeming it totally unjustifiable. In fact, it was this disaster that marked the end of hydrogen as a lifting gas for American airships. From this time on, all United States airships would fly with nonflammable helium.

The wreckage of the Roma *after it crashed at Langley Field, Virginia. Thirty-four people died in the crash.*

At the time, helium was a relatively new and very expensive gas. Hydrogen could be manufactured cheaply anywhere in the world, but helium could only be extracted from natural gas by freezing it to three hundred and twenty-eight degrees below zero. Although first observed in 1868 in the solar spectrum, helium was not found on Earth until 1895, when scientists isolated small amounts in the atmosphere and in the volcanic fumes of Mount Vesuvius. By 1903, helium was considered an excellent lifting gas for airships because it was an inert gas that could not be set afire even with a match. The problem was that no nation could make or find enough of it to fill its airships.

In 1917, the U. S. Bureau of Mines began producing helium at three installations near Fort Worth, Texas, for use overseas in World War I. This had been scheduled to be shipped to Europe when the war ended. Three years later, the bureau was making helium at a little less than ten cents a cubic foot. The American press soon made spectacular sense of the obvious, with the *New York World* writing only two days after the *Roma* burned, "There is in storage in Texas more than enough helium to have inflated the bags of the *Roma*." The army argued, comparing the $17,000 cost to fill the *Roma* with hydrogen to the $120,000 cost for helium. The press answered back, especially the *New York Globe* and the *New York Herald*, saying essentially that human life was worth any price, and "if we can't afford helium, we can't afford dirigibles."

Helium has only ninety-two percent of the lifting capacity of hydrogen, but the United States eventually came to regard its safety factor as more than sufficient compensation for the expensive process of separating it from natural gas. After the *Roma* disaster, no American airship ever flew again with hydrogen. The sadly spectacular demise of the *Roma* made this a certainty.

LEONARD C. BRUNO

Where to Learn More

Gerken, Louis C. *Airships: History and Technology*. American Scientific Corporation, 1990, pp. 239-40.

Hood, Joseph F. *When Monsters Roamed the Skies: The Saga of the Dirigible Airship*. Grosset & Dunlap, 1968, p. 49.

Mitchell, John B. *The Army Airship Roma*. Syms-Eaton Museum, 1973.

"Giant Army Dirigible Wrecked." *New York Times*. February 22, 1922, p. 1.

"Weeks Begins Inquiry into *Roma* Crash." *New York Times*, February 23, 1922, p. 1.

"Disagree On Cause Of *Roma* Disaster." *New York Times*, July 27, 1922, p. 36.

Payne, Lee. *Lighter Than Air: An Illustrated History of the Airship*. Orion Books, 1991, pp. 125-27.

Shenandoah disaster

Ohio

(1925)

Storm's destruction of the first American-built airship triggered both a revaluation of airship design and criticism of military policies.

Background

On the morning of September 3, 1925, the first American-built rigid airship was destroyed by a severe thunderstorm. The *Shenandoah,* heading west from Lakehurst, New Jersey, was in flight over Ohio. The storm that it encountered was extraordinary; the six hundred eighty-foot airship was both forced up and down and twisted like a rag doll, powerless to control its course. The *Shenandoah,* unable to endure such violent extremes, ultimately broke apart. Fortunately, twenty-nine men of a crew of forty-three survived.

In 1919, the U.S. Secretary of the Navy authorized the construction of a rigid airship (a lighter-than-air craft). This airship, called the *ZR-1,* was based on the design of a German zeppelin deployed during World War I and captured by the French after a forced-landing. American engineers modified the German design by strengthening the frame of the airship and increasing its length by thirty-three feet. This extra bulk was required because of the decision to use helium rather than hydrogen to give the airship its lift. While helium gas, unlike hydrogen, is completely safe, helium provides only 92.6 percent of hydrogen's lift. In addition to being the first rigid airship built and flown by Americans, the *Shenandoah* was the first ever to rely on helium.

Construction on the airship began at the Naval Aircraft Factory in Philadelphia in 1920. The ship made its maiden flight at Lakehurst on September 3, 1923, and one month later was christened the *Shenandoah,* a Native American term meaning "Daughter of the Stars." It was 680 feet long, with a diameter of 78.7 feet. It held 2,148,000 cubic feet of helium gas, and its five 300 HP, six-cylinder Packard engines gave it a maximum speed of 60 mph and a range of about 3,000 miles.

The Americans decided to moor their airships to masts in lieu of the German system of grounding zeppelins within hangars. These mooring masts were tall, steel-framed towers to which the airborne ship could be anchored by its nose, like a balloon tied down by a string. These mooring masts would eliminate the need for huge (roughly one thousand by three hundred feet) and expensive hangars, and enable any airport with a tall mast or tower to accommodate the airship.

Although the *Shenandoah* was experimental, the American government ambitiously planned for it to take an historic flight across the North Pole. To this effect, the government erected a network of mooring masts across the West Coast and up into Alaska. When plans for this trip were cancelled, the Navy decided to exploit the enormous public interest that had been raised in the *Shenandoah* by scheduling it for a more modest publicity trip to the West Coast, using the network of masts for stopovers.

Prior to this first extended publicity trip, the designers of the *Shenandoah* had drawn very optimistic conclusions from a storm-related accident. During January 1924, while anchored to its mast at Lakehurst, the *Shenandoah* had been blasted for four days by gale-force winds that twisted the airship on its longitudinal axis. The rolling airship eventually ripped away from the mast, leaving its nose tied fast to the tower. With a combination of skill and luck, the damaged airship just managed to ride out the storm and coast back to Lakehurst. The fact that the *Shenandoah* had survived such a test was taken as evidence that the airship was structurally sound and ready for anything.

By May 1924, the airship had been repaired. Under the direction of U.S. Navy Commander Zachery Lansdowne, the *Shenandoah* undertook a series of seaborne mooring trials, attaching itself to the U.S.S. *Patoka*. Subsequently, on October 7, 1924, the airship left for a transcontinental flight. It was during this flight that the problems peculiar to helium airships became apparent. First was the loss of performance relative to hydrogen, which though flammable provided greater lift. Second was the need to conserve the scarce and expensive helium, which compelled various expedients to minimize any releasing of the gas.

The *Shenandoah* was kept at 85 percent inflation before flight, since the helium gas expanded as the airship rose. Its "pressure height," where inflation reached 100 percent, was 4,500 feet. The ship had to remain below this altitude to prevent its automatic valves from releasing helium. Landing also led to problems in helium conservation. When the Germans landed their zeppelins, hydrogen gas was simply expelled through valves until the airship lost buoyancy. To routinely release large amounts of helium, however, was unthinkable, so the Americans adopted the stratagem of landing the *Shenandoah* at night, when the helium became colder and heavier.

For its first publicity flight across the United States, essentially a training trip, the *Shenandoah* easily could have crossed high above the Rocky Mountains. This course, however, would have meant releasing thousands of cubic feet of precious helium. The airship therefore threaded its way through many treacherous mountain passes, risking collision with cliffs. To his credit, Lansdowne steered the *Shenandoah* through the Rockies on both legs of the trip, and safely returned to Lakehurst on October 25, 1924. It is difficult today to imagine how much public interest this cross-country trip generated. In 1924, transcontinental passenger service had not been established, and only mail had been flown from New York to San Francisco. Thus, the *Shenandoah*'s trip was a scientific and engineering phenomenon.

The *Shenandoah,* nonetheless, would not fly again until 1925. Its helium was temporarily transferred to the newly christened airship, the *Los Angeles* (formerly the *ZR-3*), which had been purchased from Germany. Helium was so scarce that the United States was unable to supply two large airships with the gas.

During this interlude, Commander Lansdowne took measures to further conserve helium in the *Shenandoah.* He had eight of the ship's sixteen maneuvering valves removed, and gas-tight "jam pot" covers placed over the automatic valves. While these modifications would reduce gas loss, they also created the potential for a breakup due to excess pressure should the ship ascend too high, too quickly. Lansdowne successfully argued to the skeptical Bureau of Aeronautics, whose approval he had to secure, that the ship's maneuvering valves if necessary could be opened by hand.

A battle Lansdowne did not win concerned the scheduling of the *Shenandoah*'s next publicity flight, this time to the Midwest. A native of Ohio, Lansdowne knew very well to expect unstable weather over this region in the summer. Lansdowne attempted to have the trip postponed until fall, but the Navy granted only a two-week delay, to the beginning of September. The Navy then committed to this decision by publicizing the *Shenandoah*'s schedule, which would coincide with the many state fairs planned for that month. The Navy wanted the *Shenandoah* to get maximum exposure to the hundreds of thousands of Midwesterners who had never seen an airship.

Details of the Crash

At mid-afternoon on September 2, 1925, the *Shenandoah* set off from Lakehurst on its publicity trip to the Midwest, its fifty-eighth flight. The crew's wives, who had come to watch the departure, turned their backs on the airship as it headed west; it was considered bad luck to watch as an airship faded out of sight. The airship was greeted by Philadelphia car horns and factory whistles an hour and a half later, and soon reached the Allegheny Mountains. At around 3:00 A.M.on September 3, the crew first observed flashes of lightning, initially behind them to the east and then ahead to the west and northwest. An hour later, at 2,100 feet over Byesville, Ohio, the airship ran into strong headwinds that all but stopped its forward progress. Minutes later, the airship began to rise.

Ominously, the *Shenandoah* continued to ascend despite every countermeasure the crew attempted. The engines labored, but the airship had encountered a freak confluence of air currents in which a stream of south-moving cold air was overriding a mass of warmer air coming north. As the warm air rose, so did the airship, at a rate of six feet per second. After briefly steadying at about three thousand feet, the ship began to rise even faster, at a terrifying

The wreckage of the Shenandoah *after it broke apart over Ohio and crashed. Fourteen of the crew of forty-three died in the disaster.*

rate of nearly one thousand feet per minute. As it became imperative to release helium, Commander Lansdowne ordered that the manual valves be opened and that the automatic valves be uncapped.

When the *Shenandoah* stopped rising, at 6,200 feet, the crew prepared for the inevitable plunge. The airship began to plummet even faster than it had risen, and the crew released tons of ballast water to slow its dive. Once this measure took effect, however, another ear-popping skyward ascent began. The *Shenandoah* was now rolling and twisting like a living thing; fierce winds hammered its hull, forcing its nose to port and tail to starboard. As the crew attempted to release the fuel tanks to drop additional weight, the airship, spinning with its nose pointed thirty degrees upward, broke in two.

The *Shenandoah* split apart about a third of the way from the front. The control car, with Lansdowne among a crew of eight, was catapulted away like a stone; there were no survivors. Seconds later, four more crewmen fell with the engine car. The huge bow section, with its crew of seven, then started to rise rapidly, buoyed by some still-intact gas cells. This 225-foot bow section spun like a top for another two hours. All of its crew members, nonetheless, managed to survive by clinging to broken cables and girders until the bow section finally touched down twelve miles away from the initial break-up. The rear section of the ship itself broke into two and slowly drifted to earth, dragging along the hills and leaving a trail of injured men.

Twenty-nine men of the crew of forty-three were to survive. Events took a final morbid turn in the aftermath of the disaster, as curiosity seekers swarmed over the various crash sites and literally picked the wreckage clean. Souvenir hunters stripped everything possible from the ship's remaining parts; some of the airship's covering was later converted into raincoats and sold as *Shenandoah* Slickers.

Impact

The Court of Inquiry into the disaster concluded that the *Shenandoah* had broken apart because it was not strong enough to withstand the strains imposed on it by vertical currents. The Court left unanswered whether Lansdowne's capping of the automatic valves resulted in or contributed to the break-up.

The Court's conclusion did trigger a revaluation of airship design. To allow for severe aerodynamic loads (stresses), subsequent American airships were made much stronger, though doing so sacrificed lightness. Because all but two deaths in the *Shenandoah* were the result of the pendent control and engine cars breaking off, newer ships incorporated these functions directly into the hull. The military's decision to use helium rather than hydrogen, however, was vindicated. Since a hydrogen fire would certainly have erupted during a breakup like the *Shenandoah*'s, the survivors owed their lives to the use of helium to provide the airship's lift.

The Court of Inquiry's conclusion did trigger a revaluation of airship design. To allow for aerodynamic loads, American airships were made stronger, though doing so sacrificed lightness. Because all but two deaths in the Shenandoah were the result of the pendent control and engine cars breaking off, newer ships incorporated these functions directly into the hull.

An additional result of the *Shenandoah* crash was outspoken criticism of the American military. Most significantly, U.S. Army Colonel Billy Mitchell issued a seventeen-page indictment of the military's role in the disaster. Mitchell claimed that this and other disasters were the direct result of incompetence and even criminal negligence. Mitchell was court-martialed, convicted, and suspended from rank, duty, and pay for five years for this criticism. Even so, Mitchell achieved his purpose of focusing national attention on the military's aviation policies and practice.

LEONARD C. BRUNO

Where to Learn More

Clarke, Basil. *The History of Airships*. St. Martin's, 1961.

Hook, Thom. *Shenandoah Saga*. Air Show Publishers, 1973.

Jackson, Robert. *Airships in Peace and War*. Cassell, 1971.

"Technical Aspects of the Loss of the U.S.S. *Shenandoah*." *Journal of the American Society of Naval Engineers*, August 1926, p. 489.

Payne, Lee. *Lighter than Air: An Illustrated History of the Airship*. Orion, 1991.

Walker, J. Bernard. "Tragedy of the *Shenandoah*." *Scientific American*, November 1925, p. 301.

Wensyel, James W. "*Shenandoah*." *American History Illustrated*, February 1989, p. 24.

R-101 crash

France

(1930)

Untested, overweight, unstable, underpowered airship with a gas leak crashed in France during a severe storm, resulting in a hydrogen explosion upon impact.

Background

In 1924, the British government initiated an airship program that it hoped would prove once and for all whether very large, rigid airships were commercially feasible for long-distance, high-speed flights. The planned program was the brainchild of Sir Dennistoun Burney, a member of Parliament whose original plan, called the Imperial Airship Scheme, proposed a fleet of six airships to link Britain with far-flung possessions such as Australia, Canada, India, and South Africa. As a means of connecting the empire, airships promised to cut travel time for a voyage from Singapore to England from twenty-four days by steamer to eight days by airship. Airships would be used for mail service as well as for passenger service.

With a change in government to the Labour party, the original plan was cut back to two airships, one to be built by the government itself and the other by private industry. The former, *R-101*, came to be known as the "socialist ship," since the Labour party was leftist, while the *R-100* was called the "capitalist ship." Fierce competition ensued between the two camps, resulting in a bitter rivalry.

While construction of the *R-100* proceeded generally along normal lines, the development of the *R-101* was subject to the influence of many nontechnical factors. For instance, both design teams originally opted for diesel engines because their fuel was considered safer than gasoline. When both realized, however, that diesels would be far too heavy, the builders of the *R-100* simply switched to six proven Rolls-Royce Condor engines. The builders of the *R-101*, however, felt they could not make a similar switch because they had made public commitments to the safety of diesels. Such problems happened almost regularly with the *R-101*, as its government-sponsored program was conducted with much publicity, making it difficult to discard or alter previously announced plans.

When the *R-101* was finished five years later, actually ahead of the *R-100*, it was an incredibly heavy airship without much lifting ability. Contributing to its lifting problems were the weight of its immense diesels, plus special servo-mechanisms that controlled its rudders and elevators with power instead of by hand. It was also outfitted with a complex system of specially designed, supersensitive valves that released gas with the slightest variation in pressure (such as when the ship pitched or rolled). Its frame included some steel, which was strong but heavy, unlike the *R-100*, which only used the much lighter duralumin. But its most fundamental design departure was the use of internal rings that had no bracing wires between them.

On October 14, 1929, *R-101* made its maiden flight. By the end of November 1929, it had made seven flights, all in good weather and none at high speed, for a total of seventy hours in the air. It was an impressive sight in the air and even more so from inside the aircraft. Its palatial passenger quarters included an asbestos-lined smoking room, a dining salon, a huge lounge complete with potted plants, and even a wide, picture-windowed promenade. Despite its comfort and luxury, the ship seemed to stagger through these flights, indicating that it was dangerously over-heavy and sadly underpowered. Since it weighed 113 metric tons and had a lifting capacity of 148 tons, its disposable or actual lift was only 35 tons. This would not be enough to carry passengers on a long-distance flight. Its frustrated designers returned it to the hangar and endeavored to cut its weight.

From November 1929 to June 1930, the *R-101* remained in the hangar. Five tons of expensive equipment, including the power-steering system, were ripped out. The ship was cut in half and 45 feet of hull and an extra gas bag were added to its middle. On October 1, 1930, the *R-101* was brought out of its hangar as an entirely new ship. With a new cover and a more satisfactory disposable lift of 49 metric tons, it was 777 feet long—two feet longer than Germany's *Graf Zeppelin*. At the time, it was the largest, costliest, most complex airship ever flown. But it had not yet undergone final trials.

In the meantime, its rival, *R-100*, had made a successful round-trip from England to Canada, and the designers of the *R-101* felt compelled to fly to India as the secretary of state for air, Lord Thomson, so badly wanted. The energetic secretary had made many public announcements about flying to India for an important conference and returning to London by October 18 for another. His own sense of urgency on this matter overrode feelings of concern on the part of the people who built and flew the airship. They were especially worried since the rebuilt ship had been given no time for meaningful trial flights. In fact, it had not yet received a certificate of airworthiness, which was necessary for it to fly legally. As political pressure mounted for the flight to depart on schedule, the rules were bent and a certificate was issued on condition that full-speed trials be carried out during the voyage. Lord Thomson's famous assurance that *R-101* was "safe as a house, except for the millionth chance" would soon ring in the ears of those who saw the doomed ship depart at 6:36 P.M. on October 4, 1930.

The British airship R-101 *was very heavy and did not have much lifting ability. It also featured an unusual design element: it used internal rings that had no bracing wires between them.*

Details of the Failure

In the last frantic hours before the hurried flight to India, many wrong decisions were made that undid much of the work that had been done to lighten the ship. First, a pale blue carpet was added to cover the entrance corridor and lounge. Spreading for a length of six hundred feet and the width of a tennis court, the carpet alone added two to three metric tons of weight. While crewmen were limited to ten pounds of personal baggage each, Lord Thomson's baggage was estimated at more than one ton itself. To this were added crates of champagne, barrels of beer, and boxes of silverware for a state banquet. Finally, the decision was made to pump more fuel than was needed to reach their first stop at Ismalia, Egypt, adding nine more tons to the airship's load. This measure was taken to accommodate Lord Thomson's tight schedule, which did not allow time for refueling in Egypt.

To compound these problems, a gale was rising over France in the very direction the ship was headed. Despite the prediction of severe weather, the *R-101* lifted off at 6:36 P.M. on October 4, 1930, with fifty-four people aboard, six of whom were passengers. When the airship was released from the mast, it began to sink slightly instead of rising. Only after four tons of water ballast were unloaded did it rise sluggishly. Dropping ballast or extra weight was a standard technique used to compensate for expected gas loss during a flight. In this case, it was done immediately just to get the ship to rise.

The ship lumbered across the English Channel, flying dangerously low on an undulating corkscrew path, rising and falling as much as four hundred to five hundred feet at a time. As the *R-101* flew over the French town of Beauvais, forty miles northwest of Paris, it was pounded by winds of up to fifty miles per hour. Shortly after the ship's watch changed at 2:00 A.M., the aircraft suddenly went into a steep dive. The crew were able to use the elevators (the movable horizontal surfaces at its tail) to stop its plunge and level the ship, but immediately after it seemed to stabilize, it pitched down a second time. At 2:09 A.M., the *R-101* struck the ground at less than five miles per hour, bounced lightly, and skidded for some sixty feet before stopping.

The crash landing was witnessed by a French rabbit trapper setting snares in the woods. According to this witness, the nose-down landing was immediately followed by "a tremendous explosion that knocked me down. Soon flames rose into the sky to a great height. Everything was enveloped by them. I saw human figures running about like madmen in the wreck. Then I lost my head and ran away." Only eight people managed to get clear of the inferno, and of these, only six survived. Lord Thomson and all the ship's officers died in the flames.

Although the cause of the accident was never established officially beyond question, a court of inquiry judged that one or more of the ship's forward gas bags deflated, causing the ship to go down. More than fifty years after the crash, however, a definitive study was produced by Sir Peter G. Masefield, who offered thirteen separate but cumulative factors that doomed the airship:

The remains of the R-101 *after it crashed in France. Officials decided to go ahead with the flight despite the prediction of severe weather between England and France.*

1. The India flight as a prototype trial: committed to as early as 1924, the flight to India had to be accomplished at the earliest possible date.

2. Weak outer cover: only part of the ship had received a new cover after the extra bay was installed, and the nose and stern portions of the gas bag were original.

3. No full-speed trial: the airship was never tested when fully stressed before the actual flight.

4. Overconfidence and inadequate trials: these faults were believed to be common at the time and not unique to *R-101*.

5. Decision to go: despite an unfavorable weather report, officials still decided to go ahead with the trip.

6. Scott's judgement: Major G. H. Scott refused to admit that airships were not "all-weather" crafts and departed for India when most captains would have waited for better weather.

7. Engine starting delay: it took nearly thirty minutes to get the starboard forward engine working, and thus the ship's worst moments occurred just after the change of crews.

8. Weather: although the report predicted heavy rain and low-level turbulence, the conditions were actually much more severe.

9. Fast cruise regime: the captain ran his engines at full speed throughout the flight, stressing the ship more than it had ever been before.

10. Change of watch: at the most critical point in the flight, the more seasoned men were replaced by those with less experience.

11. Crew fatigue: much of the crew had been on continuous duty since the ship left its shed on October 1, three days before the airship took off.

12. Engines throttled back: this instinctive move to slow down the ship just as it pitched downward cost the *R-101* valuable lift and speed and contributed to its dive.

13. Calcium flares: these had been installed in the control car for use during the Channel flight, but they ignited instantly on contact with the water, starting a fire in the car that exploded the hydrogen.

Masefield summed up by saying that each of these factors contributed to the disaster and that without any one of them, the fatal chain of events would have been broken. In his final analysis, Masefield cited the weather, the cover failure, the decision to throttle back, and the calcium flares as the predominant factors.

Impact

This disaster marked the finish of airships for Britain. Perhaps as a result of the shock and the fact that the crash wiped out the elite of Britain's airship corps with one stroke, the *R-101* disaster seemed to affect the British people more than any crash since World War I. Arguments raged as to the cause of the accident, and one popular phrase, "Killed to make an Air Minister's holiday," was often heard. The British public was simply fed up with the tragedy that seemed to follow airships, and "Ban the Gas Bags" became a popular newspaper headline. Once the airship program was officially abandoned, even the successful *R-100* program was discontinued, and the $2 million airship was smashed to pieces with axes, flattened by a steamroller, and sold for scrap. Its sale brought in only $2,500.

Leonard C. Bruno

Where to Learn More

Botting, Douglas. *The Giant Airships*. Time-Life Books, 1980, pp. 122-31.

Collier, Basil. *The Airship: A History*. G. P. Putnam's Sons, 1974, pp. 166-210.

"Britain's Dirigible Horror." *Literary Digest*, October 18, 1930, pp. 10-11.

Masefield, Peter G. *To Ride the Storm: The Story of the Airship R-101*. William Kimber, 1982.

"British Airship R-101 Is Destroyed in Crash and Explosion in France." *New York Times*, October 5, 1930, p. 1.

"Weight of Rain Forced Down R-101." *New York Times*, October 6, 1930, p. 1.

Akron crash

(1933)

Crash of an advanced American airship into the Atlantic Ocean signalled the end of support for the military's airship program.

Background

During a night flight on April 3, 1933, the airship *Akron* encountered an increasingly severe storm. Just after midnight, off the New Jersey Coast, the airship became unstable and began to climb and then to descend dangerously. Pointing its nose upward in an attempt to regain altitude, the airship crashed tail-first into the sea. The impact of huge ocean waves soon the broke apart the 785-foot airship. Of the seventy-six passengers and crew members onboard, only three persons survived.

The *Akron* had been in the planning stages since 1924, and was intended as the first of the U.S. Navy's new class of American-built rigid airships (lighter-than-air crafts). After several American and international airship disasters, neither the public nor Congress was inclined to finance an airship development program, especially during a Depression. However, the dynamic Chief of the Bureau of Aeronautics, Rear Admiral William A. Moffett, persuaded Congress to authorize the building of two airships, one of which became the *Akron*. Moffett argued for the military development of the airship as a flying aircraft carrier to be used in reconnaissance.

After a prolonged study of 37 alternate designs, the contract for the airship was awarded to the Goodyear-Zeppelin Corporation, a partnership formed between the German Zeppelin company and the Goodyear Tire and Rubber Company of Ohio. Construction of what was initially called the *ZRS-4* began in late 1929, at an immense new hangar in Akron, Ohio. This hangar was itself of special design, since a conventional rectangular space of the required size would have produced internal gusting of air. The building was so large, at 1,175 feet long, 325 feet wide, and 211 feet high, that it generated a climate of its own; rain would fall inside the hangar. The most distinctive feature of the hangar was its "orange-peel" doors, which opened by folding back along the contours of the building.

Construction of the *ZRS-4*, the largest airship to date, was completed on August 8, 1931. Its overall length was 785 feet, and its eight inboard engines enabled a top speed of nearly 80 mph. Christened by Mrs. Herbert Hoover in honor of the city where it had been built, the *Akron* was a state-of-the-art ship.

After the lesson of the *Shenandoah*, which broke up during flight in a storm, the designers of the *Akron* aimed for indestructibility. Its hull structure was of immense strength, and reinforced by three keels. Since the airship would use helium, rather than flammable hydrogen, its engines could be fully enclosed in the hull. This helped reduce drag (air resistance), as did the placement of the control car directly within the nose of the ship. This positioning of the control car was, in addition, a safety advance; in the *Shenandoah* crash, most of the fatalities were due to suspended cars breaking away from the ship. In another safety advance, the Akron's propellers could swivel to provide an upwards or downwards thrust as well as a forward or reverse direction.

The *Akron* was also adapted to serve as a flying aircraft carrier. In its belly, between the two lower keels, were a large double door and an electrically powered trapeze designed to lift or lower aircraft. The airship could thus accommodate five Curtiss F9C-2 scouting airplanes. Though only the second rigid airship built by the United States, the *Akron* was technologically without equal.

The *Akron* made its first flight on September 23, 1931. For this trial flight, which was a success, the airship carried 113 people. On November 3, 1931, the airship set a passenger record by carrying 270 people on a 10-hour flight. During the next year, however, the *Akron*'s record became tarnished.

On February 22, 1932, the *Akron* smashed a tail fin while being towed from its hangar; a gust of wind had caused its ground cables to snap. In May, after two months of repairs, the airship set off on a coast-to-coast trip. On May 11, this trip was marred by a fatal accident. The airship was attempting to land at Camp Kearney, near San Diego, California. (Landing was always a problem for helium airships, since venting expensive and scarce helium gas to lose buoyancy was out of the question.) The ground crew at Kearney was inexperienced, and when the *Akron* shot upward while this crew was trying to anchor the airship by its cables, three men did not let go fast enough. As spectators watched in horror, and the press kept their cameras rolling, these men were quickly carried aloft. Two of the three men soon fell to their deaths; the *Akron* managed to take the third man onboard.

Despite the sensational publicity that this accident received, no one proposed that the new airship should be subject to a testing period to discover and resolve any operational difficulties. In the spring of 1933, the *Akron* was busily trying to recommend itself to the Navy Fleet.

The Akron *under construction in Akron, Ohio. At 785 feet in length, the airship was the largest airship to date. The specially designed hanger in which the* Akron *was built was so large that it generated a climate of its own: periodically, rain would fall inside.*

Details of the Crash

On April 3, 1933, the *Akron* set off from its Lakehurst, New Jersey base on a trip to calibrate the radio direction-finding stations that had been set up along the New England coast. The airship was flying under its new Commander, Frank C. McCord. Admiral Moffett had flown up from Washington to join the flight, as he had done many times before; it was Moffett's presence that probably prevented the flight from being cancelled when the weather turned bad. As the *Akron* took off at 7:28 P.M., under foggy conditions, its crew had no way of knowing that a cold front accompanied by severe disturbances was approaching from the west, and that a thunderstorm was developing over Washington and moving north-east.

The *Akron* first flew west as planned, reaching Philadelphia and following the Delaware River south toward Wilmington. When flashes of lightning were observed ahead, Commander McCord turned east and then, spotting additional lightning to the west, north-east. Shortly after 10 P.M., with storms clearly developing, McCord decided to head east again out to sea. He hoped to outrun the storm center and then circle back around it and return to Lakehurst.

The *Akron* dropped ballast and set its engines at full speed. By this point, the crew was uncertain of the ship's readings. McCord thought he was heading east from somewhere near Asbury Park, New Jersey, but he was more likely near Bay Head, 11 miles south. McCord was also largely uninformed about the progress of the storms, since electrical interference garbled much of a 10:00 P.M. weather report. The *Akron*, no longer sure of its position, was heading out to sea into one of the worst Atlantic storms of the decade.

The storm soon caught up with the *Akron*. The airship, increasingly in danger, was flying over the ocean through heavy rain and violent winds. McCord reversed direction and headed nearly due west, seeking the hoped-for safety of the coastline. An hour later, he saw lights below that were thought to be Asbury Park. In fact, the ship was over Barnegat Inlet, 45 miles to the south. At this stage, McCord felt he had to turn east again; he feared, erroneously, a collision with the skyscrapers of New York City.

At 12:15 A.M., once again out over the Atlantic, the *Akron* was struck by severe turbulence and started to drop rapidly. Ballast was discharged immediately, and the ship slowed its descent and almost stabilized its course. But the ship then ascended very quickly, only to begin falling again, this time at a rate of 14 feet per second. The fall continued although the engines labored against it at full force.

With the altimeter reading at 800 feet, McCord ordered the ship's elevators to be shifted in an attempt to raise the *Akron*'s nose and regain altitude. As the nose tipped up 25 degrees, there was a loud crack and a severe impact. Although everyone in the crew believed they were still airborne, the airship's lower fin and rudder had struck into the sea. The entire ship then crashed into the water, where the *Akron* was immediately rolled over by the waves and engulfed.

The *Akron* quickly began to break up. The airship carried no lifesaving equipment, and the cold waters of the Atlantic in April made it a deadly place to crash. The German tanker *Phoebus*, which had seen the ship's falling lights, sped through the heavy seas toward the fallen airship. When it arrived, an hour after the crash, there were only three survivors still clinging to debris. The remaining 73 men of the crew were casualties of the disaster, Moffett and McCord among them.

Impact

A Court of Inquiry interviewed the few survivors, but found it difficult to determine the exact cause of the disaster. The Board did conclude that Commander McCord's relative inexperience caused him to fly directly into the storm. Later studies have shown, however, that the storm was so widespread that McCord would have encountered it no matter which direction he took. Furthermore, salvaging the *Akron*'s wreckage confirmed that the airship had not suffered any structural damage prior to striking the sea. Unlike the *Shenandoah*, the *Akron* did not break up in the air.

The wreckage of the Akron *being pulled from the Atlantic Ocean. Only 3 of the 73 people on board survived the crash.*

Most experts now attribute the *Akron*'s crash to the faulty altimeter reading that led the crew to believe that the airship was flying much higher than it actually was. The storm's low barometric pressure probably caused the altimeter to be off by as much as 300 to 600 feet. The airship, flying dangerously close to the sea, had no room for the maneuver that Commander McCord attempted.

This new airship disaster, and the loss of the popular Admiral Moffett, resulted in a public outcry against further involvement with airships. To acknowledge the heavy loss of life in the disaster, Congress and President Franklin D. Roosevelt proclaimed it a national tragedy.

The *Akron*'s sister ship, the *Macon*, was the last of the U.S. military's rigid airships; it went down in a storm on February 12, 1935. Supporters had hoped that the *Akron* and the *Macon* would demonstrate the usefulness of large rigid airships for military reconnaissance at sea. This final disaster, however, underscored the lesson the public had derived from the *Akron*'s crash into the Atlantic: that the military's huge, expensive, and much-publicized airships resulted primarily in American casualties.

LEONARD C. BRUNO

Where to Learn More

Botting, Douglas. *The Giant Airships*. Time-Life, 1980.

Hovgaard, William. "After the *Akron*: Should We Continue To Build Airships?" *Forum*, June 1933, pp. 338-340.

"What Wrecked the *Akron*?" *Literary Digest*, May 6, 1933, p. 21.

"73 Lost in *Akron* Crash, 3 Survivors Here," *New York Times*, April 5, 1933, p. 1.

Smith, Richard K. *The Airships* Akron & Macon. U.S. Naval Institute, 1965.

Toland, John. *The Great Dirigibles: Their Triumphs and Disasters*. Dover, 1972.

Hindenburg crash

Lakehurst, New Jersey

(1937)

Last and largest commercial airship destroyed by hydrogen fire while landing.

Background

On May 6, 1937, the largest airship ever built, the German *Hindenburg,* had just dropped its rope mooring lines to the ground crew below after completing a flight across the Atlantic, when it suddenly exploded and burst into flames. In only thirty-two seconds, the entire ship lay crumpled and smoldering on the ground. Of the ninety-seven people aboard, sixty-two survived the fiery crash.

Since the end of World War I, German airship designer Hugo Eckener had dreamed of building what he considered the perfect airship. He hoped to be able to translate the years of experience that his firm, the Zeppelin Company, had accumulated into an airship of unparalleled size, speed, comfort, and above all, safety. The airship would have to be profitable as well. Eckener had very good reasons for wanting a new generation of airships, since by 1934, his very successful *Graf Zeppelin* had been pushed to the limits of its performance. Although it had braved the Atlantic several times, it was suited neither in size nor speed for such a route, and Eckener wanted to build an airship specifically designed for this rigorous long-distance flight.

By the end of 1934, when Eckener was about to begin actual construction of his new airship, every major nation of the world that had an airship program of its own had abandoned it or was about to do so. One by one, each country except Germany had suffered disastrous and humiliating fatal crashes. Not only was Germany uniquely untouched by serious accidents, but its record in commercial airship flying was spotless—there had never been a passenger fatality on any German airship. The *Graf Zeppelin* alone had flown over one million miles in passenger service without serious incident.

As Eckener's new ship took shape, it became apparent that it would fulfill his dream. Designated the LZ 129 and later christened the *Hindenburg* after Germany's respected war hero and president, Field Marshal Paul von

The Hindenburg *flying over New York City. The airship made the trans-Atlantic journey between Europe and the United States several times before it crashed.*

Hindenburg, its dominant feature was its immense size. At eight hundred and three feet, it was only twenty-nine feet longer than the old *Graf Zeppelin*; but with a maximum diameter of one hundred and thirty-five feet, it was fatter by far than that slim airship. Its design allowed it to hold nearly twice as much gas, giving it not only greater lifting power but making it more resistant to bending forces as well as more efficient. This leviathan incorporated the latest technology in all of its systems, best exemplified by its four Daimler-Benz diesel engines. These efficient, powerful, and lightweight eleven hundred-horsepower engines turned enormous, reversible propellers that could drive the ship to eighty-five miles-per-hour.

Traveling time on the new *Hindenburg* between Frankfurt and Lakehurst would prove to be fifty-two hours eastbound and sixty-five hours westbound—far faster than any ocean liner. And passengers would not give up any of the liner's luxury, since the *Hindenburg* was in fact a flying luxury liner. For a one-way fare of $400, a passenger could enjoy a comfortable private cabin with hot and cold running water, plus a fifty-foot long dining room that boasted an adjoining fifty-foot promenade deck with wide, sloping picture windows. The lounge on the starboard side was thirty-four feet long and was furnished with tables and comfortable upholstered chairs. Elsewhere, passengers found a writing room, another promenade deck, and spacious toilets, washrooms, and showers, as well as a separate smoking room (sealed and protected by pressurization and an airlock). In this room an electric lighter was secured by a chain for cigarette smokers, while cigar and pipe smokers had theirs lit by stewards. Passengers were strictly searched for matches when they first boarded.

Eckener's perfect airship had one weakness, however. Designed to be lifted by helium, it was forced to use hydrogen—seven million cubic feet of this flammable gas—when the United States refused to sell its scarce helium to Germany. At this time, Germany had been taken over by Adolf Hitler, and the United States prudently suspected that Germany might one day have military plans for its helium-filled airships. Hitler also recognized the very popular Hugo Eckener as an avowed anti-Nazi, and soon had him moved upstairs as the company's powerless chairman of the board. Hitler's government also became half owner of Eckener's company.

Given the politicized atmosphere, it is not surprising that the *Hindenburg*'s maiden flight, on March 4, 1936, was essentially a propaganda mission over Germany in which the airship spent four days and three nights broadcasting speeches and dropping leaflets in favor of Hitler's policies. Its first commercial flight was a March 31 trip to Rio de Janeiro carrying thirty-seven passengers. Eckener was unhappy about this long-distance flight, since the airship had not flown its test trials due to the propaganda flights. The round-trip was completed, but two of its Daimler engines failed and the ship barely made it back home. Repairs took a month to complete.

In February 1936, Germany obtained permission from the United States to make ten round-trips to Lakehurst, New Jersey, during that year. The *Hindenburg* completed all of these flights on schedule and experienced no further difficulties. The first flight to Lakehurst took two and a half days, and only two days, one hour, and fourteen minutes to return. Altogether that year, the *Hindenburg* made the ten U.S. trips plus six more to Rio, carrying more than fifteen hundred transatlantic passengers and twenty tons of mail and freight. By year's end, it had won the enthusiastic endorsement of most of the traveling public, and was beginning to interest American financiers. Eckener's dream of a perfect airship had been realized, it appeared. A year's worth of Atlantic flights had proved the ship amazingly stable. Gale force winds elicited not the slightest tremor, and it actually weathered a hurricane without the passengers being aware of how serious a storm they were flying through. Accommodations were plush, the meals served on china were sumptuous, and no one ever got seasick. Prospects for the future of the airship seemed so bright that a German-American international company was formed, with plans to put four Zeppelins into weekly service across the Atlantic.

Details of the Crash

The 1937 flying season began for the *Hindenburg* in March with the first of twenty planned flights to Rio de Janeiro. The United States had agreed to a May-through-November schedule of eighteen round-trips between Frankfurt and Lakehurst, and the first of these was booked for a May 3, 1937, departure. Despite its past full bookings, the passenger list for this flight totaled only thirty-six, but there was also an unusually large contingent of German officers and trainees. The Zeppelin company had plans for the *Hindenburg*'s sister ship, the LZ 130, to make its first flight in October 1937.

Max Pruss, a former captain of the *Graf Zeppelin*, was in command as the *Hindenburg* cast off at 8:15 P.M. from the newly completed Rhein-Main World Airport near Frankfurt. The airship headed north and reached a cruising speed of nearly ninety miles-per-hour, but encountered a strong headwind off the coast of Ireland. The ship did not fly over France or England for political reasons, and reached the North Atlantic by flying over The Netherlands and then coming down the English Channel. Its arrival time at Lakehurst was to be May 6 at 6:00 A.M., but with its speed cut to sixty miles-per-hour by headwinds, that time became impossible. Its arrival was further delayed as the weather deteriorated around Newfoundland, and the ship's speed was further reduced to thirty-seven miles-per-hour as it reached the New England coast. Its arrival time was now estimated to be around 6:00 P.M. on May 6, a full twelve hours behind schedule.

Finally, after an unusually tedious voyage during which bad weather shrouded most of the spectacular sights below, the *Hindenburg* came up the field at Lakehurst from the southwest just after 7:00 P.M., having ridden out at sea the worst of the local weather conditions. It had actually been over New York City at 3:00 P.M., but had been advised of an approaching cold front, so it hovered along the coast. This front was followed by heavy showers and a thunderstorm, but by 7:00 P.M., conditions had improved and Lakehurst told the ship to come in.

At 7:08 P.M., the *Hindenburg* emerged from the clouds at six hundred and fifty feet, roaring over Lakehurst at full speed. It was an impressive sight to the remaining spectators, as it passed over the field and made a sweeping turn to the left so it could approach the mooring from the west. As the airship returned over the field at 7:10, the shifting winds induced Pruss to make another much sharper turn to the right, for a northerly approach. With the ceiling between two thousand and three thousand feet and a light rain falling, lightning was noticed in the distant south and southwest. Pruss then valved off some hydrogen, which lowered the ship a bit, and dropped some water ballast. By maneuvering his four engines, he skillfully brought the huge ship to a complete, dead-level stop in midair at 7:20 P.M.

At 7:21, with the ship at about two hundred feet altitude and about seven hundred feet from the mooring mast, crewmen first dropped the starboard rope handling lines for the ground crew and then the port lines. The landing was proceeding normally as the ground crew continued to couple the manila rope lines to the corresponding lines on the ground. The ship was hovering between one hundred and thirty-five and one hundred and fifty feet when its

The Hindenburg *crashes to the ground in Lakehurst, New Jersey*

outer cover began to flutter and its skin seemed to be rippling. About fifteen seconds later—at 7:25 P.M.—a small tongue of flame emerged from where the skin had been fluttering and a reddish glow was noticeable from the ground. In the control car, the crew felt a shudder in the ship's frame, while at the same time, one of the crew in another part of the ship heard a muffled pop like a gas burner igniting on a stove.

Within seconds, all of these subtle phenomena were followed by a burst of exploding, flaming hydrogen blasting from the top, just forward of the upper fin. In a few more seconds, almost the entire stern was engulfed in flames and began to drop. Inside the control car, Pruss reached instinctively to drop the ballast and raise the ship, but instead made the split-second decision to let the ship's tail section crash to the ground, knowing this was the only hope of anyone possibly scrambling out alive. As the great ship's tail dropped to the ground, its nose pointed skyward and flames shot up through promenades that acted like chimneys, shooting flames out the nose "as from a blowtorch." Once this happened, the entire ship was burning and its framework quickly collapsed. From the time the first flame was noticed until the entire ship lay smoldering on the ground, only thirty-two seconds had elapsed.

The inferno that ensued killed thirteen passengers, twenty-two crewmen, and one civilian ground handler. Of the ninety-seven people on board, sixty-two amazingly survived. Life or death was simply a matter of chance. Some men in the tail walked out virtually untouched as the flames in that section went upward. A cabin boy was saved when doused by ballast water. An acrobat used his professional skills to hang from a window as the ship fell, letting go at a safe height. Others were saved by the heroic actions of the American ground crew. Captain Pruss refused to leave and went back several times to help survivors get out. The flames of the *Hindenburg* continued amidst the chaos of Lakehurst for three hours.

More than any other airship disaster, the *Hindenburg* crash remains the most mysterious and most controversial. Explanations abound as to what caused it suddenly to catch fire, but over the years they have narrowed down to two major theories: ignition of hydrogen by sabotage or by some natural source of electricity. As for sabotage, investigations by both the Germans and Americans yielded no evidence, although some argue that it was in the interest of both countries to downplay any embarrassing hint of sabotage. A book published in 1962 even gave the name of the individual on board who supposedly did the deed. But to this day, there has been no conclusive proof that the airship was sabotaged.

There are several, more natural explanations that attempt to account for the disaster, and it was one of these that the German and American investigating teams chose to endorse. Both teams of experts concluded that the airship's hydrogen was ignited probably by some type of atmospheric electrical discharge. The Americans argued that St. Elmo's Fire—a discharge of electricity that sometimes occurs as an eerie bluish glow on the prominent parts of a ship or aircraft in stormy weather—ignited the hydrogen. The Germans said that the manila ropes dropped to the ground became wet, resulting in the airship "becoming a piece of ground elevated into the atmosphere." This equalization of the static charges between the ship and the ground meant that the *Hindenburg* would itself discharge electricity into the atmosphere, a phenomenon known as "brush discharge."

Finally, both natural cause arguments assume that there was free hydrogen somewhere around the ship, available to be ignited, and point to the fluttering cover as an indicator that hydrogen was escaping. Bolstering this observation is the ship's tail-heaviness, witnessed by most just before the explosion. This lack of lift may have indicated a possible hydrogen loss. The final, official judgment was that the *Hindenburg* was destroyed accidentally by unusual but natural causes. Hugo Eckener agreed with this verdict, but Captain Pruss argued for sabotage. As the debate continues even today, a recent bit of information in favor of the natural causes argument points out that the *Hindenburg* had been painted with a different type of aircraft dope, or preparatory substance, and that it was this new kind of dope that ultimately helped create the deadly spark.

The skeleton of the Hindenburg *after it crashed. This disaster effectively ended airship travel.*

Impact

At the time of the *Hindenburg* crash, the German airship *Graf Zeppelin* was flying back from Rio. On its arrival in Germany, the ship was grounded until the cause of the crash could be determined. No zeppelin ever made another flight. The end of the *Hindenburg* marked the end of hydrogen as a lifting medium and the end of airship travel.

Despite the fact that for more than a quarter century commercial Zeppelins had carried fifty thousand passengers without a fatality, airship travel was now considered untenable. Much of the impact of the crash must surely be attributed to the fact that the disaster was the most thoroughly documented crash of its time. The sights and sounds of that terrible evening were graphically documented on film and quickly disseminated over and over again in every movie newsreel. On radio, the heartbreaking, raw emotion of newsman Herb Morrison—"Oh, the humanity and all the passengers!"—conveyed the horror and pity he felt as he watched helplessly.

The spectacular burning of the *Hindenburg* had an impact on public opinion that far exceeded the fatality count. Despite the fact that nearly two-thirds of the people on board survived, its name became linked forever with tragedy and sudden, terrifying technological disaster. Afterward, nothing could be done to erase the image of airship travel as at best, a dead end in the history of flight, or worse, an invitation to disaster. Two years after the *Hindenburg*'s last flight, an airplane carried its first paying passenger across the Atlantic.

LEONARD C. BRUNO

Where to Learn More

Dick, Harold G. *The Golden Age of the Great Passenger Airships, Graf Zeppelin and Hindenburg*. Smithsonian Institution Press, 1985.

Hoehling, Adolph A. *Who Destroyed the Hindenburg?* Little Brown, 1962.

"Sky Horror: Hindenburg Crash," *Literary Digest*, May 15, 1937, pp. 10-12.

Payne, Lee. *Lighter Than Air: An Illustrated History of the Airship*. Orion Books, 1991, pp. 218-29.

"Oh, the Humanity! Hindenburg Disaster," *Time*, May 17, 1937, p. 35ff.

Vaeth, J. Gordon. "What Happened to the Hindenburg?" *Weatherwise*, December 1990, pp. 315-22.

KEEP OFF MY REAR
I'M EXPLOSIVE!
DVG010

Automobiles

- Chevrolet Corvair (1959-1963)
- Ford Pinto (1971-1976)
- Firestone 500 tire (1972-78)
- Audi 5000 (1978-1986)

Chevrolet Corvair suspension and heater controversy

(1959-1963)

Alleged design flaws in the Chevrolet Corvair's rear suspension brought the American automotive industry under federal regulation.

Background

The Chevrolet Corvair, introduced in September 1959, was the first rear-engine automobile to be mass-produced in the United States. Alleged design flaws in the suspension were blamed for a series of accidents, some of which were fatal, and led to groundbreaking lawsuits against General Motors as the manufacturer of the car. In addition, the Corvair's direct-air heating system was criticized for drawing noxious and potentially dangerous fumes into the passenger compartment. The controversy surrounding the Corvair, in which consumer advocate Ralph Nader figured prominently, brought about the passage of the 1966 National Traffic and Motor Vehicle Safety Act, the first federal legislation to establish government standards for automotive design safety.

The Corvair was a compact car whose design was quite radical by the standards of the conservative American automotive industry. Its rear-mounted engine was air-cooled like that of the Volkswagen. Unlike its imported rival, however, the Corvair featured an innovative engine and transaxle system. It had fully independent coil spring suspension and unit-body construction that nearly eliminated the heavy iron frame of traditional American automobiles.

The design of the Corvair generated controversy even before the car was introduced to the public. No rear-engine car had yet been mass-produced in America, and there were many skeptics. In fact, both Ford and Chrysler tried to raise doubts about the stability of a rear-engine compact car in advance advertisements for the Falcon and the Valiant, the models they introduced at this time. Some more objective magazine reviews of the car, although positive in tone, raised questions about the car's high-speed handling. Nonetheless, the Corvair initially enjoyed respectable sales, and Motor Trend magazine named it "Car of the Year"—a coveted honor—in April 1960.

Details of the Controversy

The first documented Corvair fatality involving a loss of control occurred in 1960, when a California teenager named Don Lyford crossed the centerline of the road during an S-turn and was killed in a collision with an oncoming car. A Los Angeles attorney investigating the accident interviewed a police officer who claimed to have seen "six of them [Corvairs] flip out of control" in recent months.

Over the next several years, this attorney litigated scores of legal cases against General Motors involving the Corvair. None of these legal cases reached a definite conclusion until 1964, when Rose Pierini, who had lost an arm in a rollover crash, was awarded $70,000 in an out-of-court settlement. This settlement was widely reported in the press as a victory for the plaintiff, and opened the floodgate to litigation against General Motors.

In November 1965, a book was published that escalated the controversy. *Unsafe at Any Speed*, by Ralph Nader, is popularly regarded as the book that killed the Corvair. Actually, only the book's first chapter discussed the Corvair. Nader intended the book as an indictment of all the sins of the Detroit establishment, and principally of General Motors. The first sentence of the preface reads: "For over half a century, the automobile has brought death, injury, and the most inestimable sorrow and deprivation to millions of people." His thesis about the Corvair was that General Motors executives had marketed a car they knew to be unsafe because their desire for profit outweighed all other considerations. In short, Nader all but accused them of negligent homicide.

Unsafe at Any Speed, by Ralph Nader, is popularly regarded as the book that killed the Corvair. Actually, only the book's first chapter discussed the Corvair. Nader intended the book as an indictment of all the sins of the Detroit establishment, and principally of General Motors.

Under ordinary circumstances, *Unsafe at Any Speed* probably would have remained obscure. But at the time of its publication, Nader was working on the staff of Senator Abraham Ribicoff, who chaired a Senate subcommittee that was crafting a bill to establish federal standards for automotive design. Thus, the book received widespread attention, and Nader was called on during congressional hearings as an expert witness on various automotive topics, including the safety of the Corvair.

Nader's testimony, along with some well-timed publicity, set the stage for a strong National Traffic and Motor Vehicle Safety Act that was unanimously passed by both houses of Congress, and was signed into law by President Johnson on September 9, 1966. Nader's role in the passage of the bill established him as a patron saint of the new "consumerism." Nevertheless, the congressional hearings did not, as Nader had demanded, lead to the recall of all 1960-63 Corvairs. In fact, in spite of Nader's nearly ten years of effort toward this end, a comprehensive recall of the Corvair never took place.

General Motors discontinued production of the Corvair in 1969. Chevrolet's litigation problems, however, were not over. A new round of allegations, now concerning the design of the vehicle's direct-air heater, returned public attention to the Corvair's alleged flaws. Throughout the decade, General Motors had received hundreds of letters complaining of acrid fumes emitted through the heater vents, which could be noxious and even dangerous to passengers. Under continual pressure from consumer advocate groups, the

In this 1959 Ford-produced film, a Corvair (right) skids out of control while trying to keep up with a Ford Falcon. Consumer activist Ralph Nader charged that the film refuted General Motors' claim that the Corvair had good handling ability.

National Highway Traffic Safety Administration finally made a determination that the Corvair's direct-air heating system "creates an unreasonable risk of accidents and injury to persons . . . such engine fumes do in some cases contain carbon monoxide in sufficient concentrations to harm or endanger the occupants of the vehicle."

General Motors' reaction to this finding was grudging. In December 1971, General Motors mailed a letter to all known Corvair owners suggesting that they take their cars to a dealership for inspection of the heating system. Since the release of fumes depended on the condition of engine and heater duct seals, General Motors treated the entire issue as a maintenance matter and refused to accept any financial responsibility, even for the recommended inspections. The letter's additional suggestion that winter drivers turn the heater off or roll the windows down if the fumes became too noxious was not well-received. But General Motors, already embattled over the Corvair's suspension system, resolved to wait out this new tempest as the cars aged and were removed from service.

Details of the Corvair's Design

Despite attempts by the Corvair's design engineers to slim down its powertrain, the system through which power generated by the rear-mounted engine was transmitted to the axle, over sixty percent of the vehicle's weight remained concentrated on the rear axle. The Corvair had fully independent suspension (each wheel could move without affecting the movement of any other wheel), and the rear wheels were driven by "swing axles" between the wheels and transaxle. These axles had universal joints only on the side connected to the transaxle.

Most cars are designed to "understeer," so that their tendency in a high-speed turn is to go straight, and the driver must increase the rotation of the steering wheel to compensate. An oversteering car tends to increase the angle of a turn, so that the driver must adjust by reducing the rotation of the steering wheel. One of the allegations against the Corvair was that, because its front (steering) wheels were so lightly loaded, the vehicle would normally understeer but suddenly begin to oversteer during a high-speed turn that generated significant lateral (sideways) thrust. This phenomenon could get an unskilled and unprepared driver into trouble. The dispute was not over whether the phenomenon existed—even the Corvair's defenders did not deny that it did—but under what conditions the oversteering occurred, and whether the average driver could be expected to cope with it.

One strategy adopted by the Corvair's designers to offset the oversteering phenomenon was to stipulate that different air pressures be maintained in the front and rear tires. The specified cold pressure in the front tires was fifteen psi (pounds per square inch). The pressure specified for the more heavily-loaded rear tires was twenty-six psi. Although these instructions were spelled out in the owner's manual, they were routinely ignored, and anyone seeking advice on the subject at service stations often received incorrect and contradictory information. The failure to maintain the tire-pressure differential exacerbated the tendency of the car to oversteer.

The other allegation regarding the Corvair's suspension system related to the design of the rear drive axles. Nader and other critics claimed that during sharp cornering, the camber (vertical angle) of the inboard rear wheel could shift by as much as eleven degrees. Since there was no universal joint between the wheel and the axle, this alleged angle change would be transmitted to the axle. The axle, acting as a lever, would then exert an upward force on the transaxle, causing excessive body sway and greatly increasing the likelihood that the car would roll over. Critics made much of the fact that a front anti-roll bar was originally included in the Corvair design, but had been eliminated to reduce costs.

The allegation about the camber of the inboard rear wheel was fervently denied by representatives of General Motors, and was never proven by the Corvair's critics. However, there was some evidence to support the allegation. Shortly after the introduction of the Corvair, aftermarket accessories became available to improve the car's ride and handling. The best-selling of these, the EMPI Camber Compensator, was widely touted as a cure for the car's suspension ills. In the 1964 model year, General Motors adopted suspension modifications very similar to the Camber Compensator. From that year onward, the incidence of accident-related lawsuits decreased dramatically.

In the 1965 model year, the Corvair's rear suspension was completely redesigned. By that time, however, negative publicity about the car had done its damage. The Corvair's sales revived slightly in 1965, but in 1966, the year of the Ribicoff hearings, sales dropped catastrophically from 209,152 to 88,951. Virtually no model changes occurred between 1966 and 1969, the car's final year of production, and the Corvair was officially canceled on May 14, 1969. In that year, only 3,102 Corvairs were produced.

The separate allegations about the Corvair's heating system dogged the car from the start, although these complaints were overshadowed by the suspension controversy until after General Motors halted production of the car. The passenger compartment heater in most cars relies on heat drawn from the engine coolant. The Corvair's engine was air-cooled, making this approach impossible. The engine's rear location also created complications. The Corvair relied on a system of ducting and an electric fan to route some of the air used to cool the engine into the passenger compartment. If noxious fumes such as oil smoke or carbon monoxide escaped from worn engine seals—a virtual certainty given the state of the technology—those fumes were pulled into the passenger compartment.

Impact

The controversy that began in 1959 with the introduction of the Corvair gave rise a new legal concept: the liability of an automobile manufacturer for negligence of design safety. The conventional view had been that automobile drivers should look out for their own well-being and were responsible for avoiding accidents. Little or no thought was given to designing automobiles such that accidents would be survivable or less injurious. Vehicle interiors, for example, had no seat belts or interior padding, headrests, or knee bolsters. Protruding knobs and horn buttons often produced injuries that, in hindsight, seem almost ludicrously preventable; the same was true of protruding fins and hood ornaments on the exterior. Brake systems were not failsafe, steering columns sometimes broke loose and impaled drivers, and broken windshields became jagged and deadly.

Although such design characterized the industry as a whole, it was the negative publicity surrounding the Corvair that galvanized a new attitude about product safety and manufacturer responsibility. The National Traffic and Motor Vehicle Safety Act developed by Senator Ribicoff's subcommittee was largely written by Ralph Nader, who regarded the auto manufacturers as nothing less than criminals. Although in Nader's opinion the Safety Act was not nearly strong enough (for example, he held that executives should be imprisoned for non-compliance), it brought into existence the National Traffic Safety Agency, later called the National Highway Traffic Safety Administration, which drafts the regulations adhered to by all auto manufacturers today.

Ironically, that agency issued a report in July 1972 exonerating the Corvair of charges that it was exceptionally unstable in handling. The report concluded that "The handling and stability performance of the 1960-1963 Corvair does not result in an abnormal potential for loss of control or rollover, and it is at least as good as the performance of some contemporary vehicles both foreign and domestic."

JOEL SIMON

Where to Learn More

"My Six Years in the Spin, Flip and Burn Car." *Car Collector*, July 1981, pp. 17-19.

Cray, Ed. *Chrome Colossus: General Motors and Its Times*. McGraw-Hill, 1980.

Knepper, Michael. *The Corvair Affair*. Motorbooks International, 1982.

McCarry, Charles. *Citizen Nader*. Saturday Review Press, 1972.

"Car of the Year." *Motor Trend*, April 1960.

"Corvair Corsa Road Test." *Motor Trend*, January 1965.

Nader, Ralph. *Unsafe at Any Speed*. Grossman, 1965.

National Highway Traffic Safety Administration. "Evaluation of the 1960-1963 Corvair Handling and Stability." DOT HS-820, July 1972.

O'Connell, Jeffrey and Arthur Myers. *Safety Last: An Indictment of the Auto Industry*. Random, 1966.

"Safe at Any Speed." *Road & Track*, July 1966.

"Miscellaneous Ramblings." *Road & Track*, October 1966.

Sobel, Robert. *Car Wars*. E. P. Dutton, 1984.

"SCI Analyzes Ed Cole's Corvair." *Sports Car Illustrated*, November 1959, p. 23.

Ford Pinto rear-impact defect

Automobiles prone to burst into flames following rear-impact collisions led to a comprehensive recall and prolonged negative publicity and legal entanglements for Ford.

(1971-1976)

Background

Soon after its debut in 1971, the Ford Pinto became the hottest subcompact on the market. Over the Pinto's 1971-80 lifespan, more than three million of the vehicles were sold. By the middle of the decade, however, reports began to surface that the Pinto tended to catch fire after rear-impact collisions. *Mother Jones* magazine and the Center for Auto Safety maintained that design flaws in the vehicle made its fuel tank system vulnerable and hazardous, and Ford came under pressure from the National Highway Traffic Safety Administration (NHTSA). In 1978, Ford recalled 1.4 million 1971-76 Pintos for safety modifications. By the mid-1980s, at least fifty-nine people had burned to death in Pinto accidents, resulting in more than one hundred lawsuits against Ford.

The initial sales figures for the Pinto were good news for Lee Iacocca, who spearheaded the development of the vehicle as an "import-fighter." Iacocca's goal was a sensible "econocar" priced under $2,000 and weighing less than two thousand pounds. Ford and Iacocca were credited with success until the Center for Auto Safety—a consumer group run by disciples of Ralph Nader—began receiving a substantial volume of complaints about the Pinto. The Pinto fuel tanks were allegedly easily punctured in a rear-impact crash, setting off a blaze of fire that would kill or seriously burn the occupants of the car.

The Center for Auto Safety demanded government action, and in the fall of 1977 the National Highway Traffic Safety Administration initiated a "defects investigation" of the Pinto. The Pinto wasn't the first Ford vehicle to be singled out as a possible fire hazard. The Illinois Supreme Court had ruled the flange-mounted gas tanks in the 1966 Fairlane 500 unsafe, and Ford was criticized at the time for withholding reports that revealed it had been aware of the risk.

Because the Pinto was Ford's first significant entry into the small-car market, it was rushed into production. The typical forty-three month development phase for a new automobile was abbreviated to as little as twenty-five months (thirty-seven months in one historian's estimate). Tooling came early in the schedule, before the quality assurance process was completed. The $200 million in tooling installed would have discouraged any modifications of the design.

Even though Ford had been a safety pioneer in the mid-1950s, offering such innovations as seatbelts, it retreated from those standards for the subcompact Pinto. "Safety doesn't sell" was a common attitude in Detroit, and the product objectives for the Pinto, as reported by the Society of Automotive Engineers, did not include vehicle safety. A 1969 memo suggested that a safer over-the-axle gas tank design had been rejected because it would slash luggage space. Ford was also charged with employing a series of delaying tactics to fight a proposed safety rule—Federal Motor Vehicle Safety Standard 301—that could have compelled a redesign of the Pinto.

Details of the Controversy

In the fall of 1977, *Mother Jones*, a San Francisco-based investigative magazine, unleased a bombshell against Ford. A lengthy article demonstrated that the development phase for the Pinto had been accelerated, and pinpointed flaws in the vehicle's fuel-tank structure, particularly in terms of its vulnerable position. More damaging than the technical details was the article's assertion, based upon previously-concealed corporate documents, that Ford executives knew from their own tests that Pintos were prone to catch fire after rear-end collisions. Ford executives did not correct the problem, however, because doing so—at an estimated $10 to $15 per car—would cost more than what Ford estimated it would pay out in accident-related lawsuits.

Three design details bore the brunt of criticism by the *Mother Jones* article: 1) the tank sat too far to the rear, barely six inches ahead of the bumper; 2) the fuel filler tube was likely to separate and create spillage; and 3) four bolts protruding backward from the differential housing could easily puncture the tank, which was only three inches away, during a rear-end collision. Several other under-chassis components were identified as capable of piercing the fuel tank. The *Mother Jones* article also charged that the Pinto's floor pan could separate, allowing a fuel tank fire to spread into the passenger compartment, and that the vehicle's front doors tended to jam shut in a crash and prevent escape. Other critics of the Pinto noted its flimsy bumpers and generally fragile construction. Pinto was the first modern American Ford, said one, to lack rear subframe members (a steel skeleton), a cost-control and weight-saving measure.

When hit at thirty-plus miles per hour, the *Mother Jones* article claimed, a Pinto's rear end "would buckle like an accordion, right up to the back seat. The tube leading to the gas-tank cap would be ripped away from the tank itself, and gas would immediately begin sloshing onto the road." The buckled

The wreckage of a 1973 Ford Pinto that burst into flames after being hit from behind by another car. One person died in this July 1978 crash which occurred one month after Ford announced a voluntary recall of 1.4 million 1971-76 Ford Pintos.

tank would jam against the differential housing, with its protruding bolts. Add a random spark, and both cars "would be engulfed in flames." Safety expert Byron Bloch (whose credentials would be caustically questioned by Ford) insisted it was "irresponsible" to put "such a weak tank in such a ridiculous location in such a soft rear end."

Ford documents revealed that early Pintos had been tested more than forty times. Engineers reported that in a twenty-one and a half miles per hour test, the "fuel filler pipe pulled out of the fuel tank and the fluid discharged through the outlet. Additional leakage occurred through a puncture ...." Unless the car had been structurally modified, claimed *Mother Jones*, "every test at twenty-five miles per hour or more ... resulted in a ruptured fuel tank." Ford's testing continued after the Pinto was in production. A report on fuel tank integrity of 1971-72 Pintos concluded that the rear-end structure was unsatisfactory for a twenty mile per hour crash. Major revision was deemed necessary for the Pinto to withstand a thirty mile per hour collision.

Ford documents also revealed proposals to improve the Pinto's rear-end safety. These proposals included repositioning the tank, moving the filler pipe, shielding the protrusions, and installing a bladder in the tank. The most effective of the solutions might have been adopting a fuel tank that rides above the rear axle and differential. Ironically, Ford held a patent on that type of tank, and used it on the European-built Capri, which was imported to the United States. Soon after the Pinto's debut, the Capri passed the equivalent of a forty-five mile per hour rear-end crash test. British ads even boasted of the Capri tank's safety.

Ford vigorously defended the Pinto, arguing that its fuel system presented no unreasonable risk. Ford officials insisted that, in addition to meeting all applicable government-mandated safety requirements, the Pinto was as safe as other vehicles of its size and type. The Pinto did meet federal standards, but until 1977 those standards did not include a requirement that cars withstand a specified rear-end collision force.

Despite Ford's defense of the Pinto, a series of recalls followed. In October 1976, NHTSA ordered 372,584 Pinto, Bobcat, and Mustang II models recalled because of potential fuel leaks from the engines. A few 1977 Pintos were recalled because the rear-bumper nuts installed might puncture the fuel tank. The 1976 Pintos were recalled because a thirty miles per hour frontal impact could lead to fuel leakage.

NHTSA announced in May 1978 that its own investigations "demonstrate that low-to-moderate-speed rear-end collisions of Pintos produce massive fuel tank leaks due to puncture or tearing of the fuel tank and separation of the filler pipe from the tank." In the "presence of external ignition sources," the NHTSA determined, such fuel leaks could "result in fire." A dozen test thirty-five mile per hour rear-end collisions yielded two fires from fuel spillage. Moreover, government experts knew of thirty-eight cases of fuel-tank damage, leakage, and/or ensuing fires to date, resulting in twenty-six fatalities.

On June 15, 1978, shortly before a public hearing that probably would have forced a recall, Ford announced a "voluntary" recall of 1.4 million 1971-76 Pintos (the fuel tank had been redesigned beginning with the 1977 Pinto) and thirty thousand similar-make 1975-76 Mercury Bobcats. Ford did not admit to the existence of a defect, but agreed to make changes to "end public concern"—a thirty to forty million dollar public-relations gesture. Among the modifications performed for the recall, Ford agreed to install two high-density polyethylene shields around the fuel tank.

Consumer advocate Ralph Nader called Ford's planned remedy "an inadequate technical fix; the cheapest way out." Nader wanted Ford to revert to a double-wall tank, as used on bigger Fords. Retired Chrysler auto-safety director Roy Haeusler warned that many Pinto motorists were "sitting in virtual time bombs" due to unsafe tank designs.

This owner of a 1975 Ford Pinto warned other drivers to keep a safe distance when following her.

Ford doubtlessly hoped the Pinto controversy would fade after the recall, but Ford remained beleaguered in the law courts. Among the lawsuits filed against Ford was a sensational Indiana case that soon would focus national attention on the Pinto. In August 1978, three teenage girls died when their 1973 Pinto burst into flames after being struck from the rear by a van. As the Indiana case opened, Ford faced more than fifty civil suits and at least two class-action suits. Courts had already awarded plaintiffs as much as $6 million in damages. With the Indiana trial, which began in early 1980, Ford earned the dubious distinction of being the first corporation charged with a criminal offense—"reckless homicide." Although for this case Ford only faced a maximum fine of $30,000, a guilty verdict could affect twenty-three other pending civil suits.

Rather than attempting to prove a design flaw in the Pinto, Indiana prosecutor Michael Cosentino argued a narrower issue: that Ford had allowed the Pinto "to remain on Indiana highways, knowing full well its defects." Employing the legal talents of former Watergate prosecutor James F. Neal, Ford pleaded not guilty, and was ultimately acquitted of all charges. In this lawsuit, the novel principle that manufacturers of a product could be criminally responsible for injuries resulting from its use was not upheld.

Presiding judge Harold Staffeldt found "a revolt against technology" implicit in the case, and refused to establish the legal precedent.

Despite Ford's acquittal, the negative publicity surrounding the Indiana case remained extremely damaging, and many consumer advocates continued to lobby against the car. The Pinto began dropping out of the market, to be replaced in 1981 by the Ford Escort.

Impact

Had Ford executives and engineers been wholly unaware of the Pinto's potential to catch fire in rear-end collisions, the alleged defect would not have created such a public furor. After the emergence of seemingly incriminating documents, Ford officials denied having crash-tested early Pintos, and insisted that the vehicle had no inherent safety problems. Ford executives noted that their own family members drove Pintos. Ford's corporate credibility, however, was suspect; in 1974, the company had pleaded guilty to cheating on emissions tests, paying a $3.5 million fine.

The controversy surrounding the Pinto created widespread public distrust of the American automotive industry. Most compelling was the allegation that the Ford cold-bloodedly put dollars ahead of lives. *Mother Jones* declared that Ford executives decided not to modify the Pinto for safety after a "'cost-benefit analysis,' *which places a dollar value on human life,* said it wasn't profitable to make the changes" [italics in original]. After the enactment of the National Traffic and Motor Vehicle Safety Act in 1966, Ford secured an agreement from federal policymakers that such a "cost-benefit" principle would be acceptable.

A Ford analysis to determine the desirability of installing a fuel valve in certain vehicles to help prevent fires during rollovers illustrates the nature of the "cost-benefit" principle. Analysts estimated that unless changes were made, twenty-one hundred vehicles would eventually burn, causing one hundred and eighty deaths and one hundred and eighty serious injuries. Placing monetary values at $200,000 per death, $67,000 per injury, and $700 per vehicle, that came to a total of $49.5 million—the amount Ford expected to pay in lawsuits. In contrast, modifying the vehicles would cost $137 million. By forgoing the safety modification, Ford could save $87.5 million.

Allegations that American-made vehicles represent fire and safety hazards did not end with the Ford Pinto. In the early 1990s, the Center for Auto Safety charged that almost five million 1973-87 General Motors pickup trucks were fire-prone in side-impact collisions because of the "sidesaddle" placement of their fuel tanks. The National Highway Traffic Safety Administration pressured General Motors to recall and modify the affected trucks. Various civil lawsuits and thirty-six class-action suits have been filed against General Motors; in February 1993, an Atlanta jury awarded $105.2 million to the family of a seventeen-year-old killed in such a vehicle.

Shortly before, on November 17, 1992, the TV newsmagazine "Dateline NBC" had aired a simulated crash involving a General Motors truck; the truck obligingly burst into flames. Not until General Motors investigated the "Dateline NBC" story did the truth about the simulated test emerge: among other interventions, the person in charge of the simulation had placed igniters along the truck's fuel tank to ensure that it would catch fire. "Dateline NBC," which prepared an unprecedented on-the-air apology, had fallen victim to the same sort of commercial pressures and deadlines that lead to the production of defective automobiles.

Nonetheless, consumer advocates continue to lobby against both the pickups and General Motors' proposed resolution to the controversy: offering owners of affected pickups a $1000 coupon towards the purchase of a new General Motors vehicle of a comparable class. This marketing offer—whereby General Motors would more than recoup the value of the certificates in new sales profits—hardly puts questions about the corporate ethics of the American automobile industry to rest. Under the terms of their offer, General Motors would not be required to repair the affected trucks or even acknowledge that the trucks are defective. In fact, the affected trucks could be traded in as part of the new vehicle purchase—and be immediately resold.

James M. Flammang

Where to Learn More

Dowie, Mark. "Pinto Madness." *Mother Jones*, September-October 1977, pp. 18-32.

"Car Safety: Who's for Burning?" *The Economist*, January 12, 1980, pp. 30-1.

Everett, David. "GM Settles Truck Suit with Coupons." *Detroit Free Press*, July 29, 1993; pp. A1, A4.

Lacey, Robert. *Ford: The Men and the Machine*. Little, Brown, 1986.

"Domestic Safety Defect Recall Campaigns." National Highway Traffic Safety Administration. 1976 ed., p. 21; 1977 ed., p. 16; 1978 ed., pp. 12-14.

Panztur, Andy. "Pinto Criminal Trial of Ford Motor Co. Opens Up Broad Issues." *New York Times*, January 4, 1980, pp. 1, 23.

Rowand, Roger. "An All-New Ball Game for Engineers." *Automotive News*, January 21, 1980, p. 48.

Stuart, Reginald. "U.S. Agency Suggests Ford Pintos Have a Fuel System Safety Defect." *New York Times*, May 9, 1978, p. 22.

"Pinto Ruling." *Time*, February 12, 1979, p. 87.

Firestone 500 steel-belted tire failure

(1972-1978)

Seven and a half million Firestone-built tires were recalled when it was discovered their treads were likely to separate from the main structure, especially when driven underinflated.

Background

After a promising send-off in 1972, Firestone's new "500" steel-belted radial tires ran into trouble. Customers began to return failed tires in unusually high numbers, while tragic accidents were attributed to blown-out Firestone 500s. Most common of the complaints was that the outer tread—the portion of the tire that hits the road—was inclined to slither apart from the rubber carcass or main body.

Through the early decades of motoring, tire technology lagged behind other automotive developments. Not until the mid-1950s did the first tubeless tires arrive. Then came fiberglass-belted tires, adding strength to the carcass. Domestic cars soon borrowed the radial belt pattern from European automakers, and it began to elbow aside the bias-belted styles. Radials have their plies or inner core material running at right angles to the beads or sides of a tire, giving better strength, traction, and fuel economy.

By the 1970s, the new steel-belted radials seemed to promise the best of all possible worlds, mixing the strength of steel with the stability and longevity of the radial belt pattern. Industry sales began to sag by 1974, partly because radials lasted longer, creating less demand for replacement rubber.

In October 1973, *Consumer Reports* rated the Firestone 500 steel-belted radial in the top two. Neither the 500 nor any other tire showed damage in a 325-mile test at 90 miles per hour, so the magazine based its evaluation on tread life, not safety factors. No one seemed to suspect that anything within the tire might cause trouble as it aged.

Actually, someone within the Firestone organization did fear trouble almost from the start. In September 1973 (a year after production began), Thomas A. Robertson, the director of development, sent a memo to top management warning, "We are making an inferior quality radial tire which will subject us to belt-edge separation at high mileage." Nevertheless, Firestone made more than twenty-three million such radials over the next five years, and insisted the tire had no safety defects.

In 1972, Robertson had expressed concern about Firestone's relationship with automakers that installed equivalent tires (made by Firestone, virtually identical to the 500s, but lacking the Firestone label) as original equipment on new cars. "We are badly in need of an improvement in belt separation performance," he warned, "particularly at General Motors, where we are in danger of being cut off by Chevrolet because of separation failures."

In one of a series of incidents, the Louis Neal family was driving near Las Vegas in June 1974. When one of its Firestone 500 steel-belted radials blew out, their car went out of control and crashed. The parents were killed, and one child severely crippled. Firestone settled out of court for $1.4 million.

Between 1973 and May 1978, when the House Subcommittee on Oversight and Investigations opened hearings, fifteen deaths had been reported in which such blowouts were a major cause or main contributing factor. By mid-1979, at least forty-one deaths were known to have occurred. The federal government counted fourteen thousand complaints from consumers.

Consumer advocates knew about the Firestone problem by 1976, having received numerous complaints about tread separations and resultant blowouts. In 1977, the Center for Auto Safety—a watchdog group operated by activists who had worked with consumer advocate Ralph Nader—pushed for government investigation. Having discovered that half the complaints it studied involved Firestone, Clarence M. Ditlow, head of the center, wrote to Firestone president Mario A. Di Federico in November 1977. Rather facetiously, he suggested the "company should shift half of its advertising budget into quality control."

Acting upon data supplied by the Center for Auto Safety, the National Highway Traffic Safety Administration (NHTSA) surveyed 87,000 owners of new cars, asking if they had complaints about radial tires. Of the two thousand two hundred and twenty-six owners of Firestones who returned a questionnaire, 46 percent reported problems. The rate for Goodrich was 33 percent; Goodyear and Uniroyal, 32 percent. Despite a court order (obtained by Firestone) to suppress the report, these figures were "inadvertently" released, according to the agency.

Even before the large-scale 1978 recall, Firestone 500 steel-belted radials had been subject to four smaller recalls for manufacturing defects, which affected a total of 410,000 tires. But Firestone still opposed a full recall. Instead, it aimed the blame at a careless public, charging that drivers failed to keep tires properly inflated, bumped them against curbs, overloaded their cars, and drove too fast.

The NHTSA investigated for seven months, studying six thousand consumer reports. Then, in July 1978, it made an "initial determination" of a safety-related defect.

On December 5, 1978, Firestone and NHTSA agreed to a recall of the affected tires, built from September 1, 1975 to the end of 1976 (for five-rib tires) or through April 1976 (for the seven-rib style). "Failure," said the NHTSA announcement, "can result in a loss of air with a possible loss of control of the vehicle which may result in vehicle crash." The official decree covered 14.5 million tires, but only an estimated 7.5 million that remained on the road wound up entitled to free replacement. Approximately 6 million earlier tires could be exchanged at half price.

Production of the 500 continued into April 1978, but the recall did not apply to tires made after the cutoff dates, because internal construction had changed by then. Neither did it apply to nonradial (bias-belted) 500s.

Not all of the affected tires wore a "Firestone 500" label. A dozen others included the Firestone TPC (used on GM cars); Montgomery Ward Grappler; Seiberling RT 78; Atlas Goldenaire II; Caravelle Supreme; Shell Steel Radial; K-Mart Steel Radial 400; and Zenith Supreme.

In a departure from customary practice, the federal government listed no specific engineering cause for the flaw and recall. Instead, NHTSA based its decision upon the simple fact that, based on the number and nature of complaints from consumers, an excessive number of these tires had been demonstrated to be defective. Thus, a pattern of failure was deemed sufficient to prompt the recall.

Details of the Defect

Like other belted tires of the 1970s, the Firestone 500 radials carried a set of reinforcing belts beneath the tread. Whether made of fiberglass or steel, these belts encircled the tire's carcass, enhancing the rigidity of its structure. In the case of the Firestone-built radials, heat buildup within the tire appeared to cause the tread to separate from the steel-belted inner layer. When that occurred, the ultimate result could be a blowout.

Prior recalls of some 500s suggested that a serious problem lay waiting. In March 1976, a small number (two hundred) of Firestone 500s were recalled because of a possibility they were "undervulcanized, which could lead to separation in lower sidewall area." A year later came a record-size recall of 400,000 tires, because they exhibited "distortion or separation in tread area" in a high-speed indoor test. Firestone cited a temporary production problem as the reason for the defect.

One technical detail is certain: steel does not bond readily to rubber. Even when coated with brass, the twisted steel wires that form a tire's belt pattern resist becoming one with the carcass. In the early days, all domestic tire makers were having trouble producing radials.

Reps. John Moss and James Collins of the U.S. House Commerce Oversight and Investigations subcommittee examine a damaged Firestone 500 tire during that committee's 1978 hearings on the safety of the tire.

One technical detail is certain: steel does not bond readily to rubber. Even when coated with brass, the twisted steel wires that form a tire's belt pattern resist becoming one with the carcass. In the early days, all domestic tire makers were having trouble producing radials.

Because NHTSA failed to specify the precise nature of the defect, and no one else presented conclusive evidence, any analysis must be speculative. One such scenario, based upon investigation by a Georgia newspaper, was presented in *Consumer Reports*.

After examining internal Firestone documents and considering basic manufacturing processes, the *Atlanta Beacon Journal* determined that time was a significant factor in tread separation. Most such tires failed later in life. The tire could be trouble-free for thousands of miles, but a foundation for failure may have existed from the start.

To help overcome the fact that the brass-coated steel belts don't bond easily to the tire's rubber carcass, the belts were first coated with a rubber compound. This created a better base; but if that compound absorbed moisture from the air, it could set the stage for possible disaster. When the tire was vulcanized (heat and pressure applied, to harden the rubber), ammonia formed in the base rubber. Mixed with the internal moisture, this created a corrosive substance. Over time, that substance tended to corrode the brass coating of the steel wires, weakening their bond with the surrounding rubber.

A Firestone document from mid-September 1975 appeared to affirm that diagnosis. Employee P. F. Murray wrote that degraded adhesion resulted from "corrosion of the brass caused by ammonia from compounding ingredients combined with moisture." Firestone was investigating "dezincification" of belts and the effects of ammonia and moisture.

Critics charged that even if underinflation were actually the cause of failures, the company was negligent in failing to warn owners of the danger. Underinflation causes a tire to flex far more than normal. In addition to increased mechanical stress on the rubber, it can cause a tire to run hotter. If motorists had been aware that a tire inflated to appreciably less than the recommended volume would trigger eventual tread separation, they may have kept closer watch on tire pressures.

Firestone and its defenders continued to dismiss the charge of an inherent defect. Malcolm Lovell, chairman of the Tire Industry Safety Council, concurred with Firestone that the problem stemmed from underinflation.

Corporate records turned over to NHTSA, however, suggested that Firestone was hardly unaware that trouble was brewing. In addition to the Robertson correspondence, a 1973 letter from Atlas Tire—one of the companies that marketed a Firestone-built radial under its own name—stated that "Firestone is coming apart at the seams and drastic action is required." In 1976, representatives of Shell indicated they were ready to stop marketing their private-brand tire because of customer returns. Montgomery Ward warned that returns of its Grappler 8000 had reached "epidemic proportions," which "amplifies the fact we were given a bad product."

No other radial-tire line had a failure record anywhere near that of the 500 series. Firestone's claimed 7.4 percent "adjustment rate" (the number of tires returned for refund, divided by the total number sold) was twice the figure for competitors. Goodyear, Firestone's top rival, rated 2.9 percent.

When Representative Al Gore questioned that rate during congressional hearings, Firestone's counsel explained that 500s were their most expensive tire, so customers would be more likely to complain. Further investigation revealed that in reality, an industry record of 17.5 percent (not 7.4 percent) of these tires were returned to dealers.

Impact

Even if Firestone was not culpable for flawed manufacturing processes, the company's angry denials of responsibility, and of any inherent flaw, drew fire in the press. Worse yet, on learning that a recall was imminent, Firestone sold off the final 500s in southeastern states, at clearance-sale prices. *Business Week* described this last-ditch maneuver as having the appearance of a "desperate effort to unload damaged goods."

At congressional hearings in May 1978, Firestone argued that the tire had been discontinued eighteen months earlier, to be replaced by a new 721 series (also steel-belted radials). While the phaseout had indeed begun by 1977, and significant design changes were instituted between 1976 and 1977, final 500s were coming out of the factory as late as April 1978.

In a vain effort to block a total recall, John F. Floberg, Firestone's vice president and general counsel, attempted to suppress release of NHTSA's survey of radial-tire owners by obtaining a court order. (Firestone condemned the survey's validity, because half of the questionnaires went to Firestone tire owners.) That effort went awry, as the survey was released anyway, and the publicity told consumers more about the issue than they would otherwise have known. Firestone chairman Richard A. Riley, meanwhile, continued to insist the 500s were safe, and that the company had "been completely aboveboard."

During ten months of debate, noted *Business Week*, "additional revelations about the tires hardened NHTSA's position, as well as public opinion." Firestone's prior knowledge of the "unusually high volume of its customers' complaints" became evident, while the company "repeatedly tried to thwart investigation." *Fortune* magazine charged that in its defense, Firestone had "used tactics that worsened the ordeal," concluding that the company's response "may well become a classic."

Analysts predicted that the wave of negative publicity could skim 2 to 3 percent from Firestone's usual 25 percent share of the replacement-tire market. Meanwhile, the company prepared to spend $230 million for the recall. By mid-1979, about three million Firestone 500s had been replaced, but NHTSA chief Joan Claybrook and other critics expressed disappointment with the pace of the recall.

Firestone still faced a series of lawsuits charging property damage and loss of life, plus a $1.7 billion class action. By the time of the recall, about two hundred and fifty such suits had been filed. Nine verdicts had gone against Firestone, twenty-two ruled in the company's favor, and sixty-four had been settled out of court. Firestone also faced an inquiry by the Securities and Exchange Commission, which wondered if the company had adequately informed its shareholders about the 500 issue. Although the SEC apparently took no action against Firestone, in 1979 a group of about 4,000 shareholders filed a class-action suit against the company accusing it of concealing information about the 500 defect. Firestone settled with this group for $3.2 million in 1983, without admitting any fault or liability.

Although some critics went so far as to compare the Firestone 500 debacle to President Nixon's Watergate crisis, the company's woes turned out to be relatively short-lived. To help bolster its faded image, Firestone enlisted folksy actor Jimmy Stewart to help promote the new 721 series. They also offered a then-radical two-year warranty. Before long, Firestone managed to rebound near its former stature in the industry.

JAMES M. FLAMMANG

Where to Learn More

"The Big Firestone 500 Recall." *Consumer Reports*, April 1979, pp. 199-200.

"Firestone 500: Blow Out." *The Economist*, October 14, 1978, pp. 121-122.

Feldstein, Stuart A. "How Not to React to a Safety Controversy." *Business Week*, November 6, 1978, p. 65.

"The Case for Firestone." *Forbes*, November 13, 1978, p. 106.

Gray, Ralph. "Firestone Adds Batch of New Entries in Recall Wake." *Advertising Age*, February 5, 1979, p. 4.

Hinsberg, Patricia. "Analysts Assess Firestone Recall." *Automotive News*, October 30, 1978, p. 2.

Kahn, Helen. "Secret Documents Hint at Firestone Coverup." *Automotive News*, January 1, 1979, pp. 1, 23.

Louis, Arthur M. "Lessons from the Firestone Fracas." *Fortune*, August 28, 1978, pp. 45-48.

"Motor Vehicle Tire Recall Campaigns." National Highway Traffic Safety Administration, 1976 ed., p. 72; 1977 ed., pp. 79-80; 1978 ed., p. 67.

"Forewarnings of Fatal Flaws." *Time*, June 25, 1979, pp. 58, 61.

Audi 5000 sudden acceleration

(1978-1986)

Allegations of "sudden acceleration" led to public relations crises for Audi and the development of the automatic shift lock.

Background

On March 7, 1989, the National Highway Traffic Safety Administration (NHTSA), the government agency responsible for overseeing safety issues in the motor vehicle industry, released the results of their research on motor vehicle "sudden acceleration"—defined as unintended and unexpected high-power acceleration from a stationary position or very low initial speed, accompanied by apparent braking ineffectiveness. The NHTSA announced in a press release that an "exhaustive, independent study of the sudden acceleration phenomena in cars concluded that pedal misapplications that may be aggravated by vehicle design are the most probable explanation for most reports of sudden acceleration." These NHTSA test results were of particular interest to two groups: Volkswagen of America, the corporate parent of Audi; and former and current owners of Audi 5000 vehicles, as the investigation was a direct result of much-publicized allegations of sudden acceleration in Audi 5000 sedans, model years 1978 to 1986, equipped with automatic transmissions.

Victims of the Audi's alleged sudden acceleration reacted to the findings with disbelief. To these owners of runaway cars, the Audi 5000 was capable of becoming a "possessed" vehicle that could not be stopped by any driver action. In their view, the NHTSA's use of the term "pedal misapplication," which implied that drivers mistook the gas pedal for the brake pedal, deflected the blame away from the party that was responsible: Audi. Audi accident victims stated in no uncertain terms that their foot was on the brake pedal, not the gas.

Audi stood by the NHTSA's claim that the sudden acceleration problem was the result of driver error, to which poor pedal design may have contributed. This view was supported by Audi's post-accident examinations of the vehicles at issue. These examinations revealed no mechanical defect that would account for the complete brake failure described by most accident victims.

The allegations of sudden vehicle acceleration were not restricted to Audi, although a disproportionate share of the cases involved the Audi 5000s. Nor was the issue new when, in the early 1980s, Audi found itself at the center of controversy. As early as 1967, the NHTSA was researching cases of "unwanted power," as the phenomena was initially termed. Since that time, the NHTSA has conducted over one hundred investigations—involving twenty vehicle manufacturers—of the sudden acceleration problem. These manufacturers include General Motors, Ford, Chrysler, AMC, Toyota, Nissan, Honda, Subaru, Volkswagen, Mercedes-Benz, Saab and Volvo.

The NHTSA's examinations focused on very specific vehicle components that could malfunction in the manner described by the victims of sudden acceleration. Among the more commonly investigated components were: cruise control systems that might cause the throttle (which controls the air flow to the engine and thus engine speed) to stick open; computer-controlled fuel injection systems (which regulate the amount of fuel sprayed into the engine cylinders) that might affect engine performance; and electronic idle-speed control systems that could malfunction and lead to sudden engine surges.

This series of investigations prompted twenty-six government-mandated safety recalls and influenced seventy-five voluntary recalls, in which the manufacturer modified the vehicles without government intervention. In some cases (Nissan, Volvo, and General Motors), the recalls addressed a mechanical defect thought to be responsible for the sudden acceleration. For example, a loose engine mount (the device that holds the engine to the vehicle frame) caused the engines in some General Motors vehicles produced in the 1970s to twist in the engine bay, affect the throttle, and generate unwanted acceleration. Generally, however, no specific mechanical defect was discovered, and the NHTSA's final ruling in the majority of its investigations identified driver error as the apparent cause of the sudden acceleration accidents.

Details of the Controversy

The Audi 5000 sudden acceleration controversy was different from allegations involving other vehicles in two respects: the rate of reported occurrence and amount of the publicity generated. Reported incidents in the Audi reached 645 per 100,000 vehicles, nearly five times the rate of occurrence for any other manufacturer's vehicle. A larger difference, however, may have been in the press coverage of the allegations against Audi. Since no correlation exists between design and engineering changes in the Audi 5000 and the increased number of acceleration problems reported, one argument holds that the increase in reported "cases" stemmed from the level of negative publicity that the Audi 5000 received.

Publicity over the acceleration issue primarily arose from three groups: Audi itself; the Center for Auto Safety (CAS), a consumer advocacy group founded by Ralph Nader; and the NHTSA. The publicity peaked on November 11, 1986, when CBS's *Sixty Minutes* made the Audi controversy its lead story. The controversy actually began in 1982, however, long before the *Sixty Minutes* cameras rolled, when Audi responded to a preliminary NHTSA investigation of public complaints about the Audi 5000 tending to accelerate unexpectedly. Audi engineers claimed at that time that floormat interference (the floormat sliding under the pedals and interfering with braking and acceleration) was the apparent cause of the problem, and recalled the vehicles to install a metal stud that would keep the floormats in place.

Incidents of sudden acceleration continued to be reported, and Audi initiated a second recall in September 1983. In this recall, Audi placed a spacer atop the brake pedal to further differentiate the brake pedal from the accelerator pedal and thus lessen the chance for pedal confusion. This second recall indicates the stance Audi continues to take regarding allegations of sudden acceleration: the occurrence is primarily due to driver error. These two recalls were voluntary, and the NHTSA, determining that Audi's recalls had addressed the problem, ruled that no further action was necessary and closed its investigation.

In February 1986, Audi announced a third recall, in which it would increase the distance between the brake and accelerator pedals. This recall was consistent with Audi's policy of attributing the blame for sudden acceleration on driver error. In fact, the positioning of the Audi pedals was not unusual: the distance between brake and accelerator pedals before the recall, 2.4 inches, was well within the typical range of 2.3 to 2.6 inches.

It was at this point that the media attention escalated. On March 19, 1986, Robert Adams, as the Attorney General of New York, held a press conference to demand an Audi recall. Two months later, on May 28, Neil Hartigan, the Attorney General of Illinois, made the same demand, claiming that an Audi recall was necessary "to ensure the public safety and welfare of the citizens of Illinois and our sister states."

On this same day, the Audi Victims Network (AVN), a group founded in February 1986 and comprised of forty drivers who claimed to be victims of Audi sudden acceleration, held a meeting with Audi representatives. And Audi itself chose this day to announce a third recall, in which it would increase the distance between the brake and accelerator pedals. This recall was consistent with Audi's policy of attributing the blame for sudden acceleration on driver error. To support their position, Audi produced statistics revealing that a relatively large percentage (thirty-six percent) of sudden acceleration incidents were reported of vehicles that had been driven less than four thousand miles. Audi's argument was that a driver unaccustomed to the vehicle and its pedals easily might apply the gas pedal by mistake. It is worth noting, however, that the positioning of the Audi pedals was not unusual. In fact, the distance between brake and accelerator pedals before the recall, 2.4 inches, was well within the typical range of 2.3 to 2.6 inches.

In June 1986, shortly after the pedal adjust recall, Audi initiated its fourth service recall on the 5000, this time to install an automatic shift lock (ASL) device. In January 1987, Audi undertook a government-mandated recall to install the ASL on all 1978-1986 Audi 5000s with automatic transmissions. The ASL device required a driver to depress the brake pedal before shifting out of park. The reasoning behind the ASL installation was that even if the vehicle surged once the transmission was engaged, the vehicle's brakes held the car against full depression of the accelerator pedal, a brake characteristic common to all motor vehicles. Throughout these recalls, Audi strove to present itself as a company responsive to its customers' needs. This representation was hotly disputed by the AVN and the Center for Auto Safety, who viewed the sudden acceleration as deadly engineering defect that Audi was not addressing.

A fifth Audi recall was initiated in tandem with the ASL installation to replace a faulty idle stabilizer valve, a device that controls the engine speed under varying load conditions (such as towing or extreme temperatures) when the car is at idle. The stated purpose of this last recall was to eliminate the potential for engine fires, but some critics perceived it as a method of covertly fixing the sudden acceleration problem. In the letter issued to all Audi 5000 owners regarding the fifth recall, Audi indicated merely that "irregular engine idle might cause unsatisfactory car performance." Dan Howell of the CAS, however, alleged that the process amounted to a "secret recall" since Audi would not have replaced the valve unless it affected the safety of the vehicle.

The controversy over the Audi also pitted the CAS against the NHTSA. In pubic statements, the CAS encouraged Audi owners not to settle for the ASL recall as a solution to the sudden acceleration problem. The CAS had received reports that the ASL devices were ineffective against sudden acceleration. Moreover, since accident victims claimed to have had the brake pedal depressed before the occurrences of sudden acceleration, CAS Executive Director Clarence Ditlow maintained that the recall "flies in the face of the facts of virtually every runaway Audi accident ever reported." Diane Steed, the NHTSA Administrator, countered that the CAS should have encouraged Audi owners to take advantage of the recall, since "your organization has not provided any other credible explanation (or solutions)."

The general report that the NHTSA released in 1989, which concluded that there were no mechanical defects to account for the occurrences of sudden acceleration, ended the federal government's investigation of the phenomena, even if it did little to appease those Audi 5000 owners who had experienced the problem. By this point, more than 1,400 reports of sudden acceleration—resulting in 330 injuries and 7 deaths—had been received by the NHTSA and Volkswagen of America for the 250,000 model years 1978 to 1986 Audi 5000 vehicles sold.

Impact

In June 1988, a class action settlement against Audi was proposed by the Circuit Court of Cook County, Illinois. The settlement offered what amounted to a rebate on the future purchase or lease of an Audi vehicle to owners of 1978-86 automatic transmission Audi 5000s who had suffered property damage from sudden acceleration. Under the terms of this offer, victims would receive a certificate valued from $300 to $2000, depending on the model year of the affected vehicle. This paradoxical measure would benefit plaintiffs only if they continued to purchase Audi vehicles—thereby stimulating sales of the very vehicles in question. Needless to say, the CAS advised owners not to accept the settlement and thereby forfeit their right to further compensation. Audi owners, however, were never offered any other settlement.

Due to the sudden acceleration controversy, Audi's sales dropped thirty percent in one year. The vehicle was redesigned and the 5000 nomenclature dropped, but the damage had been done, and most experts agree that the Audi 5000 was a financial disaster. Even though no proof was uncovered of defective design, Audi's "Art of Engineering" reputation was tarnished—a bleak prospect for a seller of high-tech and upmarket gadgetry.

The most widespread result of the Audi controversy, nonetheless, is evident in most vehicles equipped with automatic transmissions: the automatic shift lock. Although not required by federal law, the device at least partially protects drivers from sudden acceleration and shields auto manufacturers from the public relations disaster experienced by Audi.

GREG JANICKI

Where to Learn More

Center for Auto Safety. "Audi 5000 Sudden Acceleration 1992." 1992.

———. "NHTSA Requests Audi Recall." *Impact*, Vol. 12, No. 2.

———. "NHTSA's Acceleration Report Leaves Questions Unanswered." *Impact*, Vol. 14, No. 4.

Csaba, Csere. "Audi Agonistes." *Car and Driver*, June 1987, pp. 51-57.

"NHTSA Announces Results of 'Sudden Acceleration' Study." *DOT News*, March 1989.

"The Riddle of Unintended Acceleration." *Road and Track Special Report*, 1987.

"An Examination of Sudden Acceleration." Report DOT-HS-807-376, *U.S. Department of Transportation*, January 1989.

Bridges

- Quebec Bridge (1907, 1916)
- Tacoma Narrows Bridge (1940)
- Vancouver Second Narrows Bridge (1958)
- King's Bridge (1962)
- Point Pleasant Bridge (1967)
- West Gate Bridge (1970)
- Sunshine Skyway (1980)
- East Chicago ramp (1982)
- Zilwaukee Bridge (1982)
- Mianus River Bridge (1983)
- Schoharie Creek Bridge (1987)

Quebec Bridge collapses

(1907, 1916)

Inadequate design estimates of stresses on compression members led twice to the collapse during construction of bridges across the Saint Lawrence River.

Background

Cantilevering a railroad bridge across eighteen hundred feet of the Saint Lawrence River at Quebec was a risky enterprise that failed twice, first in 1907 and again in 1916 (a cantilever bridge consists of two sections or arms that extend and meet above the middle of a stream or river). The first disaster attracted wide attention from the general, trade, and professional press, and led to major reforms in the practice of the civil engineering profession. The second accident reinforced the value of engineering analysis.

Promoters of a bridge across the Saint Lawrence had, long before 1907, dealt with serious problems and setbacks stemming as much from economics as from technology. As so often happens, economics dictated the final design despite the fact that this was to be the longest span ever constructed and budgets should have been generous. A group of civic leaders first proposed the bridge in 1852, but thirty years passed before a charter was granted. It was 1900 before financial aid was obtained from the city of Quebec and the Canadian government.

Meanwhile, the group, known as the Quebec Bridge and Railway Company, had retained the services of Theodore Cooper, at the time one of the world's foremost railway bridge experts, to advise the company and approve all plans. Bidders were asked to supply their own designs, with the result that the specifications were drawn up by the low bidders. The Phoenix Bridge Company designed the superstructure (the vertical extension from the *substructure*) and William Davis & Son, the substructure (the underlying support structure).

As the world's largest single span (1,800 feet), the bridge was to be longer than the Brooklyn Bridge and longer than each half of Scotland's Forth Bridge, which it resembled. To cut expenses, it was to be pin-connected rather than riveted, and assembled from minimal-size lattices rather than the massive tubular components that braced the Forth against wind and other loads. Without ever visiting the site, Cooper decreed that the span be two-hundred feet longer than the original Phoenix plans. The lattice-type cantilevers springing from the north and south banks of the river and the two-part central suspension span were to be constructed by traveler crane. With the substructure in place, work on the superstructure began in 1905, but progress was slow.

Details of the Collapses

On August 29, 1907 the south arm was approaching midchannel when something went wrong in a bottom chord member and the entire structure crashed into the water, killing about eighty men (the exact number of casualties reported in contemporary publications varies from seventy-four to eighty-two). There had been a warning earlier that day, when the cantilever end had deflected excessively and the engineers on the site had sent a frantic telegram to Cooper in New York. By the time his reply ordering everyone off the bridge arrived and was relayed from the Phoenix Bridge Company office, it was too late.

The cause was traced to buckling of a compression member due to inadequate bracing. One of the first reports relied on visual descriptions of the wreckage: "Of all the bottom-chord sections of the anchor arm . . . there is only one that exhibits characteristic buckling distortion, and that is the 9th chord. All others are bent and crushed, broken at the splices, cracked across, burst open, and most variously battered—but none has a well-defined buckle."

That technicality, however, was overshadowed by deeper concerns. Contemporary newspaper reports called the failure "the greatest engineering disaster" and the trade publication *Engineering News* took the matter rather personally. Its first "news report" began editorially: "Public confidence in engineers and engineering constructors and in safety and reliability of their works is an asset of the whole engineering profession. To have this public confidence receive such a blow as this at Quebec is a loss almost incalculable."

The *News* assured its readers that this "largest structure to date" had been superbly handled with "the erection planned as carefully as it was designed. The possibilities of mischance which affect ordinary erection work were eliminated. Each manipulation of every piece has been studied, . . . the general course of construction laid out so at every stage the structure was fully as safe as the finished bridge." Yet, the editors added after noting the failure of some compression members, "It is well to confess that our knowledge of the actual limits of safe stress on long steel columns of exceptional size is by no means perfect."

The on-site investigation immediately located the buckled compression member, and writers for the *News*, backing down from their praise of the design and construction planning, seized the fact that the compression

The first collapse of the Quebec Bridge (shown here) was attributed to insufficient bracing of webs of bottom chords. Approximately eighty people died in the disaster.

members were of unusual design, untested and measuring fifty-six feet long and four-and-a-half-feet wide.

It was noted that compression members were designed for a unit stress under full dead, live, and wind loads of twenty-four thousand pounds per square inch, about two-thirds the elastic limit of the metal; it was also noted that the members were carrying more than two-thirds of this amount.

Several weeks later the *News* published in full Theodore Cooper's written reply to a Canadian Royal Commission inquiry. Cooper, then seventy years old, had first declined to comment and, being disabled, had never visited the site. He had received weekly reports at his New York City home, and at one point in 1906 had stopped construction of the center post because the anchor arm was not behaving as expected.

Cooper traced the construction from its inception and answered legal questions about his commission, the supervisory provisions in the contract, and details of the design. Referring to an earlier attempt to prevent problems with the south arm, he wrote: "I believe that if prompt action had been taken to protect Chord 9 West from further deflection, which could have been done by the employment of three hours' work and $100 worth of timber and bolts, the defects and deficiencies which we now recognize in the compression chord and members could, at a later date, have been corrected and the bridge could have been made perfectly safe and sufficient for its intended purpose."

The Royal Commission report, released in February 1908, bluntly noted that it was Cooper who, as consulting engineer, "accepted and approved the defective design that originated in the office of the Phoenix Bridge Company." It found that the lattice bars in failed members were too weak for their duty, the dead load had been set too low in calculations, and "a graver error was the extraordinarily high working stress" permitted in the design.

The Commission stressed a "lack of professional knowledge of present day reaction of steel columns under load," and the editors of another trade newspaper, the *Record*, added, "The real lack is in the latticing: Engineers know little or nothing as to the proper theory of that design."

In 1908, the Canadian government decided on an entirely new bridge on the same site, naming three prominent engineers to a new board to get it designed and built. The new design was also an eighteen hundred-foot cantilever, but was heavier and stronger. It would be the first in North America to be riveted rather than pin-connected, and would also use the then-novel k-truss to add strength by bracing long compression members.

By 1911, the board awarded a contract to the Saint Lawrence Company, and by the end of 1915, some thirty-two thousand tons of steel had been placed. The construction method had been altered so that the center span would be lifted 150 feet onto the ends of the cantilever arms rather than being built out from them. An elaborate lifting mechanism was devised from double girders and vertical "chains." Jacks on one girder would raise the span in a three-stroke cycle. Between lifts, the span would be pinned to the chain.

The span, fifty-one hundred tons of nickel steel, was fabricated three miles downstream and barged to the site on September 10, 1916. After a few lift cycles, the engineers and a large party of distinguished guests went to a celebration breakfast. When the lift resumed, the span crumpled and slid into the river, killing eleven men.

Two representatives of the *Record* were on hand, and photos taken by them before, during, and after the accident show how one corner jumped from its support. The unsupported side pulled apart, one truss and then the other crumpled, and finally one lifting girder tore free, snapping mooring lines.

A Board of Engineering inquiry focused on "insufficient or improper design" of the rocker-joint castings that were supporting the corners of the span and the supposition that the support was "in an unstable condition." Again, the editors of both trade publications speculated that assumption of stresses at just over twenty-thousand pounds per square inch seemed too low for the lifting members. A year later, a fifty-four hundred ton replacement span was

The first collapse (1907) of the Quebec Bridge led to major reforms in civil engineering practices. The second, in 1916 (shown here), reinforced the value of engineering analysis.

floated in. This time, four "chairs" under the ends of the span were attached to the eyebar lifting chains to raise the span in two-foot lifts. To the cheers of thousands of spectators, the seventy-fourth lift brought the span into proper position, and when the pins were slipped through the eyebars, the final clearance was between one-eighth inch and one inch. Even the reporters cheered: "In Quebec, every bell and whistle and automobile horn was turned loose. Flags and bunting were thrown into the wind everywhere, for Quebec realized that its dream of 30 years had come true."

Impact

The main lessons learned from Quebec's double disaster were that loads must be checked, and checked again, with components tested to failure. Several reports pointed out that the mistake in the first design (dead load 30 percent too low) had been made by two designers who were overworked and under great pressure, and that their work had not been checked. The ultimate cause was placed in the designers' unwarranted extrapolation from much smaller successful bridges. The reports agreed that experiments should have been made to confirm the high design stress.

As a result, standards and codes were written to require more conservatism in design of such structures and/or testing to detect failure of full-scale compression members. And, as one observer put it, "This collapse gave a shot in the arm to the then-new American Society for Testing and Materials (ASTM)."

Rita Robison

Where to Learn More

"Quebec Bridge: Major Disaster, Major Lesson." *Civil Engineering*, August 1976, p. 65.

"Quebec Bridge Disaster: Report of the Commission." *Engineering Magazine*, April 1908, pp. 100-102.

"The Compression Members of the Quebec Bridge." *Engineering News*, October 3, 1907.

"Theodore Cooper on the Quebec Bridge and its Failure." *Engineering News*, October 31, 1907, pp. 473-76.

Holgate, Henry, J.G.G. Kerry, and John Galbraith. "Report of the Royal Commission on the Cause of the Collapse of the Quebec Bridge." *Engineering Record*, March 14, 1908, pp. 309-10.

"Giant Bridge Span Collapses; 11 Dead." *New York Times*, September 12, 1916, p. 1.

"Why the Quebec Bridge Failed." *Scientific American*, October 12, 1907, pp. 257-58.

"Report on the Quebec Bridge Disaster." *Scientific American*, April 25, 1908, p. 290.

"Further Light on the Quebec Bridge Disaster." *Scientific American*, September 30, 1916, pp. 288-89.

Skinner, Frank W. "Collapse of the Quebec Bridge." *Engineering*, September 20, 1907, pp. 391-93.

Worthen, Sanborn S. "The Bridge that Balked." *Trains*, March 1985, pp. 26-32.

Tacoma Narrows Bridge collapse

Washington State

(1940)

The failure that brought an abrupt end to the search for "a slender ribbon bridge deck" also introduced the importance of wind dynamics to bridge design.

Background

Although the newspapers were filled with important stories from throughout the world on Saturday, November 8, 1940—stories such as the RAF bombing Berlin, the Luftwaffe bombing London, and Franklin D. Roosevelt returning to Washington D.C. for an unprecedented third presidential term—there was still room on the front page of the *New York Times* for a single-column story, "Big Tacoma Bridge Crashes 190 Feet into Puget Sound."

The Tacoma Narrows Bridge was the third largest suspension span bridge in the world and only five months old. The center span, measuring two thousand eight hundred feet, stretched between two four hundred twenty-five-foot-high towers, while the side spans were each one thousand one hundred feet long. The suspension cables hung from the towers and were anchored one thousand feet back towards the river banks. The designer, Leon Moisseiff, was one of the world's foremost bridge engineers. He and his partner Fred Lienhard had earlier developed the calculations for determining load and wind forces used by bridge designers everywhere.

Following the widespread design effort of the 1930s toward "streamlining" every product from teakettles to locomotives to airplanes, Moisseiff's intent was to produce a very slender deck span arching gently between the tall towers. His design combined the principles of cable suspension with a girder design of steel plate stiffeners—running along the side of the roadway—that had been streamlined to only eight feet deep.

The $6.4 million bridge had opened with much fanfare on July 1, celebrated as a defense measure to connect Seattle and Tacoma with the Puget Sound Navy Yard at Bremerton, Washington. Owned by the Washington State Toll Bridge Authority, the bridge had been financed by a Public Works Administration grant and a loan from the Reconstruction Finance Corporation, and had been constructed in only nineteen months.

The bridge gained notoriety even before it opened, and was nicknamed "Galloping Gertie" by people who experienced its strange behavior. Forced to endure undulations that pitched and rolled the deck, workmen complained of seasickness. After the opening, it became a challenging sporting event for motorists to cross even during light winds, and complaints about seasickness became common.

Despite Moisseiff's reputation as a top-ranked engineering consultant, State and Toll Bridge Authority engineers were more than a little nervous about the behavior of the slender two-lane span, which was only thirty-nine feet wide. Its shallow depth in relation to the length of the span (eight to two thousand eight hundred feet) resulted in a ratio of 1:350, nearly three times more flexible than the Golden Gate or George Washington bridges. Engineers tried several methods to stabilize, or dampen, the oscillations—the up-and-down waves of the deck.

The first method involved attaching heavy cables—called tie-down cables—from the girders and anchoring them with fifty-ton concrete blocks on shore. The cables soon snapped, and another set installed in a second try lasted only until the early morning hours of November 7. A more successful method, a pair of inclined stay cables connecting the main suspension cables to the deck at mid-span, remained in place but proved ineffectual. Engineers also installed a dynamic damper, a mechanism consisting of a piston in a cylinder, that also proved futile as its seals were broken when the bridge was sandblasted prior to being painted.

Measurements and movie camera films taken over several months gave the engineers a good idea of how the bridge was moving in the wind. They charted the oscillations and vibrations, and discovered the movements were peculiar. Rather than damping off (dying out) very quickly as they did in the Golden Gate and George Washington bridges, Galloping Gertie's vibrations seemed almost continuous. The puzzle, however, was that only certain wind speeds would set off the vibration; yet there was no correlation between the vibration and the wind speeds.

Led by Frederick B. Farquharson, professor at the University of Washington's engineering school, the study team applied actual measurements of the bridge movements to a scale model, hoping to find ways to stabilize it. They suggested installing additional stabilizing cables, attaching curved wind deflectors, and drilling holes in the girders to permit wind to pass through. Unfortunately, the report was issued only a week before the bridge collapsed.

Interest in the phenomenon rose with the onset of the brisk fall winds pushing through the valleylike narrows that lie between the cities of Tacoma and Bremerton. The public as well as the engineers kept watch, and when the bridge finally snapped, it became one of history's most documented disas-

The center span of the Tacoma Narrows Bridge writhes under a gale force wind. Moments later, the bridge collapsed.

ters. Cameras—still and movie—recorded the collapse on film. A newspaper man was a mid-span survivor, supplying a firsthand account of the disaster.

Details of the Collapse

Witnesses included Kenneth Arkin, chairman of the Toll Bridge Authority, and Professor Farquharson. As Arkin later recalled, that morning he had driven to the bridge at 7:30 to check the wind velocity, and by 10:00 saw that it had risen from thirty-eight to forty-two miles per hour while the deck rose and fell three feet thirty-eight times in one minute. He and Farquharson halted traffic, watching while the bridge, in addition to waving up and down, began to sway from side to side. Then it started twisting.

Meanwhile, Leonard Coatsworth, the newspaperman, had abandoned his car in the middle of the bridge when he could drive no further because of the undulations. He turned back briefly, remembering that his daughter's pet dog was in the car, but was thrown to his hands and knees. Other reporters described him, "hands and knees bloody and bruised," as he crawled five hundred yards while the bridge pitched at forty-five degree angles and

For months the motions, while disturbing, had been symmetrical and the roadway had remained flat. Such twisting motions caused metal fatigue, resulting in the collapse on November 7, 1942.

concrete chunks fell "like popcorn." The driver and passenger of a logging truck told a similar tale of having jumped to the deck and crawled to one of the towers, where they were helped by workmen.

By 10:30 the amplitude (distance from crest to valley) of the undulations was twenty-five feet deep and suspender ropes began to tear, breaking the deck and hurling Coatsworth's car and the truck into the water. When the stiffening girder fell one hundred ninety feet into Puget Sound, it splashed a plume of water one hundred feet into the air. Within a half hour, the rest of the deck fell section by section, until only the towers remained, leaning about twelve feet toward each shore. Overlooking the bridge was an insurance company billboard that bragged, "As safe as the Narrows Bridge." The slogan was covered up before the end of the day. The only casualty of the collapse was the dog.

Moisseiff's first public comment was, "I'm completely at a loss to explain the collapse."

Charles E. Andrew, chief engineer in charge of construction, said that the collapse was "probably due to the fact that flat, solid girders were used along the sides of the span." He wanted to make clear that his original plans called for open girders, but "another engineer changed them." He compared Galloping Gertie with New York's Whitestone Bridge, which had been completed a year earlier. It was the only other large bridge designed with web-girder stiffening trusses, "and these caused the bridge to flutter, more or less as a leaf does, in the wind. That set up a vibration that built up until failure occurred."

The Whitestone, now known as the Bronx-Whitestone, was indeed very similar to the Tacoma Narrows. However, the stiffening trusses were twice as heavy and the deck twice as wide. Its designer, Othmar Ammann, had consulted Moisseiff, as had the designers of such structures as the Golden Gate and the San Francisco-Oakland Bay bridges. The Whitestone is four thousand feet long, with a two thousand three hundred-foot main span and three hundred seventy-five-foot-high towers. It, too, had early oscillation problems, but they had been successfully damped with a device similar to the one tried on Galloping Gertie.

After the Tacoma failure, Ammann realized the danger of the too-slender deck, and insisted that a new steel truss be superimposed onto the deck. His redesign, which also added cable stays between the towers and deck to prevent twisting, stabilized the bridge. In 1990, a tuned mass damper (a fairly recent invention) was added to the deck during a rehabilitation.

The first investigations into the collapse of Tacoma Narrows detailed how the bridge had come apart. For months the motions, while disturbing, had been symmetrical and the roadway had remained flat. The lampposts on the sidewalks remained in the vertical plane of the suspension cables even as they rose, fell, and twisted. But on November 7, a cable band slipped out of place at midspan, and the motions became asymmetrical, like an airplane banking in different directions. The twisting caused metal fatigue, and the hangers broke like paper clips that have been bent too often.

Impact

But what caused Galloping Gertie to twist so violently when other bridges had survived gale-force winds? Engineers looking into the problem of the twisting bridges were able to explain that winds do not hit the bridge at the same angle, with the same intensity, all the time. For instance, wind coming from below lifts one edge, pushing down the opposite. The deck, trying to straighten itself, twists back. Repeated twists grow in amplitude, causing the bridge to oscillate in different directions.

The study of wind behavior grew into an entire engineering discipline called aerodynamics, parallel to that of the airplane industry. *Vortex shedding* and *flutter* were added to the vocabulary. A vortex is a spiral that can be seen in the wake of a ship or, in a wind tunnel, by wisps of smoke added for the purpose. Some vortices do not affect oscillation, but others form a flutterlike

This new bridge was opened in 1950. Its four-lane deck and stiffening trusses form a box design that resists torsional forces. Excitation is controlled by hydraulic dampers at the towers and at midspan.

pattern that has the same frequency as the oscillating bridge. Eventually no bridge, building, or other exposed structure was designed without testing a model in a wind tunnel. With the development of graphic capabilities, some of this testing is now done on computers.

Dozens of papers are published each year about these subjects; nevertheless, misconceptions continue. In 1990, K. Yusuf Billah of Princeton University noticed that currently popular physics textbooks were using an incorrect version of the cause of the Tacoma Narrows collapse. He enlisted Robert H. Scanlan of Johns Hopkins University, and they "set the record straight" in the February 1991 issue of the *American Journal of Physics*. The textbooks were claiming "forced resonance (periodic natural vortex shedding)" as the cause. Instead, Billah and Scanlan pointed out, "the aerodynamically induced condition of self-excitation [vibration] was an interactive one, fundamentally different from forced resonance."

Two years after the collapse, the remains of the Tacoma Narrows Bridge were scrapped. In 1950, the state opened a new eighteen million dollar bridge designed by Charles Andrew and tested in wind tunnels by Farquharson and other engineers. Its four-lane, sixty-foot-wide deck and stiffening trusses twenty-five feet deep form a box design that resists torsional forces. Excitation is controlled by hydraulic dampers at the towers and at midspan.

The Tacoma Narrows Bridge collapse remains one of the most spectacular failures in the history of engineering. It is certainly one of the best known because of the movie film and widely reproduced still photos, but as Billah and Scanlan demonstrated, the aerodynamic details tend to become blurred. The new science of wind engineering, however, is now routinely applied to every type of structure.

RITA ROBISON

Where to Learn More

Alexander, Delroy. "A Lesson Well Learnt." *Construction Today*, November 1990, p. 46.

Ammann, O. H., T. von Karman, and G. B. Woodruff. "The Failure of the Tacoma Narrows Bridge." Report to the Federal Works Agency, March 28, 1941.

Billah, K. Yusuf, and Robert H. Scanlan. "Resonance, Tacoma Narrows Bridge Failure, and Undergraduate Physics Textbooks." *American Journal of Physics*, February 1991, pp. 118-24.

Farquharson, Frederick B. "Collapse of the Tacoma Narrows Bridge." *Scientific Monthly*, December 1940, pp. 574-78.

Goller, Robert R. "Legacy of Galloping Gertie 25 Years After." *Civil Engineering*, October 1965, pp. 50-53.

Jackson, Donald C. *Great American Bridges and Dams*. Preservation Press, 1988, pp. 327-78.

"Big Tacoma Bridge Crashes 190 Feet into Puget Sound." *New York Times*, November 8, 1940, p. 1.

Peterson, Ivars. "Rock and Roll Bridge." *Science News*, 1990, pp. 244-46.

Petroski, Henry. "Still Twisting." *American Scientist*, September-October 1991, pp. 398-401.

Vancouver Second Narrows Bridge collapse

British Columbia

(1958)

The failure of temporary construction support caused the collapse of bridge spans.

Background

On June 17, 1958, two spans of the Vancouver Second Narrows Bridge collapsed while under construction. The accident was caused by the collapse of a temporary tower supporting the partially completed bridge. Eighteen workers were killed in the collapse, which caused four million dollars in additional construction costs on the bridge.

The Second Narrows Bridge connects Vancouver, British Columbia, with its northern suburbs across Burrard Inlet, the city's harbor. The structure was built for the British Columbia Toll Highways and Bridges Authority. The main bridge, a steel cantilever truss structure, has a total length of over two thousand feet. Unlike older, simpler, and shorter bridges whose spans or sections rest independently on their piers or abutments, those of a cantilever bridge run continuously over or extend beyond the piers. The main bridge has three spans: an 1,100-foot-long center span balanced by two side spans, one 465 feet and one 466 feet long. Four steel truss spans, each 276-feet-4-inches long, make up the northern approach to the main bridge. The structure was designed by Swan, Wooster and Partners, a Vancouver engineering firm. Dominion Bridge Company was the contractor for the construction of the steel spans. The foundations and bridge piers were constructed by Peter Kiewit Sons and Raymond International.

Details of the Collapse

By mid-June 1958, the approach spans were in place and erection of the northern side span of the main bridge was in progress. The length of the span required the use of two temporary supports for construction, since the side span would not be self-supporting until its full length was in place. Each temporary support, called a falsework bent, consisted of two columns, one under each side of the span. The columns were built on temporary piers in the harbor. These piers were supported by a group of foundation piles. The load from each column was distributed to the foundation piles by a grillage—a two-layer grid of steel beams. The lower set of beams sat on top of the foundation piles. The upper layer, a set of four beams set side by side, supported the column bases.

On June 17, the first side span was supported on a permanent concrete pier at one end, and was overhanging the first falsework bent, designated "bent N4," at the other. At 3:40 P.M. that afternoon, bent N4 collapsed, plunging the partially completed span into Burrard Inlet. The falling metalwork pulled the permanent pier it was resting on out of line, which caused the adjacent approach truss to collapse as well.

Immediately after the accident, the government of British Columbia appointed a royal commissioner, Sherman Lett, chief justice of the provincial supreme court, to determine the cause of the collapse. The commissioner selected five leading engineers to investigate and report on the matter: F. M. Masters and J. R. Giese of the United States; J. R. H. Otter and Ralph Freeman of Britain; and A. B. Sanderson of Canada. Materials testing and special investigations were conducted at the University of British Columbia and testing laboratories in Vancouver.

The commissioner's report concluded that the collapse was caused by failure of the four upper grillage beams. The webs (the vertical portion) of the beams buckled laterally, causing the collapse of the falsework bent columns.

One of the errors in calculation was even discovered before the accident, but no corrective action was pursued.

Faulty design of the falsework or temporary columns led to the grillage failure. The commission discovered two major errors in the Dominion Bridge Company's grillage design calculations. The first mistake was in checking the grillage beam shear strength (the capacity of a beam to carry a load in its vertical plane; shear stress tends to tear a beam vertically, usually at supports or at points of concentrated load). The cross-sectional area of the entire beam was used in the calculation rather than just the area carrying the load. This mistake would lead the grillage designer to believe the beam strength was about twice as much as it actually was.

A second calculation, which checked the need for web stiffeners, was also incorrect. Stiffeners are metal plates welded to beam webs to give them additional stiffness and resistance to buckling. The contractor's engineer had used the one-inch thickness of the beam flanges (the horizontal elements) rather than the actual 0.65-inch web thickness. This led to the erroneous conclusion that no stiffeners were needed.

A separate investigation by Dominion Bridge Company came to the same conclusion—that incorrect calculations led to a fatally inadequate grillage design. One of the errors in calculation was even discovered before the accident, but no corrective action was pursued. The two engineers responsible for the calculations were both killed in the collapse.

Wood blocks and plywood pads had been included in the grillage to provide some bracing of the beams. Laboratory tests indicated that these blocks were only marginally effective at best. Most of the wooden blocks were not even located at the most effective bracing points. The investigation performed at the University of British Columbia also indicated that the ability of the beam webs to resist buckling was not adequately predicted by the usual design formulas for column buckling.

In his report, the royal commissioner laid the blame for the collapse on the Dominion Bridge Company. The commission found the contractor negligent for "(a) failing properly to design and substantially construct false bent N4 for the loads which would come upon it ... ; (b) failing to submit to the engineers plans showing the falsework the contractor proposed to use in the erection ... ; and (c) leaving the design of the upper grillage of false bent N4 to a comparatively inexperienced engineer, and failing to provide for adequate or effective checking of the design and the calculations made in connection with the design."

The commissioner also found that a failure in the construction process had contributed to the accident. His report pointed out that the bridge design engineers, Swan, Wooster and Partners, had a responsibility to make sure the contractor submitted the falsework plans and calculations for their approval, as required by the project contract. The engineers certainly knew that the bridge was under construction, and they had prepared the section of the project specifications that required engineer's approval of the temporary falsework structures. Commissioner Lett concluded that "there was a lack of care on the part of the engineers in not requiring the contractor to submit plans of the falsework." Ironically, the satisfactory performance of Dominion Bridge on earlier projects may have contributed to the design engineer's laxness in pursuing the falsework plans and calculations for review.

The commissioner recommended that on future large bridge projects the consulting engineers recommend allowable stresses for temporary construction support structures, and that the contractor be required to submit all construction plans and calculations for approval prior to construction. The contractor, however, would always remain legally responsible for the adequacy of construction methods and temporary structures.

Impact

After the inquiry, construction resumed on the bridge. Two concrete bridge piers were damaged in the collapse and had to be rebuilt. This required the careful removal of two thousand cubic yards of reinforced concrete. The collapsed superstructure spans were salvaged, and some undamaged members were reused. Erection of the bridge continued according to the original plan, with the notable addition of careful checking and review of all construction calculations and plans. The additional time and materials required to reconstruct the damaged portions of the bridge added four million dollars to the original contract price of sixteen million dollars.

The editors of *Civil Engineering* magazine noted that the collapse "illustrates the ever-present risks that are inherent in construction, due to human error.

Nearly twenty people were killed as failure of temporary construction support caused the Vancouver Narrows bridge spans to collapse.

The failure emphasizes the need for utilization of all possible checks on construction procedures."

Today, the leading bridge design firms continue to carefully review and check the contractor's construction plans and calculations. The collapse of the Vancouver Second Narrows Bridge was neither the first nor the last incident of mistaken temporary construction calculations leading to disastrous consequences. The accident vividly highlights the importance of independent checking of critical aspects of the construction process.

W. N. Marianos, Jr.

Where to Learn More

"What Happened at Vancouver? Probers Seek Cause of Bridge Collapse" *Engineering News-Record,* June 26, 1958, pp. 21-22.

"Faulty Grillage Felled Narrows Bridge in Vancouver, B.C." *Engineering News-Record,* October 9, 1958, p. 24.

Hrennikoff, A. "Lesson of Collapse of Vancouver 2nd Narrows Bridge" *Journal of the Structural Division, Proceedings of the American Society of Civil Engineers,* December 1959, pp. 1-20.

Masters, F. M., and J. R. Giese. "Findings on Second Narrows Bridge Collapse at Vancouver" *Civil Engineering,* February 1959, pp. 60-63.

King's Bridge failure

Melbourne, Australia

(1962)

"Brittle fracture" of supporting girders attributed to poor material quality and inadequate welding techniques caused the failure of a single-span section of a steel bridge.

Background

On July 10, 1962, on one of the coldest days of the year in Melbourne, Australia, a single-span one hundred-foot-long section of the King's Bridge gave way under the load of a crane being carried on a trailer. Completed only fifteen months before, the bridge had been designed to withstand such loading, and no distress should have occurred. Yet the girders fractured. The structure sagged about one foot under the load but did not collapse, borrowing support from the continuous concrete deck above the girders and some walls extending up from the ground to enclose the space below. Although not particularly catastrophic, the failure of the King's Bridge immediately attracted considerable attention, due to the fact that, as part of the public domain, bridges affect everyone.

At the time the bridge design began (1957), "brittle fracture" was a recognized phenomenon, responsible for the failure of major structures worldwide. And current engineering knowledge, if properly applied, should have been adequate to prevent failure. However, the responsibilities for the failure are not as clear as it might seem, and the lessons learned were varied.

The King's Bridge spans the Yarra River, as part of a freeway system in the center of Melbourne. The twenty-three-hundred-foot-long bridge consists of two separate parallel one-way bridges, each of three lanes, rising at the center of the river to allow the passage of boats. Support piers were placed at regular intervals along the bridge to support reasonable spans for steel girder construction using high-strength (low-alloy) steel. The design incorporated suspended span construction. Unlike simple span construction, where all beams extend from support to support, in suspended span construction, every other span incorporates a cantilever (overhang) from the girders of the two adjacent spans, therefore reducing the length of the center girder required to complete the span. This type of construction substantially reduced the amount of steel required for the structure.

Before the structure was designed, the general features of the bridge were prescribed by the constructing authority, the Country Roads Board (CRB) of Victoria, in what was called the outline design. This outline design was accompanied by specifications (a written document accompanying the drawings) allowing several possible combinations of materials to be used in the design of the bridge: reinforced concrete, prestressed concrete, mild steel, and high-strength steel.

Fourteen tenders were received and the main contractor was selected. "Utah" Australia was selected to build the bridge for a total cost of $5.63 million. The proposed design called for a reinforced concrete deck cast in place upon welded steel plate girders of high-strength steel. The steel to be used was that specified by the CRB in the outline design, following British Standard 968 (BS968). Although welding had been used in Australia for major bridge repairs and some new construction, welding of high-strength steel was untried. At the time, high-strength steel had been used extensively in Britain but primarily in bolted or riveted applications. Although there was much experience in the United States with welding of high-strength steel, it was based on different materials that had been tested extensively for their suitability for welding. In fact, it was not known if high-strength steel manufactured to BS968 had ever been used for welded girder construction.

Details of the Failure

The driver of the trailer described his experience on that cold night: "I had travelled about fifty yards onto the city side of the bridge when I felt something—I thought the load had shifted. Then I heard a loud noise. I stopped, saw a dip in the bridge, then jumped back in the truck and kept going. I thought the whole thing might collapse."

It was extremely fortunate that the fracture occurred at night when little traffic was on the bridge, which carried forty-five-hundred vehicles daily, and that the fractured span was surrounded by walls to support the bridge after it could no longer support itself. The failure occurred in a manner that was not catastrophic and did not result in loss of life or property damage. Yet the bridge was closed immediately and traffic was severely restricted while the cause of failure was determined and remedial measures were undertaken.

Initial inspection of the failed structure revealed that all of the four girders supporting the roadway had cracked at the transverse (crosswise) weld at the end of each cover plate. (Cover plates were added to girders in the center portions of spans, where bending forces were greatest, to help stiffen the girders.) The trailer was carrying a forty-seven-ton crane producing stresses of twenty kips per square inch (twenty-thousand pounds per square inch) in the steel girder at the point of failure. The allowable stress in the material was twenty-six kips per square inch and the material should have been able to take approximately 1.7 times the allowable stress (or forty-five kips per square inch) before failure occurred. Why then did this failure occur at less than half of this value?

The cracks extended to various depths of the girders with three girders cracking through the entire section. Upon closer examination of the failures, it was found that all of the cracks had started before the night of the failure, as evidenced by rusting over part of the crack surface. In fact six of the seven fractures exhibited paint on part of the crack surface, indicating that the cracks had been present in the shop, possibly before any stresses were applied to the girders. At the end of these cracks, in two of the fractures, there was evidence of fatigue cracking: an indication that the fractured girders had continued to serve, carrying loads in the cracked condition. This condition was identified as a multistage fracture.

It was determined that poor welding had produced notches at the ends of most of the cover plates. These notches, along with poor materials and residual stresses (stresses produced by the weld as it cooled), allowed the notch to produce a fracture with very small applied stresses. Once these cracks were present in the structure, the weakened girders were subject to higher stresses than had been anticipated in the design. These higher stresses continued to cause the crack to grow under a condition of frequently repeated loads from traffic called fatigue. The crack continued to grow at a steady pace under normal loadings from traffic until the critical condition was reached.

The final failure on July 10, 1962, was blamed on a condition know as brittle fracture. Although steel is known as a ductile material (a property where, like plastic, the material will substantially elongate under excessive loads before breaking), brittle fracture occurs under a set of conditions that can cause a sudden failure without observable deformations or warning movements of the structure. Conditions for brittle fracture include low temperatures, poor material quality (affected by both chemical composition and grain size), and stress conditions. If the structure is not redundant (designed to safely carry the applied loads by redistributing internal forces if an element fails), then brittle fracture can result in a crack propagating through the structure catastrophically, leading to sudden collapse. Had the nonstructural walls not carried the load, the failure would have most likely been catastrophic.

An initial report by a committee of investigation found the design to be in general conformance with the criteria set forth by the CRB. Although the design conditions at the ends of the cover plates were known to produce high local concentrations of stress and were considered an undesirable feature, the details were allowed by the applicable codes and standards. It was concluded that the cracking was caused during fabrication either by insufficient preheating of material before welding (preheating is used to reduce the residual stresses by allowing the weld and material to cool at the same rate) or through the use of welding electrodes that were not adequately protected from moisture (moisture introduces hydrogen into the weld and causes brittleness). In either case, the residual stresses and toughened material in the heat-affected zone (the material around the weld that is affected by the weld) produced the conditions for cracking. These cracks had not been identified by several types of testing that were supposed to verify the quality of the welds and were ultimately included as part of the permanent structure.

The orders for the steel placed by the fabricator apparently omitted the special toughness requirements specified by the CRB. Thus, the contractor was bound by contract to supply steel that conformed to specifications but had no tests to prove conformance. Apparently, the lack of adequate documentation was allowed to continue throughout the job with the knowledge of the CRB.

The committee concluded in their final report that the problems of brittle fracture were known and precautions were taken to prevent these effects; however, these precautions proved inadequate.

Impact

An inspection program was immediately undertaken to determine the structural integrity of the remainder of the bridge. The program revealed a considerable number of flaws throughout the bridge and it was concluded that defects should be assumed to exist in all girders.

Immediate attention was given to efforts to reopen half of the bridge to restricted traffic. Temporary supports were erected at points where failure involved a high level of public risk. This was relatively easy for low spans of the bridge where new supports could be added. However, the high spans over the river required some creative solutions. Two schemes were developed to repair these areas: (1) extending temporary supports up from the riverbed to the bridge deck while the bottom flanges (edges) of the girders were either repaired or removed and replaced, or (2) the addition of cables which would be prestressed to reduce the stresses in the cracked flanges of the girders. (In prestressing, a force is applied to an element to condition it to withstand its normal load more effectively). The investigation committee recommended the second scheme, and permanent repairs were made after extensive consideration of alternatives at a cost of $1.5 million.

Exemplified by the King's Bridge collapse, even minor, seemingly inconsequential details can become critical if not properly accounted for in engineering, design, fabrication, and inspection.

The final report prepared in June 1963 identified the many aspects of the contract, the design, the materials, and the fabrication that allowed the conditions that produced the failure. Ultimately, the blame was placed on the various organizations involved, detailing the exact shortcomings of each, and concluding that they were collectively responsible for the "unhappy state of affairs."

The King's Bridge was designed to be the first major structure built of high-strength steel in Australia, showing a great leap of faith in the implementation of a new technology. There was an attempt to make up for the lack of precedence through a modest testing program, yet the combination of new technology, unfamiliar techniques, and lack of local experience proved too risky for such a large project.

At the time of design, there were no provisions for the repeated nature of traffic loadings or fatigue loadings in the Australian design specifications. Today, repeated loads are a normal part of engineering consideration for many structures. As was learned on the King's Bridge, even minor, seemingly inconsequential details can become critical if not properly accounted for in design, fabrication, and inspection.

As with the welding technology, the unusual design-construct contractual arrangement, in which the main contractor was responsible for the design beyond the outline design stage, was also unfamiliar to the parties involved and, upon closer examination, contained gaps in the duties and responsibilities when compared to the practice that had been in place for some time. A final report of the investigation added that the arrangement failed to provide adequate overall supervision of the various aspects of the work.

The actual toughness requirements specified by the CRB were questioned by the committee of investigation. It was later stated that the toughness values required by the CRB's specifications were inadequate to protect the bridge from brittle fracture. Today, toughness values are selected by the designer based on the particular application but are often specified in the applicable design standards.

Another important observation by the committee was that the presence of these cracks prior to failure, especially the girder cracked over most of its depth prior to the failure, might have been observed had the span been open to view. The span that failed, being at the lowest elevation, was surrounded by walls to avoid the space beneath from becoming a dumping area. This eliminated any possibility of visual detection. The committee concluded that bridge structures should remain open to view for easy inspection. Provisions for regular maintenance and inspection are now a common consideration in the design phase of major bridges.

KIRK IAN METTAM

Where to Learn More

Duncan, J. L. "What You Can Learn from a Failure." *Engineering,* May 1, 1964, pp. 610-11.

"Report on the King's Bridge Failure, Melbourne." *Engineer,* August 31, 1962, pp. 370-1.

"Failure of King's Bridge, Melbourne." *Engineer,* October 25, 1963, pp. 661-62.

"Failure of King's Bridge in Melbourne." *Engineer,* March 20, 1964, pp. 520-22.

"Brittle Fracture of an Alloy Steel Bridge." *Engineering,* September 21, 1962, p. 375.

"Discussion and Communications re: The Development of Electric Welding with Particular Reference to King Street Bridge." *Journal of the Institute of Engineers of Australia,* June 1962, pp. 148-49.

"Lessons from Brittle Failures." *Engineering News-Record,* September 20, 1962, p. 306.

"Steel Blamed in Bridge Failure." *Engineering News-Record,* September 20, 1962, p. 139.

"Bridge Collapse Blamed on Improper Steel." *Engineering News-Record,* September 19, 1963, pp. 48-9.

Madison, R. B., and G. R. Irwin. "Fracture Analysis of King's Bridge, Melbourne." *Journal of the Structural Division, Proceedings of the American Society of Civil Engineers*, September 1971, pp. 2229-244.

Masterson, C. A. "The Development of Electric Welding with Particular Reference to King Street Bridge." *Journal of the Institute of Engineers of Australia*, June 1961, pp. 197-211.

Point Pleasant Bridge collapse

West Virginia/Ohio border

(1967)

The country's worst bridge disaster spurred research into the metallurgical properties of high- stress steel and halted use of some experimental cable designs.

Background

Ten days before Christmas. Six minutes to five in the afternoon. Shoppers are pushing rush hour traffic well beyond its normal high volume in Point Pleasant, West Virginia and Gallipolis, Ohio, towns that face each other across the Ohio River. Traffic between them crosses the great structure known locally as the Silver Bridge—the nickname *Silver* reflecting local pride that it was the first bridge in the United States to have been coated with aluminum paint. At five minutes to five, cars and trucks fill the two-lane bridge bumper to bumper, stalled in both directions by traffic lights.

At four minutes to five on that fateful December 15, 1967, all three spans of the 1,460-foot bridge lay either in the ice-choked river or on its banks. Under, on top of, and mingled with the twisted steel wreckage were as many as seventy-five vehicles, although estimates in the first news reports counted thirty-one with "a few" casualties. Later counts rose to forty-six persons dead and several dozens injured or traumatized. The first estimate of probable cause—traffic overloading—was issued by Professor Thomas Stelson of Carnegie Mellon University.

Dedicated in 1926, the Silver Bridge was forty-one years old. In addition to its glittering coat of paint, it was also known for its experimental use of eyebar chains instead of standard wire-strand suspension cables. An eyebar is a steel bar with enlarged ends, which have holes in them; pins through the holes join the bars together to form a chain. The designers, J. E. Greiner Company of Baltimore, had specified standard suspension cables, but an alternate design suggested by the contractor, American Bridge Company, Pittsburgh, proved to cost less when bids were opened.

John Edward Greiner, head of the engineering firm, was one of the most respected bridge designers of his day. In addition to the bridge at Point Pleasant, he was simultaneously designing its twin for another Ohio River crossing at Saint Marys, West Virginia, nearly one hundred miles upstream. As built, both bridges were supported by chains whose links were two eyebars side by side, joined together by pins. The design theory was that the eyebars shared equally in the 4.5 million pound pull on the cable. Each eyebar was 12 inches wide, 2 inches thick, and 25 to 30 feet long. A standard wire cable for a comparable bridge would be 7 to 8 inches thick, composed of 1,500 to 2,000 parallel wires.

Use of eyebars as suspension cables was not unique, although they had never before been used at this scale. Only a very few eyebar bridges were ever built in the United States. Three of them were under construction in Pittsburgh at the time, identical structures at Sixth, Seventh and Ninth streets. Their eyebar chain suspension systems (eight eyebars per link) were a response by the city's Department of Public Works to demands by the City Art Commission for "aesthetic" suspension bridges. Because they could not build normal anchorages fifty to sixty feet down to bedrock at or near the bridge sites, the city's engineers chose a self-anchored design.

Instead of anchoring the cables to bedrock, this technique uses stiffening girders—support elements along the side of the bridge—to resist the horizontal pull of the suspension chain. The supports at either end are thus only subjected to vertical forces. Finally, the engineers chose cables composed of eyebars rather than wire because they would form a better connection with the plate-girder stiffening members. These plate-girder stiffening members were simpler and more aesthetically pleasing than the conventional truss form.

In developing the eyebar scheme for the Point Pleasant Bridge, the designers stretched the idea and used the eyebar chain not just as a suspension element but as a part of the stiffening truss as well—an arrangement labeled bizarre by later critics. Only one other bridge was designed with this feature: the Florianopolis Bridge in Brazil, designed and constructed in 1926 by the firm of Robinson & Steinman, New York City. This bridge, however, had the eyebars only on the main span stiffening trusses rather than on the trusses for all three spans like the Silver Bridge.

Another deviation from standard bridge design practices occurred in the anchorages of the Point Pleasant and Saint Marys bridges. The design, called ingenious at the time, was a reinforced concrete trough at each end of the bridge. These were supported on reinforced concrete piles, and the weight of the concrete resisted the 4.5 million pound pull of the cables. At the same time, the top of each anchorage served as an excellent highway approach to the bridge.

The last complete inspection had occurred in 1951, and even then, according to highway department records, the eyebars had been observed "through binoculars."

Details of the Collapse

The collapse occurred without warning. "It was all over in seconds," said one survivor. Another described his experience: "The old bridge was bouncing up and down like it always does. Then, all of a sudden, everything was falling down. My feet touched the damned bottom of the river."

A policeman on the scene reported, "The whole bridge is in the river," and soon wrecking trucks and cranes swarmed over the river banks to retrieve smashed vehicles. Cars that had been flung eighty feet down into the water or onto land had been hit by chunks of concrete from the superstructure. One trailer truck floated several hundred yards downstream.

Later, during studies by the National Transportation Safety Board and other government bureaus, a witness testified to hearing a "loud clanking;" five minutes later, another witness added, he saw a large nut with part of a bolt lying in the roadway. A third witness also saw this.

Both towers crumbled, but the anchorages held. Early speculation suggested that the cause of the collapse was failure of one of the towers. This was tempered by a warning from Edward J. Donnelley, partner in the J. E. Greiner firm, who pointed out the obvious: "It is difficult to separate the primary cause of failure from secondary causes."

The collapse occurred without warning. "It was all over in seconds," said one survivor. Another described his experience: "The old bridge was bouncing up and down like it always does. Then, all of a sudden, everything was falling down."

Subsequent investigations underscored that warning. Engineers from Greiner, the governors of both states, the West Virginia Road Commission, and the Ohio Highway Department converged on the site. President Lyndon B. Johnson ordered a federal inquiry. News reports reflected nationwide shock that a major bridge could collapse without warning, and at least one publication sent in a "special investigative reporter."

The magazine *Popular Science* and its investigator, Alden P. Armagnac, used large headlines to promise "startling surprises" and suggest that blame for "our worst highway disaster" could be laid to "radical features" of its design. The article dwelt on the novelty of the eyebar chain, quoting one engineer about the safety and stability of the bridge depending on the chains—"if one chain link goes, it all goes." The steel used for the high-tension eyebars came under the author's scrutiny. They had been fabricated from a then newly developed heat-treated carbon steel and were its first use in the United States.

Another "startling surprise" noted by the magazine article was the "rocker tower," an innovation during the 1920s. All bridges must have a means to expand and contract with temperature variations, and previous suspension bridges either had flexible towers that could swing with the movements or saddle/rollers for the cables. The Silver Bridge was the second of its size designed so that the towers, supported on rocker shoes that allowed the towers to rotate or move horizontally as necessary, could adjust to the varying cable lengths.

Impact

The immediate effect of the Silver Bridge collapse was to spur an intensive safety inspection of bridges across the country. West Virginia alone inspected five hundred, closing the twin at Saint Marys and two others spanning the Ohio River. Short of dismantling Saint Marys, there was no way to determine if it suffered from the same flaws as its twin; it was eventually dismantled and replaced.

By order of President Johnson, three task forces were established: one to examine the effectiveness of inspection practices; one to replace the bridge; and one, under the newly created National Transportation Safety Board (NTSB), to determine the cause of the failure.

The inspection teams had to work with debris hauled from the river and its banks and laid out in a nearby field. Many of the pieces were so badly damaged that there seemed to be no way of finding the cause of the collapse. Eventually, however, the critical eyebar—on a single link in the North eyebar chain—was located; it contained a hairline crack and was soon subjected to a battery of metallurgical tests.

It was not until April 1971, more than three years after the collapse, that the exhaustive detective work resulted in a final report by the Federal Highway Administration (FHWA). The initial failure was a hairline crack that the report said "could not have been detected by an inspection method known in the state of art today without disassembling the eyebar joint. The crack caused the eyebar to fail, starting a chain reaction with one structural member falling after another."

The federal investigators had also pondered the effects of metal fatigue—failure of a metal element by cracking caused by repeated or cyclical stress—a phenomenon not too well known at the time. "No doubt the metal fatigue had resulted from repeated changes in loading and vibration of this crucial eyebar, transforming the steel slowly from a strong coarse-grained metal into a fine-grained brittle one."

In July 1971, *Civil Engineering* magazine, published by the American Society of Civil Engineers, published two conclusive articles about the Silver Bridge disaster. One was by Daniel Dicker, professor of civil engineering at Columbia University, who analyzed the mechanism of the collapse. In the second article, Charles F. Scheffey, FHWA director, explained that agency's final report. Their opinions, though not identical, were very similar.

Dicker reviewed the design, noting that the total design live load for the cables was 1,400 pounds per lineal foot and a concentrated load of 42,000 pounds applied at one point. The heat-treated carbon steel, he said, had an ultimate stress of 105,000 pounds per square inch but this strength had been achieved at the sacrifice of ductility, or the ability of the steel to stretch. The bridge's large motion under live load, described as vibrations by witnesses, indicated to Dicker that fatigue failure should have been expected.

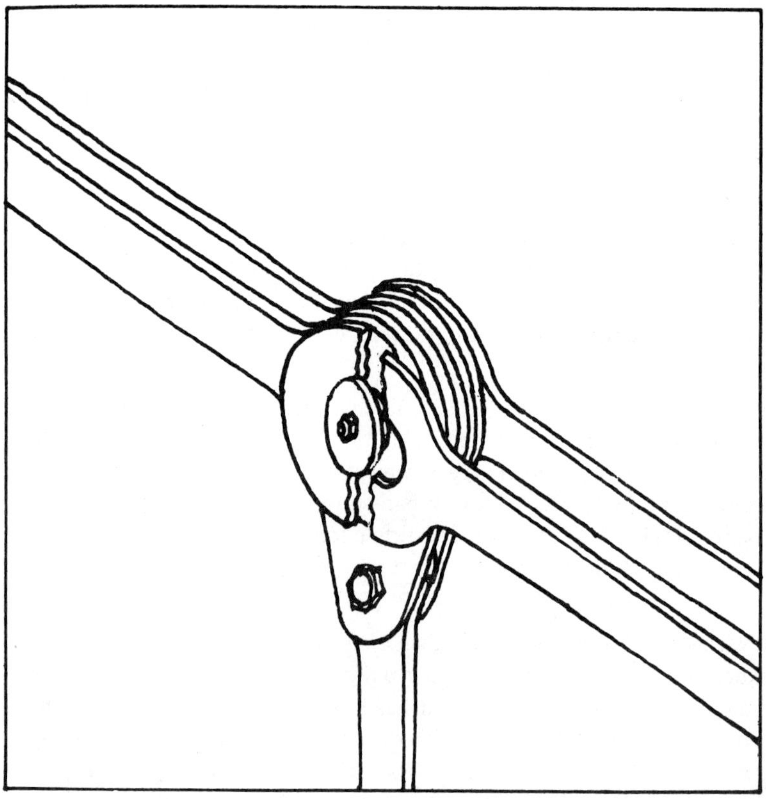

This drawing depicts the single cracked eyebar that led to the collapse of the Point Pleasant Bridge in 1967.

Dicker singled out as the cause extremely high stress concentrations at the point where the eyebar failed, caused by eyebar-to-pin friction. He concluded, "Apparently the pin cap and nut were forced off the pin . . . then two large trucks passed, causing the vibrations."

The FHWA view was that "stress-corrosion and/or corrosion-fatigue were contributing factors." Extremely high stress steels, according to the metallurgical studies, are susceptible to this particular type of corrosion (not the familiar rusting) when exposed to air pollutants such as sulfur. The FHWA report was released without resolution of the controversy, arguing "which mechanism was responsible has little bearing on actions required."

Both reports alarmed bridge owners and engineers everywhere. *Civil Engineering*'s editor issued a call for research into the properties of steels, stating that there was also a need to reexamine the design concepts and the "way we qualify materials for stress, similar to the way we qualify materials for shipbuilding."

Various task groups had already recommended an upgrading of structural safety inspections, encouraging development of new inspection standards. Many county and municipal departments were eager for these, according to the FHWA and AASHO (American Association of State Highway Transportation Officials, now known as AASHTO, with transportation added to the name).

Two years after the disaster, a new cantilever bridge was completed in record time about a mile south of the Silver Bridge site. The new structure, designed by E. Lionel Pavlo, was a modification of his bridge over the Mississippi at Vicksburg, Mississippi, under construction at the time. In 1971, Pavlo designed the replacement for the Saint Marys Bridge, also a cantilever rather than a suspension bridge.

The far-reaching consequence of the Silver Bridge collapse was a systematic nationwide inspection of all American bridges for the first time in history. States were required to conduct programs under guidelines set by the Bureau of Public Roads. In addition to the Ohio River structures, many other bridges were closed or had their posted speed and load limits reduced drastically. Many commentators expressed fears that bridges designated as "historic" would disappear under aggressive modernization programs. They were more right than wrong.

Rita Robison

Where to Learn More

Armagnac, Alden P. "Our Worst Bridge Disaster: Why Did It Happen? Silver Bridge at Point Pleasant, W. Va." *Popular Science*, March 1968, pp. 102-05.

"Causes of Silver Bridge Collapse Studied," *Civil Engineering*, December 1968, p. 87.

Dicker, Daniel. "Point Pleasant Bridge Collapse Mechanism Analyzed." *Civil Engineering*, July 1971, pp. 61-66.

Langley, Marcus. "Ohio River Bridge Disaster." *Engineering*. October 1971, pp. 763-64.

Mair, George. *Bridge Down: A True Story*. Stein and Day, 1982.

Plowden, David. *Bridges: The Spans of North America*. Viking Press, 1974, p. 238.

Scheffey, Charles F. "Pt. Pleasant Bridge Collapses: Conclusions of the Federal Study." *Civil Engineering*, July 1971, pp. 41-45.

Varney, Robert F., and John G. Viner. "Vibration Studies Relating to the Failure of the Point Pleasant Bridge." *Public Roads*, June 1971, pp. 161-66.

West Gate Bridge collapse

Melbourne, Australia

(1970)

Negligence and other human errors were to blame for collapse of a steel span under construction in Melbourne, Australia.

Background

Failure of the West Gate Bridge in Melbourne, Australia, differed from most bridge disasters in that it occurred during construction while engineers were trying to correct a severe defect that was later acknowledged as having provided ample warning. It was also unusual because many of the workers experienced death and injury not on the bridge, but below it while they were eating lunch inside the construction shacks.

The technology used to construct the bridge was reasonably new, but after a long inquiry, the official report blamed not the technology but the engineers for not paying enough attention to their calculations and, in some cases, not performing structural calculations at all.

On October 15, 1970, at ten minutes before noon, a 368-foot-long box girder span collapsed, killing thirty-five men and injuring another eighteen. The box girder was being erected in two half-widths, and this was the first half to be erected. At 11:00 A.M., the half-span had buckled, prompting a flurry of activity that included removing some thirty bolts.

The buckle, according to official reports, had been caused by heavy blocks of concrete that were used in an attempt to correct the camber (difference in elevation between the sides) of the bridge, which was incorrect because of the fairly new erection sequence that had been adopted. The Royal Commission of Inquiry eventually fixed the blame on design engineers, although the contractor did not escape censure: "The consultants . . . failed in their duty to prevent the contractor from using procedures liable to be dangerous."

The bridge is part of a three-mile-long roadway project linking Melbourne with its suburbs. The bridge and its approaches total eighty-five hundred feet, rising from ground level on the west bank of the Yarra River onto a concrete viaduct, across the river in five steel spans 192 feet above the water, and finally onto another concrete viaduct that reaches the east bank and descends to ground level. The project was designed to carry four lanes of traffic in each direction at speeds of seventy miles per hour.

The central twenty-eight hundred foot bridge was designed as a five-span steel continuous box girder supported by stay cables radiating from two towers. In such a design, the bridge consists of a series of steel boxes trapezoidal in shape. Each box is 13 feet 2 inches deep, 52.5 feet long and 83.5 feet wide at the top flange, with a cantilever bracket (used to build extensions between the arms of the bridge) at each side extending out another 10.5 feet. The usual construction method is to fabricate the boxes on the ground and to raise and bolt them together in the air, creating a cantilever bridge (one that has two sections or arms extending from opposite banks, joining in the middle above the water).

At the time of the bridge's construction, box girder bridges had become fashionable and cable-stayed bridges were quite new to the engineering profession. They differ from the traditional lattice-girder type, such as the Sidney Harbor bridge, in the same way that a modern airplane differs from the strut-and-wire, cloth-covered planes of the first two decades of flight. Both types of structures have a "skin" that supports local loads while resisting bending, shearing, twisting, and other loads. In a box girder bridge, the low profile helps control the aerodynamic stability. The main drawback is that the plates that make up the skin are subject to distortion during fabrication, and buckling is difficult to predict.

The mere fact that the West Gate Bridge was designed with recently developed technology was not the problem. Several other factors, however, compounded the situation: (1) the concrete approaches to the bridge were curved, which introduced a camber that had to be matched by the steel portion; (2) the bridge had to clear the river by 192 feet; (3) the design engineers were half a world away in London; and (4) the contractor had no experience with steel construction, yet was obligated to follow a dubious erection procedure established by a failed predecessor.

On October 15, 1970, at ten minutes before noon, a 368-foot-long box girder span collapsed, killing thirty-five men and injuring another eighteen. Many victims were in shacks below the bridge, eating lunch.

When the Lower Yarra Crossing Authority had contracted with Maunsell & Partners, Melbourne, to design and supervise the entire project, that firm had brought in Londoners Freeman Fox & Partners to design the steel bridge. The designs were completed within a comparatively short time between July and October of 1967. Haste was not considered to be a factor in the disaster, but may have led to the failure by all engineers to check the calculations thoroughly enough at various design stages.

Of the three construction contracts let in early 1968, the steel contract was troubled from the start. The low bidder, known as World Services and Construction (WSC), was a joint venture of Werkspoor-Utrecht and Wescon of the Netherlands. By 1970, the WSC work was so far behind that the Authority re-assigned the steel contract to the concrete contractor, John Holland Company (JHC) of Victoria, Australia.

Details of the Collapse

The Royal Commission later noted that the disaster began with the method that the original contractor, WSC, had chosen to erect the steel side spans. The new contracter, JHC, adopted this method as well, but instead of creating a series of full-width boxes, fabricated an entire half-width span on the ground next to the piers. The span was then raised and moved on rolling beams to its final position. The other half-span was fabricated and placed in the same way, with the two bolted along a centerline splice. On paper, the method worked. Each half-span would be fabricated with the same camber, the unsupported flange of the inner box would be stiff enough to prevent compression buckling, and the half-spans would be pulled together along their centerlines despite some outward bowing.

In practice, even special supports could not maintain the proper camber when the halves fabricated for the first span were erected. Attempts to correct the camber only caused a buckle. The site engineers tried removing some bolts and allowing the deck plate to slip. This did flatten the buckle.

The problem seemed to be solved, but the method's success led to its repeated use when the next span proved to be far out of alignment. This time, however, JHC workers added eight-ton blocks of concrete to help correct the camber. When only a one-inch difference remained, jacks completed the task of raising the low side and lowering the high side. A buckle again developed, the concrete blocks were removed, and an attempt was made to eliminate the buckle by removing some bolts from the transverse seams. Later, the Royal Commission was unable to ascertain that any calculations had been made by the contractor or by the consulting engineers that indicated the stresses induced by these remedial efforts. A representative of the consulting engineer was, however, seen directing the work.

The commission's reconstruction of the failure indicated that removal of the bolts had created a shallow V in the center of the span, shortening it and pulling one end off its roller bearings. It then fell, crushing the construction shacks located immediately below. When that end fell, the other end snapped up, fell back, and knocked over the river pier. The derrick slid down the sloping deck; tanks of fuel oil and other flammables fueled a fire that broke out in the wreckage.

Impact

The three-person Royal Commission conducted eighty sessions to interview fifty-two witnesses and review 379 exhibits before producing its lengthy report. Although designers Freeman Fox claimed that they and their calculations were not at fault, the commission decided otherwise.

The commission concluded: "Every feature of the West Gate Bridge had been used before, but not necessarily in combination or on such a large scale. Willingness to cope with imponderables is one thing that separates engineers from scientists, but it is imponderables that can provoke failures." The report noted that the box girder bridge resembled a piece of aircraft structure, yet bridge builders were not able to test their structures in the same way that

Two thousand tons of twisted metal and concrete fell into the mud as the West Gate Bridge collapsed during construction.

engineers in the aerospace industry do. Only after one full span and much of the second was built were defects observed.

As for division of responsibility, the commission could not decide who, if anyone, was the actual designer of the bridge. Even within Freeman Fox, individuals denied being responsible for other individuals' work, or even checking on it. The Melbourne contractors, WSC, claimed that they never received proper calculations from London—and that their own calculations suggested changing to thicker steel for some components. The commission found no evidence that Freeman Fox had completed a computer run to check the WSC calculations.

The report also criticized labor and communication troubles that had caused the authority to terminate the WSC contract and turn the job over to JHC. On taking over the contract, JHC was committed to erecting the bridge the way WSC had proposed. Holland did prepare a detailed procedure manual, but fieldwork outstripped the manual by the time of the accident. Instructions for obtaining the correct camber while the half-span was still on the ground were lacking. Freeman Fox also authorized departures from the manual, such as unbolting the critical transverse seams. Lines of responsibility were further confused when the London firm replaced its three top men on the site.

Although the commission noted that Maunsell and Partners of Melbourne was responsible for all contracts except steel, its resident engineer was criticized for not "interfering enough" in the steel construction. The report also concluded that labor should have done more to cooperate with JHC when that company took over the steel contract. Another union objected when the men assembling the half-boxes were assigned to erect them and work in the air, and other labor problems escalated when JHC was forced to take on WSC laborers who beonged to different unions.

The report was most emphatic, however, in blaming the designers. The reasons for failure, it said, "are to be found in the acts and omissions of those entrusted with building the bridge. . .[who] failed altogether to give a proper and careful regard to the process of structural design. They failed also to give a proper check to the safety of the erection proposals put forward by the original contractor, WSC. In consequence, the margins of safety for the bridge were inadequate during erection; they would also have been inadequate in the service condition had the bridge been completed."

RITA ROBISON

Where to Learn More

Kozak, J. J., and C. Seim. "Structural Design Brings West Gate Bridge Failure." *Civil Engineering*, June 1972, pp. 47-50.

"At Least 19 Are Killed in Melbourne as a New Bridge Collapses." *New York Times*, October 15, 1970, p. 5.

"Flaws Reported in Bridge in Australia Before It Fell." *New York Times*, November 7, 1970, p. 9.

Phillips, Mike. "Design of the Bridges That Failed." *Engineering*, October 23, 1970, pp. 419-21.

Trumbull, Robert. "Death Toll Rises to 31 in Melbourne Bridge Collapse." *New York Times*, October 16, 1970, p. 82.

Sunshine Skyway bridge collapse

Florida

(1980)

Part of a bridge and its approach spans collapsed after the Sunshine Skyway bridge was struck by a freighter.

Background

In the early 1950s, Florida's postwar development boom was beginning to spill over to its west coast, putting pressure on local roads that skirted the coves and bays of the Gulf of Mexico. Former fishing villages were being transformed into retirement centers and many small cities were expanding rapidly.

One of the largest water barriers was Tampa Bay, with St. Petersburg on its west side, Tampa on the north, and Sarasota just to the south of the fifteen-mile wide-mouth. Tampa, noted for its cigar production, was also a major port of call for ships sailing under the flags of all maritime countries.

In 1953, the state of Florida opened the Sunshine Skyway as a direct traffic link between St. Petersburg and Palmetto, a village just north of Sarasota. The Skyway was part of a new course, Route 19, and crossed the mouth of the bay on low-level causeways that rose to meet the main spans of a cantilever truss bridge that cleared the single shipping channel by 140 feet.

The Sunshine Skyway was designed by the New York City firm now known as PBQD (Parsons Brinckerhoff Quade & Douglas) and built for $12.5 million by Hardaway Constructors of Columbus, Georgia. In 1960, seven years after the initial bridge opened, state highway department engineers modified PBQD's original plans to construct a $23.5 million twin causeway-bridge. The most significant change was to substitute concrete foundation piles for the original steel piles, which had exhibited some corrosion problems in the sea-salt atmosphere. The second structure enabled state officials to separate northbound and southbound traffic, diverting southbound vehicles onto the new route.

Even though the designers had taken precautions to protect the bridge by adding flared collision walls to the pier bases adjacent to the channel, ships struck the Skyway regularly. Eight times in nearly three decades, these accidents were serious enough to go on record.

Details of the Collapse

On May 9, 1980, the ninth accident killed thirty-three people dropping some thirteen hundred feet of the southbound high-level bridge and its approach spans into Tampa Bay. At 7:30 A.M., during a sudden squall, a freighter heading into the bay struck the reinforced concrete pier at the end of the cantilever truss, one span away from the channel. According to witnesses, the bad-weather conditions included sixty-mile-per-hour winds and "a driving rainstorm." On impact, the concrete deck and beams of the main span "draped the bow" of the ship, a 606 foot phosphate carrier named the *Summit Venture.* The ship was operating under Liberian registry and was empty at the time.

Many of the victims were passengers on a cross-country bus that dropped 140 feet into the water. There existed other victims in three cars and a pickup truck. One of the survivors, truck driver Wesley McIntire, said, "The whole bridge started to sway. I could see the ship and the end of the bridge was breaking off. I just slid and I hit the ship and then I dropped into the water."

One of the survivors said, "The whole bridge started to sway. I could see the ship and the end of the bridge was breaking off. I just slid and I hit the ship and then I dropped into the water."

Four passengers were in a car with survivor Richard Hornbuckle of St. Petersburg, who reported, "I was driving on the bridge about 20-25 mph when the bus passed me, also going pretty slow. They reached the top just as the freighter hit. There was an impact, then the girders vibrated and the bridge collapsed. I saw the bus and several cars fall into the water."

Because at that time the Skyway carried an average of some seventeen thousand vehicles a day, a huge traffic jam quickly developed on the causeway and approaches to the channel bridge. Below, boat owners congregated to look for survivors. Two seagoing tugs were summoned to keep the *Summit Venture* from hitting the bridge again, although eventually the freighter was able to steam away under its own power. Highway officials rushed to inspect the northbound span, which appeared undamaged. They declared it safe and diverted all traffic there, restoring the original two-way pattern.

By the end of the first day, divers, ignoring fears that a loosened forty-foot span of the bridge might fall on them, recovered eighteen bodies, including that of a child. They reported that most of the dead were in the bus, a Chicago-to-Miami Greyhound, and that there were nine more bodies in three cars.

The Coast Guard named a formal board of inquiry, and the National Transportation Safety Board named a panel to investigate the accident, as well as its causes and possible preventive measures. Steve Plotkin, a state bridge maintenance engineer, testified that the bridge had been undergoing a $2 million repair at the time of the accident, but "it was like a toothpick against a sledgehammer. The ship weighed more than the twelve thousand tons of steel in the bridge superstructure." (Empty, the *Summit Venture* displaced twenty thousand metric tons of water.)

Within days, claims and counterclaims were being tossed around concerning the Skyway and its condition. According to Arthur Goodale, a civil engineer quoted in an "exposé" published by the *Clearwater Sun*, construction shortcuts and inferior materials were as much to blame as the impact itself. He apparently didn't think much of the concrete pilings of the southbound bridge. "The ship should not have brought down the bridge. And I don't think it did. I think it was all a matter of time before it came down."

Two state engineers promptly disagreed. One of them, Dewey Oliver, explained, "We ran into trouble with steel pilings and that's why concrete was used. We wanted to eliminate a corrosive problem."

The first repair estimate came quickly from Florida's bridge maintenance engineer, Jack Roberts, whose $20 million figure was nearly as high as the original cost of the entire southbound causeway-bridge. The state had insured the project for 80 percent of replacement cost after a 5 percent deductible, and Florida state officials looked to Washington for 100 percent federal emergency funds to cover their share.

Cleanup started immediately when Thomas Alberdi, Florida Department of Transportation chief engineer, called in the original contractor, Hardaway Constructors. The firm worked on a cost-plus contract while Tampa officials estimated loudly that the port was losing $50 million a day while vessels were trapped there. The Coast Guard explained that it would take time to clear the channel.

Impact

The accident drew worldwide attention to the need for protecting bridge piers against ship collisions. One result is that the main spans of bridges built in the 1980s and 1990s are considerably longer than their predecessors. Wherever possible, they are constructed with one or both main piers placed well outside the shipping channel, on a man-made island if not on natural land.

The U.S. Coast Guard issued warnings about the vulnerability of bridge piers "near" shipping channels as well piers adjacent to them. In Canada, officials issued "take notice" warnings, and other countries followed suit. Experts estimated that the impact of a thirty-nine thousand-ton ship at 10 knots would be equal to the impact of a jumbo jet traveling at 120 knots.

Rather than repair the southbound bridge, Florida DOT officials began planning a new bridge with a far longer main span over the shipping channel. Their planning involved extensive risk analysis and a commission to Greiner Engineering Sciences of Tampa for research into and design of pier protection systems for the replacement bridge. The researchers concluded that the piers had to be designed against ship impacts ranging from one thousand kips at

In three decades, the Sunshine Skyway had been struck eight times by ships. The ninth collision, in 1980, resulted in the collapse of the main span and the deaths of thirty-three people.

the low-level trestles near land (a kip is one thousand pounds) to 12,000 kips for the main channel piers. These directives evolved into a design for man-made islands surrounding the main span piers and dolphins (huge pieces of concrete) surrounding the piers supporting the approach spans. Greiner engineers also designed an elaborate warning system for motorists that senses impacts and other structural aberrations.

The Florida DOT engineers decided on a single central bridge, 95 feet wide and 175 feet above the channel, carrying two lanes of traffic and a full emergency lane in each direction. It is a spectacular cable-stayed concrete segmental bridge with a 1,200-foot-long central span, 540-foot-long back spans, and towers that rise 242 feet above the roadway. Unlike many of the world's cable-stayed bridges, the Skyway's cables are arranged in a single plane, anchored in the central median. When completed in 1987, the Skyway was the longest span of its type.

On either side of the main span are twin trapezoidal box girder structures that descend to the twenty-six-foot level of the trestle-type causeways. Figg & Muller Engineers of Tallahassee, designed the cable-stayed portion, while the low level approaches were designed by PBQD.

Cost for the new Sunshine Skyway ran more than $220 million and included demolishing the center span of the damaged causeway. This left the approaches, 4,000 feet on one side and 8,000 feet on the other, to be outfitted as the world's largest recreational fishing piers.

RITA ROBISON

Where to Learn More

"Ship Wipes Out Cantilever Truss." *Engineering News Record*, May 15, 1980, pp. 12-13.

"Engineers Seek to Shield Bridges from Ship Hits." *Engineering News Record*, February 3, 1983.

Galambos, Charles F. "Bridge Safety." *Public Roads*, September, 1984, pp. 41-47.

Garcia, A. M., and Rita Robison. "Sunshine Skyway Nears Completion." *Civil Engineering*, November 1986, pp. 32-35.

Mair, George. *Bridge Down: A True Story*. Stein & Day, 1982.

"Florida Bridge Collapse Linked to Faulty Work." *New York Times*, May 31, 1980, p. 8.

"32 Killed as Ship Wrecks Bridge in Florida." *New York Times*, May 31, 1980, p. 1.

Highway ramp collapse

East Chicago, Indiana

(1982)

Poor quality scaffolding and numerous safety violations led to the fatal collapse of a portion of a highway ramp during construction.

Background

On April 15, 1982, shortly after 11:00 A.M., thirteen workers were killed in an event that the Indiana Department of Transportation (IDOT) called "the worst construction accident in Indiana history." Two sections of a highway ramp under construction collapsed without warning. While no concrete was being poured at the moment of collapse, slabs of set concrete were being placed. The cause of the accident was later determined to lie within the falsework, or scaffolding, rather than in the design of the elevated ramp itself.

An eyewitness described the disaster as happening in slow motion. The first section collapsed without warning; three minutes later he heard a "big boom" and the second section collapsed, catapulting workers into the air, according to news reports. In addition to the thirteen who died, fifteen workers were injured severely as the ramp crashed onto the banks of an adjacent industrial canal. Some went down with the rubble; others were caught underneath. The body of one was found head down in a pillar of concrete that had hardened around him.

Some seventy-five workers were on the site of the Cline Avenue Extension, also known as the Riley Road Interchange, part of a highway designed to link the steel mills lining the Lake Michigan shore with the city of East Chicago, Indiana. The ramp, which is twenty-eight feet wide and allows for two lanes of traffic and a shoulder, rises over railway tracks and industrial sites.

The ramp had originally been designed as a precast segmental concrete bridge by Figg & Muller Engineers of Tallahassee, Florida. In this type of bridge, huge sections are cast in a special yard separate from the site, transported by truck, and lifted by crane. The sections are box-shaped, with a bottom slab and top slab connected by sloping webs. Their ends fit together and are "glued" either by epoxy or by concrete poured into the joint.

Tendons—cables within the box shape—apply most of the holding force, allowing several segments to span between piers.

Superior Construction Company of Gary, Indiana was named low bidder, at $13.5 million, for the contract. Its engineers, however, sought and won permission from the DOT to change the construction method to cast-in-place concrete. The subcontract for installing the reinforcing steel and posttensioning cables was let to Midwest Steel Company. The change meant that a substantial amount of falsework would have to be constructed in order to form the complex concrete box-shaped superstructure. Superior Construction purchased falsework designed and manufactured by Anthes Industries, Mississauga, Ontario, Canada.

Details of the Collapse

The job went smoothly at first. Piers for the concrete ramp were completed, and the superstructure was being built in sections known as units. The sections were each 180 feet long and terminated in mid-air 42 feet beyond a pier. The concrete for each unit was cast in place (poured) in two steps. First, the bottom slabs and webs were cast, then the top slab. The ramp was designed as a posttensioned structure—one in which much of the strength is derived from steel cables that are tensioned (tightened) after the concrete is hardened. In this project, the top slabs were posttensioned after the concrete had cured to a prescribed strength.

Prior to the posttensioning, all loads were being carried by the falsework, the system of steel and wood forms erected to hold the freshly cast concrete. Erecting falsework is a separate and important discipline within the construction industry, as it must be done with great care to ensure the safety of the workers and the integrity of the finished project. Here, the plywood boxes of the formwork rested on longitudinal stringers supported by transverse crossbeams, which in turn were supported by shoring towers located next to the piers and at one-third points of a span.

The shoring towers were prefabricated from welded steel frames joined by bolting on tubular braces. At the top, U-shaped heads met the crossbeams, and at the bottom, the tower legs were supported by individual concrete pads that had been formed with only a small amount of wire mesh instead of heavier steel reinforcing bars. Actually, the tower legs rested on sand jacks, ten-inch-square boxes of two-by-four lumber and sheet metal on top of the concrete pads. These devices enabled the workers to remove the towers by lowering the jacks after a unit was completed and post-tensioned.

A ramp rather than a straight elevated highway, this project had an uphill slope of 3.6 percent—a 3.6-foot rise every 100 feet. The U-shaped heads had to have wedges inserted above them to maintain the slope. The fact that no wedges were found in the debris was one of four major findings listed by later investigators as contributing to the accident.

The original construction drawings had shown towers supported by a guy-wire system such as that used with telephone and other utility poles. The contractor, however, received permission to substitute the tubular X-bracing, even though it provided less stability.

At the time of the collapse, concrete was being placed for the top slab of unit 4, up to about twenty feet from pier 408. The diaphragm form over this pier was about half full of fresh concrete, and the unit 4 tendons had been attached to unit 3. For a few minutes after unit 4 crashed down, unit 5 stood "teetering," according to witnesses, then also collapsed. The domino effect took all but 20 feet of one 180-foot span and the full 180 feet of the other.

Impact

The Indiana DOT lost no time in calling in investigators. Teams came to the site from the Federal Occupational Safety and Health Administration (OSHA), the National Bureau of Standards (NBS), and the Federal Highway Administration (FHWA). Forensic engineering specialists arrived from Concrete Technology Laboratories (CTL), in Skokie, Illinois. These specialists are trained to reconstruct disasters much like medical specialists are trained to perform autopsies.

In addition to the thirteen who died, fifteen workers were severely injured. Some workers went down with the rubble; others were caught underneath. The body of one was found head down in a pillar of concrete that had hardened around him.

Some investigations involved full-scale load tests on the concrete pads and sand jacks, which were found capable of sustaining loads of 160 kips (1,000 pounds per square inch) when the design load of the tower leg was 81 kips. Two separate structural analyses were required, because when the bottom slab was cast, the weight of the concrete was transferred to the stringers and crossbeams; when the top slab was cast, the load was distributed to the towers by continuous-beam action of the U-sections. The structural analysts concluded that the collapse had not been triggered by failure of stringers, crossbeams, U-heads, shoring towers, or sand jacks, even though some of the jacks were loaded "critically."

Lawsuit activity began almost immediately. An IDOT inspector who had been injured in the accident filed a $15 million suit against the general contractor. He charged that the scaffolding had been improperly installed and maintained, that guy-wires were not fixed to the falsework, and that the job had been improperly inspected by others. Two ironworkers also filed $15 million suits.

Early speculation about the cause of the collapse centered on the falsework; this was confirmed late in August by a CTL interim report. After a four-month investigation, the forensic engineers pinpointed the area where the collapse started, one of the steel supporting towers. They cited "differential movement at the base of the towers," and "transfer of forces through the system of upper jackscrews, subledgers and ledgers or both." The movement, they reported, could have resulted from cracking of the concrete pads or silt settlement beneath them.

Poor quality scaffolding and numerous safety violations led to the fatal collapse of this portion of highway ramp during construction. Workers comb through the rubble looking for victims of the disaster.

Concrete Technology Laboratories found that cracked pads had been reported months before, and that "differential movement at the base of the tower would induce forces in diagonal members. These could exceed the buckling capacities of the individual members." The report noted that a large horizontal force hinted at instability at the top of the towers.

In October 1982, NBS and FHWA together released a 215-page report shortly after the Indiana OSHA cited the two construction companies and IDOT for sixteen safety violations. One of the violations, according to the charges, was that Midwest Steel had not provided adequate escape routes for the workers.

The IDOT had been most at fault, said the NBS-FHWA report, for not responding to prior reports of cracking in the five-by-five-by-one-foot concrete pads supporting the falsework. This cracking "most likely" had triggered the collapse and occurred because the pads lacked an adequate margin of safety to support the falsework and ramp.

The report also found that the specified wedges had not been used between the stringers and crossbeams at the top of the falsework to compensate for the slope of the roadway. Another critical omission was the inadequate stabilization of the tops of the shoring towers against longitudinal movement that developed when the concrete pad cracked and the tower frame dropped slightly. The report also cited poor quality welds in supports for the falsework crossbeams at the top of the towers, which were inadequate to resist the forces resulting from the longitudinal movement.

The report did not specifically blame any person or company. The researchers did conclude, however, that had any one of the four deficiencies not been present, the collapse would probably not have occurred.

RITA ROBISON

Where to Learn More

Carino, Nicholas J., H. S. Lew, and William C. Stone. "Investigation of East Chicago Ramp Collapse." *Journal of Construction Engineering and Management*, March 1984, pp. 1-18.

"Probers Eye Concrete Ramp Collapse." *Engineering News-Record*, April 22, 1982, pp. 45-46.

"Fallen Ramp Probe Widens." *Engineering News Record*, April 29, 1982, p. 12.

"Report Eyes Ramp Failure." *Engineering News-Record*, August 26, 1982, p. 29.

"Ramp Failure Cause Detailed." *Engineering News-Record*, October 28, 1982, pp. 15-16.

"12 Workers Die in Collapse of Unfinished Bridge Ramp." *New York Times*, April 16, 1982, p. 14.

"U.S. Cites Defects in a Collapsed Road Ramp." *New York Times*, October 21, 1982, p. 23.

Zilwaukee Bridge failure

Saginaw, Michigan

(1982)

The near-collapse of a bridge segment during the construction of a concrete-span freeway bridge delayed its completion for years and added millions of dollars to the project's cost.

Background

Though it did not result in any fatalities or injuries, the near-collapse of a portion of the Zilwaukee Bridge during its construction was notable for the controversy that it generated, for its enormous cost, and for the unusual techniques applied to repair it. Designed to replace a drawbridge over the Saginaw River which was a perennial bottleneck for traffic on I-75 through Michigan's bay region, the Zilwaukee Bridge was supposed to be completed in 1983 for a total cost of $76.8 million. When a segment of the bridge suddenly shifted beneath a load imposed by construction equipment in August 1982, state officials and contractors began a lengthy dispute about who would assume financial responsibility for repairing the damage. After painstaking repairs at a price tag of over $38 million, the bridge was finally opened to traffic in 1988.

One and one-half miles in length, the Zilwaukee Bridge consists of two parallel structures—one each for the north- and southbound lanes. A cantilevered bridge, Zilwaukee was assembled from 1,592 precast concrete box girders. Concrete piers divide the bridge into twenty five spans on the northbound side and twenty six on the southbound. The precast girders, averaging 73.5 feet wide and eight feet long, and weighing roughly 140 tons, were attached one by one to cantilevered sections of the existing structure. Delivered by truck across previously completed spans, the girders were lowered into position by a steel truss launching girder that held them in place while grout and temporary prestressing were applied. The launching girder itself spanned unfinished sections of the bridge, resting upon cantilevered segments.

Details of the Accident

During construction of a bridge, the loads imposed on cantilevers while they are anchored at one end and not at the other are often far greater than the loads imposed after completion. To prevent mishaps, the movement of tons of material and equipment must be carried out with precision, and the anchored side of the cantilevers must be strong enough to handle the massive loads. On a Friday night in late August 1982, a crane operator working on a section of the bridge stretching over pier North-11 had just lifted a 157-ton box girder from a truck when the section began to sag, twisting to the side and deflecting five feet downward at its free end where normal deflection should have been only six inches. The other end of the section, which was attached to an expansion joint on the other side of the pier, rose about three feet. A temporary thirty five-foot spacer block positioned within the joint in order to immobilize it had cracked under the load, freeing the section to begin moving.

Predictably, initial concern centered around the possibility that the entire section would collapse. A state bridge design engineer, Adrianus Van Kampen, addressed the issue, stating that "if the prestressed cantilever can be returned to position, construction can continue. There is no danger that it will collapse; our figures show it is stable." Not all, however, were so optimistic.

The Michigan Department of Transportation (MDOT) quickly sent experts from its research laboratory to assess the damage, and they found that the movement of the superstructure had two major consequences. First, a section of the footing supporting the pier had sheared nearly completely through, shifting the pier slightly out of plumb. Second, the movement of the superstructure had damaged the bearings upon which it rested at the top of the dual 104-foot pier columns.

The project had been a troubled one and was already more than two months behind schedule when the accident occurred. Not long after the mishap, accusations about its cause were being hurled by several parties. Involved in the dispute were the bridge designers, BVN/STS Consulting Engineers of Indianapolis, Indiana; the builders, Stevin construction of the Netherlands in joint venture with Walter Toebe Construction of Wixom, Michigan; and the overall administrators, the MDOT.

Some time prior to the mishap, Stevin engineers found that they had to double the weight of the launching girder in order to manipulate the precast box girder sections. The new launcher weighed one thousand tons, and Stevin claimed that its original specification had been based on faulty calculations supplied by the MDOT. Early speculation therefore pinned the accident on the weight of the launcher.

Teams of investigators swarmed to the site. The MDOT brought in Howard, Needles, Tammen, and Bergendoff (HNTB) of Kansas City, Missouri to report on the cause and design repairs. The joint venture contractor retained both T.Y. Lin International of San Francisco and Wiss, Janney, Elster and Associates of Northbrook, Illinois. Meanwhile, the *Detroit News* retained a civil engineering professor from the University of Detroit to make a separate investigation.

The professor, Filomeno N. Almeida, declared in sensational front page stories that the MDOT had underestimated the damage and that the entire project was suspect. He cited evidence of water in conduits carrying post-tensioning tendons, exterior cracks, and voids in the joints between the segments. He also declared that workmanship, adequate at the start of the project, had deteriorated in the area where the accident occurred. MDOT Engineering officials disputed each claim.

An engineering professor investigating the mishap declared that workmanship, adequate at the start of the project, had deteriorated in the area where the accident occurred.

A year later, the MDOT imported yet another expert, Peter Gergely of Cornell University's structural engineering department. Gergely recommended increasing the anchorage capabilities of twenty eight pier footings that were carrying the greatest loads. This was done by enlarging them and installing reinforcing steel to connect the existing footings with new concrete surrounding them. This procedure was estimated to increase repair costs by $400,000.

In the summer of 1983, the state reached a settlement with its contractors stipulating that the state pay $4.8 million dollars in claims and purchase several million dollars in specialized equipment from the contractors for future use, while the contractors pay $29 million dollars in claims.

The next year, the "official" cause of the accident was finally published in a report from the Government Accounting Office (GAO), which blamed the original designers for "failure to recognize or adequately consider the effective load bearing capacity of the expansion joint as designed." The report noted that the designers approved a total load of 63,700 kips per foot (a kip is one thousand pounds) but had failed to account for a thirty-ton work platform at the end of span, stating that "handling manual procedures were approved which did not specify [the] requirement that the platform be removed."

In another part of the report, the GAO maintained that the contractor misplaced one leg of the launching girder by four feet, adding 2600 kips per foot to the load. This report only raised more disputes. H. Hubert Janssen of Indianapolis, consultant to ZCE (as the BVN/STS design firm was now known) insisted that two legs, not one, had been out of place and that the additional load was 5400 kips per foot. He said that the work platform should have not been there at the time and that the state, which was supervising the job, should have known this. However ZCE went out of existence when its contract with MDOT expired in February 1984, and was supplanted by HNTB.

Impact

The repair of the damaged section presented several challenges. Before repairs to the footing could begin, the base of the pier would have to be stabilized to prevent vibration from causing a catastrophic collapse, and the only way to repair the bearings at the top of the columns was to lift the superstructure first. The MDOT estimated that the repairs would cost between $1 and $3 million.

An elaborate monitoring system was installed that would warn of imminent collapse by detecting motion in the structure, and a system of cables was stretched between the broken superstructure and an 1100-ton counterweight placed below the broken expansion joint to prevent further shifting.

The first priority was to repair the crippled pier, which was supporting some 5500 tons while bearing on the fractured footing. It was feared that unstable soil conditions might cause the pier to shift while the new footing was installed, so a complex job of freezing the soil beneath and around the pier was undertaken. Described by Paul W. Steinberger, engineering manager of Geocentric, the freezing subcontractor, as "the most complex freezing job that's ever been done in this country," it involved drilling an intricate network of two hundred refrigeration pipes beneath the footing.

Once the soil was stabilized, holes drilled through the frozen earth ninety feet to bedrock were used as forms for three new foundation columns on each side of the footing. Steel reinforcing cages and concrete filled the holes to ground level and the outer parts of a new footing were placed around the new columns. The inner part of the footing would be installed once the superstructure was raised from the pier, allowing it to slip back into its vertical position.

A structural steel frame bearing on the new footing was then erected. Fitted with twelve six hundred-ton hydraulic jacks at the top, the frame was used to lift and carry most of the superstructure's load while the new footing, eleven feet thick, was cast around the base of the columns. After the concrete had cured, the damaged bearings at the top of the column were replaced by new ones coated with teflon and stainless steel to permit rotation and sliding. Then the superstructure was lowered to rest on the pier, and a set of hydraulic rams at the expansion joint slowly drew the superstructure back into alignment.

General construction of the Zilwaukee Bridge resumed in March 1984, with S. J. Groves and Sons as contractor. Instruments were installed to monitor any unwarranted movement, and it was stipulated that all parties agree that each construction procedure was appropriate before any part of the work could proceed. Five years behind schedule and nearly 75 percent over budget with a total cost of $132.6 million dollars, the Zilwaukee bridge was finally opened to northbound traffic in December 1987 and southbound traffic nine months later.

RITA ROBISON

Where to Learn More

Arnold, C. J. "Salvaging the Zilwaukee." *Civil Engineering*. April 1986, pp. 46-49.

"Great Zilwaukee Bridge Mistake." *Construction Contracting*. April 1983, pp. 14-15.

"Box Girder Wrenched Out of Shape." *Engineering News-Record*, September 9, 1982, pp. 10-11.

"More Problems Cited for Zilwaukee Bridge." *Engineering News-Record*, February 3, 1983, p. 13.

"Zilwaukee Woes are Still Mounting." *Engineering News-Record*, August 2, 1984, p. 13.

Mianus River Bridge collapse

Greenwich, Connecticut

(1983)

Collapse of one bridge span was caused by several factors, including lack of proper maintenance over twenty-five years.

Background

On June 28, 1983, three people were killed and another three were severely injured when a one hundred-foot-long section, three lanes wide, fell out of the Mianus River Bridge in Greenwich, Connecticut. The disaster happened suddenly, without warning to the motorists traveling east on I-95. It triggered lengthy engineering and scientific investigations that ended in a lawsuit against the company that had designed the bridge twenty-five years earlier. More important, it triggered extensive inspections of bridges in all fifty states, changing inspection procedures in each.

Interstate 95 carries a heavy load of commuter traffic as well as through traffic in southwestern New England, a load estimated to be about one hundred thousand vehicles per day. In June 1983, the seasonal buildup of tourist and vacationer traffic had already begun, putting additional pressure on the Connecticut portion of I-95, which is owned and maintained by the state's Department of Transportation (ConnDot) and is known locally as the Connecticut Turnpike.

Parallel to the Long Island Sound shoreline, the turnpike stretches across numerous coves and river outlets, crossing some three hundred bridges in the state and dozens in Fairfield County (which borders New York State to the west) alone. The bridge over the Mianus River in the town of Greenwich also crosses marshland and tidal flats along its banks. The structure is 2,656 feet—about half a mile—long, and is seventy feet above the river at mean tide. The design, by Tippetts-Abbett-McCarthy-Stratton (TAMS), was typical of thousands of bridges built during that era.

Details of the Collapse

When the collapse occurred, a one hundred-foot span on the eastbound side fell seventy feet into the river and onto one of its banks. The span consisted of two 9-inch-deep plate girders, four stringers, and several crossbeams. The girders, made up of plates fastened together in an I-shape with top and bottom flanges, are parallel with the roadway and form its primary support. Crossbeams perpendicular to the girders support the stringers (again parallel to the roadway) which support the deck, a 7.5-inch-deep concrete slab that is in turn topped by two inches of asphalt paving. The entire section weighed about five hundred tons.

When a one hundred-foot span fell seventy feet into the river and onto one of its banks, three people were killed and another three were injured. The collapsed section weighed about five hundred tons.

Three features of this original design are important, as they figured in later controversies about the cause of the failure. The first is that the bridge is skewed, or oriented on the diagonal between its east- and west-end abutments. The skew in this case is fifty-three degrees. Skewing a bridge reduces the total amount of steel and concrete required but may introduce some structural weaknesses.

The second feature concerns the two-girder design. Using two rather than three or more girders is a popular way to cut total costs, but if one girder fails, so does the other.

The third feature is called the pin-and-hanger design. The span in question was supported conventionally on a pier at one end, but at the other was hung from the adjoining span, a cantilever anchored at the following pier. The girders of both spans were shaped somewhat like a reverse Z to fit each other at their ends, and a steel strap was fastened to the topmost cantilever girder and to the girder of the hung span below. The fasteners were steel pins, each with a 1-inch diameter hole along its axis that received an 11 1/4-inch-long bolt with washers and a welded nut at each end.

As is common in many bridge failures, public demand for information conflicted with the engineers' need for thorough investigation. Newspaper, radio, and television reporters raced to get the story, eager to quote any "authority" they could locate. Local residents said they had heard unusually shrill noises coming from the bridge for several days and insisted that their previous reports of such noises had not been acknowledged by police or highway authorities.

To reassure an alarmed public, Governor William O'Neill immediately ordered an inspection of all other turnpike bridges and the installation of a temporary bridge over the gap left by the failed span. Similar to the Bailey bridges used by the military during two wars, this was delivered on ten trucks and bolted together within days. Workers assembled the ten-foot-deep side trusses that were then lifted by crane onto the existing piers. They attached a deck to the trusses, providing a twenty-four-foot-wide roadway that was elevated three feet above the piers. This left room for construction of the new span while traffic proceeded on the temporary one. The temporary span, including ramps on either end, actually weighed one-half as much as the permanent span, so the total weight on the piers was kept well within safety limits.

Three people were killed when a one-hundred foot section of the Mianus Bridge fell into the water below. The collapse was due to several factors, including a lack of proper maintenance.

Before traffic could be routed over the remaining spans, however, they were thoroughly inspected and strengthened by temporary "slings" of steel. These were formed by placing girders on top of the deck, threading steel rods through the deck to reinforce the hung span below.

Speculation about causes of the collapse differed widely. Some people thought that the heavy traffic had somehow weakened the bridge. Others said its skew was at fault. Still others blamed ConnDot's maintenance practices, which had allowed parts of the bridge to rust. ConnDot officials bristled. J. William Burns, head of the agency, said that the Mianus Bridge had been inspected and found safe in September 1982, less than a year before. That pronouncement drew attention to the agency's inspection program. As it turned out, all these speculations were, in part, true.

At the time, ConnDot's inspection program fielded only twelve people in six teams to look at all 3,425 bridges in the state. The policy, according to its officials, was to inspect all bridges once in two years. If a bridge was found to be "in danger," it was to be inspected more often. With such a heavy work load, inspectors rarely used X-ray or other sophisticated detection equipment.

ConnDot called in three New York City consulting engineering firms, one for inspection, one for a detailed forensic analysis (much like an autopsy), and the third to design a replacement. Officers from TAMS were quoted widely that the pin-and-hanger design was not unique for its time (1958), but had fallen out of favor in the 1960s because it had become a "maintenance nuisance."

Other investigations were begun by the local and state police, the Federal Highway Administration, and the National Transportation Safety Board (NTSB). Professor John W. Fisher of Lehigh University, who specializes in metal fatigue, was also consulted. He looked into the fact that the remaining hung spans were wobbling and the failed pin had come loose despite being constrained from sliding. This alarmed officials in other states, who dispatched inspectors to their pin-hanger bridges. In Massachusetts, for instance, inspectors found two broken hangers on the eighteen hundred-foot Harvard Bridge over the Charles River in Boston.

The first consultant's report to ConnDot about the Mianus River Bridge was written by Hardesty & Hanover, New York City engineers. It cited three factors: (1) rust buildup on the fourteen-inch washers behind the hangers and on the rear of the hangers themselves; (2) increase in high bearing stresses of the hanger on the pins as the links moved away from the guide webs; (3) out-of-plane deformations of the girders under live loads because of the heavy skew.

In September, hearings were started by the NTSB, and ConnDot released an internal report calling for more staff and equipment, and improved training and supervision in its bridge inspection division. Even so, the report concluded, "inspectors could not have been expected to detect the imminent collapse of the bridge."

That conclusion did not account for facts that were being uncovered in separate probes. One ConnDot inspector, Jerry White, was arrested for forgery and later convicted of altering, after the collapse, his September 1982 Mianus report. He said he had submitted warnings about the bridge, but ConnDot testified it had logged in no such notes. Actually, the original report, which ConnDot had hailed as finding the bridge "safe," had rated the bridge only "fair. . . . [A] potential exists for major maintenance."

In the NTSB hearings, which took four days, twenty-six witnesses failed to agree about the cause of the collapse. They disagreed about the significance of the rust. The forensic engineer, Lev Zetlin of Zetlin-Argo Structural Investigating, New York City, downplayed the rust, citing instead the skewed design. Others focused on ConnDot's maintenance practices.

A year later, in June 1984, the official NTSB report did blame ConnDot maintenance and inspection programs, concluding that the probable cause was corrosion that caused undetected displacement of the pin-and-hanger suspension assembly. ConnDot contested those conclusions, pursuing the theory of design deficiencies. Yet the agency began to follow NTSB recommendations that included reopening drains that had been paved over; preparing inspection guide manuals; reviewing inspection reports in greater depth; establishing quality control for inspections; and placing catwalks on bridges to facilitate inspections.

As late as February 1985, forensic engineer Zetlin was still blaming the skew as he released a five-volume report that cost $800,000 to produce. Corrosion of the pins accelerated, but did not cause the accident, he reported, adding that bridge inspectors could not have predicted the accident because the phenomenon of lateral movement of the strap was not widely known and that the movement, only 1/25th of an inch a year, was too small to be detected.

Other engineers disputed the skew theory, saying that if it had been the cause, the pin shifting would have been gradual rather than the actual sudden release. ConnDot, however, sued the original designers in the state's superior court. The suit claimed that because of design negligence, TAMS should reimburse Connecticut for the $25 million cost of settlements with the three survivors and the families of the three dead victims, replacement of the one hundred-foot collapsed section, and miscellaneous items such as lost toll revenues and police overtime.

During the nineteen-day trial, the jury considered 489 pieces of evidence. They heard expert witnesses on both sides dispute what caused the hanger to slip off its pin. The state's forensic consultant blamed the original designers. Weight of the traffic, he said, put pressure on the pin-and-hanger assemblies, causing the hangers to creep slowly off the pins. He demonstrated his theory with a videotape of a laboratory model.

Professor Fisher disagreed, declaring that the collapse was initiated not when the hanger shifted off the pin, but when the edge of the pin, loaded too much by the shifting hanger, failed. His conclusion was that expanding rust had led to the collapse. Other witnesses also zeroed in on the rust. One said that the real cause was that the movement went undetected, and another insisted that the problem could have been found fifteen years earlier "if the inspectors had noticed."

One expert told the jury that paved-over drains and missing gutters let water and road salt flow directly onto the steel. His and other testimony, some coming from the state's own witnesses, backed up the defense contention that poor maintenance, not design, had caused the tragedy. Indeed, one state inspector noted that accumulations of bird droppings up to six inches deep had prevented inspectors from walking the beams. Instead, they used binoculars.

On August 14, 1986, the jury exonerated the original designers, ruling that the bridge met the design standards that prevailed in the 1950s. It is standard practice in engineering that designers are not to be held responsible for subsequent discoveries about materials or details that are not known at the time. The jury rejected the state's claim for monetary damages.

Reconstruction of the bridge was accomplished in stages so that traffic could continue without undue interruption. The design, by Frankland & Leinhard, New York City engineers, replaced the original two lines of steel girders on each roadway with seven lines. The eastbound structure was widened fourteen feet. Piles and substructures for the twenty-four spans were completed early in 1988, and in August of that year Walsh Construction Company of Trumbull, Connecticut, was low bidder (at $53 million) for the third phase of construction, which completed the new bridge late in 1991.

Impact

The collapse of the Mianus River Bridge was not the first to inspire better maintenance procedures—this has been a standard response for many decades. Even before the final jury decision in 1986, ConnDot had embarked on a ten-year infrastructure renewal program. Expected to cost $5.5 billion over the decade, the program includes inspection and maintenance of all bridges and highways. Repairs and replacements are being made throughout the state, including small towns that cannot otherwise afford such work.

Other states were prompted by Connecticut's problem to inspect their pin-and-hanger bridges, then make plans to eliminate the problem. As of 1992, nearly all have been strengthened by one means or another. One common method has been to eliminate the pin-and-hanger assembly and replace it with a stronger, bolted connection.

Nearly all states have adopted more stringent inspection programs, contracting out much of this work to private consulting engineering firms. In this way, state transportation departments can have their thousands of bridges inspected by trained professionals without having to build up enormous staff. Connecticut has been a leader in offering such contracts, and continues the practice.

Bridge maintenance has risen on the list of priorities, and all states have written procedures into their maintenance and operations manuals so that inspectors and crew personnel alike are aware of its importance. As one highly regarded bridge authority once expressed it, "The best protection for any bridge is to send out a crew with mops and brooms. Keep it clean and painted. Keep the gutters cleaned out. Attend to problems while they are very small."

Rita Robison

Where to Learn More

"New Mianus Bridge Report Disputes Earlier Study." *Civil Engineering*, April 1985, p. 10.

"Designers Cleared in Bridge Collapse." *Civil Engineering*, October 1986, p. 10.

"Failed Pin Assembly Dropped Span." *Engineering News-Record*, July 7, 1983, pp. 10-12.

"Girder Flaw Eyed in Span Collapse." *Engineering News-Record*, July 14, 1983, pp. 10-11.

"Fallen Span Had Design Ills." *Engineering News-Record*, August 4, 1983, p. 12.

"Collapse Review Continues." *Engineering News-Record*, September 22, 1983, pp. 11-12.

"Hearings on Collapsed Span Focus on Rust, Skewed Plan." *Engineering News-Record* September 29, 1983, pp. 35-6.

"State Blamed for Failure." *Engineering News-Record*, July 5, 1984, pp. 27-8.

"Collapse's Cause Questioned." *Engineering News-Record*, February 28, 1985, p. 12.

"New Bridge for Mianus." *Engineering News-Record*, August 11, 1988, p. 14.

Failla, Kathleen S. "Anatomy of Fracture: Mianus River Bridge." *Engineering News-Record*, February 19, 1987, p. 48.

Tomasson, Robert E. "Study Says Bad Design Was a Factor in Bridge Collapse." *New York Times*, July 30, 1983, p. 23.

Zetlin, Lev and James. M. Hinckley. "Learning from Failure." *Consulting Engineer*, June 1985, pp. 52-7.

Schoharie Creek Bridge collapse

New York State

(1987)

Two long-standing bridge engineering problems—scour and lack of redundancy—converged to destroy a New York State Thruway bridge.

Background

At opposite ends of the state of New York, New York City and Buffalo are linked by the 550-mile long New York Thruway. For thirty-one years, the toll road had been operated without structural problems in any of its eight hundred nineteen bridges when disaster struck on April 5, 1987. The center span of the 540-foot Schoharie Creek Bridge collapsed without warning at 10:45 A.M., killing ten people.

The highway, known formally as the Governor Thomas E. Dewey Thruway, crosses the Schoharie about forty miles west of Albany, near the town of Ft. Hunter. At thirty to forty-five feet wide, the "creek" could more aptly be called a river. That April day, it was raining, but in the rural area flooding from the storm seemed manageable even though the Schoharie was running thirty feet deep under the eighty-foot-high bridge, a good twenty feet more than usual. Thruway officials remembered that the Schoharie and other bridges in the area had survived a devastating flood in 1955, but they discounted the importance of some flood control measures being taken nearly forty miles upstream.

Operators of the Blenheim Gilboa Pumped Storage Power Project, fearing damage to the reservoir, had released flood waters into the Schoharie. The flow was later reported at a record 65,187 cubic feet per second, while the previous record had been 55,000 cubic feet per second during the 1955 flood that had left Ft. Hunter under five to six feet of water. This time the town stayed dry because of a new levee that had been constructed along its north side. The levee kept the creek within its banks, but this intensified the flow. Witnesses later said that the water "seemed to boil over on the rocks below and tear at tree trunks leaning out over the banks of the creek."

The high flow caused state highway officials to close a bridge about a half mile downstream from the Thruway, but other bridges in the region were left open to traffic. On one of them, also just downstream from the Thruway, a fifteen-year-old boy became the main eyewitness to the collapse at 10:45 A.M. "All of a sudden I heard this noise. Water splashed up high, and the road and the concrete just fell in," he told reporters. He saw a truck plunge straight into the creek, a small car that went in upside down, and a large white car that was later found downstream with its top ripped off. At least two other cars also disappeared. Nearby residents said that they heard "an explosion," and ten minutes later "it sounded like a bomb going off" when the second span fell. The two spans were swept some eighty feet downstream, and the others dropped when their piers failed within a few hours.

State trooper Peter Persico had crossed the bridge a short time before, but turned back after witnesses called the police. He found several cars and trucks parked on the partly collapsed bridge, "their drivers gawking," and with other troopers, he evacuated the bridge just before the second section collapsed. By nightfall, police had located the two lost cars and a tractor trailer, but the rescue efforts were curtailed by high water and the threat of further collapse.

Thruway officials were astounded. Chief engineer Daniel S. Garvey said, "We were just totally shocked. We're nonplussed about the whole thing. We don't understand it."

James A. Martin, deputy executive director, said that the thruway's bridges were designed to withstand one hundred-year storms. He ordered an immediate inspection of all 819 Thruway bridges, and one hundred were inspected within the first day. By the end of that day, three separate investigations had been arranged and Senator Daniel Patrick Moynihan had flown in by helicopter, promising to ask the federal Department of Transportation for $10 million to rebuild the bridge. (The Schoharie Bridge had only been insured for its construction cost of $8.8 million, with a $2.5 million deductible.)

The bridge, which in 1987 carried an average of sixteen thousand vehicles per day, was constructed in 1952 of five spans. The reinforced concrete deck sections each rested on a pair of steel girders about seven feet high and one hundred to one hundred twenty feet long. The transverse floor beams were fifty-five feet wide. Below, four piers consisted of two columns supported by a single shallow spread footing. The columns, nine feet wide at the bottom and tapering to seven feet wide at the top, were connected near their tops by a beam, giving the pier an H shape. Two piers were planted in the creek bed, two on the banks, and the superstructure also rested on two abutments.

The bridge had been rehabbed five years earlier and had been inspected last in April 1986, according to Robert C. Donnarums, Thruway deputy chief engineer. In the yearly inspections before that, no problems with the piers had been found, though one inspector, in 1977, had recommended replacement of the riprap that seemed to be missing on the upstream sides of the bridge. Riprap consists of large broken stones that are piled around the pier bases to protect them. Rehab plans made between 1981 and 1982 called for

Nearby residents reported hearing a loud explosion as the center spans of the Schoharie Creek Bridge collapsed without warning, killing ten people.

repairs and repaving of the four-lane deck, but included no mention of the riprap. Donnarums added that now, after the collapse, investigators "would pay close attention" to the flow of water around the bridge supports.

Despite the existence of rehab plans, Thruway and state officials soon learned that their records were in disarray. Construction details were lacking, and they had no idea how many other Thruway bridges were of the same steel plate, two-girder design. When they did locate the records, they found that no underwater inspection had been made, even though it had been urged at the time of the rehab.

Details of the Collapse

Investigators initially blamed the failure on the bridge's lack of redundancy (lack of back-up, load-bearing systems in case of component failure). Each of the five spans was supported independently, and the two-girder design provided no alternative load paths in case one horizontal member failed. The lack of frame-action reinforcement in the concrete piers was also considered as a factor contributing to the suddenness of the collapse. But within two or three days of the collapse, investigators began to focus almost exclusively on hydraulics—the behavior of flowing water—and its effect on pier foundations. Dr. David Axelrod, state health commissioner who also headed the State Disaster Preparedness Commission, said there was "a whole series of unusual factors" about flows in Schoharie Creek.

One investigation uncovered a report—written by a state engineer who had reviewed the bridge design before its construction—warning that the proposed design did not take into account the frequent occurrence of ice jams in the Schoharie Creek and the possibility of extremely severe floods. The report further claimed that the 540-foot length of the bridge would be too constrictive of the river and recommended lengthening the span to 775 feet. This report had been disregarded, and the construction of the bridge, along with the erection of the levees, had narrowed the river channel sufficiently to greatly increase the stream's velocity. The rapid flow of water, in turn, had caused a great deal of erosion. After the collapse, it was found that scour, the erosion of the streambed by the force of the currents, had washed nine feet of the streambed out from under the upstream side of one of the piers. When the pier finally tipped into this hole, the spans of the bridge above disengaged from their supports and fell into the river as well. Had the bridge been constructed with concern for redundancy—with one continuous span across the river rather than five independent spans—some experts maintained that the girders would have merely sagged at the point of the pier failure instead of collapsing. The slump of the bridge would have provided a visual warning that the bridge was in danger of collapsing in time to close it and prevent any loss of life.

While the lack of redundancy and reinforcements contributed to the severity of the collapse, scour was determined to be the primary cause. Scour was not unknown previous to the Schoharie Bridge failure, nor was it unique to that particular bridge. During that same spring of 1987, scour caused seventeen other bridges to fail in the northeastern United States alone. Two years later, the Hatchie River Bridge on U.S. Route 51 in Tennessee failed when material was lost from around a pier because the channel shifted laterally. Scour is not easy to track, however. One simple reason is that the same water that scours out the streambed promptly refills the hole with loose silt and gravel. Without probing, even a diver cannot see what has happened, and in time, scour recurs.

One witness saw a truck plunge straight into the creek, a small car that went in upside down, and a large white car that was later found downstream with its top ripped off. At least two other cars also disappeared.

Two forensic engineering firms were brought in to determine the cause of failure: Wiss, Janney, Elstner Associates of Chicago, and Mueser Rutledge of New York City. The National Transportation Safety Board also made a full investigation. Their reports led the New York State Commission of Investigation to fix blame on the Thruway Authority (NYTA) as early as August, four months after the collapse. The commission cited evidence that NYTA did not follow up on a critical recommendation of one of its engineering consultants that it reinstall riprap around the piers and that a staff inspector failed to look at the footings during annual inspection. They also contended that NYTA did not carry out its insurance company's recommendation that an underwater inspection be undertaken.

The commission concluded that "while the bridge design and construction may well have been deficient, with proper inspection and maintenance, this bridge would not have collapsed." The commission then recommended a complete overhaul of the state's bridge inspection program, urged bridge inspectors to review drawings before site visits—a practice not rigorously enforced—and urged the state legislature to intervene by funding bridge repairs for needy localities.

The final verdict came in April 1988, just a year after the collapse, when the National Transportation Safety Board accepted a staff report that also blamed the NYTA for not maintaining the riprap. The report cited "ambiguous plans and specifications for the original construction, an inadequate inspection and inadequate oversight by NYDOT and FHWA."

Impact

Reconstruction of the bridge began soon after the collapse. Traffic was jamming all other available roads through the area, and Thruway officials hastened to retain the New York City firm of Hardesty & Hanover, who redesigned the bridge for quick reconstruction as well as for durability. The fourteen plate girders are continuous over the piers and are spaced a conservative eight feet apart. Designing the girders to be very stiff allowed the contractor to pour the concrete deck span by span rather than wait for all steel on the entire bridge to be in place.

Due to the traffic problems, the new pier foundations had to be designed and constructed before the cause of the collapse was determined. The foundations were specifically designed to "make the system independent of scour." Steel-plated boxes, called sheetpile cofferdams, extend from the surface to twenty-three feet below the river bottom—eight feet below the foundation bottoms—to permit the construction crew to pour the pier foundations in dry conditions. The replacement bridge has deeper foundations and steel H-piles that extend down to bedrock and rigidly connect to the foundations. With proper riprap, the design protects the pier bases from scour.

The Schoharie tragedy prompted new concerns and new regulations, both at the national level and within various states. In 1988, Charles H. Thornton of Thornton-Tomasetti, a New York City engineering firm, published what he called a "powerful plea for better bridge inspection practices" in *Civil Engineering*. After having studied the Schoharie disaster, he recommended a bridge inspection program, specifying the degree of experience and familiarity with bridge design required of an inspector, the need for periodic evaluation of each bridge's complete structural and foundational system, and numerous other measures to ensure that this type of failure did not occur again.

In 1989 a study found that 494 of the 823 U.S. bridge failures from 1951 to 1988 occurred as a result of hydraulic conditions, primarily scour of foundation material. Most of these collapses, however, were on local road systems and did not generate national news as did the Schoharie Creek Bridge failure.

But the search for scour had burgeoned. In 1988, FHWA required initial screening to identify bridges most likely to be vulnerable to scour damage and evaluation for setting priorities. Each state was to develop its own plan. While FHWA provided procedures for predicting potential scour depths and designing countermeasures, it underscored a need for more research, especially measuring devices. These are just now coming out of various laboratories.

The state of New York was among the first to bring in an independent task force to devise its plan. The task force devised a simple, straightforward method for scour screening and priority procedure—evaluating nearly twenty thousand bridges for the priority list. Interim countermeasures were installed at several sites, including riprap and guide banks to withstand scour action, underpinning supports, and closing bridges during storms.

New York's Department of Transportation, which once, according to one critic, had to "rummage through boxes in the basement" to locate specific bridge records, created the post of state bridge inspector general to bring order into the archives and the inspection process. The inspector general is also charged with reviewing new bridge designs.

RITA ROBISON

Where to Learn More

"Can You Top This? (Schoharie Creek Bridge Reconstruction)." *Civil Engineering*, January 1988, pp. 61-2.

"Thruway Blamed for Collapse." *Engineering News-Record*, August 13, 1987, pp. 13-14.

"Schoharie Report Offers Guidance on Inspection." *Engineering News-Record*, December 10, 1987, p. 11.

Green, Peter. "Feds Blame State Agency for Schoharie Failure." *Engineering News-Record*, May 5, 1988, p. 16.

Huber, Frank. "Update: Bridge Scour." *Civil Engineering*, September, 1991.

Lane, Kate. "Schoharie Bridge Probe Yields Inspection Changes." *Engineering News-Record*, May 14, 1987, p. 11.

Murillo, Juan A. "Scourge of Scour." *Civil Engineering*, July 1987, pp. 66-69.

Thornton, Charles H., et al. "Lessons from Schoharie Creek." *Civil Engineering*, May 1988, p. 46.

Buildings and Other Structures

- Boston Molasses (1919)
- Knickerbocker Theatre (1922)
- Bronx apartment house (1936)
- Texas Tower radar station (1961)
- Ferrybridge cooling towers (1965)
- Ronan Point (1968)
- Skyline Plaza (1973)
- Hartford Coliseum (1978)
- Willow Island cooling tower (1978)
- Kemper Arena (1979)
- *Alexander Kielland* oil-drilling rig (1980)
- MGM Grand Hotel (1980)
- Harbour Cay condominium (1981)
- Hyatt Regency Hotel (1981)
- *Ocean Ranger* oil-drilling rig (1982)
- L'Ambiance Plaza (1987)

Molasses spill

Boston, Massachusetts

(1919)

A giant, five-story-high steel tank suddenly fractured, releasing millions of gallons of deadly molasses that engulfed people, animals, and property.

Background

Boston's inner harbor was memorialized on January 15, 1919, by one of the most bizarre structural failures ever to occur anywhere. The disaster temporarily put the city's working-class North End on the map and made larger headlines than the Paris Peace Conference and events in revolution-torn Russia. On that mild winter day, without warning, a massive tidal wave of sweet, sticky death gushed from a fractured steel tank, leaving twenty-one dead, over one hundred and fifty injured, and many buildings crushed under nearly 12,000 tons of thick, brown, sugary molasses.

The huge fifty-foot high, ninety-foot diameter steel tank had been used by the U.S. Industrial Alcohol Company for storing up to 15,000 tons of molasses. It loomed above and was sandwiched between several buildings—a freight depot, firehouse, stables, offices, and even an elevated railway.

The tank was ordered from Hammond Iron Works in 1915 by the Purity Distilling Company on authorization of U.S. Industrial Alcohol. The treasurer of Purity ironically a Mr. Jell, ordered it without consulting an engineer. The only requirement used in making the order was that the tank have a factor of safety of three for the storage of molasses weighing 12 pounds per gallon (50 percent heavier than water). Later, at the Massachusetts Superior Court hearings, Mr. Jell said the Hammond Iron Works were "experienced and reputable [tank] makers" and that he relied on their skill and experience. It was the largest tank Hammond had ever built.

The sides of the circular structure consisted of seven rings of steel sheets, each about seven-feet high, varying in thickness from .687 inch at the bottom to .312 inch at the top. The tank's flat bottom rested on a cement sand cushion, with a concrete slab foundation. The slab was supported on concrete piles. The tank's roof, conical in shape, was made up of curved steel sheets supported by framing (beam rafters).

All the steel sheets used in construction of the tank actually proved less thick than shown on the Hammond drawings used to obtain the building permit. For instance, the bottom ring—the most stressed part of the structure—was supposed to be .687 inch; in actuality it was only .667 inch. The steel thicknesses for the other six rings were likewise 5 to 10 percent less than shown on the permit plans. The bottom ring (ring one) had a twenty-one-inch diameter manhole opening cut out of it.

The steel sheets that made up the rings were spliced together in two ways. The sheets in the bottom rings had butt joints and splice plates. The sheets in the other six rings had lap joints. Both types of joints were riveted together with six rows of rivets—three on each side of the joints. The rings were connected to each other by a lap joint using a single row of rivets.

The tank was completed early in 1916 and tested only by running six inches of water into it. On a dozen occasions during the tank's three years of service it contained a maximum of around 1.9 million gallons (for periods up to twenty-five days). The content that dark day in January was some 2.3 million gallons—near capacity. It had been in the tank for four days. Months later, at the legal proceedings, several recalled that the seams of the tank were "weeping" (leaking molasses), but no one seemed to be concerned.

Details of the Disaster

It was 12:40 P.M., still lunch hour, when eyewitnesses said they heard sounds like machine-gun fire and then saw a torrent of molasses, two stories high, exploding out over adjoining property. Six children were immediately swallowed up.

The tank had fractured and burst open. A 2.5-ton section of the lower part of the tank was thrust out onto a playground 182 feet away. Another section of the structure wrapped itself around and completely sheared off one column of an adjacent elevated railway. Traveling at about thirty-five miles per hour, the molasses swept over and through everything in its path. At its most destructive moments, the gooey wave was 15 feet high and 160 feet wide.

Ralph Frye, then a reporter for *The Boston American* covering the event, recalled, "What looked like a moving wall of volcanic lava filled the street and was moving relentlessly toward me. Everything it overtook—horses, automobiles, people—disappeared." People and animals lost their lives either by drowning or by being struck by wreckage. Several horses had to be shot. Occupied houses were demolished, while the cellars of others were filled with molasses. Horse-drawn and motor vehicles were marooned or floated away. Rescue crews found one man and his wagon embedded in a mountain of molasses, man and horse both frozen in motion.

By midafternoon the flood had settled. It covered more than a two-block area and the general vicinity looked like it had been hit by a cyclone. Buildings were destroyed or bulldozed off their foundations, rails from the elevated

When a giant, five-story-high steel tank suddenly fractured, millions of gallons of deadly molasses were released, engulfing people, animals, and property.

railway were dangling in the air, and the tank itself lay on the ground, a heap of crumbled junk metal.

Police and firemen used huge hydraulic siphons around the clock to pump molasses out of flooded cellars. It was nearly a week before all the bodies were recovered and months before signs of the disaster disappeared.

During the trial that followed, the Purity Distilling Company suggested three likely causes for the failure: (1) an explosion; (2) vibrations from trains using the adjacent elevated track; and (3) the collision of a runaway trolley freight car. The explosion theory got the most play, but in the long run all three were easily dismissed as the cause.

Examination of the wreckage showed two principal fractures in the steel plate of the lowest or bottom ring, one being a vertical line through the manhole and extending up into the second ring, part way up its lap joint, which was located directly over the manhole. (The weakest part of the second ring—a lap joint—was placed right above the weakest part of the first ring: the manhole opening.) The other major fracture was a diagonal break, again through solid metal, the top of its line being near the same lap joint (in the second ring).

The second ring was separating from rings one and three by a tear in the single-riveted lap joints which connected them. Also, its vertical lap joint directly opposite from the manhole location was completely fractured.

The tank originally thought to have a safety factor of three in reality had none. Its weakest elements were its connections, especially the lap joints of the second ring, and the reduction of effective material by cutting out a twenty-one-inch manhole opening without adequate reinforcement in its bottom ring, the section resisting the most liquid pressure.

In the rings that had vertical joint failures, allowing them to be torn apart, the connections failed by shearing off rivets or by actually severing the steel plates between rivets. Some joints had a combination of both separations.

The tank originally thought to have a factor of safety of three in reality had none. Its weakest elements were its connections, especially the lap joints of the second ring, and the reduction of effective material by cutting out a twenty-one-inch manhole opening without adequate reinforcement in its bottom ring, the section resisting the most liquid pressure. The use of thinner material than called for on the permit drawings did not help matters.

The tank was improperly designed, detailed, and constructed, and calculations were never checked by an experienced engineer. Its failure was due entirely to structural weakness.

Impact of the Collapse

Immediately after the event, the Boston Building Department started requiring that all calculations of designers be filed with their plans, and that stamped drawings be signed. The catastrophe likewise probably influenced the adoption of engineering certification laws in all states not having such legislation, as well as the requirement that all plans for major structures be sealed by a registered professional engineer before issuance of a building permit.

The Boston molasses disaster resulted in countless lawsuits, and litigation took nearly six years. Over three thousand witnesses—six million words of testimony filling 40,000 pages of court records—appeared before the Massachusetts Supreme Court rendered its opinions: the tank as built was underdesigned.

The total settlement for 125 damage claims was small by today's standards, reportedly exceeding one million (in 1925 dollars).

The Boston molasses disaster heightened the engineering industry's attention to detail in analyzing the impact of large stress increases at openings in metal plates and in designing highly stressed connections—rivets, bolts, welds—that do not have redundancy (that is, those that have no backup connectors to help out in the case of failure.)

RICHARD WEINGARDT

Where to Learn More

Bluthardt, Robert. "Wave of Death." *Firehouse,* June 1983, pp. 86-88, 136.

Brown, Burtis. "Details of the Failure of a 90-Foot Molasses Tank." *Engineering News-Record,* May 1919, pp. 974-76.

Frye, Ralph. "The Great Molasses Flood." *The Reader's Digest,* August 1955, pp. 63-67.

Harding, Priscilla. "The Great Boston Molasses Disaster of 1919." *The American Legion Magazine,* December 1968, pp. 12-15.

———. "Bursting of Boston Molasses Tank Found Due to Overstress." *Engineering News-Record,* January 1925, pp. 188-89.

Knickerbocker Theatre collapse

Washington, D.C.

(1922)

A theater roof collapsed due to inadequacies in design, fabrication, and construction.

Background

On the evening of January 28, 1922, the eighteen hundred-seat Knickerbocker Theatre in Washington, D.C. was comparatively empty. A phenomenally heavy snow—approximately two and a half feet—had kept many theatergoers at home, and less than four hundred people were present. As the orchestra leader raised his baton to signal the start of the overture, a dusting of plaster fell from the ceiling onto his shoulders. A crack, several inches wide, opened in the north wall of the building next to the stage. Seconds later, the roof and the eastern two-thirds of the balcony collapsed, killing ninety-five people and injuring over a hundred others.

Open for four years at the time of the collapse, the Knickerbocker Theatre had been planned by R. W. Geare. An experienced local architect, Geare had been retained to supervise its construction and hired M. S. Rich to develop the theater's structural design. The main walls of the theater were built of hollow tile combined with structural steel framing and reinforced concrete roof and floor slabs. The curving north "Columbia Road" wall was built of hollow clay tile, with the vertical and continuous cells covered with brick, stone, and stucco, so as to resemble a series of three story buildings and thus harmonize with the residential character of the neighborhood. Windows were pierced through the entire thickness of the wall and closed by cloth screens inside the glass of the sash. Thirty-five to forty feet high, the hollow tile walls were approved only after a questionable change in the building code.

Construction began in 1916. The successful bidder for the steelwork, C. A. Schneider's Sons, stipulated that trusses be substituted for the plate girders and heavy beams called for in Rich's design. This change would save time and money since the amount of steel required would be reduced and materials were in short supply because of World War I. The steel subcontractor hired J. H. Ford of Union Iron Works to redesign the theater. Geare approved the substitution, subject to approval by the city building department, but did not check the redesign himself or ask Rich to review the changes. Authorities later characterized Ford's redesign as "so extensive as to be substantially independent of the original design."

During construction, general contractor Frank L. Wagner secured authority to strengthen the north wall by inserting a course of reinforced concrete near the top of the wall. When the steel was partially erected, Geare noted the unstayed condition of the column supporting the main roof truss and ordered that a strut be installed from the top of the column to an adjacent column at the level of the bottom chord of the truss. Geare also noted at the time that all field connections were bolted but workmanship "did not appear to be first class in all respects." In the summer of 1921, a steel cantilever was bolted to the front of the balcony to allow the addition of two rows. However, no action was taken to strengthen the supporting steel in the balcony.

Details of the Collapse

Three investigative committees were formed: a joint Army-Navy committee, appointed to investigate for the district attorney; a committee formed by the Associated General Contractors (AGC); and another committee formed by the American Association of Engineers (AAE). These committees considered and rejected several theories about the cause of the collapse. Initially, examiners attributed the collapse to cracking in the north "Columbia Road" wall and the failure of adjacent roof beams. Investigators also suspected that local foundation settlement of the walls had caused the gradual unseating of the theater's roof supports. Another theory held that the main building columns had failed due to inadequate splicing, aggravated by the load of the balcony addition. Geare had noted during construction that the major column was inadequately braced, but although inspectors discovered deformities in several columns, they found the major column upright and unbent.

The joint Army-Navy Committee pursued the aforementioned theories as well as others and issued a report citing twenty-one major and minor causes. The committees of the AGC and AAE dismissed the reasons described above as the primary causes of the collapse and instead attributed it to the failure of the main roof truss that terminated in the curved north wall.

The AGC and AAE committees attributed the failure of this truss to several flaws in the building's design. First, the steel roof members were overstressed because the actual loads greatly exceeded the design loads. The actual combined live and dead load of roof structure and snow at the time of the failure was estimated at seventy-five pounds per square foot, two hundred twenty-five percent greater than the roof's calculated capacity of twenty-three pounds per square foot. Also, the height of the main truss had been reduced by nine inches during the redesign. This alteration significantly lowered its capacity to carry bending loads and caused it to deflect excessively.

The committees also discovered that the main truss bearing was poorly designed, leading to uneven loading of its two channels and a reduction of the channels' load-bearing capabilities. In addition, contractors failed to anchor the trusses to the walls. Normal seasonal expansion and contraction of the dissimilar metal and masonry materials could cause the walls to be pushed out by the steelwork in summer and possibly allow the steel members to unseat themselves by contraction in cold weather.

Finally, the steel structure suffered from an overall lack of adequate bracing. Committee members noted in particular the omission of adequate bracing at the junction of column and truss, and the placement of a heavy, concentrated load (the main roof truss) on a high, relatively thin, hollow tile wall that was pierced with numerous openings and inadequately braced.

Based on these findings, the collapse probably occurred in the following manner. Expansion and contraction over a period of time displaced the curved north wall outward. The end of the main truss had also moved but was still on its bearing, although some damage to the improperly designed bearing was probable. On the occasion of the peak snow load on January 28, the truss deflected excessively. In a condition similar to that demonstrated by bending a rubber eraser on its long axis, the lower chord extended while the upper chord resisted the load as it attempted to shorten proportionately. Thus, the overall length of the truss between bearings might have been shortened to such an extent that the truss fell from its bearings and brought down the roof.

Two other events were taking place simultaneously. First, as the bottom chord of the truss bowed, the light tile walls were pushed outward. With neither the strength nor the mass to counter this stress, the walls may have cracked or moved enough to unseat the truss and bring down the roof. Second, the overloaded bearing deformed, causing an eccentricity in the truss. Without adequate bracing, a truss in such circumstances buckles and falls.

The roof structure of the Knickerbocker Theatre, without any snow, was overloaded by at least one hundred fifteen percent. With all its defects, the collapse of the roof was inevitable.

Each of the problems described above contributed to the failure. The roof structure, without any snow, was overloaded by at least one hundred fifteen percent. With all its defects, the collapse of the Knickerbocker Theatre was inevitable.

The AGC report placed principal responsibility for the failure on the architect and the steel designer, although it stated that the owner shared this responsibility due to his failure to provide full and proper plans and insist on adequate field inspection. The report also concluded that the owner "had induced the architect to cut corners" to save money. The AGC found the local Building Department jointly responsible with the architect for its failure to notice and prevent violations of the building code, particularly the omission of anchorage of the steel to the perimeter walls. The AAE report agreed generally with the AGC, but found the design to be carelessly done and stated that to expect plan checking by the building inspector to the detail necessary to prevent these defects would relieve the architect of that function and responsibility entirely.

In April 1922, a federal grand jury indicted Geare, Ford, Donald Wallace (the general contractor's general foreman), Richard G. Fletcher (the concrete/masonry foreman), and Julian R. Downman (the former assistant building inspector) for manslaughter. In May, Ford entered a plea of not guilty, arguing that the statute of limitations had run out since the offense did not take place within three years of the indictment. The court found him not guilty, and the other criminal indictments were dismissed as well. In August 1926, in a suit brought by the surviving families of twenty eight of those killed in the collapse, Chief Justice Martin of the District of Columbia Court of Appeals, found the structural steel contractor not liable to third parties after the completion and acceptance of the work. The suit against the architect was also dismissed.

Impact

The investigating committees made a number of recommendations intended to avoid the recurrence of this type of failure. First, they suggested that fees charged for building permits be sufficient to make plan checking and inspection possible. Although fees have grown steadily over the past seventy years, building departments are neither adequately staffed nor do they attempt to perform detailed engineering review of project drawings. Plan review is normally restricted to zoning review, egress, fire safety and other special concerns. On-site inspection by building inspectors is quite limited and often carried out on an exception basis. Building officials refuse any responsibility for technical architectural or engineering review, abdicating these responsibilities to the architect, engineer, or testing and inspection agencies. In many areas, the building officials are not qualified by experience or training to perform these reviews, nor is this ability a requirement for their appointment.

The second recommendation was that owners provide adequate plans and specifications. The Commissioners of the District of Columbia issued a revised building code on July 22, 1926, that required owners to employ an inspector, approved by the District Engineer, when constructing large buildings, theaters, places of public assembly, and reinforced concrete buildings of twenty thousand square feet or larger or containing more than one hundred tons of structural steel.

Third, the committees recommended that architects and engineers be responsible for carrying out inspections. The current American Institute of Architects model contracts limit the architect's duties during construction to "observation." These contracts specifically state this does not include inspection or supervision; they also include a provision requiring the owner and contractor to review the specifications and plans and promptly inform the architect of any errors, omissions, or inconsistences. Most major architects and engineers maintain insurance coverage, often for millions of dollars, to protect themselves against liability for their errors and omissions. The American Society of Civil Engineers also recommends against contracts including inspection of work by the responsible design engineer, to avoid liability from litigation. The inspection function is often either omitted, or where required by local or state law, of limited scope or liability, and is performed by third party engineers or testing agencies having no responsibility for the overall project or its planning.

A fourth recommendation was that building codes be revised and standardized. Although many states have adopted standard building codes, and several well regarded model codes, such as BOCA (Building Officials Conference of America) and SSBC (Southern Standard Building Code), are widely used, building codes remain a local option, varying widely from one place to another.

A final recommendation, and one that led to the most positive result of the Knickerbocker Theatre disaster, was that architects, engineers, and contractors be registered. Despite the absence of a federal law, licensing laws governing the practice of architecture and engineering exist in all fifty states and most possessions of the United States. These laws establish minimum education and experience qualifications for licensing and standards of practice to assure a knowledgeable, ethical, professional practice.

NORMAN J. GLOVER

Where to Learn More

"Indict Structural Details in Knickerbocker Reports." *Engineering News-Record,* March 23, 1992, p. 502.

"Two Committees Charge Knickerbocker Collapse to Twisting Failure of Truss End." *Engineering News-Record,* March 30, 1922, pp. 273-76.

Godfrey, E. "Buckling Wall of Theater Causes Ceiling Collapse." *Engineering News-Record,* May 25, 1922, p. 873.

"Knickerbocker Theatre Roof Collapses under Weight of Snow, Bringing Down Balcony on Audience." *New York Times,* January 29, 1922, p. 1.

"Reinspection of All Theatres in New York City Will Be Made As Result of Knickerbocker Theatre Collapse in Washington." *New York Times,* January 31, 1922, p. 3.

"Knickerbocker Theatre Sued for $280,000 by Estates of 28 Victims of Collapse." *New York Times,* January 25, 1923, p. 12.

"Trial Called to Fix Responsibility for Knickerbocker Theatre Disaster in January, 1922." *New York Times,* November 1, 1923, p. 12.

Bronx apartment house collapse

New York City

(1936)

A multistory apartment complex under construction collapsed without warning, causing multiple deaths and injuries.

Background

At 3:30 P.M. on June 19, 1936, an apartment house under construction in New York collapsed in a deadly shower of bricks, mortar, and structural steel. Collapse of the center wing of this seven-level structure resulted in the deaths of eighteen men and injured many others. At the time, the structure was hailed as a progressive, modern construction project because innovative materials, techniques, and equipment had been used to build it. Ironically, theories related to the cause of the crash immediately began to center around the materials used and the quality of the workmanship.

The apartment building, located at 112 East Mosholu, was the first "all-electric" multifamily dwelling in the Bronx and was designed for the use of electric ranges and refrigerators. Built for an estimated cost of $250,000, the building occupied a site measuring 200 by 115 feet. It was located on Mosholu Parkway between 206th Street and Van Cortlandt Avenue. Construction plans called for the building to house about sixty families in apartments ranging from small one-and-one-half-room suites to larger accommodations of up to four-and-one-half rooms. The exact date of the start of construction is not known, but building plans were filed with the Bronx Department of Buildings on January 18, 1936. The Building was essentially completed at the time of the collapse; a topping-out ceremony was held on Tuesday, June 16, just three days before the failure. The ceremony was elaborate even by New York standards and included speeches and comments by a variety of city officials and business leaders praising the modern design of the complex.

Architecturally, the apartment building consisted of three wings separated by light wells. The center wing supported an ornamental penthouse tower that many believe was the location of the start of the collapse. Records indicate that the collapse occurred without warning as masons worked to finish off the tower. Materials for completion of the tower were stacked in the vicinity and probably contributed to the ultimate collapse. While the masons were working inside the building and on scaffolding, the tower collapsed, taking with it the top three levels of the exterior wall. Several workmen were killed instantly, while others were trapped inside the jumbled mass of building materials and died later from their injuries.

A seven-level structure (six stories plus basement), the apartment building was constructed of a variety of structural materials including brick bearing walls, wood joists, and steel girders. With the exception of the wood floors, this structural concept is not much different from many of the apartment buildings, midrise hotels, and small commercial buildings constructed today. While materials used in 1936 were in general less strong than those in use today, the technology and construction expertise for using these materials were well known at the time. The apartment structure did not exhibit any stability problems during the construction phase up to the time of the crash. However, many workmen questioned at the scene of the accident were quick to point out a variety of problems including "skimping" of materials and the "speed up" construction process used. Many of these complaints were later found to be all too accurate.

Details of the Collapse

When the masons mounted their scaffold on the afternoon of June 19, little did they know that it would be the last brick and mortar some of them would ever place. The scene was not unlike a modern construction project that you might see while driving down the street of your hometown. It was in the final stages of construction and a variety of workers including bricklayers, heating contractors, and plumbers were scrambling to put the finishing touches on the building. Near the end of the day, many workers were already crowded around a table on the third floor waiting to receive pay from twenty-year-old Louis Lansafane. Lansafane's father would be among the unfortunate workmen killed in the tragedy. Inside, plumbers on the third floor were punching through a masonry pier to fit their pipes. At the same time, masons were working on the same part of the structure but on the upper floors. Carpenters were working throughout the structure and on the roof. One of them, Edward Grewing, would be credited with saving several lives. When the collapse started, he instinctively picked up his tools before running for shelter in a back stairwell, and thus was one of the few to have cutting tools in hand before rescue teams arrived. He worked frantically with his saw to free two of his fellow workmen injured in the collapse.

The collapse started when the decorative, tower-like penthouse fell inward and crashed violently through the roof, sending an estimated one hundred tons of debris crashing all the way down inside the building to the basement. Workers were caught up in the thundering landslide of construction material that sent many of them to their deaths below. The resulting rubble formed an enormous pyramidal pile of wreckage that extended upward to the third floor. The wood-and-steel-girder floors were no match for the massive weight of the falling masonry. Floors progressively collapsed under its weight and impact much the way a child's tower of blocks collapses when the top block being placed falls on the one below, then on the one below that, and so forth. A ragged three-story-high V was torn out of the exterior wall. Scaffolding on the center wing crashed to the street below. Fortunately, the two outer wings held under the tremendous force of the crash, preventing what could have been an even more horrendous disaster.

It was determined that faulty reinforcement of the ornamental tower and improper installation of structural steel support beams were the direct causes of the collapse. Contributing or underlying causes included faulty materials and poor workmanship. As is the case with many classic building failures, there was more than one type of defect at work resulting in the collapse.

Immediately following the collapse, large clouds of dust emerged out of the hollowed-out center section of the building and drifted over the nearby parkway. The construction site became one of extreme confusion, with workers trapped throughout the collapsed structure.

From across the parkway, a Works Progress Administration (WPA) foreman and several workers ran from their own construction project to help with the rescue operations. Responding to the emergency, arriving firefighters braved the shaky roof to obtain a vantage point for directing the operation of other rescuers below. Ladders were brought to the site and used along with the remaining scaffolding to gain access to trapped and injured workers. Under emergency police escort, a large construction crane was rushed to the scene all the way from 137th Street and Lincoln Avenue. The crane was intended to be used to lift construction debris to gain access to trapped workers. However, the remaining parts of the exterior walls were so unstable that first the crane was used to pull them down.

Crowd control soon became a problem as more than thirty-thousand people crowded into the area by early evening. Over one hundred police and sixty firefighters from all over the New York area were called to the site to control the crowds and help with the rescue operation. Emergency crews continued their rescue efforts throughout the night by stringing lights in the trees along the parkway.

There was little argument that the failure of the penthouse tower was the trigger mechanism for the rest of the collapse. More specifically, it was determined that faulty reinforcement of the ornamental tower and improper installation of several structural steel support beams were the direct causes of the collapse. Contributing or underlying causes centered on a combination of factors including faulty materials and poor workmanship. As is the case of many classic building failures, there was more than one type of defect at work resulting in the collapse.

Impact of the Collapse

A grand jury investigation of the collapse lasted for an entire month. District Attorney Samuel J. Foley produced the results of "a chemical analysis" that showed that too much sand was present in the mortar, which would have reduced the overall strength of the masonry. Since sand is cheap relative to the lime and cement components of mortar, the implication was that corners had ben cut to save on costs. Foley also intended to show that rushing the project to completion resulted in poor quality control. The investigation revealed that steel beams salvaged from another structure had been used in the Bronx apartment and that many of them were rusted and had holes cut in them as a result of their previous use. In addition, some of the beams used were lighter (and therefore had less load-resisting capacity) than called for in the original design specifications.

Three major findings came out of the grand jury investigation of the collapse: (1) there was evidence of some improper design of the structure; (2) many of the materials were secondhand or faulty in nature; and (3), the fast-paced construction process had resulted in faulty workmanship, especially in the masonry work. Based on these conclusions, District Attorney Foley was able to indict six individuals on criminal charges including Louis Steinberg, the builder and developer, and his son, Max. Others indicted were Harry C. Rutkins, the architect; Michael C. Thiede, a city plan examiner; Vincent A. Valentine, the masonry contractor; and Alfred F. Heine, masonry inspector from the Bronx Bureau of Buildings.

The trial that followed confirmed the grand jury charges. It was also shown that building code violations and faulty plans were not reported to the Bronx Bureau of Buildings even though it was demonstrated that the masonry inspectors knew of the problems. All six men were convicted of second-degree manslaughter. The Steinbergs and Valentine received sentences of two to five years. Heine was given six months in jail, while Thiede and Rutkins received suspended sentences.

The collapse was not the work of a single individual or the result of a lone defect. It was a result of a series of blunders and criminal activity that occurred throughout the entire construction phase. Architectural press reports of the day concluded that the disaster was a result of greed and the pressure of the speculative building process. It was also concluded that licensing of any of the individuals involved would not have prevented the disaster. But the collapse fueled ongoing arguments for licensing architects and engineers and for developing other ways to insure that experienced individuals were placed in charge of major construction projects. This failure and others of that era did support the growing call for better procedures, quality control, and construction monitoring that eventually became the licensing and permit process of today.

M. Kevin Parfitt

Where to Learn More

"Bronx Collapse Brings Criminal Charges." *Architectural Record*, September 1936, p. 172.

"Building Collapse Kills Seventeen." *Engineering News-Record*, June 25, 1936, p. 930.

"Competence versus Inspection." *Engineering News-Record*, June 25, 1936, p. 928.

"4 Killed, 12 Hurt as Building Falls." *New York Times*, Saturday, June 20, 1936, p. 1.

"Guilty as Charged: Collapse of Bronx Apartment House." *Architectural Record*, May 1937, p. 6.

"Same Old Story." *Engineering News-Record*, April 8, 1937, p. 530.

"17 Men Died When This 'Modern' Structure Collapsed." *Architectural Record*, August 1936, p. 80.

"Six Convicted of Manslaughter in Bronx Apartment Collapse." *Engineering-News Record*, April 8, 1937, p. 532.

"Six Men Sentenced in Building Crash." *Engineering News Record*, May 13, 1937, p. 720.

"16 Dead Removed from Wreckage of Bronx Building." *New York Times*, Sunday, June 21, 1936, p. 1.

Texas Tower radar station collapse

North Atlantic

(1961)

A manned North Atlantic radar station collapsed owing to a nonrigid design that could not withstand the action of wind and waves.

Background

A violent winter storm on January 15, 1961, dealt the fatal blow to an already weakened outpost of America's early warning radar system—and to the twenty-eight men who were working to repair it from earlier damages. The tragedy's costly lessons led to better designs for offshore platforms such as North Sea oil rigs.

Time magazine described the tower's collapse in its January 27, 1961, issue; the cover story that week was the inauguration of John Kennedy as president of the United States. It was the midst of the Cold War.

The government had decided in 1954 to build a group of five radar islands in the North Atlantic, off America's coastline from New York to Maine. They were called Texas Towers because of their similarity to oil-drilling rigs in the Gulf of Mexico. Plans for two of the structures were subsequently dropped, but towers number 2, 3, and 4 were built. Texas Tower number 2 (TT-2) was the first to be completed, in 1955; TT-4 was the last, with erection accomplished by the end of 1957.

They were the first offshore platforms to be built in the stormy North Atlantic. TT-4 had the further distinction of being placed in deeper exposed water than any previous framed structure; it was positioned sixty miles east of Sandy Hook, New Jersey, where the sea was one hundred and eighty-five feet deep. It was also erected on a sandy ocean floor, while the other two radar towers were founded on rock.

The twenty-one million dollar Texas Tower number 4 was a sight so impressive that ocean liners sometimes swung off course to give their passengers a view of it. The triangular steel platform measured two hundred feet on a side; a six-story building could have fit between the platform and the water below. At each corner, twelve and a half-foot-diameter hollow steel legs supported the tower and provided fuel storage areas. Lined up along the west side of the platform were three orange-and-white-checked radomes (radar-housing spheres), the tallest of which towered another sixty-three feet above the top of the platform.

In order to properly repair the joints, the whole assembly should have been towed back to the manufacturing facility. However, pressure to get the radar units in place to guard against potential Soviet air attack led the Air Force to approve on-site repairs. Working underwater on an assembly that was being buffeted by waves was not conducive to making structurally sound repairs.

Besides housing the equipment necessary for the station's operations, the platform contained living quarters for the seventy-man crew, including libraries, gymnasiums, hobby shops, pool table-equipped recreation rooms, and an excellent dining hall. The men served sixteen-month tours consisting of forty-five days aboard alternating with fifteen days ashore; to maintain morale, the facility was made as comfortable and pleasant as possible. The tower had been designed to yield to the action of wind and waves, so it constantly swayed and creaked, not only adding a new dimension to pool games but also unnerving the men over a period of time.

The leading builder of offshore platforms at the time was the DeLong Corporation, but its bid on the TT-4 project was rejected in favor of a lower-cost offer from the engineering firm of Moran, Proctor, Mueser & Rutledge and contractor J. Rich Steers. Because of the unprecedented challenges posed by TT-4's location, the selected engineers departed from the traditional design favored by DeLong and designed it in a significantly different way from any other existing offshore platform. The normal procedure was to weld the first stages of the legs to the platform, then float the assembly like a barge to its destination, extend the legs down to their anchoring positions on the ocean floor, and jack the platform up to the proper height. In the case of TT-4, however, the entire leg assembly was constructed first, then floated horizontally for five hundred miles from its manufacturing site in South Portland, Maine, to its final destination, and righted to a vertical position. The four thousand-ton platform was then floated in between the legs (a tricky, close-fitting maneuver), and jacked up to its final position. Finally, the three hundred and fourteen-foot-long legs were embedded into the ocean floor.

This procedure ultimately undermined the stability of the structure. The process of floating the leg assembly horizontally (with two legs on the water and one leg riding one hundred and thirty-five feet in the air) and then tipping it upright subjected the assembly to very different loads and stresses than those it had been designed to withstand (i.e., supporting the platform). The effects of these stresses were magnified by the fact that horizontal braces connecting the legs were not welded on, but were connected by pins that permitted some movement. Additional horizontal braces connecting the midpoints of the major leg braces were welded, but diagonal supports between the four levels of horizontal bracing were pinned for flexibility.

Further weakening the assembly, the top layer of leg braces was left incomplete so that the platform could be moved into position. One end of each brace was pinned to a leg, and the brace was lashed to the leg; the free end would ultimately be swung out and attached to another leg. Rough weather during the five-day voyage tore two of these braces loose from their lashings; the resulting excessive motion damaged the pin connections.

The Texas Tower radar station was so impressive in appearance that ocean liners sometimes swung off course to give their passengers a view of it. A six-story building could have fit between the platform and the water below.

In order to properly repair the joints, the whole assembly should have been towed back to the manufacturing facility. However, pressure to get the radar units in place to guard against potential Soviet air attack led the Air Force to approve on-site repairs. Working underwater on an assembly that was being buffeted by waves was not conducive to making structurally sound repairs.

There was yet another flaw in the design of TT-4. It had been engineered to withstand thirty-five foot-high waves accompanied by one hundred and twenty-five mile per hour winds. During its lifetime, it was subjected to winds of at least that magnitude and waves twice that high. Studies conducted after the tower had been in place over a year found that waves ten to twelve feet high produced more movement in the supports than thirty-foot waves. The smaller waves occurred closer together, allowing two of them to break simultaneously on separate portions of the tower, exerting a different pattern of stress than a single, larger wave.

The unproven design and improvised repairs set the stage for the tragedy of TT-4. The scene was further framed by the forces of nature. Within a year of the tower's completion, hurricane Daisy struck in September 1958, inflicting $500,000 damage on the tower. It ripped loose some of the bracing connections that had been reconstructed underwater during erection. Because of rough seas, the connections could not be repaired until the following May, allowing the motion of the weakened tower to inflict further wear on other joints.

During the fall of 1959, platform movement became excessive, and crews were assigned to tighten all of the connections and inspect the entire structure. They found badly worn pin connections in the bracing at twenty-five feet and seventy-five feet below the water level. To counteract the sway allowed by these loose connections, workers installed a new set of cross-braces between the legs above the water level.

The reinforcement project was completed in August 1960; the following month hurricane Donna struck. The most powerful storm yet to hit the tower, it ripped off a maintenance bridge hanging below the platform, magnified the damage to the underwater bracing, and loosened connections on the new above-water braces.

Details of the Disaster

The crew of TT-4 now called the station Old Shaky, and it was not a term of endearment. In their letters and phone conversations with relatives, they expressed increasing fear for their safety. On October 3, 1960, the Air Force asked J. Rich Steers to evaluate the tower's integrity and perform repairs; Moran, Proctor, Mueser & Rutledge would help as a subcontractor. The $500,000 contract was not confirmed in writing, however, until the following January 4.

In the meantime, by mid-November 1960, the tower's gyrations became so alarming that most of the crew were evacuated. Fourteen servicemen remained aboard TT-4 to provide power and heat for fourteen civilian repairmen. Workers reported that they had to strap themselves to their bunks to sleep, or be tossed out. The terrible lurching sent men to their knees in prayer. Fuel was leaking from the storage tanks in the tower's hollow legs. By early January, the swaying was too severe to allow the welders to work.

Increasing activity by Soviet fishing vessels deterred thoughts of abandoning TT-4. Leaving the tower's equipment unattended would have been tantamount to handing it to the enemy.

On January 7, the first inspection under the new contract was made, revealing a broken brace one hundred and seventy-five feet below the water level. Steers announced that the work required to repair this major damage was outside the scope of its current contract and asked for further negotiations. Apparently other repair work was suspended pending the negotiations as well; steel cable that was supposed to strengthen damaged upper-level braces was not installed as scheduled.

As renegotiation began on January 12, it was agreed that conditions were so severe that the tower should be evacuated on February 1. Repairs would start in May when winter weather subsided.

On the evening of January 14, a new storm began to batter Old Shaky. By dawn sixty-mile per hour winds buffeted the tower with sleet and snow. Captain Gordon Phelan, TT-4's commanding officer, awaited orders to abandon the tower. Days earlier as he was leaving on a mission, Major William Sheppard, the Texas Tower squadron commander, had delegated authority to order the evacuation to Phelan—a fact that apparently was not clearly relayed to the captain.

At 10:30 A.M. on January 15, following a sudden loud noise, the tower's motion intensified. At seventy-two miles per hour, the wind bordered on hurricane force by mid-afternoon; large cracks were sighted in the leg braces. At four o'clock Phelan conferred with superiors ashore; helicopters would be dispatched for evacuation during the first lull in the storm. At six o'clock Phelan telephoned his wife to say the tower was breaking up and the aircraft carrier *Wasp* was on its way. Workers scrambled onto the platform's deck to clear space for the helicopters and to throw confidential papers overboard. At 7:30 P.M. the tower's glowing image disappeared from the *Wasp*'s radar scope.

When the first rescue vessels reached the tower's location at 8:00 P.M., the only signs of the structure were a few bits of floating debris and the smell of diesel oil. All twenty-eight of the occupants had perished.

Impact

The spectacular collapse of Texas Tower number 4 captured the attention of the popular press. *Time* and *Newsweek* each devoted nearly a page to the event, and *Life* ran a six-page photo story featuring dramatic pictures of rescue attempts. In April 1962, *Reader's Digest* ran a condensation of a vivid account of the tower's final days, the full version of which appeared in *True, the Man's Magazine.*

The Air Force immediately launched an investigation of the disaster, shielding it behind claims of national security. The designs of the radar towers had never been classified, however, and within weeks public pressure produced disclosure of the findings. Colonel William M. Banks, the highest regional officer in charge of the Texas Towers, was court-martialed on charges of dereliction of duty. He was acquitted.

The Senate Armed Services Committee's Subcommittee on Preparedness held public hearings on the tower's design and construction. The final report cited the nonrigid design, innovative erection methods, and the decision to repair the damaged leg assembly on-site during the structure's construction as major factors underlying the collapse. Referring to the nearly continuous history of problems with TT-4, the subcommittee concluded that "an unbroken chain of error and mistakes in judgment had ended in stark tragedy."

Offshore platform designers learned from this costly failure to avoid pinned connections and rely on rigid, welded structures. Platform managers learned to attend quickly to structural weaknesses, lest the forces of wind and waves amplify them to dangerous levels.

As for America's early warning system, airborne radar units became available shortly after TT-4 was built. Texas Tower number 4 would not be reconstructed. The two remaining towers continued their radar surveillance of the North Atlantic until 1963, when they were dismantled in favor of the more effective airborne units.

Loretta Hall

Where to Learn More

"Air Force Marks Tower Damage." *Engineering News-Record,* January 26, 1961, pp. 28-29.

"Broken Lower Brace Probable Key to Tower Collapse." *Engineering News-Record,* February 2, 1961, pp. 22-23.

"A Case of Too Little Too Late." *Engineering News-Record,* May 18, 1961, p. 37.

"The Death of 'Old Shaky.'" *Newsweek,* January 30, 1961, pp. 33-34.

"Death on Old Shaky." *Time,* January 27, 1961, pp. 15-16.

"DeLong Sounds Off on Texas Tower No. 4." *Engineering News-Record,* May 11, 1961, pp. 18-19.

"Radar Tower Collapses in Atlantic." *Civil Engineering,* February 1961, p. 88.

Ross, Steven S. *Construction Disasters.* McGraw-Hill, 1984, pp. 355-67.

"Taps Too Late for Tower 4." *Life,* January 27, 1961, pp. 75-80.

"Texas Tower Tilts Down into Place." *Engineering News-Record,* October 25, 1956, pp. 40-44.

Wylie, Evan McLeod. "Tragedy on Tower Four." *Reader's Digest,* April 1962, pp. 94-102.

Ferrybridge cooling towers collapse

Pontefract, England

(1965)

Design that inadequately accounted for wind forces led to the collapse of three giant water cooling towers, spurring important research into wind dynamics.

Background

On November 1, 1965, at the Ferrybridge power station near Pontefract, England, three of the largest natural draft cooling towers in the world collapsed during a gale force wind. The cooling towers at Ferrybridge, operated by Britain's Central Electricity Generating Board, were structurally complete but not in operation. The wind stretched the reinforced concrete membranes of the towers beyond their structural capacity, causing the concrete to vibrate at a high-pitch and ripple into waves. The upper rim and the side sections ultimately buckled into the center of the towers.

Cooling towers play a critical role in power generation by providing low-temperature water for use as a coolant within the power plant. The eight towers at the Ferrybridge power station were "natural draft" towers, which cool water by spraying it through cooler air into large concrete tubes. The hourglass shape of natural draft towers and the warming of the air by the water draws the air up through the concrete tube without the aid of fans, like smoke up a chimney. Differences in water and air temperature and heat losses during evaporation cool the water. The cooled water then pools at the base of the tower and is recirculated to the power generating station.

The cooling towers at Ferrybridge rose 375 feet above the collection pools, spanned 290 feet in diameter at their bases, and weighed 8,000 tons. They were "thin shell" towers, with concrete structural membranes only five inches thick. In contrast, the shell of an egg as large as a Ferrybridge tower would be 25 inches thick. The eight towers at Ferrybridge were grouped into two rows of four, staggered into a checkerboard pattern. They were spaced closer together than towers in previous designs to take advantage of a "pillar of support," a localized area of superior foundation material.

Britain's Central Electricity Generating Board, an acknowledged leader in thin shell concrete cooling towers, did not venture into innovative design or construction with the Ferrybridge towers. Although these cooling towers were the largest to date, the increased size represented only a minor extrapolation from similar towers constructed two year before. The Board had contracted the design and construction of Ferrybridge towers to Film Cooling Towers (Concrete) Ltd., at an estimated cost of $700,000 to $840,000 each. Film Cooling Towers (Concrete) Ltd. had subcontracted the design to the consulting firm of Messrs. C. S. Allott and Son. The series of design and construction contracts left some ambiguity as to which parties were ultimately responsible for the design of the towers.

Details of the Collapse

A strong wind blew against the cooling towers at the Ferrybridge power station on November 1, 1965, creating a windward and a leeward row of towers. This wind was later estimated to have traveled at 76 to 84 miles per hour at 33 feet above the ground, and at 93 to 104 miles per hour at the upper rim of the towers. As the wind pushed past the towers in the windward row, it was funnelled through the spacing between them onto the windward faces of the second row of towers.

As the wind pushed past the towers in the windward row, it was funnelled through the spacing between them onto the windward faces of the second row of towers. The resulting vibration of the enormous concrete towers produced a high pitched whine that was reported by observers to resemble "someone rubbing a finger round the rim of a wineglass."

This pressure caused the leeward towers to lean away from the windward towers, stretching their windward sides in the way that leaning from the waist stretches the muscles on one side. At the same time, the updraft through the towers created a force that tended to lift the towers off their foundations. This lifting action reduced the effective weight of the towers and increased the stretching of the concrete shells. The resulting vibration of the enormous concrete towers produced a high pitched whine that was reported by observers to resemble "someone rubbing a finger round the rim of a wineglass."

The wind eventually stretched the windward sides of three of the second row cooling towers beyond the strength of the steel reinforcing in the concrete shells. The towers lost their structural integrity and could no longer hold their shape. Initially, the upper rims of the three cooling towers buckled into the tower centers. The windward sides collapsed into the tower centers in turn, dragging in the adjacent side sections of the structures.

Three leeward towers collapsed in this fashion. Although the failures were sudden, they were not simultaneous. The first tower failure occurred at 10:30 A.M., the second about ten minutes later, and the third at about 11:30 A.M. A spectacular photograph was taken from the windward side of one of the towers during its collapse, showing the overstretched windward side of the tower dragging and twisting the adjoining sections inward.

Impact

The Central Electricity Generating Board formed a committee with expertise in concrete, wind dynamics, and civil engineering structures to investigate the causes of the failure. When records and inspection of the collapsed structures revealed no significant deficiencies in construction and materials, the committee reviewed the design of the towers. It was noted that the engineers had enlarged the height and diameter of the towers, in relation to the prototype design, without any corresponding thickening of the concrete shells. However, wind tests by Britain's National Physical Laboratory had indicated that five-inch thick shells could withstand winds up to 200 miles per hour. Therefore, the committee directed their investigation to the engineer's methodology in calculating wind pressure requirements.

The designers of the cooling towers had relied on data about specific wind speeds at predetermined heights above the ground in determining the structural requirements. The wind speeds were assumed to be static (uniform in speed throughout their duration). The design engineers then used equations to adjust this static wind data for tower height, the impact of temporal fluctuations in the wind (dynamic effects), and other conditions. These calculations formed the basis for designing the steel reinforcing in the concrete shells and in other structural features of the cooling towers. This approach to designing the thin shell towers was standard engineering practice at the time.

The committee found that the collapse of the cooling towers was related to various shortcomings in this process of calculating wind pressures: (1) the designers ignored the impacts of clustering the towers or locating them near other large power station structures; (2) the designers incorrectly interpreted a report by the National Physical Laboratory; and (3) the designer's equations to calculate the wind pressure at the tops of the towers generated a figure 24 percent less than what the equations specified in the British Standard Code of Practice would have produced. The cumulative effect of these and other deficiencies was the use of far less steel reinforcing in the walls of the towers than required. This led to the structural collapse of the towers at wind speeds that the committee estimated would occur every five years, far less than the design life of the towers.

The disaster at Ferrybridge had significant repercussions on the power station itself and on the structural engineering community as a whole. At the Ferrybridge power station, the five towers that did not collapse were inspected; crack lines revealed that they had also been damaged by the gale force wind on November 1. The engineers determined that these towers did not have to be demolished, but could be altered to make them structurally sound. A second layer of concrete was added to the outside of each of these towers, doubling the thickness of the shells, to provide the additional strength required. The three towers that collapsed were rebuilt using a similar "double-skin" design. Nonetheless, regulations were instituted to prevent anyone from approaching the cooling towers during gale-force winds, and these regulations have remained in effect over 25 years.

The most significant impact of the design failure at Ferrybridge, however, involved theoretical advances in the field of structural dynamics. The investigating committee recommended that the entire approach to designing cooling towers be reassessed, as current engineering practices did not adequately address the dynamic impacts of wind. The engineering community responded with extensive analytic and experimental research. The analytic approaches applied fundamental engineering concepts and the relatively new finite element method of structural analysis to predict the effects of dynamic wind pressures on cooling towers. Experimental researchers built models resembling the towers at Ferrybridge, installed them in wind tunnels, and tested the structural response. The research confirmed the investigations conducted by the committee, and led the way to many advances in cooling tower design.

The results of the research initiated by the collapse of the cooling towers at Ferrybridge ultimately were incorporated into international structural design methods and building codes. In this way, the structural failure at Ferrybridge led to an expanded understanding of wind dynamics and helped to prevent future engineering disasters—vividly illustrating the role that failure can play in the evolution and advancement of technology.

DAVID N. FORD

Where to Learn More

"Towers Tumble—Grave Trouble for CEGB?" *Electrical World*, November 29, 1965, p. 77.

"Spacing Cited in Cooling Tower Collapse." *Electrical World*, February 7, 1966, p. 51.

"Cooling Tower Failures at Ferrybridge." *The Engineer*, November 12, 1965, pp. 789, 805.

"Ferrybridge-'C' Cooling Towers." *The Engineer*, December 24, 1965, p. 1049.

"Ferrybridge Cooling Towers." *The Engineer*, August 26, 1966, pp. 291-92.

"What Lessons from Ferrybridge?" *Engineering*, September 2, 1966, pp. 399-403.

"High Wind Levels British Cooling Towers." *Engineering News Record*, November 25, 1965, p. 45.

"Cooling Tower Failure Is Puzzle." *Engineering News Record*, March 3, 1966, p. 28.

"Design Blamed for Tower's Fall." *Engineering News Record*, September 8, 1966, p. 20.

Gardner, Noel J. "Response of Cooling Tower to Turbulent Wind." *Journal of the Structural Division, Proceedings of the American Society of Civil Engineers*, October 1968, pp. 2057-75.

Sykes, J. H. M. "Ferrybridge Collapse Report Cites Wind-Force Design." *Electrical World*, December 26, 1966, pp. 42-43.

Ronan Point tower collapse

London, England

(1968)

Inadequate connections between walls and floors caused the progressive collapse of one corner of a prefabricated high-rise apartment building.

Background

At a quarter to six on the morning of May 16, 1968, the residents of the Ronan Point apartment building in London awoke to the sound of an explosion followed by the cascading collapse of an entire corner of the twenty-four-story structure. The British government acted quickly to inspect and reinforce hundreds of similar prefabricated concrete high-rise apartment buildings in order to restore public confidence in the construction technique widely used to address the post-World War II housing shortage that still affected the country.

In this particular neighborhood, over a fourth of the dwellings had been destroyed by enemy attacks. An $11.5 million complex of nine identical apartment buildings was planned for the area; opening in the spring of 1968, Ronan Point was the second to be completed. Its upper twenty-two stories provided 110 "flats," as the British refer to apartments.

In the quest to rebuild European cities as quickly as possible, a variety of systems had been developed to construct large buildings from concrete wall and floor panels that were mass produced at central locations and brought to the construction site for assembly. This prefabricated construction (also referred to as system building) was very efficient in terms of minimizing cost, labor force, and construction time.

The Borough of Newham selected Taylor-Woodrow-Anglian, a London contractor, to build its nine-tower complex using the Larsen and Nielsen prefabrication system developed by Danish engineers. The system had been widely used throughout Western Europe, Turkey, and Hong Kong for fifteen years; Taylor-Woodrow had built three thousand such apartment units in the British Commonwealth since 1962. In a decision that would later raise questions of conflict of interest, the contractor insisted that Philips Consultants be selected as the project's structural engineer; Philips was a wholly owned subsidiary of one of Taylor-Woodrow-Anglian's two parent companies.

As with other prefabrication systems, the Larsen and Nielsen method featured one-story-tall wall panels and room-size floor slabs, which were stacked and cemented together like a high-rise gingerbread house. In a typical interior joint, two floor panels would rest atop opposite edges of a wall panel; the gap would be filled with concrete, and another wall panel would be set directly above the lower wall. Exterior wall panels were formed with a lip that would extend past the end of the floor slab as it rested on half the width of the wall; the lips on the upper and lower walls were offset so they would overlap and could be bonded together with concrete.

What differentiated the various systems was primarily the design of the interpanel joints. A row of concrete teeth was cast into the edges of the Larsen and Nielsen floor panels to increase the surface area contacted by the concrete grouting used to bond the wall and floor sections together. Two strands of steel reinforcing bar were installed horizontally between the floor panel edges (in the case of an exterior wall, there would be only one floor panel and one reinforcing bar).

On a previous project, a government engineer concerned about insufficient connection between the walls and floors had required the addition of steel bolts running vertically through the wall panels. At a joint, the end of the bolt would extend through a tie plate in the lower edge of the upper wall; a nut atop the bolt could be tightened to press the tie plate firmly into contact with the concrete poured between the wall and floor panels. Apparently Philips Consultants was not convinced of the need for the bolts, which were also included in the Ronan Point design; a postcollapse inspection would reveal that the bolts were uniformly left untightened.

Another difference of opinion between Philips Consultants and a government engineer involved the loads for which the building should be designed. The initial design called for the building to withstand an imposed floor loading of thirty pounds per square foot and a wind loading of seventeen pounds per square foot. Philips Consultants was told to increase the values to forty pounds per square foot imposed floor loading and twenty-four pounds per square foot wind loading.

After modifying the design to meet the borough's specifications, Philips Consultants submitted a document certifying the proposed building's structural stability. Perhaps out of a desire to save time, or possibly simple complacency, that certification was accepted without an independent review of the design details and calculations. The tower was built, and 250 tenants moved into their new homes.

Details of the Collapse

Ivy Hodge was one of Ronan Point's new tenants. When she awoke on May 16, 1968, Ivy put on her robe and slippers and went to the kitchen to heat a kettle of water. Her friend Charley Pike had been kind enough to hook up her gas stove when she moved into her eighteenth-floor flat a month before. Since the local gas pressure sometimes dropped low enough to extinguish pilot lights, allowing gas to leak when the pressure returned, she had promptly taken the safety precaution of disconnecting the pilot.

Everything looked and smelled normal this morning as she filled the kettle and set it on the stove. But when she struck a match to light the burner, she was knocked to the floor by an explosion. She suffered only second degree burns on her face and arms, but her apartment sustained fatal damage.

Two exterior wall panels of her corner living room were blown outward and fell to the ground. The disappearance of those two wall sections left the six floors above Ivy's apartment without vital support for their weight. Walls, floors, and furniture from above came crashing down into her living room. The resulting shock to the floor and supporting walls under her apartment exceeded their loading capacity. Floor by floor, the corner of the building crumbled to the ground.

It was primarily living rooms that were destroyed as the concrete panels fell, but adjacent bedrooms also sustained some damage. Fortunately, most of the tenants were still in bed, out of the path of worst destruction. Even so, over a dozen residents were injured, five fatally.

When the dust settled, all that remained of Ronan Point's southeast corner was twenty-one tiers of floor slab remnants, their two interior edges still connected to neighboring panels, dangling toward a pile of rubble below. The corner wall panels had disappeared completely.

Unfortunately, Charley Pike's good intentions were not matched by knowledge or experience. He had connected the stove to the building's gas line with a brass nut that did not meet British Gas Board standards. He compounded the error by overtightening the nut, which was weak enough to crack slightly. The result was a slow leak.

High-rise apartment buildings were relatively new fixtures in England in the late 1960s; building codes had not been revised to reflect their characteristics—particularly with respect to wind loads, which logically have a greater effect on tall buildings than short ones. Although prefabricated construction was officially encouraged as a solution to the housing scarcity, building codes had not been modified to address system building specifically. Furthermore, since it was virtually unknown in frame or brick buildings, Britain's building codes did not take into account the possibility of progressive collapse (in which failure of one component triggers failures in adjacent components, allowing damage to spread through a structure rather than being contained to the area of the initial failure).

Recognizing the inadequacy of official guidance regarding the structural stability of a category of buildings that housed tens of thousands of people, the government moved quickly to avert public panic. Within days, the British

Following a gas explosion in one of the apartments, inadequate connections between walls and floors resulted in the progressive collapse of one corner of the prefabricated Ronan Point apartment building.

home secretary appointed an inquiry committee consisting of a lawyer and two prominent engineers.

The investigation was directed at Ronan Point specifically, but also at prefabricated high-rise buildings in general. In the specific case, the tribunal found that the design and construction of the joints between wall and floor panels were inadequate to allow the components to act as an integral system when subjected to an overload.

Impact

The impact of the Ronan Point Tower collapse extended to more than the eighty families evacuated from that particular building. The investigative committee's interim report, released three months after the disaster, asserted that all prefabricated buildings employing load-bearing walls could be susceptible to progressive collapse. It went on to say that another Ronan Point-type disaster could be triggered by some event other than a gas explosion—such as ground settlement, accidental physical damage, or heat rising through a structure that was on fire.

The investigative committee's interim report asserted that all prefabricated buildings employing load-bearing walls could be susceptible to progressive collapse. It went on to say that another Ronan Point-type disaster could be triggered by some event other than a gas explosion—such as ground settlement, accidental physical damage, or heat rising through a structure that was on fire.

Some thirty thousand apartments in roughly five hundred systems-built structures over six stories high were suddenly suspect. Local authorities were advised to tell residents of all high-rise apartments whether their buildings were prefabricated, and to inspect such buildings for structurally sound joints between wall and floor panels. Gas supplies were to be cut off in any buildings failing inspection.

Examining that many buildings took time. On November 15, 1968, the *New York Times* reported that fourteen hundred families were being evacuated from twenty-nine apartment towers in London. The six thousand evacuees were housed at government expense during a year-long, $2.4 million structural reinforcement program.

The tribunal's final report, issued in early November, found fault with specific individuals and actions involved with Ronan Point's design and construction. It also addressed broad criticism at architects and structural engineers, in general, for inadequate attention to the design of prefabricated buildings. Numerous members of the Institution of Structural Engineers responded with pointed rebuttals and objections to the report's recommendations.

In addition to its provisions for inspecting and reinforcing existing buildings, the report also recommended changes in the way future systems structures would be designed. Specifically, the problem of progressive collapse should be addressed by designing structurally sound joints and by positioning load-bearing walls in such a way that there would be alternative paths for transferring loads (thus, if one wall section failed, the load above it could be borne by neighboring sections).

In an attempt to account for explosive forces within future buildings, the report recommended requiring them to withstand higher wind loads than current codes required. Critics argued that static and dynamic loads would affect a structure differently, expressing doubt that designing a building to tolerate a constant pressure from a strong wind force would ensure that it could withstand the quick burst of pressure from an explosion.

The events at Ronan Point influenced both design philosophy and building codes around the world. But in the Borough of Newham, a damaged building required attention. In the year following the disaster, the damaged corner was reconstructed, and brackets were added to reinforce the connections between floor slabs and exterior wall panels. Tenants moved back into the building. Within five years, however, ominous wall cracks began appearing and the building was evacuated again.

Rehabilitating the building to meet new building codes would have been prohibitively expensive, so the building was condemned to destruction. A public-minded architect named Sam Webb, suspecting shoddy workmanship, convinced the government to disassemble Ronan Point rather than leveling it with explosives. The strategy paid off. The joints between walls and floors were found to be loosely filled with rubbish, not firmly packed with concrete.

This frightening discovery raised further questions about the safety of similar buildings. Eventually, hundreds of such structures were condemned and demolished.

The Ronan Point collapse touched off wide-ranging debates about the stability of prefabricated construction, the adequacy of building codes, and the advisability of supplying high-rise apartments with gas for cooking. Ultimately, system building and gas fuel were sanctioned—provided the design of each structure accounted for the possibility of explosion and progressive collapse, and the craftsmanship was adequate.

LORETTA HALL

Where to Learn More

"Anatomy of Collapse." *Engineering*, November 15, 1968, p. 707.

"Failure of a High-Rise System: How Safe Should the Structure Really Be?" *Architectural Record*, November 1968, pp. 169-170.

"Gas Turned Off: British Fear Explosions in High-Rise Prefab Buildings." *Engineering News-Record*, August 22, 1968.

"The Implications of the Report of the Inquiry into the Collapse of Flats at Ronan Point, Canning Town." *The Structural Engineer*, July 1969, pp. 255-84, and September 1969, pp. 387-89.

Levy, Matthys, and Mario Salvadori. *Why Buildings Fall Down*. Norton, 1992, pp. 76-83.

Skyline Plaza collapse

Bailey's Crossroads, Virginia

(1973)

Premature removal of temporary wood shoring and formwork caused the collapse of a high-rise apartment building under construction.

Background

Early on the afternoon of Friday, March 2, 1973, a tall concrete apartment building under construction near Bailey's Crossroads, Fairfax County, Virginia, suffered a dramatic collapse of a major portion of the structure. Twenty-three stories plus four basement levels came crashing down, along with a large construction crane and an adjacent parking garage also under construction. Fourteen people lost their lives and another thirty-four were injured.

Investigators from the Occupational Safety and Health Administration (OSHA), U.S. Department of Labor, arrived at the job site within hours and began the task of examining the damaged building. They were especially interested in whether or not any violations of OSHA's safety regulations had caused or contributed to the collapse.

The Center for Building Technology of the National Bureau of Standards (NBS) was asked to provide technical assistance. The OSHA and NBS teams were on site gathering data for two weeks following the disaster. They reviewed photographs of the project and statements made by workers present when the incident occurred. Using a computer program bureau engineers modeled the structure in three dimensions to calculate the stresses in the building's structural components at the time of the collapse.

The Bureau issued a report of findings that concluded that the most probable cause of the collapse was the failure of the concrete floor slab at the twenty-third floor due to premature removal of temporary wood shoring and formwork supporting it.

Building A-4 was one of several buildings in the Skyline Plaza complex. Structural plans prepared by the design engineer for Building A-4 indicated eight-inch-thick floor slabs made of reinforced concrete with a design strength of three-thousand pounds per square inch, a typical concrete specification. The floor slabs were shown supported on rectangularly shaped reinforced concrete columns of varying strengths and spacings, again a typical arrangement for this type of structure.

According to a note on the structural plans, the temporary wood shoring and formwork installed to support each concrete floor as it was poured was to remain in place until at least two complete floors above it had been constructed. This would allow the new concrete sufficient time to harden and become strong enough to carry the loads from above without the aid of the wood supports, which could then be removed.

Concrete reaches its design strength within a month of its placement. About 70 percent of this "twenty-eight-day" strength is achieved by the seventh day after pouring. In its first few days, fresh concrete is considerably weaker, hence the need for supporting formwork during the early "curing" period.

Governing OSHA safety regulations also contained minimum time requirements for leaving forms in place: at least ten days (of temperatures exceeding fifty degrees Fahrenheit) for concrete floor spans of over twenty feet, which Building A-4 contained.

Details of the Collapse

The pace of construction of Building A-4 had been running at about one floor built per week. However, near the end of February, it had been accelerated to almost twice that rate. The portion of twenty-fourth floor slab being poured against its supporting wood forms on March 2 was being installed above a section of twenty-third floor slab that was only four days old, rather than the usual seven or so.

Construction workers who provided eyewitness accounts of the events leading up to the collapse gave contradicting descriptions of the stage of construction underway on that Friday morning, especially in regard to the status of the shoring supporting the twenty-third floor slab.

Some said all of the formwork at the twenty-second story was still in place (as had been required by the structural plans). Others stated that the shoring was already in the process of being removed (even though the twenty-fourth floor slab had not yet been completely cast) and that only some of it remained in place at the time of the collapse. Still others claimed that all of the shoring at this level had been removed because the wood was needed elsewhere in the building.

Shortly after lunch, some of the workers observed severe deflections or bends (six inches to two feet) in both the twenty-third and newly poured twenty-fourth floor slabs. The deflections grew until the slabs failed, dropping onto the twenty-second-floor slab below. This still-young slab at the twenty-second floor, cast on February 20, was not strong enough to withstand the impact load of the falling material from above, and so it also failed. The collapse continued in this manner, with each successively lower floor slab hit by the moving weight of more and more concrete, and unable to support it.

Skyline Plaza experienced a type of failure known as progressive collapse, that is, a major sequential catastrophe initiated by a comparatively smaller, localized triggering mechanism. The question was just where the trigger had occurred and why.

Several potential sources of failure were investigated and later dismissed as not responsible. The NBS team's computer model of Building A-4 was created to determine numerically where a stress concentration high enough to initiate failure could have been produced. Concrete strengths were assumed based on age, temperature history, and field test data. The analysis was performed both with and without shoring.

The results from the computer program indicated that without shoring, significant overstresses were generated in the twenty-third-floor slab at four specific column support locations. The type of stress involved was what is known as shear stress, or more exactly, punching shear. This type of failure is typically characterized by strong concrete columns punching through comparatively weaker, thin horizontal slabs that have been loaded beyond their capacities.

Analysis of Building A-4's twenty-third-floor slab indicated that the actual shearing stresses to which the slab was probably subjected at the four overstressed locations exceeded its approximated capacity by 30 to 100 percent.

The NBS investigation also determined that there were significant problems with the wood shoring system itself on this project. The plan sheets submitted by the concrete contractor to the county for the shoring system showed no lateral cross bracing. Such bracing was required by the OSHA regulations. Some photographs taken of the project during construction showed a limited number of cross braces in place, poorly fastened at best, and certainly not up to OSHA's standards.

Additionally, photographs indicated that the shoring posts being used were less than effective supports, as much of the wood was observed to be in a deteriorated condition or "out of plumb," and without adequate end bearing surface areas.

The Skyline Plaza collapse served to call attention to several deep-seated problems present in the construction industry in the United States:

1. Project general contractors (and their subcontractors) are typically responsible for the design and placement of temporary supports such as shoring and formwork, even though they may not be completely aware of the loading that may occur during construction. They are not particularly knowledgeable about the capacity of the structure, build, nor necessarily of the degree to which time affects material strength. Project structural

Skyline Plaza experienced a type of failure known as progressive collapse, that is, a major sequential catastrophe initiated by a comparatively smaller, localized triggering mechanism.

engineers, on the other hand, more aware of the loads and capacity of the structure, are normally not involved in the design of temporary systems such as formwork and shoring.

2. Regardless of how adequate a design may appear on paper, its implementation in the field is subject to the level of quality control present during construction. Although some independent inspection takes place, general contractors are for the most part responsible to ensure that the intent of the plans and specifications is carried out in the built structure. The degree to which they succeed varies widely, and is at least partially a function of their skills and experience in management and decision making.

3. Although usually desirable from an economic standpoint, the practice of constructing buildings faster and faster can be detrimental. Speeding up the construction schedule on Building A-4 to the point that wood shoring only recently placed was already being moved somewhere else did not allow the concrete enough time to gain sufficient strength to support itself before form removal.

Impact

On March 4, 1973, two days after the collapse, the entire portion of Building A-4 east of the failure zone was demolished, leaving about half of the building intact. The missing half was later replaced; the structure was completed and occupied.

Two months later, a Fairfax County Grand Jury indicted the general contractor (who was also the owner), the concrete subcontractor and the job superintendent. The concrete subcontracting firm, also charged with willful violation of OSHA regulations, was ultimately ruled responsible, and fined. Charges against the general contractor were dismissed; the superintendent was found not guilty. Fines assessed the concrete subcontractor amounted to less than $20,000, despite an estimated $8 to $10 million in building damage, not to mention the human injuries and deaths.

An injured construction worker later successfully sued the building's architect and engineer for negligence and was awarded $500,000. The court blamed the designers for not visiting the job site to inspect the progress of the work.

Unfortunately, no sweeping revolution of the industry has yet taken place. Many of the problems that this particular collapse, as well as others like it, served to underscore are still with us today.

No sweeping revolution of the industry has yet taken place. Many of the problems that this particular collapse, as well as others like it, served to underscore are still with us today.

Current building codes do not have specific requirements for analyzing abnormal loading or for dealing with the avoidance of progressive collapse, either during construction or in completed structures. However, today's codes do have more restrictive detailing requirements for reinforced concrete construction than were present in the codes of 1973, especially in high wind and seismic zones, where there is increased attention to connections. This provides for greater structural integrity overall, and consequently for better resistance to collapse in an indirect way.

Much of the added conservatism in current building codes may be attributed to the efforts of the Portland Cement Association, whose studies of concrete structures have produced recommendations later codified by the American Concrete Institute and ultimately adopted in building codes. The association's "alternative path" methodology contains recommendations for tying building elements together to provide increased structural integrity and greater resistance to partial collapse.

In summary, the construction industry is still far from perfect in terms of its ability to translate a good design from a set of plans into a well-built, safe structure. However, improvements have been made in building codes and their enforcement, which are helping to make collapse as destructive as Skyline Plaza's less likely. If this trend continues, and further strides in education take place, hopefully such failures can be made a thing of the past.

LINDA BUCHACA

Where to Learn More

Breen, J. E., ed. *Summary Report—Research Workshop on Progressive Collapse of Building Structures.* Sponsored by the National Science Foundation, the National Bureau of Standards, and the U.S. Department of Housing and Urban Development, 1975.

"Bailey's Crossroads Synopsis." *Civil Engineering,* November 1975, pp. 59-61.

Cohen, E., N. Burns, and A. Meenen. "Skyline Towers Parking Structure Investigation," Prestressed Concrete Institute, 1974.

Leyendecker, E. V., and G. S. Fattal. *Investigation of the Skyline Plaza Collapse in Fairfax County, Virginia,* National Bureau of Standards, Building Science Series 94. U.S. Government Printing Office, 1977.

"Six Killed As a Crane Drops through Virginia Building." *New York Times,* March 2, 1973.

Ross, S. S. *Construction Disasters: Design Failures, Causes and Prevention.* McGraw-Hill, 1984.

Schousboe, I. "Bailey's Crossroads Collapse Reviewed." *Journal of the Construction Division, Proceedings of the American Society of Civil Engineers,* June 1976, pp. 365-78.

Schultz, D. M., et. al. *Design and Construction of Large-Panel Concrete Structures,* Portland Cement Association, Reports 1 through 6 and supplemental reports A through C. U.S. Department of Housing and Urban Development, U.S. Government Printing Office, 1975-1979.

Winter, G., and A. H. Nilson. *Design of Concrete Structures,* 8th ed. McGraw-Hill, 1972.

Hartford Civic Center Coliseum roof collapse

Hartford, Connecticut

(1978)

Collapse of a space frame roof caused by a lack of bracing of top chord truss members.

Background

The initial flurries of a New England snowstorm that was destined to drop 4 1/2 inches of wet snow over much of the East Coast by daybreak were not enough to deter some five-thousand fans from turning out for an evening college basketball matchup between the University of Connecticut and the University of Massachusetts at the Hartford Civic Center Coliseum on January 18, 1978. Less than five hours after the spectators had departed, however, the accumulated snow proved enough to bring down the space frame roof of the five-year-old arena, sending more than 1,400 tons of steel, concrete, and other debris crashing onto the arena's ten-thousand seats. Although no one was injured in the predawn collapse, investigators noted that if the roof had failed just a few hours earlier, several thousand people would have been killed.

The same night smaller roofs collapsed in two other Connecticut towns, killing one person. Three days later, during a second snowstorm, another space frame roof collapsed over an auditorium at C. W. Post College in Brookville, New York. This sudden rash of failures—and particularly the failure of the prominent and previously acclaimed Hartford arena roof—temporarily shook the public's faith in space frame roof systems and eventually led engineers who design them to regard their computer models with a more jaundiced eye.

Design of the Hartford Civic Center complex, which was at the heart of the city's downtown redevelopment program, began in 1970. The Philadelphia-based Kling Partnership was the architect for the complex, working with Hartford-based associate architect Danos & Associates. The $75 million project included shops, offices, and convention space as well as the arena with its 2.4-acre space frame or space truss roof. At the time the roof was constructed in 1972, it was the second largest space frame roof in the country, second only to the roof erected over the activities center at Brigham Young University in Provo, Utah.

The twenty-one-foot-deep, $1.4-million space frame, designed by the Hartford office of Norfolk, Connecticut-based Fraioli, Blum, & Yesselman Engineers, consisted of offset upper and lower grids made from horizontal steel members that were connected every thirty feet by steel diagonal members. The upper horizontal members were braced by additional steel diagonals, and the main diagonals were braced at their mid-points by horizontal steel bars. When completed, the 300-by-360-foot roof structure looked like a series of linked inverted pyramids.

Bethlehem Steel Corporation of Bethlehem, Pennsylvania, the project's steel fabricator and erector, shop-welded and bolted some of the space frame's connections before shipping it in sections to the site. Crews assembled the entire space frame structure on the ground, installing all heating and cooling ductwork and electrical and plumbing components. Then, over a two-week period, workers hoisted the massive roof up almost one hundred feet with jacks located atop temporary steel towers. The roof was then lowered into place at its final location eighty-three feet above the arena floor, supported on four massive concrete pylons located at the corners forty-five feet in from the edges of the space frame. The arena's three-inch-thick composition roof was supported not on the top chord of the space frame, as was typical, but on three-foot-high steel posts or struts welded to top chord members.

Part of the attraction of the space frame was that it was expected to save as much as half a million dollars in construction costs in comparison to a more conventional roof system. Since the roof was unconventional in design, project designers asked the city of Hartford for additional funding to pay for a computer analysis of the structure. That analysis, which convinced the architects, engineers, and city officials that the structure was safe, proved to be the space frame's undoing.

Details of the Failure

Witnesses to the collapse of the Hartford arena roof said the structural failure broke the predawn silence on January 18 with a thunderous crack that shook the windows in nearby buildings. Although exactly how and where the roof began to fail is a matter of some dispute, most investigators agree that individual top chord members of the truss, which were already overstressed, began to buckle under the added weight of the accumulated snow and ice. As these members failed, the load they carried was distributed to adjoining top chord members, which also buckled, eventually causing lower chord members to buckle and bring down the whole structure. A major shortcoming of the space frame system was that there was no redundancy built into it: when specific individual members failed there were no additional members in the structure that were able to carry that extra load.

Several different theories about exactly what caused the roof to collapse and the sequence of the failure were put forward by several teams of investiga-

Project designers relied heavily on computer analysis of the safety of the Hartford Civic Center. The conclusions of the analysis were tragically flawed, as this photo of the Center's collapsed roof shows.

tors. The first to release its findings was the New York City-based structural engineer Lev Zetlin Associates (LZA), hired by the city of Hartford immediately after the collapse. Hartford also hired two other consultants—Loomis and Loomis Incorporated Consulting Professional Engineers of Windsor, Connecticut, and Buck & Buck, Hartford—to inspect other sections of the civic center complex for structural soundness.

After a complete structural analysis of the roof and an examination of the bolts and welds used, LZA placed the blame for the failure of the space frame squarely on the shoulders of its designers. Noting that the roof loads were 1.5 million pounds more than the design loads used in the engineering calculations, LZA investigators claimed that the roof had begun to fail the moment it was jacked into place.

The space frame's greatest areas of vulnerability were at the top chord members on the exterior edges of the frame structure. Those members were particularly at risk because, due to their location at the outside of the frame, they were not braced at their midlengths by diagonal supports in the same manner that the interior upper-chord members were. According to LZA, those top chord members were overloaded by as much as 852 percent on the arena's east and west faces. Investigators from LZA said that these inadequately braced top-chord members of the truss buckled laterally under compression loads, beginning on a north-south line three bays from the eastern edge of the roof.

Lev Zetlin calculated that the roof was designed to support only 69 to 74 pounds per square foot, and that at the time of collapse the load was between 66 to 73 pounds per square foot, including snow loads of up to 19 pounds per square foot. Consultants for the firm said that the roof should have been designed to support a load of at least 84 pounds per square foot, and optimally, design loads should have been 140 pounds per square foot, with a safety factor of 1.67.

But another theory, put forward by Loomis and Loomis, blamed the collapse on torsional buckling of the diagonal members of the space frame, meaning that truss members twisted and bent rather than just simply bowing under pressure. Many of the truss members were cruciform in cross section, having been made up from four steel angles welded together back to back, and some types of cruciform sections are susceptible to torsional bucking. Although Loomis and Loomis was first hired to check other civic center buildings and later became a subcontractor to LZA, the firm undertook its own analysis of the failure at its own expense.

Loomis and Loomis said the collapse began near the support pylon at the roof's northwest corner. As that corner caved in, the force knocked the space frame off the other three pylons and the roof folded along a north-south line near the center.

Still another theory, released several months later, blamed the collapse on the failure of the weld that connected a scoreboard to the roof. That theory, developed by Hannskarl Bandel, a structural consultant hired by the architect's insurance company, suggested that because the bottom chords of the truss had already been stretched by the weight of the snow on the roof, the failure of the weld holding the scoreboard was enough to release 150,000 pounds of force and trigger the collapse.

Although the discrepancies in theories in the finer points of the collapse sequence may never be totally reconciled, it is undisputed that the roof gave off many clues to the impending disaster during its short lifetime and that these clues were largely ignored by designers and city officials. During erection, the contractor recorded a deflection of 8.4 inches in the bottom chord of the space truss, already higher than the expected deflection of 7.35 inches even though the roofing had not been added. The problem of the space frame's deflection was also evident when the fascia panels could not be bolted onto steel framing supports on the exterior of the truss as planned and had to be refitted and welded instead. By April 1975, the space frame had deflected by more than 12 inches, instead of the 8.5 inches predicted by the computer model. Lev Zetlin's report concluded that the city, the construction manager, the general contractor, the designers, the inspectors, and the steel contractor were all at fault for the collapse due to their inadequate inspection and lack of quality control efforts.

Impact

The failure of the Hartford arena roof, although it caused no injuries, served as a great embarrassment to the city, which counted on the civic center complex to help revitalize the city. As a result of the collapse, the World Hockey Association's New England Whalers, who called the arena home, were forced to move to Springfield, Massachusetts, and play in a much smaller arena. Other groups scheduled to use the arena during the two years it remained closed were also inconvenienced, including the Ringling Brothers, Barnum & Bailey Circus, which filed a $1 million lawsuit against the city claiming breach of contract in February 1979.

Redesign of the arena roof began almost immediately after the collapse under Ellerbe Architects of Minnesota. Although the arena's seating capacity was expanded to 14,500 seats from the original 10,000, the new roof that covered it was far simpler than the original space frame structure. The redesigned roof consisted of two parallel vertical trusses that rested on the original concrete pylons, raised up about twelve feet to accommodate the arena's increased volume. Six secondary trusses framed into the main trusses, and tertiary trusses framed into the secondary trusses. The new roof measured 320-by-440 feet, and weighed 1,200 tons and used 360 pieces, compared with the original, which weighed 200 tons more and used 4,400 pieces. The coliseum reopened on February 7, 1980.

The collapse of the Hartford Civic Center Coliseum roof, considered an engineering marvel when it was built, shook public confidence in wide-span roofs, especially space frame roof structures. Former president Gerald R. Ford refused to believe computer models that indicated that a space frame roof system constructed over a museum built in his honor in Michigan in 1980 was safe. Instead, Ford had the roof covered with water-filled boxes to test the space frame's deflection.

In the spring of 1978, Travelers Insurance Company paid the largest individual claim in its 114-year history—$11.3 million—to Hartford, in addition to the $1 million it had already paid the city to begin cleanup after the collapse. The repairs and expansion of the arena were later paid for with this insurance money, plus a $5 million state grant and a $14 million federal grant.

Although no criminal wrongdoing was determined—on July 26, 1979, the state closed its investigation into the collapse, not finding sufficient evidence to warrant a grand jury probe—financial liability for the collapse was placed on the building team. In August 1978, Hartford charged twenty-two firms and individuals—including the architects, engineers, contractors, and fabricators—with breach of contract and negligence. This action followed the filing of a $14-million lawsuit by Travelers against the same parties. The action was resolved out of court in March 1984, when Hartford's city council and Travelers accepted a $12.5 million settlement, netting $1.8 million for the city and $10.6 million for the insurance company.

The collapse of the Hartford Civic Center Coliseum roof, considered an engineering marvel when it was built, shook public confidence in wide-span roofs, especially space frame roof structures. Skepticism of the stability of these structures was so high immediately after the Hartford collapse that former president Gerald R. Ford refused to believe computer models that indicated that a space frame roof system constructed over a museum built in his honor in Grand Rapids, Michigan, in 1980 was safe. Instead, Ford had the roof covered with water-filled boxes to test the space frame's deflection.

The collapse provided a prominent example for the architectural and engineering community of the dangers of relying too heavily on a computer model in order to pare down a structure to the slimmest, least expensive, and least redundant solution possible. In the case of the Hartford arena, a structure designed to be just strong enough to satisfy requirements allowed no margin for error; and when errors occurred, the roof came tumbling down.

VIRGINIA KENT DORRIS

Where to Learn More

"Design Flaws Collapsed Steel Space Frame Roof." *Engineering News-Record*, April 6, 1978, pp. 9-11.

"Roof Collapse Suit Settled." *Engineering News-Record*, April 5, 1984, p. 16.

Levy, Matthys, and Mario Salvadori. *Why Buildings Fall Down*. W. W. Norton, 1992, pp. 68-75.

McQuade, Walter. "Why All Those Buildings Are Collapsing." *Fortune*, November 19, 1979, pp. 58-62.

Newman, Bruce. ". . . And the Roof Came Tumbling Down." *Sports Illustrated*, January 30, 1978, p. 20.

"Raising the Roof." *Newsweek*, February 12, 1979, p. 17.

"Civic Center Investigation Closed." *New York Times*, July 28, 1979, p. 20.

Petroski, Henry. *To Engineer Is Human*. Doubleday, 1992, pp. 198-200.

Ross, Steven S. *Construction Disasters*. McGraw-Hill Inc., 1984.

Smith, E. A. and H. I. Epstein. "Hartford Coliseum Roof Collapse: Structural Collapse Sequence and Lessons Learned." *Civil Engineering*, April, 1980, pp. 59-62.

"The Night the Roof Fell In." *Time*, January 30, 1979, p. 28.

Cooling tower collapse

Willow Island, West Virginia

(1978)

A cooling power plant tower collapsed as it was being built when construction loads were placed on concrete that had not been allowed to reach adequate strength.

Background

Over a period of four years the United States federal government had investigated four construction fatalities at the Pleasants Station power plant at Willow Island in West Virginia. These accidents would be eclipsed on April 27, 1978 by one of the worst construction disasters in North America. On that date, fifty-one workers pouring concrete 170 feet aboveground would fall to their death as the support for the wet concrete failed. Several of the workers had successfully leaped to an adjacent safety net, but still lost their lives as the collapsing portion of the structure violently dragged the safety net with it.

The scope of the Willow Island projects consisted of construction of two coal-fired power plants near the Ohio River in West Virginia. Each of the two units was to have a separate cooling tower. The first cooling tower was constructed without incident. The second cooling tower met with disaster.

Power plants produce excessive or waste heat that is dissipated with the aid of cooling towers that are located nearby and to which they are connected by pipes. This waste heat is absorbed by cold water, and the heat that it picks up is then rejected into the atmosphere by a cooling tower which has air flowing over the heated water. Cooling towers thus create conditions in which hot water evaporates in the presence of moving air. This is known as evaporative cooling.

There are many different types of cooling towers, but this one being built was a fairly common, natural-draught hyperbolic concrete cooling tower. This means that it was a large, chimney-like tower whose distorted hourglass shape allowed air to circulate inside naturally (without fans) since the warmer air inside would naturally rise. At the time of failure, the tower had reached one hundred and sixty-six feet high. The first cooling tower had already been completed and had a base diameter of three hundred and fifty-eight feet with a height of four hundred and thirty feet. Concrete was the material of choice to construct these tall towers since it is strong and can be shaped economically into nearly any shape.

Concrete is composed of a predetermined mixture of cement, sand, stone, and water. Initially, when these ingredients are mixed together, a thick liquid is formed. The cement and water experience a chemical reaction called hydration, which gives concrete its strength. The sand and stone are merely fillers in the mix. Hydration and this strength gain do not occur instantly and actually take several weeks. In general, normal concrete has reached 25 percent of its strength after one day, 50 percent of its strength in five days, and 100 percent in about four weeks. These guidelines are approximate since weather conditions have an effect on the rate of strength gain. If these rates of strength gain are too slow for the contractor, chemical additives can be used to provide early strength gain. These chemical additives, called accelerators, are mixed into the concrete batch. Accelerators help a contractor to get the job done quicker.

After the concrete is mixed, it is poured into steel or wood frames that will produce the desired shape and support the concrete until it gains the appropriate strength. These frames are called formwork and perform a function similar to an ice cube tray. Ice cube trays not only hold water until it hardens to a solid, but can also be reused.

Formwork is very costly. In fact, the formwork costs are two to three times more expensive than the cost of the concrete. For this reason, a contractor will purchase a limited amount of formwork to construct only a portion of the structure and then reuse the formwork for the remainder. For example, in a high-rise structure, new formwork can be used on the first, second, and third floors. After the first floor concrete has reached an adequate strength, the first floor formwork is removed and is reused to construct the fourth floor. In a similar manner, the second floor formwork is reused to construct the fifth floor, and so on. This method is called leapfrogging because of its similarity to the children's game. Leapfrogging saves on formwork costs but can have catastrophic results if the formwork is removed too early. It is also essential that the concrete on the few levels below where the workers are pouring the new concrete be sufficiently hard so that the formwork attachments that are secured to it will not pull out when more concrete (and thus more weight) is added above it.

Details of the Collapse

The formwork system used to construct the Willow Island cooling tower had been successfully employed by the same company to build thirty-six towers in seven years. The method utilized a procedure similar to that of the leapfrogging system.

The walls of the cooling tower were poured in five-foot-high sections called lifts. The cast aluminum forms into which the concrete was poured were used simultaneously on four consecutive lifts. After the lowest lift had obtained adequate strength, the formwork would be slid five feet upward to begin the next lift. The lower three levels of formwork would support the concrete at each level as well as the construction loads from the level being constructed. This process provided for the pouring of one lift each day. The weak link in the chain was the reliance on the three lower levels of formwork attached to relatively new concrete to support the heavy construction loads of the level being added above.

Disaster struck on April 27 while the concrete for the twenty-eighth lift was being placed. The twenty-seventh lift had been placed only twenty hours earlier, providing relatively weak concrete for connections of the formwork to the concrete. At the time of the collapse, a bucketful of concrete was being lifted upward to the twenty-eighth lift. The lifting system as well as the weight of the bucket were supported fully by the formwork. This bucket was very large since it held one cubic yard (two thousand, five hundred pounds) of concrete.

The collapse was observed by workers at ground level. No accounts by workers at the top were available, since every worker at that level perished in the fall. Those on the ground said that the interior work platform peeled away "like someone opening a can" and it, along with the exterior platform, fifty-one workers, and tons of concrete, were pulled inside the tower and collapsed into a heap of jumbled debris. The formwork ripped loose simultaneously in both circular directions and met directly across from the starting point.

Those on the ground said that the interior work platform peeled away "like someone opening a can" and it, along with the exterior platform, fifty-one workers, and tons of concrete, were pulled inside the tower and collapsed into a heap of jumbled debris. The formwork ripped loose simultaneously in both circular directions and met directly across from the starting point.

The branch of the federal government responsible for enforcing safety laws is the Occupational Safety and Health Administration (OSHA) by authorization of the Department of Labor. The Department of Labor requested the expertise of the National Bureau of Standards (NBS), a technical branch of the federal government, to investigate the cause of the collapse. The investigation was complicated because the 170-foot fall left behind an indistinguishable mass of debris. The NBS (now called the National Institute for Standards and Technology) investigation focused on three possible causes of the collapse: 1) Failure of the structure that supported the concrete as it was lifted, 2) failure of the formwork system, or 3) failure of the concrete that supported the formwork.

A structure that lifts and supports the bucketful of concrete is called a hoisting system. The investigation of the hoisting system concluded that the components of the system were not the cause of the failure. Beams, cables, and bolts used in the hoisting system were designed and fabricated properly. Conclusions also vindicated the formwork system.

By process of elimination, all other potential causes of failure were ruled out and suspicion pointed to failure of the connection of the formwork to the concrete. The NBS needed to analyze the structural integrity of the cooling tower by comparing the strength of concrete needed (demand) to the strength of the concrete (resistance) at failure. The strength of the concrete that was needed was determined using sophisticated computer programs. The strength of the concrete at twenty hours (age at collapse) was determined from field records kept by a testing company. Comparison of demand to resistance showed that the twenty-hour-old concrete could not sustain the construction loads. The NBS concluded that the formwork had been moved and reloaded without allowing the concrete to reach its proper strength.

Based on the investigation by the NBS as well as their own, OSHA issued twenty citations against the contractor. Ten of these were classified as "willful," which left the contractor open to potential criminal liability. In October 1978, OSHA recommended to the Justice Department that criminal action be initiated based on the ten willful violations. Some of these included improper anchoring of several parts of the scaffolding system, failure to provide critical instruction to key personnel, failure to make tests of the concrete to determine if adequate strength was obtained, and inadequate training and instructions for employees.

Criminal charges against the contractor were subsequently dropped. Prosecuting attorneys had a heavy burden of proving that the contractor had a reckless disregard for OSHA regulations and safety. Although criminal charges did not stick, the contractor did pay substantial amounts in damages to the families of employees who died as well as a fine imposed by OSHA.

Families of the victims of the collapse were outraged by the dismissal of criminal charges. Many people believed that the OSHA investigation had not gone far enough because OSHA was trying to cover up for failure to enforce their own regulations. OSHA was also accused by West Virginia of being uncooperative with its state commission investigating the accident.

The workers and their families were a tightly knit group. Many of them lived in a trailer park next to the construction site. Almost all families were affected to some degree. One worker on the ground watched in horror as four of his brothers, three uncles, two brothers-in-law and two cousins fell to their death.

The families were disturbed both by the failure of the contractor to follow good technical procedures, and by what they perceived as the sacrifice of safety to economics. Workers were paid for only eight hours for each concrete lift which encouraged them to finish the job as quickly as possible.

Impact of the Collapse

As a result of its investigation, the NBS recommended additional guidelines for existing Occupational Safety and Health Administration regulations. These guidelines stipulated that OSHA should require the use of an independent specialist in cooling tower construction to ensure conformance with their regulations; that test results from concrete samples should be consulted prior to removal of formwork or any other operation that relies on concrete strength; that time of formwork removal and movement should be deter-

The cooling tower collapsed as it was being built when construction loads were placed on concrete that had not been allowed to reach adequate strength. The twisted framework of the scaffolding can be seen inside the structure (right).

mined by an engineer; and that employees should be made aware of all critical aspects of the formwork system.

The Willow Island cooling tower collapse became one of the driving forces for change in safety requirements for construction workers.

In the first half of this century, Congress enacted the Contract Work Hours Standards Act to provide safety and health standards for various types of employees. Congress then amended the act in August 1969 to include employees in the construction industry. This amendment became known as the Construction Safety Act. The purpose of the Construction Safety Act is to provide construction workers with a safe working environment and to reduce the severity and frequency of construction accidents. The Occupational Safety and Health Act authorizes the Secretary of Labor to adopt and enforce the Construction Safety Act. A review of the standards in effect at the time of the collapse however, revealed ambiguities, gaps in coverage, and a lack of details. It became apparent after the accident that changes in the Construction Safety Act were necessary.

OSHA later drafted a revised version of the Construction Safety Act and made it available for comments from industry in June 1988. The new standards became effective in August 1988 and included two major changes.

The first of these changes shifted more responsibility for formwork decisions to the contractor. The old standards had required that the engineer or architect determine when formwork could be removed, thus allowing the contractor to be a spectator rather than a participator in the plan. But the architect and engineer could not be expected to safely make such a decision, since they were generally not aware of the construction procedures at a job site. OSHA determined that it would be safer as well as more practical to place this responsibility on the contractor. The new wording in the standard reads "Forms shall not be removed until the employer determines that the concrete has gained sufficient strength."

The second major change resulted in elimination of what was known as Table Q-1. This table provided the contractor with a schedule for removing formwork, but it lacked the accuracy and additional safety measures that could be obtained by physical testing of the concrete. The new act requires that concrete be tested using recognized methods to determine that adequate strength has been achieved.

Where to Learn More

Department of Labor. *Concrete and Masonry Construction Safety Standards*, subpart Q, 29CFR part 1926, August 15, 1988.

Department of Labor. *Concrete and Masonry Construction Safety Standards, Final Rule*, vol. 53, no. 116, June 16, 1988.

"Form Collapse Kills 51 on Cooling Tower Job." *Engineering News Record*, May 4, 1978, pp. 10-11.

Lew, H. S. "West Virginia Cooling Tower Collapse Caused by Inadequate Concrete Strength." *Civil Engineering*, February 1980, pp. 62-67.

Morrison, Allen. "Willow Island Aftermath: The Limits of OSHA." *Civil Engineering*, March 1980, pp. 68-73.

Ross, Steven. *Construction Disasters*, McGraw-Hill, 1984, pp. 377-88.

Kemper Arena roof collapse

Kansas City, Missouri

(1979)

Roof of award-winning arena collapsed when fatigued bolts failed during a rainstorm.

Background

On the first June weekend of 1979, members of the American Institute of Architects (AIA) began to arrive in Kansas City, Missouri, for the professional association's annual conference. Before the meeting got into full swing on Monday, June 4, many of the visiting architects took advantage of the opportunity to visit the R. Crosby Kemper Jr. Memorial Arena, a structure that had been recognized as one of the nation's finest buildings by the AIA with an award just three years earlier.

Just after the architects' conference concluded for the day on Monday, however, a section equal to about a third of the area of the prestigious building's roof crashed to the floor during a violent rainstorm. The collapse of this high-profile project, only seven months after the spectacular collapse of the roof at another modern arena in Hartford, Connecticut, sent shock waves through the architectural community. Although early speculation blamed a severe wind and rainstorm for the catastrophe, investigations revealed that the collapse was actually caused by a combination of engineering and natural factors.

The $23.2 million, seventeen-thousand-seat Kemper Arena, designed by architect Helmut Jahn of Chicago-based architect-engineer C. F. Murphy Associates, was completed in 1974 and used by the Republican party for its national convention in 1976. The building was widely hailed as structurally innovative, and became an instantly recognizable landmark, largely due to the way in which its 324-by-310-foot roof was constructed. The arena roof was supported not on columns or piers located inside the building, as is typical, but was suspended from three massive "space trusses" shaped like giant, three-dimensional croquet wickets that rose up and over the building.

Those space trusses, which are 81 feet high and span 360 feet horizontally, run in the north-south direction and are spaced 99 feet apart. The horizontal sections of the triangular trusses are made up of tubular steel members as large as 4 feet in diameter. The trusses have three equal 54-foot-long sides, and in each truss, the single top chord and two bottom chords are connected by diagonal elements. The vertical legs of the wicket-shaped trusses are also triangular in cross section. The bottoms of these vertical legs rest on reinforced concrete conical footings that in turn rest on underground piles.

The building was widely hailed as structurally innovative, and became an instantly recognizable landmark, largely due to the way in which its roof was constructed. The roof was supported not on columns or piers located inside the building, as is typical, but was suspended from three massive "space trusses" shaped like giant, three-dimensional croquet wickets that rose up and over the building.

The roof itself was constructed of seven trusses—each composed of two 99-foot long trusses—that ranged in depth from 5 feet to 9 feet, and laid out perpendicular to the space frame trusses above. These seven secondary trusses were connected by a series of long-span joists, spaced 9 feet on center. This network of secondary and tertiary trusses supported a concrete slab roof that was reinforced by metal roof decking.

The roof structure was suspended from the three massive, space frame trusses by forty-two steel pipe hangers, made from ten-inch-diameter pipe. Six of these hangers were connected to each of the seven secondary steel trusses in the roof below. The hangers were bolted at the top to the two bottom-chord members of the three triangular space frame trusses above. The hangers were welded at the bottom to a pair of 1-inch-thick steel base plates, cushioned with a 1/4-inch-thick layer of resin-injected plastic. The plates, which had four holes drilled through them, were bolted to the top flanges of the secondary trusses with four 1 3/8-inch high-strength A490 steel bolts.

High-strength steel performs well under tension, and is appropriate for static or constant loading conditions, but it is a brittle material with low ductility, so it may perform poorly under dynamic loading conditions or when structural movement or bending is involved. At the Kemper Arena, the bolted connections in the hanger system needed to resist two type of stresses: tension stresses that resulted from the weight of the roofing materials, mechanical systems, rain and other loads on the roof; and horizontal stresses imposed by wind blowing against the building walls causing the connections to rock and sway. Although six of the hangers were fastened to the roof structure at the top and bottom with hinged connections, the majority of the hangers were hinged only at the top and were attached with fixed connections at the bottom.

Details of the Collapse

On the evening of July 4, while a rainstorm dropped about four inches of water on Kansas City, the Kemper Arena stood empty except for a maintenance worker and several security guards. The maintenance worker reported that he heard a loud cracking sound at about 7:00 P.M. and walked into the arena to investigate. There he noticed two streams of water pouring down on either side of the scoreboard hanging from the ceiling. As the scoreboard broke loose and a large portion of the ceiling began to give away, the maintenance and security people turned and fled the building. No one was hurt.

Two days after the failure, Kansas City hired James L. Stratta, a Menlo Park, California-based engineer and fellow of the American Society of Civil Engineers who was active in the investigation of earthquake-related failures, to investigate the collapse. It took Stratta only two months conclude the investigation and issue his report.

Stratta attributed the roof failure to fatigue weakening of the high-strength bolts that connected the hangers to the roof trusses. The bolts were reduced to between 20 percent and 25 percent of their original strength by repeated and frequent rocking of the roof under wind loading during the five-year period that the building had been standing. Stratta's report cited data contained in the 1974 *Guide to Design Criteria For Bolted and Riveted Joints*, written by John W. Fisher and John H. A. Struik, a document that describes how high-strength bolts that were tightened and loosened five times failed at one-third their original design load.

Stratta also noted that the bolts—installed by Kansas City-based steel subcontractor Kansas City Structural Steel, working under general contractor J. E. Dunn Construction Company, also of Kansas City—may not have been tightened sufficiently during construction to bring the metal plates into full contact with the steel trusses. Two connections that were tested at the University of Missouri at Columbia after the collapse showed that a torque or twisting load of just 400 to 750 pounds was needed to loosen the bolts rather than the 2,500 plus pounds that would have been required if the bolts had been installed correctly.

The Kemper Arena roof collapse began when two of the bolts in a connection at a central hanger in the southern half of the roof failed at between 30,000 and 40,000 pounds each, rather than at the 160,000 pounds they were expected to carry. After those two bolts failed, the load they had carried immediately was shifted to the other two bolts in the plate in the same connection. Those two bolts were suddenly subjected to double the tension stresses as well as new bending stresses imposed when the steel plate they connected began to move, and they also quickly failed. Once the first hanger failed, the load it had carried was shifted to adjacent hangers. Those hangers were not able to carry the extra load and quickly failed.

The fatigued bolts finally snapped when a greater-than-anticipated pool of water accumulated on the Kemper Arena roof. The 129,000-square-foot roof had been constructed with only eight 5-inch-diameter drains, designed so that the building would be able to collect and hold rainwater during a sudden downpour and then release it gradually into the city's sewer system instead of releasing it quickly and overloading the system. To prevent excessive water buildup on the roof, scuppers (openings) were created in the arena's exterior parapet walls two inches above the top of the roof so the water could simply run out and down the sides of the arena if it reached that level.

The rainwater that fell on June 4, however, was pushed to the south side of the arena roof by wind gusts of up to seventy miles per hour. In addition, an atmospheric condition called the Bernoulli effect created a suction that

The roof of the award-winning Kemper Arena collapsed when fatigued bolts failed during a violent rainstorm.

allowed even more water to accumulate on the south end of the roof. As the amount of water on the roof increased, the roof structure began to deflect, allowing more water to build up before it reached the height necessary to drain out of the scuppers in the parapet wall. Eventually, the water on the roof reached a depth of nine inches near the drains.

Impact

Immediately after the investigation results were reported, Kansas City officials began preparations to repair and reopen the Kemper Arena. The three primary exterior space trusses that supported the roof from above were not damaged in the collapse, and officials opted to reuse them in the reconstructed arena.

The new roof was again hung from the three space frame trusses, but each hanger was created from a single 2 1/4-by-6 1/2-inch steel bar 10 1/2 feet in length. The hanger bars were fastened to the bottom chords of the space trusses above with a steel pin connection. The bar in turn was passed down through the secondary truss to the bottom chord and was welded to the truss diagonals. The new, more ductile connections were designed by Stratta working with Bob D. Campbell & Company, a Kansas City-based engineering company. The connections, which were tested by the University of Missouri, were shown to be able to carry 600,000 pounds each, well over the 210,000 pounds that the original connections were designed to carry even before fatigue. In addition, the center of the roof was raised by thirty inches to improve water runoff, and the number of drains was increased from eight to twenty.

The Kemper Arena reopened in March 1980, looking very much as it had when it first opened six years earlier, after the $5.2 million reconstruction job was completed. Two months later, Kansas City charged C. F. Murphy Associates Inc. with twenty-two counts of negligence, including failure to design for appropriate wind forces, water ponding, and drainage. The city also charged Kansas City Structural Steel Company with four counts, including improper tightening of the bolts. The suit claimed $8.97 million in damages and lost revenue. That lawsuit, as well as additional suits filed against Kansas City by a circus, an ice show, and a souvenir vendor—all alleging loss of income and breach of promise for canceled contracts—were settled just two days into a court trial in the fall of 1983.

The Kemper Arena collapse raised concern within the architectural and engineering communities over the use of A490 high-strength bolts in light of their apparent lack of resistance to dynamic loading (loadng from moving entities such as water and wind). After the failure, a number of engineers reported checking structures to make sure that the connections that used the bolts were functioning as had been anticipated, even in situations in which dynamic loading was not expected to be a factor. Largely because the Kemper Arena roof failure did not cause any injuries or loss of life, the incident was virtually forgotten soon after the facility was repaired when Kansas City was struck by a far more devastating disaster: the collapse of two overhead walkways at the Hyatt Hotel that killed 113 people.

VIRGINIA KENT DORRIS

Where to Learn More

"Failed Bolt Connections Bring Down Arena Roof." *Engineering News-Record*, June 14, 1979, pp. 10-12.

Goldberger, Paul. "Kansas City Arena Loses Roof in Storm." *New York Times*, June 6, 1979, p. 1.

"Kemper Arena Reopens with New Roof Connections." *Engineering News-Record*, March 13, 1980, p. 18.

Levy, Matthys, and Mario Salvadori. *Why Buildings Fall Down*. W. W. Norton, 1992, pp. 57-67.

McQuade, Walter. "Why All Those Buildings Are Collapsing." *Fortune*, November 19, 1979, pp. 58-62.

"Prizewinning Arena Collapses." *Time*, June 18, 1979, p. 74.

Ross, Steven S. *Construction Disasters*. McGraw Hill, 1984.

Alexander Kielland oil-drilling rig collapse

North Sea

(1980)

Faulty welding was responsible for a crack in one of the legs supporting an offshore oil-drilling rig; the crack led to the collapse of the leg and the fatal capsizing of the rig.

Background

In 1976, the French-built *Alexander Kielland* drilling rig was completed. The drilling rig was to travel from site to site in the North Sea in an attempt to locate oil beneath the sea floor. If a test drill well was successful and oil was located, the semisubmersible drilling rig would cap the drilling hole and make way for a permanent drilling rig and continue its nomadic search for more oil. Weighing in excess of twenty million pounds, the drilling was designed to be semisubmersible. A semisubmersible drilling rig is towed on top of pontoons. On reaching its destination, the pontoons are filled with water, sinking a portion of the rig under water, making it more stable.

The *Alexander Kielland* drilling rig resembled a five-legged stool. The five legs of the rig were constructed of steel members that were each supported on a pontoon. A 130-foot-tall steel tower was supported on the stoollike platform. This tower was designed to support the oil-drilling mechanism.

Before it even left the fabrication yard, the *Alexander Kielland* was considered obsolete, since more advanced rigs were already operating in the North Sea. But rather than scrapping the drilling rig, the owners (Stavanger Drilling, a Norwegian company) decided to modify it to become a "flotel." A flotel is a floating hotel. Drilling crews need rest and relaxation after completing their rigorous twelve-hour shift. Transportation to land was impractical, and the constant transient search for oil dictated living arrangements that were mobile. The *Alexander Kielland* was retrofitted to accommodate the living requirements of up to 366 workers. Even though the flotel lacked architectural beauty, it offered the comforts of a plush hotel. A movie theater, game rooms, and saunas were a few of the comforts offered to the workers.

The structure was eventually leased by the Phillips Petroleum Company. On March 27, 1980, the *Alexander Kielland* was semisubmerged and anchored to the floor of the North Sea, halfway between England and Norway. The North

Sea was rough on that day, with winds in excess of sixty-five miles per hour and twenty-foot-high waves, but this was far less ferocious than this sea could deliver. At half past six in the evening, some of the crew were dining while others were engaged in various forms of recreation. Twenty minutes later, the *Alexander Kielland* would be floating upside down, two-thirds of its predominantly Norwegian crew killed.

Details of the Disaster

After two loud banging noises followed by an ear-shattering cracking sound, the *Alexander Kielland* started leaning precariously to one side. A large wind gust soon completed the overturning. Only the workers who were near exits and made a decision to abandon ship had a chance of survival. Many originally thought that the leaning had been initiated by sabotage or a collision with a submarine. These notions were soon dispelled as it became apparent that it was caused by the structural failure of one of the five supporting legs. With one supporting leg gone, the oil rig became an unstable four-legged structure.

The response to the distress signal of the *Alexander Kielland* was immediate. Nearly fifty ships, supplemented with the aid of helicopters, searched for survivors thrown from the rig. The rescue parties represented at least five different nations. Rescuers knew that time was of the essence since the frigid waters with temperatures below fifty degrees Fahrenheit would kill a person within twenty minutes.

Before it even left the fabrication yard, the Alexander Kielland *was considered obsolete, since more advanced rigs were already operating in the North Sea. But rather than scrapping the drilling rig, the owners decided to modify it to become a "flotel," or a floating hotel.*

After the initial search for those thrown into the sea, the search turned to looking for lifeboats as well as searching for survivors who might still have been trapped in the capsized drilling rig. Cameras lowered into the sea showed that all the windows had been broken in the violent overturning and no air pockets existed to sustain any trapped worker. Eighteen hours after the capsizing of the rig, the search for survivors ended with 123 of the 212 workers dead.

Investigation of the sunken oil rig pointed to failure of a strut that braced one of the five legs. When the strut failed, the leg that depended on it also failed. Survivors of the disaster said they could see the broken leg floating away as they plunged into the sea. This caused the entire drilling rig to become unstable. Ironically, the *Alexander Kielland* was to have been taken out of service two months earlier. However, delays in the construction of the *Henrik Ibsen*, its replacement, provided the *Alexander* with an extended life that led to disaster.

The *Alexander Kielland* was towed in its capsized position to the coast of Norway. It sat at that location for more than three years while debates over its fate continued. Families of victims whose bodies were never recovered requested the drilling rig be uprighted before being demolished. Several attempts at righting the drilling rig were unsuccessful. Approximately three years after its collapse, the *Alexander Kielland* was successfully uprighted by an elaborate and expensive plan (the Norwegian government spent $34 million on this final attempt alone). Surprisingly, only six of the thirty missing bodies were found. In November of 1983, the *Alexander Kielland* was towed to sea and blown up; it sank to the sea floor a quarter mile below.

The *Alexander Kielland* disaster happened two weeks after the British government released a report about concern of the structural integrity of these drilling rigs. Concerns centered around the structures' ability to withstand the constant pounding of the waves. The *Alexander Kielland* sank before any action could be taken to remedy the problems addressed in the English report.

The five legs of the *Alexander Kielland* were braced with steel struts. These struts allowed for distribution of the load forces to all the legs. This load distribution system can be understood by examining the construction of a bar stool. A bar stool constructed of four legs that are in no way interconnected will tend to be rickety and unstable. If the legs are not independent but instead are interconnected by short members between the legs, the stool is no longer unstable. These short members are struts and their function is to have the legs act as a unit. The struts between the legs of the *Alexander Kielland* performed the same function. The *Alexander Kielland* capsized when one of its legs collapsed. The leg collapsed when the strut that braced it pulled loose. The strut failed as a result of a poor quality weld being subjected to the repetitive loading of the ocean waves.

Welding provides an efficient means of rigidly connecting two members together. Great care must be taken when using welded connections. Poor quality welds may induce cracks in the members or leave imperfections in the weld itself. The basic procedure for obtaining quality welds includes the four following steps:

1. Members to be welded must be properly fabricated so that they fit snugly against each other.

2. Surfaces of the member to be welded must be preheated and the proper welding material must be used. These decisions must be made by a welding specialist and serve to reduce the possibility of cracking during welding and during the service life of the members.

3. Proper welding techniques must be followed by workers to avoid substandard welds.

4. The welded area must be inspected to guarantee that the weld is of acceptable quality and that there are no cracks. Inspection can be done with visual techniques such as magnetic particles or dye penetrants. These materials align themselves in a crack making detection of surface cracks possible. However, inspection systems making use of ultrasound and X-rays are much more reliable for detecting flawed welds.

Preexisting cracks in structural members and welds will not necessarily produce a failure. The crack leads to failure when it is subjected to repetitive

The four remaining legs of the Alexander Kielland *are shown here (upper left); the fifth supporting leg collapsed because of a fatigue crack. The rig capsized, killing 123 of 212 on board.*

loads. These repetitive loads, known as fatigue loads, will increase the size of the crack with each cycle of loading until the crack has grown to a catastrophic size. The effects of fatigue loading can be demonstrated by taking one of the ends of a paper clip, bending it, and then pushing it back to its original position. The bending will not cause failure of the paper clip unless the bending is continued through many cycles. This is known as fatigue failure. If cracks existed in the paper clip prior to bending it, the number of cycles to failure would be reduced significantly. A member with a preexisting crack from a poor-quality weld coupled with the repetitive loading of the sea waves would have a very limited life.

Although the *Alexander Kielland* was never used as an oil-drilling rig, it was fully outfitted to perform that function. It is important that an oil-drilling rig be properly aligned with the well cap at the bottom of the sea. This is accomplished by installing a hydrophone on the drilling rig. A hydrophone is an electronic device that receives signals from the well cap located on the sea floor. These signals assist the drilling rig in determining its proper position. The hydrophone was attached to the *Alexander Kielland* by cutting a fourteen-inch-hole in the leg strut and welding on a supporting plate. The weld used to attach the support plate was poorly done as well as having insufficient strength. The welding produced a three-inch-crack in the strut before fabrication was ever completed. This was evidenced by paint being found on the inside surface of the crack, which would not be the case if the cracks had developed while the *Alexander Kielland* was at sea.

The crack soon propagated after the oil rig was put to sea. Each wave of the sea water pounded the structure, and increased the size of the crack. When the crack reach two-thirds of the circumference of the strut, it failed, initiating the capsizing of the drilling rig.

It is interesting to note that in 1961, an offshore structure used as a United States Air Force radar station collapsed. The members comprising the tower (the "Texas Tower") were assembled with boltlike connections rather than welded connections. Part of the reason this tower collapsed was that the force of the water caused the bolt holes to enlarge, allowing movement in the joints of the structure. Because of this, designers switched over to welded connections of the type used in the *Alexander Kielland*.

Impact of the Collapse

The capsizing of the *Alexander Kielland* confirmed suspicions of researchers that the design and construction of drilling rigs was seriously inadequate. England's Department of Energy released revisions to its drilling design rules one year after the *Alexander Kielland* disaster. This revised code required design changes for new oil-drilling rigs as well as a reevaluation of existing rigs. The code changes addressed the reality that although drilling rigs were designed to withstand two million cycles of loading from the ocean waves, the drilling rig would probably be subjected to fifty times this amount.

Other rule changes included that all semisubmersible drilling rigs meet inspection and other requirements used for seagoing vessels. Had this requirement been in place, the three-inch crack in the strut most likely would have been found prior to launching of the *Alexander Kielland*.

Where to Learn More

Apple, R. W., Jr. "137 Feared Lost in Collapse of North Sea Platform." *New York Times*, March 29, 1980, p. 1.

Apple, R. W., Jr. "Loss of North Sea Rig Is Traced to Mysterious Crack." *New York Times*, April 6, 1980, p. 4.

"Kielland Collapse at OTC." *Engineering News-Record*, May 15, 1980, p. 14.

"Test Reveals Faults in Offshore Design Methods." *Engineering News-Record*, March 5, 1981, pp. 14-15.

"Weld Fractures Capsized Hotel Rig." *Engineering News-Record*, April 6, 1981, p. 51.

Henskey, R. D., Jr. "Scores Missing as a North Sea Oil Platform Collapses." *New York Times*, March 28, 1980, pp. 1-3.

Lewis, P. and I. Mather "Another Strike for the Cruel Sea." *Maclean's*, April 7, 1980, pp. 29-30.

Nielson, J. and J. LeMoyne. "Death on an Oil Rig," *Newsweek*, April 17, 1980, p. 49.

MGM Grand Hotel fire

Las Vegas, Nevada

(1980)

Second most deadly hotel fire in United States history led to a nationwide reevaluation of fire codes.

Background

Sparks from a short-circuit started a fire in the MGM Grand Hotel in Las Vegas, Nevada on November 21, 1980. Despite having passed inspections, the building's fire protection systems could not prevent the fire from engulfing the world's largest gambling hall in smoke and flames. Thick black smoke filled the air ducts and escape stairwells in the twenty-one floors of guest rooms. Scenes reminiscent of Hollywood disaster films became reality as guests were awakened by screams and smoke. Some raced to the roof or ran down smoke-filled stairwells. Others called for help or huddled by broken windows as smoke billowed into their rooms. Eighty-five people died and over six hundred were injured, primarily due to smoke inhalation rather than the fire itself. The disaster dramatically accelerated the updating of fire code regulations for both new and existing high-rise buildings.

The MGM Grand Hotel is an enormous facility, including a casino, two 1,000-seat theaters, forty shops, five restaurants, 2,076 guest rooms, a jai alai fronton, and a sports arena. A "T" shaped tower of guest rooms rises above two entertainment levels, each as large as twenty football fields. The fire and its impact were concentrated near the casino on the upper entertainment level and in the guest tower. Of the fire protection systems designed and built into the MGM Grand, the four most important during the fire were the building egress system, the fire suppression and alarm system, the system of fire zones, and the heating, ventilation and air conditioning (HVAC) system.

The hotel's building egress system was intended to provide safe routes of passage out of the building to safety. People in the casino and adjacent spaces could exit the building directly onto the street through several large sets of doors. Hotel guests could exit through one of the two stairwells in each of the three branches of the T-shaped guest tower.

The fire suppression and alarm system was designed to detect and control a fire. The fire suppression system was based on automatic water sprinklers located in selected portions of the building; the heat of a fire would set off these sprinklers. The minimum guidelines of the fire code in effect at the time of hotel's design did not require that the entire facility be equipped with sprinklers. Thus, the casino, deli, most floors in the guest tower, and many other areas were not protected by sprinklers. An especially important area that was not protected was the casino's "eye in the sky." This network of walkways above the casino's ceiling allowed security personnel to monitor gambling operations unnoticed. The alarm system was designed to alert the hotel's security office of an outbreak of fire. The security office would then verify the alarm and use the building's alarms and public address systems to alert occupants to any danger.

A system of fire zones isolated distinct sections of the hotel and casino facility to prevent the spread of fire and smoke and to protect occupants and property. Fire zones are formed by enclosing parts of a building with fire-resistant construction. These enclosures must effectively seal off the many openings between building parts and any gaps between construction materials through which fire and smoke could pass. Important openings in fire zone enclosures at the hotel include access panels into elevator and air shafts, structural joints for building movement, and the junctures of structural steel and drywall.

The heating, ventilation, and air conditioning (HVAC) system at the hotel was designed to stop the flow of air into the building during a fire. Fire dampers—hinged louvers similar to those on attic exhaust fans in many residences—are fundamental to this strategy. The fire dampers in large HVAC systems are typically installed where air crosses fire zone enclosures and adjacent to air distribution fans. Fire dampers are usually held open by a link designed to melt in the heat of a fire, thereby closing the louvers and stopping the flow of air. In this way, fire dampers allow the passage of air through the building under normal circumstances while maintaining the integrity of fire zone enclosures.

Three factors caused the fire protection systems at the MGM Grand Hotel to perform poorly during the November 21 fire: the original construction did not fulfill the intent of the designers; the contents of the casino level were unusually dangerous in a fire; physical and operational modifications since the opening of the hotel had reduced the effectiveness of the fire protection systems.

The construction deficiencies impaired the building's ability to suppress a fire. Potential smoke paths across fire zone enclosures were left open, or were covered with materials that could not resist fire or stop smoke. Moreover, materials with inadequate flame-spread and smoke ratings were used to build ceilings and attic areas, and the elevator shafts were not adequately vented. The decorations, furnishings, and finishes on the casino level contained high amounts of synthetic materials. These materials can easily catch fire and burn, and produce large amounts of smoke. Some fire dampers had been bolted open or had their "melt-down" links replaced with metal wire, preventing the dampers from functioning. In addition, several areas designed for 24-hour use—spaces where the fire code did not require sprinklers, on the assumption that any fire would be detected quickly—were in fact closed down during low-use periods. One such area was the deli, where the fire started.

The MGM Grand Hotel was thus a facility whose faulty and inadequate fire protection systems had been crippled by modifications. Despite the numerous flaws in its fire protection systems, the hotel was not in gross violation of the fire design and construction standards that prevailed at the time. The building underwent all required fire reviews and inspections before opening, and passed a fire inspection just six months before the disastrous fire.

Details of the Fire

About 7:10 A.M. on November 21, 1980, an MGM Grand Hotel employee noticed sparks coming out of a gaming board in the closed-down deli. A chef tried to extinguish the fire, but soon was forced to flee the area. The fire fed on fresh air supplied by the HVAC system and the abundant combustible materials, generating smoke that was laden with unburned fuels. These flammable gases collected in the "eye in the sky" above the casino, and were infiltrating the casino space itself as the fire department entered. Fire fighters later reported that they noticed a stratified layer of smoke six to eight feet from the ceiling. Within approximately twelve seconds, the smoke had dropped to four feet above the floor, as the fire threw open the glass doors at the west end of the hotel and took on the appearance of a fire storm.

The inadequacies in the hotel's fire protection systems aided the progress of the fire. The fire incapacitated the alarm system, preventing those in the tower from receiving any warning. The fire zone enclosures failed, allowing smoke to enter the hotel's air ducts, elevator shafts and stairways and migrate throughout the building. The HVAC system, rather than depriving the fire of oxygen, provided it with fresh air. The chimney effect created by the vertical passageways and the HVAC system drew the smoke quickly up into the guest tower hallways and rooms. The building egress system, however, proved to be the most deadly factor. The only path of exit from the guest rooms was down the stairwells to the street. Once hotel guests entered these stairwells, self-locking doors prevented them from returning to the hallways. Thus, as smoke filled the stairwells, many guests became trapped. Only the automatic sprinkler system performed as designed, preventing the fire from spreading into the guest tower or beyond the casino-level areas that were not equipped with sprinklers.

Most of the guests and employees on the casino level were able to flee out of the building ahead of the fire. Some guests in the hotel tower, rather than attempting to evacuate down the stairwells, escaped to the roof. Many other guests remained in their rooms and broke or opened windows to gain access to fresh air. The fire department was able to control the fire by around 8:30 A.M., and evacuated the majority of the survivors by 11:00 A.M.

The decorations, furnishings, and finishes on the casino level contained high amounts of synthetic materials—materials that caught fire easily and produced large amounts of smoke.

Impact

The most important impact of the MGM Grand Hotel disaster was the nationwide revision of local fire codes. The fire brought two issues to the forefront of fire code discussions: (1) the danger of smoke over and above that of fire, and (2) the failure of fire protection systems to evolve with fire protection capabilities.

Given that over ninety percent of those who died in the disaster were overcome by smoke rather than fire, several communities increased the emphasis on smoke control in their fire protection regulations. These changes represented a shift in priorities from saving buildings to saving lives. Smoke detectors were developed that would signal the HVAC system to exhaust air out of a fire area and pressurize the surrounding spaces to contain the circulation of the smoke. Although fire protection professionals are not in complete agreement over the feasibility of all smoke control methods, many cities now require that all high-rise buildings be equipped with smoke control systems.

Fire-protection systems in buildings fail to evolve with fire protection capabilities for two reasons: the continuous improvement of fire protection

Helicopters were used to rescue nearly 1,000 people from the fast-moving MGM Grand Hotel fire.

knowledge and technologies, and the long life of buildings. In the eight years between the design and construction of the MGM Grand Hotel and the November 21 fire, for example, the following advances in fire protection for high-rise buildings were incorporated into local fire codes: smoke detectors, refuge centers, direct connection of fire alarms to fire departments, communication systems for use by firefighters during a fire, and smoke exhaust systems. Thus, the extended life-span of buildings causes many facilities to remain in use long after their fire protection systems are outdated.

Fire codes can require existing buildings to be upgraded to comply with newly adopted guidelines, but such retroactive codes, while increasing safety, carry large costs for building owners and communities. The higher cost of "retrofitting" and operating high-rise buildings can ultimately reduce the attractiveness of a city to developers and businesses. In 1980, few fire codes were retroactive. Since the MGM Grand fire, several cities have adopted retroactive codes. The legality of several of these fire code provisions has been challenged in court.

Nonetheless, the MGM Grand Hotel fire illustrates how a technological disaster can act as a catalyst for societal change. Such disasters often lead to a reevaluation of the relative values placed upon different aspects of the community—human life, buildings and facilities, the prospect of economic expansion. The nation's second most deadly hotel fire provoked a flurry of discussion about alternative approaches to and priorities within fire protection. The fire code reforms that resulted point to the large-scale political significance of the MGM Grand Hotel disaster.

DAVID N. FORD

Where to Learn More

Brannigan, Vincent M. "Record of Appellate Courts on Retrospective Fire Safety Codes." *Fire Journal,* November 1981, pp. 62-72.

"MGM Grand Hotel Victim of Old Code." *Engineering News Record,* November 27, 1980, pp. 10-11.

"MGM Fire Rekindles Smoke Debate." *Engineering News Record,* December 4, 1980, pp. 14-15.

"Compromised Codes Threaten Life Safety." *Engineering News Record,* January 29, 1981, pp. 26-30.

"MGM Fire: Short Circuit to Disaster." *Engineering News Record,* June 11, 1981, pp. 13-14.

"Cities Stiffen Fire Codes." *Engineering News Record,* January 14, 1982, pp. 16-7.

"Fire at the MGM Grand." *Fire Journal,* January 1982, pp. 19-32.

Harbour Cay condominium collapse

Cocoa Beach, Florida

(1981)

A five-story concrete residential building collapsed during construction owing to a combination of design errors and construction problems.

Background

On Friday, March 27, 1981, shortly after 3:00 P.M. as concrete was being poured on the roof, a five-story condominium building under construction collapsed suddenly without warning, killing eleven construction workers and injuring another twenty-three workers. Within a matter of minutes the 242-by-52-foot structure was reduced to a pile of rubble, leaving thirty-four of the thirty-six construction workers in the building dead or injured. So complete was the collapse that the only portion of the original structure left standing was the frame of a stairwell and scattered columns projecting above the debris.

The building, part of the Harbour Cay development located in Cocoa Beach, Florida, was to be the first of a series of apartment buildings within the complex. The project was being built during a construction boom within Brevard County as the market for retirement condominiums was exploding. Investigation of the collapse identified a number of major design flaws along with several construction defects that compounded and eventually led to the devastating collapse. It was also apparent that the level of construction activity in Brevard County far outpaced the ability of the local building department to control and monitor the design and construction of a project of that size.

The building's owner-developer, Univel of Cocoa Beach, also served as the general contractor. The design team consisted of local architect William Juhn and a retired NASA engineer, Harold A. Meeler. Meeler, who was hired by Univel through an oral agreement, in turn hired Augustus M. Allen, another retired NASA structural engineer, to perform structural calculations.

The $5.5 million project was a cast-in-place concrete structure with plan dimensions of 242 feet by 58 feet. The five-story structure, supported on piles, did not have any basement. Column spacing in the long direction varied from 23 feet 4 inches to 27 feet 8 inches on center, and in the short direction from 18 feet 4 inches to 22 feet 2 inches, resulting in a plan configuration of nine bays by three bays. A cantilever balcony projected over the west face of the building. The floor framing consisted of an 8-inch-thick flat-plate steel reinforced concrete slab with a story-to-story height of 8 feet 8 inches. The slab was tapered down to 6.5 inches at the balcony. This is a common framing system for residential construction because it is relatively inexpensive and allows for a very thin floor construction.

Details of the Collapse

From eyewitness accounts and subsequent investigations it is apparent that the collapse originated at a slab-to-column interface somewhere in the center of the fifth floor slab. As the slab cracked at one interior column, the redistribution of forces within the flat slab overstressed the adjacent slab-column interfaces, and a progressive collapse occurred across the fifth floor. As the fifth floor slab cracked and fell below to the fourth floor, it overstressed the fourth floor slab, and the process was repeated until each level formed a pile of rubble on the ground. The cause of the initial failure is described by the term *punching shear*. A number of factors combined to result in a top slab that did not possess adequate strength.

The cause of the initial failure was punching shear. A slab without adequate strength will allow a column to punch up through, much like a sharp pencil punching through a taut sheet of paper above it. Instead of the floor being too heavy or thick for the columns to hold it up, in this phenomenon the floor is too thin, and the columns pop through it.

Punching shear around the perimeter of the supporting columns is a primary mode of failure of a flat-plate reinforced concrete slab. A slab without adequate strength will allow a column to punch up through, much like a sharp pencil punching through a taut sheet of paper above it. Instead of the floor being too heavy or thick for the columns to hold it up, in this phenomenon the floor is too thin, and the columns pop through it. The narrower the dimensions of the columns, the stronger the slab must be to resist this tendency of the column to punch through. The three primary factors determining the capacity of the slab-column interface to resist punching shear are the compressive strength of the concrete (this is its capacity to resist a crushing force, and is measured in pounds per square inch that can be resisted without failing), the effective depth of the concrete slab measured from the top layer of reinforcing steel in the slab to the bottom surface of the slab, and the size of the columns.

In the Harbour Cay project, the concrete columns supporting the fifth floor measured ten inches by eighteen inches, and the specified concrete compressive strength was four thousand pounds per square inch. The reinforced concrete structure was to be designed in accordance with the *Building Code Requirements for Reinforced Concrete* (ACI 318-77), as published by the American Concrete Institute (ACI). A slab designed according to ACI requirements for bay size, floor loads, column dimensions, and concrete strength for this structure should have been eleven inches thick instead of the actual eight inches. From the examination of the original design calculations after the collapse, it was evident that the engineers failed to check their design for punching shear.

We've mentioned that the dimension between the bottom of the slab and the reinforcing steel within the slab has a key role to play in the capacity of the slab to resist punching shear. The design for this project called for the steel to be located 6.3 inches from the bottom of the slab. Bent metal wire supports, known as metal chairs, are normally used to set the reinforcing steel at the correct distance from the bottom of the wood form before placement of concrete. In this case the chairs used by the concrete contractor were set too low, resulting in a height of approximately 5.3 inches from the bottom of the slab to the reinforcing steel. Also, in some areas, instead of providing an 8-inch-thick slab, the contractor only poured a 7.5-inch slab. Thus, a slab that was significantly underdesigned due to design errors was further weakened by construction faults.

In accordance with the building code requirements in Florida, the structural engineers submitted drawings and calculations for review and approval of the building department. Seventy-nine sheets were submitted. Although Meeler signed and sealed each sheet, by his own admission he only prepared a single calculation sheet related to the foundation design and did not thoroughly review the other calculations. Further, Cocoa Beach city engineer Raymur H. Murray was quoted as saying, "In this case, I skimmed through the calculations and felt that the engineer had taken an adequate approach to the design, so I accepted the calculations as signed and sealed by a professional engineer." To compound matters, the city relied on Meeler for construction inspections without performing any of its own. Prior to the collapse, extensive cracking and excessive deflections or bending of the floor slabs were noticed. These were brought to the attention of Meeler and Allen during their field inspections; however, their construction reports filed with the building department did not mention the cracks.

Impact of the Collapse

An unfortunate series of errors, oversights, and miscalculations accumulated in this case to result in a construction collapse that cost eleven workers their lives. But the tragedy raised two important issues.

First, if the design had followed the applicable design code and the contractor had then built according to the design drawings, the collapse would not have occurred. This collapse was caused purely by human error, it was not an act of nature.

Second, why should the failure of one element so easily have propagated across a structure and resulted in a total collapse with eleven deaths and twenty-three injuries among the thirty-six construction workers on the site? Structural engineers know very well the brittle nature of concrete when improperly designed. Unlike other materials, badly designed concrete can lead to a brittle and sudden failure of the sort observed in this case. Qualified structural engineers know how to properly detail and reinforce concrete so that in extreme situations it can undergo large deformations without fracture. This property, called ductility, is required to prevent sudden failures, and is of primary importance to a well-designed structure. These deformations offer a warning to those nearby that something is wrong, and warnings can mean the difference between life and death. The previous analogy of a sharp pencil punching through a taut piece of paper would represent a very brittle failure. If we replace the paper with a thin piece of (ductile) plastic, we would expect the plastic to stretch and deform before allowing the pencil to punch through.

In its August 1982 report documenting its investigation of this failure, the National Bureau of Standards (NBS) concluded that, "a check for punching shear capacity omitted in the design calculations resulted in a slab too thin for the design loads." It also was critical of the nonuniform consistency of the concrete produced onsite. The NBS researchers, however, confined their judgments to matters that were solely technical, and did not comment on the obviously implicit issues concerning culpability and liability.

After the collapse, Florida's building officials instituted a new requirement for the construction of structures more than three stories tall. The mandate states that a professional engineer, called a "Threshold Inspector," must be on site while structural work is being performed.

The Harbour Cay condominium was not rebuilt, but another condominium now stands on the site.

LEO ARGIRIS AND ERIN A. ZENO

Where to Learn More

U. S. National Bureau of Standards. *Investigation of Construction Failure of Harbour Cay Condominium in Cocoa Beach, Florida.* August 1982.

"Collapse of Flat Slab Condo Kills 11 Construction Workers." *Engineering News-Record,* April 2, 1981, pp. 11-12.

"Florida Collapse Blamed on Design." *Engineering News-Record,* June 18, 1981, pp. 39-40.

"Collapse Report Cites Shear." *Engineering News-Record,* October 22, 1981, pp. 13-14.

Montgomery, Paul. "10 Are Dead and 14 Injured in Florida Building Collapse." *New York Times,* March 28, 1981, p. 8.

Hyatt Regency Hotel walkways collapse

Kansas City, Missouri

(1981)

The collapse of two suspended walkways caused by a simple design change resulted in the worst structural failure to date in the United States.

Background

On the evening of July 17, 1981, some 1,500 people had gathered in the glass-walled, four-story atrium at the Kansas City Hyatt Regency Hotel for the hotel's weekly tea dance. As the Steve Miller Orchestra played swing music, some attendees danced on the floor of the open lobby while others watched from three pedestrian walkways that crossed the space overhead. Just as the band struck up Duke Ellington's "Satin Doll," a sound like a clap of thunder rang out as two of the overhead walkways collapsed, showering glass, concrete, steel, and people down on the dancers below. It took rescue personnel using heavy equipment twelve hours to reach the last victims trapped under debris. The collapse, which has since become known as one of the greatest structural tragedies in the history of the United States, killed 114 people and injured nearly 200 others.

The Hyatt walkways collapse was especially disturbing because it was caused by the failure of a connection later described by engineers as simple enough to be used as a design problem in an undergraduate engineering school class. The connection that failed was not built as originally shown in architectural drawings, but instead was constructed according to shop drawings, prepared by the steel fabricator, that depicted an alternative design. Investigators blamed the failure on that design change, which effectively doubled the stress on the connection.

The tragedy focused new attention on the design of engineering details by fabricators and materials suppliers and raised questions about the responsibility of a project's engineer-of-record for the oversight of such details. Two engineers eventually lost their professional licenses as a result of their role in the walkway collapse.

The Kansas City Hyatt Regency, designed by the joint venture architect Patty Berkebile Nelson Duncan Monroe Lefebvre Architects Planners and the structural engineering firm of Gillum Colaco Associates, opened about one year before the walkway collapse. The hotel included a forty-story guest-room tower and a four-story wing containing restaurants and meeting rooms; these structures were connected by a large, open atrium lobby. Dramatic atriums had become a trademark of Hyatt hotels during the preceding ten years, following a trend begun with the opening of architect John Portman's Atlanta Hyatt Regency, which contained a spectacular twenty-two-story-high open atrium.

The Kansas City Hyatt atrium contained three hanging walkways, each 120 feet long—consisting of four spans of 30 feet each—and slightly more than eight-and-a-half feet wide. The walkways spanned the atrium at the second-, third-, and fourth-floor levels. Each walkway was connected to balconies at either end. The walkways' box beams were welded to embedded plates to form a rigid joint on the south end while the north end connection was constructed as an expansion joint; sliding bearings allowed the beams room to expand or contract in reaction to changes in temperature without causing stress or cracking.

The walkways were constructed of lightweight concrete on metal decking. Each walkway was supported from underneath along its length by 16-inch-deep, wide-flange steel beams, and across its width by 8-inch-deep steel box beams, welded beneath the walkways at 30-foot intervals. The hanging rods that suspended the walkways from the roof supports attached to both ends of the box beams at these seams. The box beams were made from pairs of 8-inch steel channels that were welded together at their open ends. The hanging rods, which were also made of steel, measured 1 1/4 inches in diameter.

The fourth- and second-floor level walkways were constructed one above the other along the west wall of the atrium, while the third-floor level walkway was built about thirteen feet away toward the center of the room. To support the weight of the walkways, the ends of the hanging rods were passed through holes drilled through the welded box beams and secured from below by a washer and nut. No stiffeners, plates, or other reinforcements were used to further secure the connections. Each walkway was hung from three pairs of steel rods.

The design change that occurred between completion of the original design drawings and construction of the hotel concerned the way in which the walkways were connected to the steel hanging rods. The original drawings called for both the fourth- and second-floor level walkways to be hung from the same continuous steel rods. The rods, which would have been about forty-five feet long, would have been attached at the top to the atrium ceiling, passed through the box beams at the fourth-floor level walkway, and continued down to pass through the box beams at the second-floor level walkway. The weight of each walkway would have been supported by the nut-and-washer connection under the box beams at each level.

Instead of using continuous rods, however, contractors hung the walkways from two separate sets of rods. As constructed, each transverse box beam of the upper walkway had two holes drilled through either end of it: one hole two inches from the end and another hole six inches from the end. One set of rods passed through the outside holes, suspending the fourth-floor bridge from the atrium ceiling and supporting its weight with only a nut and washer underneath each beam. Another set of beams passed through the inner holes, held in place by a nut and washer connection above the fourth-floor beams, and supported the second-floor walkway from the bottom by another nut and washer connection. This change in design diverted the weight of the second-floor walkway from the ceiling supports to the connections of the rods and beams above it.

Henry Petroski, a civil engineering professor at Duke University who studied the failure, suggested that the situation could be compared to one in which two men hang separately by their hands from the same rope. If their hands were strong enough to support their own weight, the men would not be in danger of falling. But if instead of grasping the rope, the man at the bottom held onto the legs of the man hanging above, the man at the top would suddenly be supporting twice his weight. If the person at the top was just barely strong enough to support the double load, any small weight addition would be enough to make him loose his grip.

The design was probably changed because the use of such a long steel rod in the detail would have been somewhat tricky, perhaps involving casting a long rod in which the lower half was threaded like a screw so that a bolt could be twisted almost halfway up its length to support the upper walkway. Engineers later suggested that the best way to accomplish the original design would have been to use a sleeve nut, a device that could connect two separate threaded rods below the fourth-level walkway to form one continuous rod.

Details of the Collapse

Initial speculation about the tragedy focused on the possibility that such a large crowd of people had congregated on the walkways to watch the dance contest that the weight capacity was exceeded. Others theorized that people dancing on the walkway had set in motion a vibration that caused the collapse.

Investigations into the collapse were launched by several groups, including Failure Analysis Associates of Palo Alto, California, Packer Engineering Associates of Chicago, Lee Lowery, a professor at Texas A&M University and Rex Paulsen of Fay Engineering Company in Denver. Probably the most prominent investigation was undertaken by the National Bureau of Standards (NBS), which simulated the conditions of the collapse in the laboratory and tested the original materials from the collapsed walkways as they became available.

The report produced by NBS investigators, led by Edward O. Pfrang, the chief of the agency's structures division, was released in February 1982. NBS blamed the change in the design detail for the catastrophe and concluded that as built, the walkways could support only 27 percent of the combined dead

The third walkway of the Hyatt Regency remained intact, while two other suspended walkways collapsed primarily due to a change in the design of the suspension rods.

load—the weight of the materials of the bridge—and live load—the weight of the people standing on it—that they should have been designed to carry under the Kansas City building code. Using video footage shot by a local television crew filming the tea dance, investigators were able to estimate that only sixty-three people were standing on the second- and fourth- level walkways at the time of the collapse, a number that would have created a live load of less than 10 percent of what the walkway should have been designed to support. Each of the hanging rods should have been able to support 68,000 pounds to meet the design load requirements, but the box beam connections collapsed under a load of just 18,600 pounds.

NBS found that the collapse of the walkways began when the welding of one of the fourth-floor box beams split apart, and the nut and washer supporting the bottom of that walkway slipped through the hole. The load of both walkways was then transferred to the remainder of the fourth-level walkway connections and these box beams, too, fractured, unable to sustain the weight. The fourth-floor walkway collapsed onto the one below it, sending both walkways crashing down onto the lobby floor.

Although the report concluded that the connections as originally designed would have withstood the forces that caused the collapse, it asserted that that design was found lacking as well. The original hanging rod design would have supported only about 60 percent of the design load requirement under the building code, the investigators said. The investigators also reported that they had some minor reservations about the quality of the welding of the channels that comprised the steel box girders, but indicated that materials and workmanship did not play a significant role in the disaster.

Impact

Despite objections from investigating engineers and lawyers representing victims of the catastrophe, the hotel's owners, Crown Center Redevelopment Corporation, removed the only remaining atrium walkway within days of the collapse. A later inspection of that structure, however, uncovered evidence that box beams in that walkway had also deformed near the rod-to-box-beam connections, indicating that the third walkway was probably overstressed also. The side walls of those box beams had bowed out, and the bottom walls had deformed by as much as 1/4 inch where the washer and nut pulled up against the bottom of the beam. The hotel's owners replaced that walkway with a new seventeen-foot wide causeway at the second-floor level, supported on ten columns built up from below. The hotel remained closed for eleven weeks, and was reopened in the fall of 1981. Following the collapse, the owners of nearly sixty other Hyatt hotels with atriums of various designs inspected their structures for safety as well.

At least 150 lawsuits were filed in circuit court after the collapse seeking total compensatory damages of $1 billion and punitive damages of about $500 million. Virtually all of the suits were eventually settled out of court. In 1983, a grand jury in Kansas City cited a lack of evidence in deciding not to issue indictments against the parties involved for criminal negligence.

Firefighters (upper right) continue rescue efforts as others look on. The collapse of the walkways killed 114 people and injured nearly 200 others.

In February 1984, however, the Missouri Board for Architects, Professional Engineers and Land Surveyors filed a complaint against the building's chief structural engineer, Jack K. Gillum, and the project engineer, Daniel M. Duncan, both then officers with St. Louis-based GCE International, for gross negligence in the design and analysis of the walkways. The complaint launched a protracted investigation and hearing during which 450 exhibits were examined and 5,000 pages of transcript were produced.

During the hearing, which was closely watched by the engineering community, Gillum and Duncan argued that the failed connection was designed by the project's steel erector and fabricator, Havens Steel Company of Kansas City. The engineers' attorney affirmed that it had become common practice within the construction industry to delegate the task of designing steel connections to the steel fabricator because the fabricator could best select the most economical way of building such details. An engineer for Havens, in turn, testified that the detailing on the Hyatt project had been subcontracted to WRW Engineering of Kansas City. Neither Gillum and Duncan nor Havens Steel admitted changing the connection design, and each blamed the other for the alteration. However, the shop drawings submitted by Havens were stamped and initialed by the architectural joint venture, the structural engineer and the contractor.

Much testimony at the hearing centered on a collapse of part of the atrium's roof during the construction of the hotel in 1979, long before the walkways collapsed. Although no one was hurt in the roof collapse, it prompted the building's owners to ask for a review of the structure, essentially giving the engineers a second chance to catch the mistake. Evidence presented showed that Duncan submitted a report at that time indicating that he had checked the suspended bridges, and that Gillum told the project's construction manager that every connection in the atrium had been reviewed.

In November 1985, Gillum and Duncan were found guilty of negligence for having failed to review the steel shop drawings. Missouri administrative law judge James B. Deutsch, who presided at the hearing, said he found evidence of deliberate fraud on the part of the engineers for not thoroughly checking the connections when the roof collapsed in 1979. In January 1986, the Missouri Board for Architects, Professional Engineers and Land Surveyors voted unanimously to permanently revoke the engineering licenses of Gillum and Duncan, and to revoke the certificate of authority for their firm, GCE International.

The Hyatt tragedy attracted widespread public attention because of the large loss of life. The collapse also forced the engineering community to take a new, closer look at the responsibility implied when an engineer affixes his professional seal to a set of project documents. The judge in the Hyatt case made it clear that while the courts would allow the engineer to delegate the job of designing a detail to another party, that engineer could not delegate the responsibility for that detail to the other party as well.

More than a decade later, the lines of responsibility between a structural engineer-of-record and a project fabricator are still a subject of controversy. Although the Hyatt case seemed to indicate plainly that the structural engineer could be held responsible for errors made by a fabricator, many structural engineers and architects do not want to be held liable for errors made by others and have tried to "write away" responsibility in project specifications and other contractual documents. Many firms also require the fabricator's own engineer to seal shop drawings that come out of that company.

Although all of the issues raised by the Hyatt walkways collapse may never be completely addressed, the tragedy did succeed in drawing the engineering profession's attention to long-standing problems that had gone unresolved. The collapse inspired countless professional seminars, conferences, and papers and is at least partially credited with inspiring the American Consulting Engineers Council to institute its peer review process and the American Society of Civil Engineers to publish its *Quality in the Constructed Project* manual in 1988. Probably no U.S. engineer will ever design a hanging

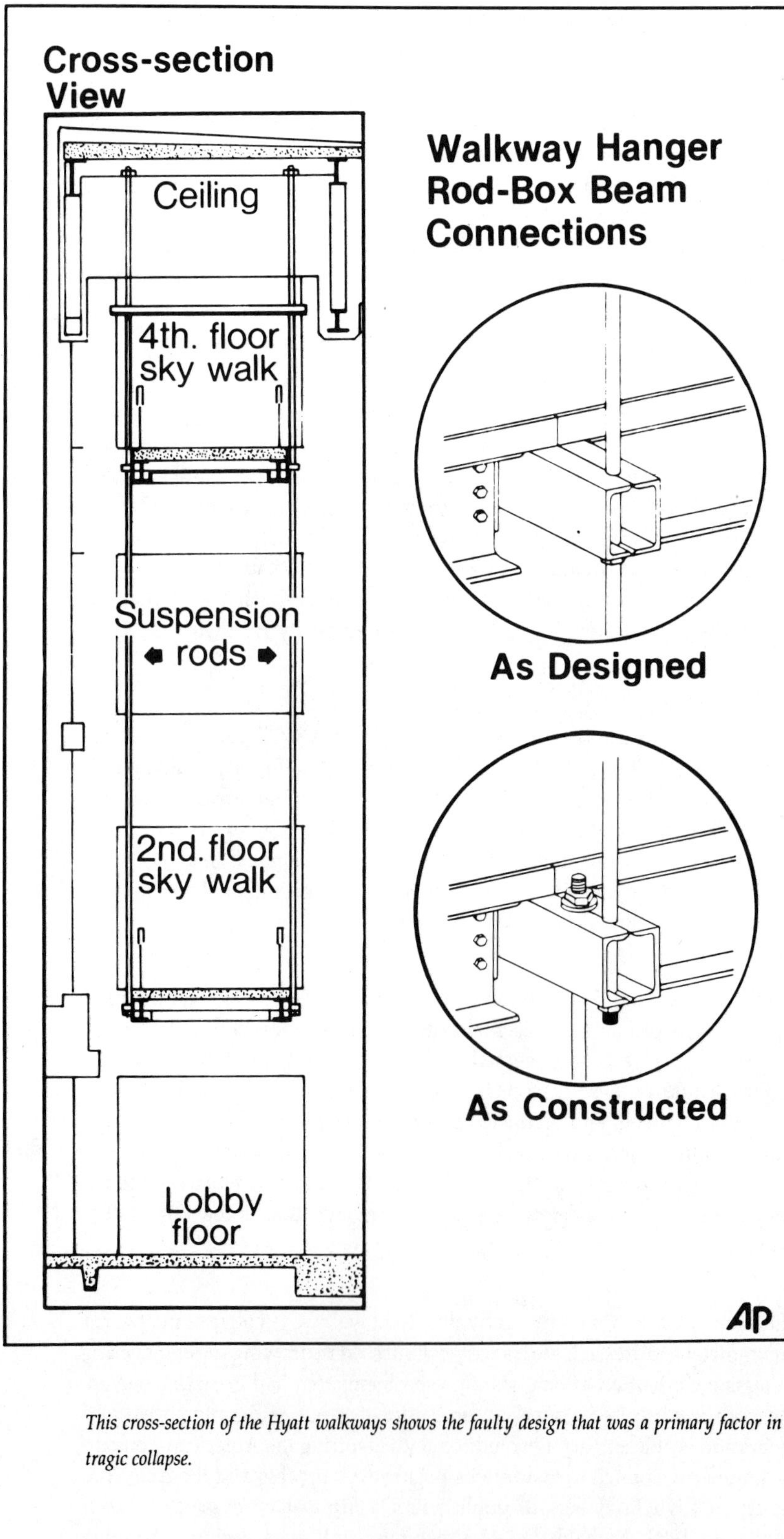

This cross-section of the Hyatt walkways shows the faulty design that was a primary factor in the tragic collapse.

walkway structure again without thinking of Kansas City and checking each detail.

VIRGINIA KENT DORRIS

Where to Learn More

Alm, Rick. "Hyatt Engineers Lose Licenses in Missouri." *Engineering News-Record,* January 30, 1986, p. 11.

"Hyatt Hotel Engineers Cited for 'Negligence'." *Engineering News-Record,* February 9, 1984, p. 14.

"Hyatt Hearing Traces Design Chain." *Engineering News-Record,* July 26, 1984, pp. 12-13.

Marshall, R. D., et al. *Investigation of the Kansas City Hyatt Regency Walkways Collapse,* U.S. Dept. of Commerce, National Bureau of Standards, 1982.

"Death Trap in Kansas City." *Newsweek,* July 27, 1981, pp. 30-31.

Petroski, Henry. *To Engineer Is Human.* St. Martin's Press, 1985.

Pfrang, Edward O., and Richard Marshall. "Collapse of the Kansas City Hyatt Regency Walkways." *Civil Engineering,* July 1982, pp. 65-68.

Ross, Steven S. *Construction Disasters.* McGraw-Hill, 1984.

Ocean Ranger oil-drilling rig sinking

North Atlantic

(1982)

A broken window on an offshore oil-drilling rig precipitated a catastrophic chain of events that caused the sinking of the rig and the death of all on board.

Background

On February 15, 1982, a small broken window in the control room of the *Ocean Ranger*, a mobile offshore drilling rig, eventually caused the ocean vessel to capsize, resulting in the death of every crew member. Though such a conclusion may seem ridiculous, it is no different than concluding that a spark causes a dynamite explosion. In the case of the *Ocean Ranger*, a broken window was the spark, and the formula for the dynamite was control operators not properly trained to handle emergency situations, inadequate lifeboats, substandard emergency protective clothing, freezing temperatures, and a fierce storm.

The cause of the broken window will always be a mystery since the entire crew was killed. It could have been broken by the heavy winds, high waves, or possibly it occurred during drilling operations taking place earlier that day. Regardless of how it happened, the missing window allowed some of the control panels of the *Ocean Ranger* to be drenched with ocean water. The drenching caused a short-circuit in the ballast control panel. This not only made keeping the rig floating in a stable manner impossible, it also caused the rig to lean precariously forward.

Built in 1976 in Hiroshima, Japan, the *Ocean Ranger* was a semisubmersible oil-drilling rig. The rig was owned by the New Orleans-based Ocean Drilling & Exploration Company (ODECO), Mobil Canada.

The *Ocean Ranger* was an odd-looking structure. The working platform was supported on eight columns, making the oil-drilling rig look like an eight-legged table. The eight legs were supported in two groups of four on top of multicompartment tanks called ballast tanks. The four-hundred-foot-long ballast tanks would be filled with air to keep the structure floating or water to help the rig maintain its stability. If heavy loads were being placed on the front of the ship, it would cause listing in the front. The listing would be countered by allowing water to flow into the rear ballast tanks resulting in leveling of the rig. When the load was removed, the ballast tank was pumped out to avoid listing at the rear. An experienced control room operator would be assigned to operate the ballast system since it was a crucial part of the oil rig's stability.

The ballast control panel was designed specifically for the *Ocean Ranger*. The panel consisted of an abundance of electrically operated switches that opened and closed valves on the ballast tanks. In the event power was lost, the switches could be operated by the use of a manual control rod. However, without power, the gauges would not work and the position of the valves could not be determined.

Details of the Disaster

The weather on February 14, 1982, was relatively calm during the day. The *Ocean Ranger*, as well as two other mobile offshore drilling rigs, were drilling for oil 160 miles off the coast of Newfoundland, Canada. These drilling rigs, which ranged from three hundred to four hundred feet in length, were operating in waters in excess of 250 feet deep. Due to the potential instability of the drilling rig and other potential hazards, each drilling rig was required to be accompanied by a rescue ship. The accompanying rescue ship typically doubled as the supply boat and drifted within two miles of the drilling rig. In the absence of survivors from the *Ocean Ranger*, the sequence of mishaps was reconstructed from radio transmissions received by these other five ships.

The lifeboats the crew boarded in the early morning were not designed for the fierce environment they were about to battle. The launch of the boats required a seventy foot drop into the ocean. This drop, coupled with high winds, made the safe use of these emergency boats a fantasy.

The unfortunate chain of events began at dinnertime on February 14. Internal communication was overheard that the crew was mopping up water and cleaning up broken glass in the control room, and also that the crew members were getting shocks from a control panel. As the evening passed, the weather turned fierce. The drilling rigs were pummeled by fifty-five-foot-high waves, eighty-mile-per-hour winds, and subzero temperatures. Conversations throughout the evening between the *Ocean Ranger* and the other ships showed no cause for alarm until around 9:00 P.M. At that time, an internal transmission on the *Ocean Ranger* requested an electric technician report to the control room because valves for stabilizing the drilling rig were opening and closing on their own.

Shortly after midnight, the *Ocean Ranger* radio operator sent out a distress message. The message, which was repeated ten to twenty times in a half hour, transmitted that the drilling rig had a severe list that required immediate assistance. Less than two hours later, the crew of the *Ocean Ranger* proceeded to life raft stations to begin abandoning ship because the listing of the oil rig had reached the point of no return. As the crew boarded the lifeboats, they most likely believed that in a short time they would reach safety. Unfortunately, the lifeboats offered these crew members little chance of survival.

The lifeboats the crew boarded in the early morning were not designed for the fierce environment they were about to battle. The launch of the boats required a seventy foot drop into the ocean. This drop, coupled with high winds, made the safe use of these emergency boats a fantasy. Some investigators have concluded that some of the lifeboats experienced extensive damage from bouncing off the side of the Ocean Ranger during launching.

Probably only one of the lifeboats successfully survived the launch and had a chance for rescue. This lifeboat had sustained damage to the bow during the launch and the crew on board had to bail out water that was pouring in through the hole. Since the boat was floating, the men had life jackets, the boat's motor was operating, and the rescue boats were approaching, the crew might have been saved had the rescue ships possessed adequate life saving equipment.

Sensing trouble, the *Seaforth Highlander*, the rescue ship assigned to the *Ocean Ranger*, started moving toward the distressed ship. Rescue ships typically stay within a mile or two of their assigned ship. This distance is close enough for emergency response, but still far enough away to avoid collision and tangling of anchor lines. Due to the severe weather, the *Seaforth Highlander* had drifted eight miles away from the *Ocean Ranger*. On hearing the distress signal, the rescue ship proceeded full speed ahead, but did not arrive until an hour later.

The crews on the rescue ships for the other two mobile offshore oil-drilling rigs had been monitoring the *Ocean Ranger* situation. At 1:15 A.M., their services were desperately needed. Within one to two hours, both rescue ships were in the vicinity of the sinking oil rig; but the *Seaforth Highlander* was the only rescue team that saw any living members of the *Ocean Ranger* crew. Just before 2:00 A.M., the *Highlander* sighted distress flares. As the crew approached the flares, they spotted a lifeboat with a hole in it heading in their direction. The lifeboat and the rescue ship maneuvered to be side by side. The rescue ship tossed lifelines to the lifeboat. The lifeboat was fastened to the rescue ship with one of these lines. The rough sea was violently tossing the two boats. Several men exited the covered part of the lifeboat to board the rescue ship, causing the boat to capsize and throwing the men into the sea. The *Seaforth Highlander* threw inflatable life rafts to the men in the sea. Although the life rafts were thrown over within minutes and landed right next to the men, they made no effort to grab for them. The bitter cold sea had practically paralyzed the crew. The *Seaforth Highlander* could do little else since the ship was not equipped with baskets to scoop people out of the water. After many attempts to pull the crew out of the ocean, the *Seaforth Highlander* had to move on in search of other crew members, but in vain.

The search for survivors continued for a week after the capsizing, by air and sea. Drift plots were calculated to estimate the farthest distance that an object would drift from its original position under existing weather conditions. The search within the drift plot boundaries located several lifeboats and bodies, some over forty miles away from the capsizing. The search was called off on March 1, 1982.

Eighty-four people were killed on the *Ocean Ranger*. Only twenty-two of the bodies were ever recovered.

Eighty-four crew members perished as the Ocean Ranger's *ballast control panel short-circuited, causing a dangerous list of the rig.*

Impact

The window that broke in the control room was directly to the side of the ballast control panel. The incoming water short-circuited the panel, which meant the system could only be operated by manual controls. In all likelihood, not a single person on the *Ocean Ranger* knew how to operate the ballast control panel manually. Even if an operator was experienced with the manual controls, the system was not realistically operable since all information gauges were also shorted out. A radio transmission from the *Ocean Ranger* stated that the ballast valves were opening and closing on their own. The erratic behavior of the ballast control board was causing the ballast tanks in the front of the rig to fill with water. The *Ocean Ranger* was tilting ten to fifteen degrees forward in rough seas. This angle of tilt would normally not be an immediate problem. However, at this angle, the holes in the drilling rig where the anchor chains were kept started to flood, drastically increasing the list. The *Ocean Ranger* was now totally out of balance and started to slide into the North Atlantic as the crew abandoned ship.

One of the most disturbing aspects of the *Ocean Ranger* disaster was that the damage could have been limited to only the loss of property. All crew members probably reached the lifeboats safely. Unfortunately, the lifeboats were not designed to withstand the severe pounding of the seventy-foot drop from the rig into the ocean. Also, many crew members floating in lifejackets, could easily be spotted with the assistance of the emergency lights attached to the lifejackets. But without protective equipment from the cold ocean, the jackets were merely keeping dead bodies afloat. Those who successfully kept their lifeboat afloat did not fare any better. The rescue boats were not equipped to pull survivors from them under the harsh storm conditions.

As a result of the capsizing of the *Ocean Ranger*, the National Transportation Safety Board instructed the United States Coast Guard to require that:

1. An operating manual be provided for semisubmersible mobile offshore units that includes include guidance for countering accidental flooding of ballast tanks, as well as for preventing flooding into the anchor chain compartments;

2. Control room operators be licensed in ballasting procedures for mobile offshore drilling units;

3. Exposure suits to protect those in the ocean be provided for 150 percent of the persons on offshore drilling rigs where water temperatures may go below sixty degrees;

4. Vessels assigned for rescue purposes to the mobile offshore drilling rigs have capabilities for retrieving disabled persons from the water under adverse conditions;

5. The suitability of approved lifeboats, life rafts, and lifesaving equipment be evaluated.

One month after the disaster, the families of the victims brought a 1.7 billion dollar lawsuit against ODECO and Mobile Canada, the owners and operators of the *Ocean Ranger*, respectively. By mid-1984, all but a handful of the families had reached out-of-court settlements with the companies involved.

Where to Learn More

"Capsized Rig's Design O.K." *Engineering News Record*, February 25, 1982, p. 12.

Clugston, Michael. "An Aftermath of Sorrow and Anger." *Maclean's*, March 1, 1982, pp. 28-31.

———. "The Search of the *Ocean Ranger*." *Maclean's*, May 3, 1982, pp. 17-18.

———. "How the *Ocean Ranger* Sank." *Maclean's*, February 21, 1983, p. 14.

———. "A Preventable Tragedy." *Maclean's*, August 27, 1984, p. 45.

Giniger, Henry. "Hope Fades for 84 on Rig; 18 on Soviet Freighter Die." *New York Times*, Wednesday, February 17, 1982, p. 3.

Joyce, Randolph, "The Cruel Sea." *Maclean's*, March 1, 1982, pp. 26-27.

———. "The *Ocean Ranger*'s Mystery Deepens." *Maclean's*, March 8, 1982, pp. 24-25.

LeMoyne, James, et al. "The *Ocean Ranger*'s Night of Death," *Newsweek*, March 1, 1982, p. 48.

National Transportation Safety Board. *Capsizing and Sinking of the U.S. Mobile Offshore Drilling Unit Ocean Ranger*, February 15, 1983, Report no. NTSB-March 83-2.

"Wreck of the *Ocean Ranger*." *Time*, March 1, 1982, p. 29.

"Sea Extracts Its Price in Hunt for Oil." *U.S. News and World Report*, March 1, 1982, p. 6.

L'Ambiance Plaza collapse

Bridgeport, Connecticut

(1987)

An apartment building collapsed during lift-slab construction, probably caused by an inadequate design for connecting the floor slabs to the column-mounted lifting mechanisms.

Background

Shortly after lunch on April 23, 1987, the twin towers of the half-completed L'Ambiance Plaza apartment building fell like a gigantic stack of concrete pancakes, killing twenty-eight workers and injuring sixteen. The second-worst construction accident in American history focused attention on the importance of designing for construction loading as well as the static conditions of completed structures.

The state of Connecticut had financed the building as part of an economic revival effort for the city of Bridgeport. The $17 million project would provide two hundred and eighteen apartments in thirteen aboveground stories atop three levels of underground parking.

In the interests of economy and (ironically) safety, the lift-slab technique was chosen to build the structure. Once the first level of concrete flooring was poured on the foundation, slabs for the remaining floors and the roof were poured in layers atop the base. After each slab had hardened sufficiently, it was posttensioned to increase the strength of the slab by tightening high-strength steel wires embedded in the concrete.

When all of the slabs were ready, they would be lifted in groups of about three by jacks positioned atop each of the building's steel columns. As lower-level slabs reached their destination heights, they were welded to the columns; upper-level slabs became working platforms from which column extensions were erected to permit raising floors to higher levels.

This procedure of pulling buildings up by their own bootstraps is economical, because the concrete is formed and poured essentially at ground level. It reduces the amount of work that must be done on scaffolding, diminishing the risk of workers falling or being hit by falling tools or other material.

Posttensioning allows for longer spans (fewer columns) and thinner slabs, reducing materials costs.

Texstar Construction Corporation was selected to lift the slabs for L'Ambiance Plaza. It had already built eight hundred lift-slab structures without a fatality, although there had been a few mishaps during the thirty seven-year history of the procedure.

Two incidents within the first six years of lift-slab construction helped formalize essential safety measures. In 1954, a 250-ton slab fell sixteen feet while being lifted at a site in San Mateo, California; this demonstrated the need for cribbing (supporting the slab with temporary posts until attachment to the columns was completed) and sway bracing (keeping the unstable structure from shifting sideways by anchoring it with temporary cables). The necessity of sway bracing was underscored two years later when a partially completed eight-story garage in Cleveland, Ohio, leaned eight feet from vertical before being winched back into position.

A succession of problems had plagued L'Ambiance Plaza even before the structure's ultimate collapse. In one instance, an automobile jack was braced against one column and used to move another column back into vertical. In another, workers attached jacks to four slabs at one end of the building while attaching jacks to only three slabs at the other end; when lifting operations began, one slab cracked and was subsequently patched. There were also difficulties with fitting columns together, problems with jacks, and incidents involving deformed shearheads.

Shearheads, consisting of steel channel sections cast into the concrete slab to form a frame around each column, were critical to the structure's integrity. They provided attachment points for the lifting rods that linked each slab to the jacks above. They were also the connection points at which slabs were welded to the columns. Just a year before the L'Ambiance Plaza collapse, a shearhead failure at another Texstar project in Stamford, Connecticut, had allowed a slab to fall fifteen inches, injuring a worker.

Details of the Collapse

Work seemed to be proceeding normally on April 23, 1987. L'Ambiance Plaza's two rectangular towers were offset, so that they shared half a wall. Columns in both wings extended through the ninth-floor level, and a stack of upper-floor slabs sat parked at the top awaiting extension of the columns. Another set of three slabs—the ninth, tenth, and eleventh floors—had just been raised to the top of the west tower, and workers were preparing to secure the set's bottom slab to the columns.

Suddenly a loud metallic bang rang out. With an ominous rumble, cracks spread through the ninth floor slab as through a shattering sheet of ice. The slab collapsed, pulling those above down with it. As the upper floors dropped, each lower floor crashed in succession in a kind of collapse known as "progressive collapse." The attached east tower was wrenched into failure as well. Within five seconds a heap of crumbled concrete punctuated with twisted steel was all that remained.

Suddenly a loud metallic bang rang out. With an ominous rumble, cracks spread through the ninth floor slab as through a shattering sheet of ice. The slab collapsed, pulling those above down with it. Within five seconds a heap of crumbled concrete punctuated with twisted steel was all that remained.

The catastrophe apparently began at column E4.8, which stood in front of slab cutouts for an elevator shaft centered on the twin towers' shared wall. Posttensioning wires curved around the cutouts. The building's original design had included an additional column nearby; a redesign, however, eliminated the extra column, leaving E4.8 the most heavily loaded column in the building.

Inexplicably, no changes in the jacking plan accompanied the redesign. Hydraulic jacks placed atop the columns to lift the slabs had a capacity of 178 kips (178,000 pounds); at some locations, superjacks with a capacity of 300 kips were substituted. The jack atop E4.8 was the smaller type.

As workers prepared to secure the crucial package of three 320-ton slabs prior to the collapse, they evidently realized that the slabs were not quite level. Raising them a fraction of an inch at column E4.8 put additional strain on the jack at that location. A superjack atop next-in-line column E3.8 was apparently used to help in the leveling effort.

What failed, however, was not the jack atop E4.8, but its connection to the ninth-floor slab. Sections of ninety-degree-angle steel were welded to opposite sides of each shearhead, extending into the channels of the column's steel I beam. A threaded rod connected to the jack above extended through a slot in the horizontal section of each of these "lifting angles." A nut screwed onto the end of the rod provided a surface on which the lifting angle rested.

As the leveling was attempted, the lifting angle at column E4.8 twisted upward, allowing the nut to slide toward the column. The rod swung out of its slot, slamming into the column with a thunderous 75,000 pounds of force.

Without support at that location, the seven-inch-thick slab began to sag and crack. The sudden loss of connection between the slab and the jack overburdened jacks at neighboring columns. In the ultimate wreckage, the surrounding columns bent toward E4.8 like accusing fingers.

Several other factors may have contributed to the difference in severity between the L'Ambiance Plaza shearhead failure and the episode in Stamford. Construction of concrete shear walls in strategic locations, which would provide stability to the column-and-slab skeleton, improperly lagged several floors behind slab placement. The builders had not installed any cribbing or sway bracing, further contributing to the structure's instability.

Moreover, inadequate safety factors in Texstar's shearhead design and construction increased the likelihood of that component failing. O'Kon & Company, the Atlanta-based engineering firm that designed L'Ambiance Plaza, reviewed the shearhead connection design only for postconstruction loading, not for loads imposed during the lifting operation. Although American National Standards Institute specifications called for a minimum

The L'Ambiance Plaza collapsed during lift-slab construction, probably caused by an inadequate design for connecting the floor slabs to the column-mounted lifting mechanisms.

safety factor of 2.5 in jacking assemblies, subsequent laboratory tests of this shearhead design revealed a safety factor between 1.1 and 1.3. That is, the shearhead and lifting angle assemblies began to deform under a load of 160 kips; they would have had to support 375 kips to meet the 2.5 safety factor.

Impact

Precise details of what caused the disaster have never been established conclusively.

A team from the National Bureau of Standards (NBS) investigated the incident on behalf of the federal government. Upon release of its report, the U.S. Occupational Safety and Health Administration (OSHA) announced record fines totaling $5.11 million against the prime contractor, TPMI/Macomber, and the lift-slab subcontractor, Texstar. Each entity was assessed 238 willful violations for inadequate lifting angles. They and two other subcontractors were also cited for inadequate lateral support and inadequate inspection, testing, and correction of weld problems.

In late October 1987, all of the known victims of the accident jointly filed suit against the city of Bridgeport and two of its employees who were responsible for issuing the building permit and inspecting the project. Attempting to avert an unwieldy number of separate lawsuits lingering interminably in the court system, a two-judge panel engineered a mediated $41 million settlement among all parties.

For the defendants, the agreement had mixed consequences. On the positive side, OSHA dropped all but $430,000 of its proposed fines, settling for recovery of the investigation's costs. On the other hand, over twenty parties involved in the ownership, design, and construction of the structure were assigned some level of responsibility for the disaster. Some felt that a longer, more extensive investigation might have exonerated them.

For the plaintiffs, the settlement meant a quick resolution of their claims. Payments and annuities for survivors and families of those killed in the collapse totaled $30 million. In addition, it was decided that L'Ambiance Plaza would be rebuilt; ownership of the completed building would be transferred to the plaintiffs, who would subsequently receive a portion of the future earnings of the facility.

For the construction industry, the settlement meant that the exact causes of the catastrophe might never be fully identified. While further investigation was not barred by the settlement, the lack of court proceedings to apportion liability removed much of the impetus for additional analysis.

Within days after the accident, the city of Bridgeport hired Thornton-Tomasetti, a New York City-based structural engineering firm, to conduct an investigation. Four years later their report suggested that the catastrophe had been triggered by the loss of a wedge holding the slab in place at column E4.8, resulting in a sudden sagging that initiated the collapse sequence. The report also charged that the NBS team's misidentification of elements of the debris when they were moved from the site hindered efforts to analyze the accident.

The American Society of Civil Engineers' (ASCE) annual convention in October 1991 included three sessions relating to the L'Ambiance Plaza disaster. The most heated discussions involved Neil M. Hawkins, an engineer who had worked as a consultant to Texstar. He argued that the collapse had originated when pressure from dirt filled in behind an adjacent retaining wall caused poorly designed lower-level slabs to crumble. He disputed the laboratory tests showing inadequate shearhead safety factors, and charged that OSHA had rushed to blame Texstar without allowing other investigators prompt access to the debris.

The spectacular failure with its heavy loss of life generated some changes in construction procedures and law. For example, OSHA directed contractors to meet the 2.5 safety factor criterion on all lift-slab jobs. In response to the fact that almost half of those who died at L'Ambiance Plaza had been working on tasks unrelated to the jacking process, OSHA also obtained an injunction banning the presence of nonessential workers underneath slabs during the lifting operation at a subsequent Texstar job site. Two weeks after the collapse, an ironworkers' union in Massachusetts adopted a policy prohibiting its welders from working underneath concrete slabs as they were being lifted.

The L'Ambiance Plaza collapse was the third major structural or construction failure in the state of Connecticut in a decade. Based on recommendations from a governor-appointed committee, the state enacted a new construction safety law in 1988. Key provisions included improving the qualifications, visibility, and power of building inspectors and requiring that building plans and specifications for large projects be reviewed by an independent engineering consultant.

The disaster underscored the need for designers to evaluate connections as well as the major elements of a structure. It also focused attention on the fact that temporary loads and stresses during construction must be analyzed when determining the adequacy of a design. Because of construction accidents such as the one at L'Ambiance Plaza, ASCE formed a committee to write a new standard on structural design loads during construction.

The collapse of L'Ambiance Plaza inserted an exclamation mark into the story of modern construction practices. It made the point that small details can be the most critical elements in large projects.

LORETTA HALL

Where to Learn More

Godfrey, K. A., Jr. "After L'Ambiance Plaza." *Civil Engineering,* January 1988, pp. 36-9.

Korman, Richard. "Changes Asked in Safety Law as L'Ambiance Pact Is Signed." *Engineering News-Record,* December 8, 1988, pp. 13-14.

———. "Column May Help Locate Cause of Fatal Collapse in Bridgeport." *Engineering News-Record,* April 30, 1987, pp. 10-12.

———. "Flawed Connection Detail Triggered Fatal L'Ambiance Plaza Collapse." *Engineering News-Record,* October 29, 1987, pp. 10-12.

———. "Mediated Settlement Seeks to Close the Book on L'Ambiance Plaza." *Engineering News-Record,* November 24, 1988, pp. 10-11.

———. "L'Ambiance Plaza Won't Rest Easy." *Engineering News-Record,* November 4, 1991, pp. 14-16.

Korman, Richard, and Kathleen S. Failla. "Lack of Temporary Bracing Studied." *Engineering News-Record,* May 7, 1987, pp. 12-13.

Levy, Matthys, and Mario Salvadori. *Why Buildings Fall Down.* W. W. Norton & Company, 1992, pp. 173-82.

Vogel, Shawna. "The Domino Building." *Discover,* January, 1988, p. 88.

Chemical and Environmental Disasters

- Sudbury, Ontario (1883 to present)
- Toyama, Japan (1912-1955)
- DDT (1939 to present)
- Love Canal, New York (1942-1953)
- Hanford (1944-1988)
- Minamata, Japan (1955 to present)
- Agent Orange (1961-1971)
- Times Beach, Missouri (1971)
- Methylmercury seed (1971-1972)
- Flixborough, England (1974)
- Seveso, Italy (1976)
- Bhopal, India (1984)
- Institute, West Virginia (1985)

Ecological disaster

Sudbury, Ontario

(1883 to present)

The area around Sudbury, Ontario has been deforested by emissions from years of nickel and copper open roast smelting.

Background

Nearly two billion years ago, a large meteorite struck the earth near what is now Sudbury, Ontario. The meteorite's impact melted the earth's crust, and metals became concentrated in rich sulfide ores near the surface. In 1883, workers on the first trans-Canadian railroad cut into a surface bedrock deposit of nickel ore, marking the beginning of commercial mining in the region. Today, the area around Sudbury supplies approximately a fifth of the world economy's requirements of nickel, along with copper, and some iron, cobalt, gold, silver, sulfur, and selenium.

The valuable metals at Sudbury exist in an ore mixed with sulfur in chemical compounds called *sulfides*. To obtain the metal, one must remove the sulfur, which traditionally has been accomplished by burning the ore at a high temperature to combine it with oxygen. This process is called *smelting*.

In the first forty years of mining at Sudbury, the mining companies smelted the ore in large open "roasts." The roasts consisted of a layer of timber cut from the local forests, over which were layered piles of the sulfide ore. The wood was set ablaze, raising the temperature of the ore to the point where it, too, would ignite. Once the ore itself was burning, the fire was self-sustaining and smoldered for several months. After the fire had subsided, the miners removed the cooled, crude copper and nickel metal and shipped it to a refinery for processing into pure metals. At its peak in the early twentieth century, the mining companies in Sudbury operated as many as thirty open roasts, which burned day and night.

Details of the Disaster

The open roast method was crude but inexpensive—not only because it did not require complicated furnaces, but the heavily forested area supplied cheap and plentiful wood to fuel the smelting. Although economically cheap, the open smelting process was very expensive environmentally. The timber cut to fuel the fires was *clear-cut*, that is, wide tracts of forest were completely stripped of trees. This accelerated soil erosion and destroyed forest ecosystems.

The open roast beds emitted a toxic cloud which contained not only gaseous sulfur dioxide, but also an acidic mist and metal particles. These toxic clouds were completely uncontrolled, and wafted over neighboring land, where the heavier metal particles settled to the earth and the sulfur dioxide fell as acid rain.

The chemical composition of the ores found in Sudbury constitutes a recipe for environmental disaster: a variety of metals found in the ores are toxic to plants and animals when the metals are concentrated in soil and in lake water. The sulfur which accompanies the metals forms sulfur dioxide during the smelting process, which then forms acid rain when the exhaust gases mix with water in the atmosphere.

The toxic chemicals produced in and emitted from the open roasts killed the surrounding vegetation, which accelerated the erosion of soil from the underlying bedrock. Once the soil washed away, the toxic clouds scarred even the granite bedrock.

Although the open roasts were inexpensive, they were also somewhat inefficient in extracting metals from the ore, and the mining companies began to replace them with smelting furnaces. In 1928, the Canadian government banned the use of open roasts entirely.

The new smelters vented exhaust gases through smokestacks built above the furnaces. The smokestacks, however, emitted pollutants higher into the atmosphere and the surrounding vegetation was still damaged, lakes were acidified, and toxic contaminants were still spread throughout the area.

The sulfur dioxide emissions reached their peak in the late 1960s, when the Sudbury area was the world's largest *point source* (a localized, discrete, and fixed contaminant emission source) of sulfur dioxide.

Hoping to alleviate the persistent, severe local smog and the severely damaged areas (closest to the points of emission), engineers devised taller smokestacks. The highest of these is the "superstack," a single smokestack 1,250 feet (381 meters) tall, erected in 1972 at the largest iron ore recovery facility located in Sudbury at the Copper Cliff mine.

Although the tall smokestacks have greatly improved the local air quality, areas far downwind of the smelters—where the smelter exhaust falls along with rain—pay the price in the a form of regional acid rain problem. Scientists believe that acid rain is caused when gaseous sulfur dioxide mixes with water vapor in the atmosphere. When the water condenses, the sulfur dioxide enters a solution with the water, and forms a dilute acid. The greater the concentration of sulfur dioxide in the air, the more acidic the rain becomes, which condenses from it.

As a result of sulfur dioxide and other toxic emmissions, the area around the International Nickel smelter in Sudbury, Ontario, has been transformed into a moonscape.

In addition, the sulfur dioxide emitted from mining operations around Sudbury has decreased since its peak in the 1960s, due in part to economic changes. During the 1980s, the world's metals markets experienced a dramatic decline. As the value of nickel dropped, the mining companies in Sudbury scaled back their operations, and the number of people employed in the mines fell in ten years from about one third of the workforce in the area to about one sixth. The decreased production of metal was accompanied by a decrease in the production of pollution. Nevertheless, the amount of pollution produced by the mining activities at Sudbury remains high, although its damage is more widely dispersed.

Impact

The environmental cost of metals mining and processing on the Sudbury area has been enormous. Sudbury is one of most extensively studied areas of environmental damage in the world.

Scientists have been able to trace the impact of mining on the Sudbury ecosystem by examining soil samples for pollen. The layers of soil deposited on the bed of lakes in the area form a record of the amount and kinds of pollen which fell into the lake water each year. When scientists looked at layers dating from the beginning of mining activity in the late nineteenth century, they found that the amount of pollen decreased by two-thirds over the next several decades. This decrease corresponded to the period of open roast use showed that the surrounding vegetation was severely damaged during that period. They also found a dramatic increase in smoke particles, which indicates the clear-cut and pollution-damaged forests frequently caught fire and burned, adding to the area's pollution. There was an increase in pollen, however, beginning in the early 1930s—about the time open roast use decreased and the construction of the first tall smokestacks began.

The ecosystems nearest to the smelters have been the most severely affected. Scientists estimate that during the open roast era, forty square miles around the roasts were completely stripped of vegetation, and another 140 square miles outside of this barren inner zone lacked conifers—evergreen trees—common to the region.

Scientists estimate that during the open roast era, forty square miles around the roasts were completely stripped of vegetation, and another 140 square miles outside of this barren inner zone lacked conifers—evergreen trees—common to the region.

It was found that even outside the 140-square mile area, conifers suffered from stunted growth, high death rates, and needle damage within an additional 710-square mile zone. Furthermore, it was found that in a larger zone comprising 1,582 square miles, conifers, although surviving and not stunted, nevertheless had damage to their needles from the pollution. Even today few such trees grow near the smelters.

In the early 1980s, additional studies showed that although the damage was far less widespread than in the open roast era alone, it nevertheless was striking: in a zone nearly two miles (3.2 kilometers) in diameter around the smelters, there is no forest, and only one percent of the ground is covered by vegetation. In a ring from two miles to five miles (eight kilometers) outside of the smelters, there are forest remnants only where there is water and ground sheltered from the pollution which occurred over the years. In a ring from five miles to twelve miles (19.3 kilometers) surrounding the smelters, the forest is continuous, but lacking in a natural diversity of species. Today, there are few evident effects on terrestrial ecosystems located more than twelve miles from the smelters.

In addition to the effects on the forests from soil contamination and air pollution, the lakes near Sudbury have been devastated by the area's mining legacy. Lakes near the smelters are polluted with metals from the smelter stacks, and their waters are acidified by acid rain. The combination of pollutants completely killed the plant and animal life in several lakes in the area, and dramatically reduced the life in many others. Exacerbating the acidification of the lakes is the fact that limestone, which acts as a buffer for acidity in water, is lacking in the lakes around Sudbury.

The acidification of the lakes has affected animals which depend on the lakes, as well. The common loon, for example, breeds on freshwater lakes, where it collects fish, frogs, insects, and aquatic plants to feed its young. The decline in the amount of these valuable food sources deprives young loons' diets of animal protein which they need to thrive. Moreover, common loons nest on the shores of lakes, and the parents are too heavy to fly to other lakes and return with additional food to supplement their chicks' diets. Consequently, the principal source of food for loon chicks born on pollution-acidified lakes is nonnutritious algae, and many chicks eventually die of malnutrition.

Many scientists argue that the tall smokestacks merely redistribute the large emissions of sulfur dioxide from the areas near the smelters to those downwind. The effects can be seen in distant lakes, for although the metals, which are relatively heavy, do not travel far, the acid precipitation from the smokestacks affects lakes more than 25 miles (40 kilometers) downwind. Although the damage to Sudbury's environment is extensive, there are a few hopeful signs. In the 1980s, the province put 4,600 people to work restoring the area's ecosystem. The workers replanted trees, removed tree stumps, spread fertilizer to assist in plant growth, and spread lime to reduce the acidity of the soil.

The mining companies, prompted by tough new laws designed to curb air pollution, have also taken steps to clean up the exhaust from the smelters. For example, the company which operates the superstack spent $500 million on a new smelter and systems to remove acid rain-causing sulfur dioxide from the superstack's emissions. The tighter controls have been in place now for more than a decade, and scientists believe that many of the lakes have partially recovered. They caution, however, that the recovery will probably remain partial unless even stricter controls are enacted.

L. Carol Ritchie

Where to Learn More

Alvo, R. "Is the Laughter of the Loons to be Stilled on Our Acid Lakes?" *Canadian Geographic*. June-July, 1987, pp. 46-50.

"Reclaiming Land and a Positive Image." *Environment*. October, 1987, p. 21.

Freedman, B. *Environmental Ecology: The Impacts of Pollution and Other Stresses on Ecosystem Structure and Function*. Academic Press, 1989, pp. 30-41.

Nriagu, J., ed. *Environmental Impacts of Smelters*. J. Wiley, 1984.

"Nickels from Heaven." *Scientific American*. December 1985, pp. 82-3.

Walmsley, A. "Case Study: An Acid Test." *Maclean's*. September 17, 1990, p. 62.

Cadmium poisoning

Toyama, Japan

(1912-1955)

Mining runoffs containing toxic cadmium collected in drinking and irrigation water and poisoned Japanese villagers, causing degenerative Itai-Itai disease.

Background

The degenerative effects of cadmium-polluted water used for decades for drinking and irrigation of Japanese rice fields were finally linked in the 1960s to a familiar disease called Itai-Itai. Once the disease was officially identified by the Japanese government as being caused by downstream runoff from mining processes, legislation was passed to prevent further contamination and to compensate some of the victims.

Middle aged women suffered from Itai-Itai ("ouch-ouch") disease as early as 1912, decades before the cause—cadmium poisoning from mining effluents—was pinpointed. Although men, young women, and children were also exposed, ninety-five percent of the victims were women over fifty years old who had ingested low levels of cadmium in rice and drinking water for many years. They usually had given birth to several children and had lived more than thirty years within three kilometers of the lower stream of the Jinzu River in Toyama prefecture, Japan.

The disease started with symptoms similar to those of rheumatism, neuralgia, or neuritis. Then came bone lesions, osteomalacia (bone softening) and osteoporosis (bone adsorption), along with renal disfunction and proteinuria (excess protein in the urine). As it escalated, pain in the pelvic region caused the victims to walk with a ducklike gait. Next, they were incapable of rising from their beds because even a slight strain caused bone fractures. The suffering, often with unbearable pain, could last many years before it ended with death. At autopsy one patient had twenty-eight fractures to the ribs and seventy-two total in the whole body. Overall, an estimated 199 Itai-Itai victims had been identified by 1965, and at least 162 had died by December 1992. Only gradually had the causes been correlated with the effects.

Gold had been mined on the upper stream of the Jinzu River at least by 720 A.D., and a working mine was developed in 1589. The Kamioka Mining

Station was modernized in 1887 to become a leader in smelting lead and zinc and producing sulfuric acid. The Kamioka mining activities were increased in the 1910s and again in the 1930s and 1940s to support Japanese military expansion. As output increased, farmers downstream noticed that rice plants near their irrigation outlets turned yellow and died. They demanded compensation for crop damage, which the Mitsui Mining and Smelting Company began paying in 1890; and sedimentation facilities were installed in 1932.

Even so, the number of victims increased during and after World War II. During peak production approximately four thousand tons of ore were mined daily to yield annual outputs of about forty thousand tons of zinc, ten thousand tons of lead, and one hundred tons of cadmium. As the ore was processed, cadmium, lead, and zinc were systematically released into the wastewater from the flotation processes used to separate and concentrate the ores and were discharged to the Takahara River, an upper tributary of the Jinzu River. Downstream, farmers using the water for drinking and crop irrigation also received the suspended particles of cadmium it contained. Cadmium is taken up by rice plants more easily than most of the other heavy metals and can reach levels that are toxic to human beings before the vegetation itself shows any signs of damage. Consequently, the farmers tried to minimize the amount of cadmium reaching the plants by passing the river water through small sedimentation pools before it was released into their paddy fields; however, the procedure was largely ineffective. In 1955 the mining company built a dam in the mountains to create a waste sedimentation lagoon to remove most of the solids from the wastewater before it was released into the river. The cadmium levels in the water were dramatically lowered; but nearby soils, already high in cadmium, continued to receive more from airborne emissions, and local families continued to eat the contaminated rice.

Details of the Disaster

While the symptoms of Itai-Itai disease were first observed early in the twentieth century, only in 1946 did a local physician, Dr. Noboru Hagino, study and name the disease. He first reported his findings in 1955. Other investigators first thought the condition was due to some type of nutritional deficiency such as vitamin D, protein, or calcium; however, based on the geographical distribution of patients, Dr. Hagino had concluded by 1957 that heavy metal mining pollution was the cause. The patients' bones and tissues showed high levels of cadmium, zinc, lead, and copper, which were also found in the mining wastewater. With a half-life of thirteen years in the human body, cadmium replaced calcium and accumulated, especially in pregnant women.

In 1961 cadmium poisoning was unofficially reported as the cause of Itai-Itai, disease supported by a 1942 report on osteomalacia at a cadmium factory in France. Then Toyama prefectural authorities initiated measures for pollution control and began to investigate the causes of the disease. The Japanese Ministry of Health and Welfare and the Ministry of Education did not begin investigating the connection between Itai-Itai disease and the environmental pollution of the area by cadmium until 1965, ten years after the suspected causative agent was announced, and neither group acknowledged the most likely cause, mining wastes. Moreover, because no new cases appeared after 1955, it was difficult to correlate the victim's cadmium intake from water and foods, and efforts were focused on determining the relationship between exposure, rice content, proteinuria, and cadmium excretion. In 1966 a national nutrition survey reported that the average daily intake of cadmium in Japan was about 0.060 milligrams, while the intake of cadmium in food in the endemic area was estimated at 0.6 milligrams per day, or ten times the normal intake. About half of this was estimated to come through the intake of cadmium-contaminated food. In the opinion of some experts, the intake of cadmium from food and water had been very much higher in the earlier years of the twentieth century, when the disease was developing and before sedimentation lagoons were in place.

Since 1968, many investigations have also been conducted to evaluate the effects of environmental cadmium contamination in other areas of Japan where mines extract sulfide ores of copper, lead, and zinc. A second incident of cadmium contamination of the environment resulted from the Ikuno mine owned and operated by the Mitsubishi Metal and Mining Company. The wastewater was released into the Ichi and Maruyama rivers, which drained to the Seto and Japan seas, respectively. The concentrations of cadmium in the rice paddy soils ranged from about ten to twelve parts per million in the vicinity of the mine to between one and three parts per million in the downstream watershed draining into the sea. The cadmium concentrations in unpolished rice produced from 1972 to 1979 in these river basins ranged between 0.1 and more than 1.0 parts per million cadmium and correlated with the distribution patterns of cadmium in the soils.

Japanese governmental associations did not begin investigating the connection between Itai-Itai disease and the environmental pollution of the area by cadmium until 1965, ten years after the suspected causative agent was announced, and they did not acknowledge the most likely cause, mining wastes.

Similar cases were also reported in other mining districts downstream from the Hosokura mine on the Namari and Nikazama rivers in Miyagi prefecture, the Usui and Yanase rivers in Gumma prefecture, the Okudake River in Ōita prefecture and the Sasu and Shiine rivers on Tsushima Island in Nagasaki prefecture. Around the latter in 1964, soil, rice, and water contained even higher levels of cadmium than those along the Jinzu River; and the victims had the same symptoms. Even though individuals downstream from these mines were identified with typical Itai-Itai disease symptoms, they were never recognized as official victims because the doctors in the Ministry of Health and Welfare (Environment Agency after 1971) Itai-Itai Study Group lacked unanimity in their diagnosis. As a result, only the patients within the endemic area of the Jinzu river basin have been compensated, and that was after political intervention.

In 1968 a member of the House of Councilors asked the minister of health and welfare to clarify the relationship between Itai-Itai disease and pollution from the mine. Then the disease was officially declared to have resulted from the following process: renal tubular dysfunction first caused by cadmium poisoning, followed by the development of osteomalacia, which resulted in Itai-Itai disease, with inductive factors such as pregnancy, lactation, changes in internal secretions, aging, nutrition, and deficiency in calcium. At the same time the Ministry of Health and Welfare announced that effluents from Kamioka mine were the only practical source. Proclaiming that national health should take precedence over industrial profits, for the first time in Japan, responsibility was laid directly on a large enterprise for the pollution it caused.

Impact

On March 9, 1968, after seven years of negotiations with Mitsui Mining and Smelting Company, twenty-eight patients and surviving family members filed a damage suit against the company. The decision of the Gifu District Court on June 30, 1971, favored the plaintiffs and was appealed by the company; but on August 9, 1972, the Nagoya High Court upheld the compensation order, seventeen years after Dr. Hagino first drew attention to the problem. The survivors and their families were the first pollution victims in Japan to win a lawsuit against a major company. The victims won because in 1939 Article 109 of the Mining Act had imposed strict liability on mining facilities for damages caused by their activities. The plaintiffs only had to prove that cadmium discharged from the mine had caused their disease, not that the company was negligent. As epidemiological proof of causation sufficed as legal proof in this case, this principle set a precedent for other pollution litigation as well. Still, only patients within the Jinzu river basin were acknowledged and compensated, not those in other areas where victims had been identified, such as Gumma, Miyagi, Ōita, and Nagasaki prefectures.

Besides losing their health, the farmers' livelihoods were also jeopardized. Even after the discharges of cadmium were lowered, the distribution of cadmium in surface soil remained a problem. Where paddy soil contained six parts per million cadmium, concentrations increased as much as two hundred times in rice roots and twenty times in polished rice grains. In 1969 the government initially set a surveillance level at 0.4 parts per million cadmium in unpolished rice. However, because much of the contaminated farmland produced grain in excess of that level, in 1970 the Foodstuffs Hygiene Law officially increased the maximum allowable standard to 1.0 parts per million cadmium for unpolished rice and 0.9 parts per million for polished rice. Otherwise, the extensive contamination of Japanese rice paddy soils would produce huge quantities of unmarketable rice. Normally, polished rice from all areas of Japan averaged 0.07 parts per million cadmium.

Because the kidneys are considered to be the most critical organ in chronic cadmium processing, the maximum allowable limit was set to protect an individual from the renal dysfunction that often precedes Itai-Itai disease. However, many scientists have contended that the existing Japanese cadmium standards of 1.0 and 0.9 parts per million cadmium for unpolished and polished rice do not adequately protect individuals against cadmium-induced renal tube damage. This position is supported by the joint FAO-WHO Expert Committee on Food Additives. In 1972 the committee proposed a provisional tolerable weekly intake of between 0.4 and 0.5 milligrams of cadmium per person, equivalent to about 0.065 milligrams of cadmium per person per day. Accordingly, the only way a person weighing fifty kilograms who consumes five hundred grams of rice daily could stay within the FAO-WHO guidelines would be if the rice contained less than 0.1 parts per million cadmium and the person had no other cadmium intake.

In an effort to restore the farmland, Japanese authorities instituted a program wherein, each year, the most highly contaminated soils in a small area are exchanged with uncontaminated soils, and less contaminated soils are rehabilitated through the addition of lime, phosphate, and a cadmium sequestering agent, ethylenediaminetetraacetic acid (EDTA). As of December 1990, about 4,340 hectares (66.7 percent) of the approximately 6,510 hectares of the most highly contaminated farmland had been restored. In the remaining contaminated areas, families continue to eat home-grown rice and other farm produce; however, the effects of cadmium poisoning can be mitigated by treating the victims with massive doses of vitamins (B_1, B_{12}, D), calcium, and various hormones. Studies are being conducted to discover the best methods to achieve more rapid excretion of cadmium, but with decreasing cadmium exposure and improved standards of nutrition, Itai-Itai disease should no longer occur. The human and economic costs involved in reaching that point have been very high.

The Study Group for Global Environment and Economics estimated the total annual cost in 1989 for health indemnities, court judgments, crop losses, and soil restorations for the Jinzu river basin at ¥2.5 billion ($17.5 million) compared with an estimated ¥603 million ($4.2 million) for pollution control. The group concluded that it would be much more cost effective to implement pollution control early to prevent damages. Thus, beside a legal precedent to establish fault by the polluter, economic considerations may well encourage more forthright efforts to prevent this type of environmental pollution and human health degradation in the future.

Frank M. D'Itri

Where to Learn More

Kjellström, T. "Itai-Itai Disease." *Cadmium and Health: A Toxicological and Epidemiological Appraisal*, vol. 2, edited by L. Friberg, C. G. Elunder, T. Kjellström, and G. F. Nordberg. CR Press, 1986. pp. 257-90.

Kobayashi, J. "Pollution by Cadmium and the Itai-Itai Disease in Japan." *Toxicity of Heavy Metals in the Environment*, part 1, edited by F. W. Oehme. Marcel Dekker, pp. 199-260.

Kogawa, K. "Itai-Itai Disease and Follow-Up Studies." *Cadmium in the Environment*, part 2, edited by J. O. Nriagu. Wiley, 1981, pp. 1-37.

Study Group for Global Environment and Economics. *Pollution in Japan-Our Tragic Experience*, 1991, pp. 43-8.

Tsuchiya, K., ed. *Cadmium Studies in Japan: A Review*. Kodansha and Elsevier/North-Holland Biomedical Press, 1978.

WHO. "Environmental Health Criteria for Cadmium." *Ambio*, vol. 6, 1977, pp.287-90.

DDT insecticide contamination

(1939 to present)

Contamination of the ecological system resulting from the use of DDT has exposed the potential destructiveness of pesticides.

Background

DDT is the popular name for the chemical compound dichloro diphenyl trichloroethane, which was widely used as an insecticide from World War II until the early 1970s. With its prolonged toxicity, DDT can contaminate the entire food chain, destroying much plant, aquatic, bird, and animal life. DDT contamination extends to human beings, and a recent study links DDT to breast cancer in women. The publication of Rachel Carson's *Silent Spring* in 1962 was primarily responsible for raising public consciousness in the United States about the ecological threat posed by DDT and other pesticides. Civic activism and lawsuits—in the face of resistance from chemical manufacturers and federal policymakers—led to the formation of the U.S. Environmental Protection Agency in 1970 and the banning of most uses of DDT. DDT, however, is still used in many countries, including Mexico.

The proper chemical name for DDT is 2,2-bis (p-chlorophenyl)-1,1,1-trichloroethane. The compound has the empirical formula of $C_{14}H_9Cl_5$. DDT is a colorless crystalline powder that is slightly aromatic. It is insoluble in water, moderately soluble in alcohols, and highly soluble in some oils and fats. Its melting point is 109 degrees celsius. DDT was first synthesized in 1874 by the German chemist Othmar Zeidler. In 1939, the Swiss scientist Paul Müller discovered the insecticidal properties of DDT. He received the Nobel Prize in Medicine in 1946 for his discovery of DDT's effectiveness in controlling insect-borne human diseases.

DDT was used widely and successfully during World War II to eliminate insect populations carrying typhus and malaria. DDT was also useful in preventing yellow fever, river blindness, elephantiasis, and bubonic plague. After World War II, DDT was primarily exploited for agricultural purposes. Before DDT became available, farmers and gardeners used arsenicals (compounds containing arsenic) to control pests. Arsenicals are very costly to manufacture, and DDT, which acts as a nerve gas on insects, was both cheaper and more effective. DDT was used extensively against beetles, moths, lice, butterflies, flies, and mosquitoes. By 1959, 124 million pounds of DDT were being manufactured annually for disease control, agriculture, gardening, and animal husbandry. DDT has been used throughout the world, but most heavily in tropical zones.

Details of the Controversy

DDT has proven to be a significant threat to the entire ecological system. DDT poisoned the habitat and food supplies of large populations of plant, marine, bird, and animal life. In humans, where DDT's low solubility in water and high solubility in fat result in great bioconcentration, the primary danger lies in DDT's chronic effect (its effect over time).

The publication of Rachel Carson's *Silent Spring* in 1962 was largely responsible for alerting the public to the widespread ecological threat posed by DDT. Born in 1907, Carson attended the Pennsylvania State College for Women and later Johns Hopkins University, where she earned a master's degree in biology. Carson taught at Johns Hopkins and the University of Maryland before taking a position with the U.S. Bureau of Fisheries (which later became the U.S. Fish and Wildlife Service) in 1936. A respected scientist and author, Carson wrote two books about the ocean and shorelife before her publication of *Silent Spring*.

The theme of *Silent Spring* is the interdependence of life on earth, such that polluting the ecosphere with pesticides or other toxic compounds results in unintended and unforeseeable consequences. The book documented the effects of DDT and other pesticides in reducing the populations of birds, fish, and certain mammals when indiscriminantly applied to control specific insects. Carson did not denounce all use of pesticides, but rather advocated the selective use of pesticides based on a full understanding of the consequences of their use.

Silent Spring begins with "A Fable for Tomorrow," which describes a village with no birds, fish, or wild animals. Another chapter describes the toxic effects of pesticides, especially DDT, on organisms. A later chapter—"And No Birds Sing"—details the effect of DDT on birds. In this chapter, Carson describes the spraying of elm trees with DDT to slow the destruction due to Dutch elm disease. Robins feed on the earthworms underneath the trees, which have ingested the poisonous leaves that fall to the ground. The robins become sterile or die from the poison, resulting in a rapid reduction in their numbers. The book's last chapter, "The Other Road," discusses the use of alternative pest control methods, such as predator insects, bacteria, sterilants, crop rotation, and ecological biodiversity (allowing many plants and animals to interact).

Although naturalists in the United States had noticed a reduction in the local birdlife population, most did not suspect the cause until *Silent Spring* was published. Many citizens then became concerned about the environmental effects of pesticides and turned to the legal system, Congress, and the

Executive Branch of the federal government to force a reevaluation of pesticide use.

Impact

In 1962, the year *Silent Spring* was published, President John F. Kennedy, Secretary of Agriculture Orville Freeman, and Secretary of the Interior Steward Udall turned their attention to ecological hazards posed by pesticide use. Their purpose was to weigh benefits in agricultural productivity against the large-scale environmental contamination. Many members of the Agriculture and the Interior departments, however, resisted attempts to reevaluate or restrict the usage of DDT and other pesticides. Many of these federal employees had close alliances with the pesticide manufacturers, who aggressively attacked Rachel Carson and other ecological "fanatics." Thus, many policymakers in the federal government continued to defend practices that obviously resulted in indiscriminate poisoning of the environment.

In 1964, Congress acted to amend the Federal Insecticide, Fungicide, and Rodenticide Act (FIFRA), tightening label guidelines and requiring that safety information be provided. In 1970, the Environmental Protection Agency, which consolidated the major programs and agencies that dealt with environmental pollution, was created by President Richard Nixon. The Environmental Protection Agency was authorized to regulate the introduction of new agricultural chemicals (and other compounds) into the marketplace, establishing safety guidelines and testing procedures. Disagreement continues to exist between the Agriculture Department and the EPA concerning the appropriate parameters for the risk/benefit analysis of a pesticide.

Since the banning of DDT, other chlorinated compounds and organophosphates have been developed. Many of these compounds are much more toxic than DDT, but lose their toxicity quickly and break down into relatively harmless products.

In 1969, a year before the EPA was established, federal regulations had begun phasing out the use of DDT. At the end of 1972, EPA Administrator W. D. Ruckelshaus announced a ban on almost all uses of DDT. Citizen's lawsuits had been responsible before this date for the cessation of DDT use in many local and state jurisdictions; Michigan and Arizona, for example, had banned the use of DDT in 1968.

Since the banning of DDT, other chlorinated compounds and organophosphates have been developed. Many of these compounds are much more toxic than DDT, but lose their toxicity quickly and break down into relatively harmless products. While farmers applied pounds of DDT per acre, only grams per acre of the newer pesticides are required. Moreover, the newer pesticides are often designed to destroy only specific insects or weeds.

Nonetheless, many problems with pesticide use remain. In *Silent Spring,* Rachel Carson predicted the increasing resistance of insects to DDT and other pesticides. This prediction was based on scientific evidence available to her at the time—the late 1950s and early 1960s—and proved accurate. In the 1930s, before DDT, American farmers lost about one third of their crop to insects, weeds, and disease. Currently, in the 1990s, farmers still lose the same proportion of their crops, with pesticide costs exceeding $4 billion. Now, however, the entire ecosystem contains toxic compounds that may effect future organisms in unknown ways.

The long-term effects of DDT on humans have not been fully determined. Research being carried out world-wide shows that DDT is found in all human blood and fat, and shows up in high levels in mother's milk in the tropics. There are instances of people being exposed to relatively large amounts of DDT without exhibiting obvious symptoms of poisoning. For example, the small city of Triana, Alabama, is downstream from the U.S. Army's Redstone Arsenal. Olin Chemical Co. manufactured DDT at this site from 1947 until 1970, pouring tons of DDT-laden wastewater into Indian Creek, which runs through Triana. The local fish and water bird population diminished significantly. Triana residents eating the Indian Creek fish, however, accumulated high levels of DDT and other chlorinated compounds without exhibiting noticeable effects.

There is evidence, however, that high DDT levels may affect the behavior of children, increase suicide among older people, and either cause pancreatic cancer or increase the chances of its occurrence. Most recently, a study by Mary Wolff of the Mount Sinai School of Medicine in New York City ties DDT levels in women to breast cancer. An increase in breast cancer was known to have coincided with the widespread application of DDT, but this study, published in 1993 by the *Journal of the National Cancer Institute*, is the first to establish a statistical link.

JULIAN LARSON

Where to Learn More

Carson, Rachel. *Silent Spring*. Houghton Mifflin, 1962.

Garabrant, D. H. et al. "DDT and Related Compounds and Risk of Pancreatic Cancer." *Journal of the National Cancer Institute*, 1992, pp. 764-71.

Graham, Frank, Jr. *Since Silent Spring*. Houghton Mifflin, 1970.

Lewis, Richard L., ed. *Sax's Dangerous Properties of Industrial Materials*, 8th Edition. Van Nostrand Reinhold, 1992.

Silent Spring Revisited. American Chemical Society, 1987.

Sittig, Marshall. *Handbook of Toxic and Hazardous Chemicals and Carcinogens*, 2nd Edition. Noyes, 1985.

Stranahan, Susan Q. "The Town That Ate Poison." *National Wildlife*, June/July 1984, pp. 16-19.

"The Joy Ride Is Over: Farmers Are Discovering That Pesticides Increasingly Don't Kill Pests." *U.S. News & World Report*, September 14, 1992, pp. 73-74.

Whorton, James. *Before Silent Spring: Pesticides and Public Health in Pre-DDT America*. Princeton University Press, 1974.

Love Canal toxic waste contamination

Niagara Falls, New York

(1942-1953)

A Niagara Falls, New York, neighborhood, built on top of a toxic waste dump, had to be abandoned in 1980 after it was determined to be a threat to public health.

Background

Love Canal is probably the best known, and most infamous, of the nation's toxic waste sites. The name has come to symbolize chemical contamination of areas and neighborhoods.

Between 1942 and 1953, the Olin Corporation and the Hooker Chemical Corporation buried over twenty thousand tons of deadly chemical wastes, including dioxin, in Love Canal, located near Niagara Falls, New York. Many of the compounds dumped there are known to cause cancer, miscarriages, birth defects, and other illnesses and disorders.

In 1953, Hooker Chemical donated the land to the local board of education for a token payment of one dollar, but did not clearly warn of the dangerous nature of the chemicals buried there, even when a school, homes, and playgrounds began to be built in the area. In 1976, after years of unusually heavy rains raised the water table and basements became flooded, problems became apparent in the neighborhood. Homes began to reek of chemicals; children and pets came home with chemical burns on their feet and hands, and some pets even died, as did trees, flowers, and vegetables.

Soon, people in the neighborhood began to experience an extraordinarily high number of serious and unexplainable illnesses, including higher-than-normal rates of cancer, miscarriages, and deformities in newborns. In the spring of 1978, alarmed by the situation and frustrated by the lack of action on the part of local, state, and federal authorities, a twenty-seven-year-old housewife named Lois Gibbs took matters into her own hands. She organized her neighbors into the Love Canal Homeowners' Association, and began a two-and-a-half-year-long fight to have the government relocate them to another area. In the process, she turned Love Canal into a household name across the country, and helped focus nationwide attention on the growing problem of toxic waste disposal, which culminated in passage of the so-called Superfund Law in December of 1980.

Details of the Disaster

As the number of people experiencing health problems in the neighborhood rose, it became increasingly apparent that the community might not be safe for human habitation. Under pressure from neighborhood activists and as a result of extensive coverage in the local and national media, government officials finally began to recognize the need for action. In August 1978, the New York State health commissioner, Dr. Robert P. Whalen, recommended that pregnant women and children under the age of two be evacuated, saying that there was "growing evidence. . .of subacute and chronic health hazards, as well as spontaneous abortions and congenital malformations" at the site. When the state tested the air, water, soil, and homes for toxic chemicals later that month, over eighty different compounds were found, many of which were thought to be capable of causing cancer. Chemical pollution of the air was measured at two-hundred-and-fifty to five thousand times the levels considered safe.

The 1978 study found an unusually high miscarriage rate of 29.4 percent in the neighborhood, with five of the twenty-four children born in the area listed as having birth defects. State health officials estimated that women in the area had a 50 percent higher-than-normal risk of miscarriage. Another report found that in 1979, only two of seventeen pregnant women in Love Canal gave birth to normal children. Four had miscarriages, two had stillbirths, and nine had babies born with defects.

Epidemiological studies of the affected population revealed an alarming pattern of illness among exposed residents. For example, on Ninety-sixth Street, where fifteen homes were located, eight people developed cancer in the twelve-year period between 1968 and 1980: six women had cancerous breasts removed; one man contracted bladder cancer; and another developed throat cancer. In addition, a seven-year old boy experienced convulsions and died of kidney failure, and a pet dog had to be destroyed after developing cancerous tumors.

Some of the most alarming health data were gathered by Dr. Beverly Paigen of the Roswell Cancer Institute, who found a much higher incidence of illness among people who lived in houses that were located above moist ground or wet areas—those most prone to contamination by rising ground water. On March 21, 1979, testifying before the House Subcommittee on Oversight and Investigations, she described the tragic history of several families who lived in one such house located directly over liquid wastes that were seeping out of the ground. Among the four families, there existed three cases of nervous breakdown, three hysterectomies due to uterine bleeding, cancer, or both, and numerous cases of epilepsy, asthma, and bronchitis.

Dr. Paigen also found a significant excess of childhood disorders among youngsters born to residents of wet areas, including nine instances of birth

Section of the fence surrounding Love Canal where a residential community had been built over a dumpsite containing over twenty thousand tons of deadly chemical wastes.

defects among the sixteen children born in such areas between 1974 and 1978. She determined that the overall incidence of birth defects was 20 percent; the miscarriage rate was estimated at 25 percent, compared with just 8.5 percent for women moving into the area. She also reported that eleven out of thirteen hyperactive children lived in wet areas, and that 380 percent more asthma occurred there than in the "dry" areas of Love Canal. The incidence of urinary disease and convulsive disorders was almost triple that of dry areas, and the rates of suicides and nervous breakdowns almost quadruple. Significantly, Dr. Paigen found that Love Canal residents suffering from illnesses ranging from migraines to pneumonia to severe depression reported marked health improvements when they moved out of the area and away from the contamination. Eventually, New York State authorities termed the area "a grave and imminent peril" to the health of those living nearby. Several hundred families were moved out of the neighborhood, and the others were advised to leave; the school was closed and a barbed wire fence placed around it.

In May 1980, further testing revealed high levels of genetic damage among neighborhood residents, resulting in an additional seven hundred and ten families being evacuated at a cost estimated to run between $3 million and $60 million. In October 1980, President Jimmy Carter declared the neighborhood a disaster area. In the end, some sixty families decided to remain in their homes and reject the government's offer to buy them out. The total cost for the cleanup has been estimated at $250 million.

Twelve years after the neighborhood was abandoned, the state of New York approved plans to allow families to move back into some parts of the area, and homes were permitted to be sold.

Impact of the Disaster

The major impact of the massive publicity generated by Love Canal was to draw national attention to the dangers of toxic chemicals and hazardous waste, and to pressure Congress and the White House to pass laws to address the problems caused by such pollution. Public outrage was particularly fueled by revelations that Hooker had engaged in a coverup of the dangers of the situation at Love Canal. When Hooker deeded the land to the Niagara Falls Board of Education for one dollar, the company did not warn the board about the lethal nature of the chemicals it had buried there. But Hooker did take pains to protect itself: the legal document transferring the property disclaims liability for any deaths or injuries that might occur on the land and specifies that the school board assume responsibility for any claims that result from exposure to the buried chemicals. In its letter agreeing to donate the land to the new school, Hooker's executive vice president adopted the tone of the good corporate citizen, stressing the company's desire to serve the needs of its community.

Even when a neighborhood began to be built in the area, Hooker failed to issue warnings about the dangers there. On June 18, 1958, a company memo observed that "the entire area is being used as a playground," and that "3 or 4 children had been burned by material at the old Love Canal property." But concerned about legal repercussions, Hooker remained silent. Ten years later, roadworkers building a highway near Love Canal uncovered leaking drums full of toxic chemicals that Hooker analyzed and discussed in a March 21, 1968 internal company memo. Again, no action was taken to alert the public about the presence of these dangerous chemicals.

Indeed, the chemical industry continued to downplay the threat to public health documented at Love Canal long after evidence was uncovered. For example, on October 14, 1979, the late Armand Hammer, then chairman of Hooker's parent company, Occidental Petroleum, said on NBC's "Meet the Press" that the Love Canal problem had "been blown up out of context." And when Hammer chaired Occidental's annual stockholder's meeting in May 1980, the company rejected a resolution calling on the firm to adopt policies designed to prevent future Love Canals. Even the vice president of the Chemical Manufacturers Association, Geraldine Cox, suggested during a July 2, 1980 television interview on the "MacNeil/Lehrer News Hour" that Love Canal residents were comparable to hypochondriacs with imaginary or exaggerated illnesses.

By 1980, shocking information was being released concerning thousands of other toxic waste dumps scattered across the country that were potential Love Canals. The U.S. Environmental Protection Agency (EPA) estimated that there existed thirty-two thousand to fifty thousand hazardous waste disposal sites in the United States, of which twelve hundred to two thousand might pose "significant risks to human health or the environment." And new chemical waste sites were being discovered at a rate of two hundred a month.

At the time, the EPA estimated that only about 10 percent of the one hundred and fifty million tons of hazardous wastes being generated each year were disposed of in a safe and legal manner; 90 percent were being dumped illegally or disposed of in a way that posed a potential threat to humans or the environment. The agency called the situation "the most serious environmental problem in the U.S. today." (More recent estimates put the amount of hazardous waste being produced each year at 300 million tons—roughly a ton for every man, woman, and child in the country.)

On May 16, 1979, Assistant Attorney General James Moorman testified before the House Subcommittee on Oversight and Investigations that toxic waste dumping was the "first or second most serious environmental problem in the country." He pointed to a lack of effective environmental regulations and a lack of enforcement of existing anti-dumping laws.

Since Love Canal, and partly in response to the uproar it caused, laws designed to protect the public from toxic chemicals have been passed or strengthened. In November 1980, a provision of the Resource Conservation and Recovery Act (RCRA) went into effect, theoretically requiring that toxic wastes be tracked "from cradle to grave." And on December 11, 1980, President Carter signed into law a diluted version of the Environmental Emergency Response Act—the so-called Superfund Law—that creates a

Drums containing contaminated sediment from Love Canal. Revelations that the Hooker Chemical Company had engaged in a coverup of the site's lethal effects prompted widespread public outrage.

fund to pay for the cleanup of hazardous waste sites and makes owners and operators of waste disposal sites, as well as producers and transporters of hazardous materials, liable for cleanup costs. But while these laws have greatly reduced the improper disposal of dangerous chemicals, weak government enforcement has significantly weakened their effectiveness. For example, as of the beginning of 1993, only 149 of 1,256 priority Superfund sites had been cleaned up.

Despite the apathetic response of the Reagan and Bush administrations to environmental concerns, the new laws and regulations have eliminated many of the abuses of the past, and have led the chemical industry to begin recycling much of its waste (a reported half of the thirty-eight billion pounds produced in 1991). Nevertheless, toxic waste disposal remains a critical problem. In May 1993, the Agency for Toxic Substances and Disease Registry estimated that there existed over sixty-four thousand areas in the United States contaminated by hazardous wastes that require cleanup; that some forty-one million Americans were living within four miles of a hazardous waste site and might be at some risk; and that at least two hundred and fifty thousand people had been affected by "acute release events," such as waste releases, spills, or accidents. Numerous studies show that people living near such waste dumps have a much greater risk of bearing children born with birth defects, of contracting cancer, and experiencing other serious health effects.

It may well prove impossible to completely clean up the nation's dump sites at any price, but the anticipated costs for steps that must be taken in the next few years are staggering. One EPA study estimated that it would cost over $44 billion just to clean up the most dangerous sites, with the public having to pick up half the tab. Other projections put the figure for addressing Superfund waste sites at over a trillion dollars—which would require half a century to accomplish.

LEWIS G. REGENSTEIN

Where to Learn More

Blumenthal, R. "Fight to Curb 'Love Canals.'" *New York Times*, June 30, 1980.

Brown, M. H. "Love Canal Revisited." *Amicus Journal*, Summer 1988, pp. 37-44.

———. "A Toxic Ghost Town: Ten Years Later, Scientists Are Still Assessing the Damage from Love Canal." *Atlantic*, July 1989, pp. 23-26.

Gibbs, Lois. *Love Canal: My Story*. State University of New York Press, 1982.

Kadlecek, M. "Love Canal—10 Years Later." *Conservationist*, November-December 1988, pp. 40-43.

"Fleeing Love Canal." *Newsweek*, June 2, 1980, p. 56.

Regenstein, L. G. *Cleaning Up America the Poisoned*. Acropolis Books, 1993, pp. 136-45.

Hanford radioactive waste reservation

Washington State

(1944-1988)

Dumping of radioactive waste from a plutonium-producing reservation has contaminated the environment along the Columbia River.

Background

The Hanford Reservation is an environmental disaster that is already fifty years old, may be getting worse, and may only be resolved in thousands of years, when the radioactive life of the wastes is exhausted. The beginnings of the disaster were dramatic. During the bitterest moments of World War II, the United States military and scientific brain trust conceived of the Manhattan Project, in which the atomic bomb was developed that obliterated Nagasaki, Japan, and concluded the war. The Project was accorded the highest praise; extraordinary efforts in science, technology, engineering, mobilization of people and equipment, and defense strategy terminated the war, redirected global defense, focused the future of energy on the atom, and literally changed the world. The project behind the Project was the development of the Hanford Engineering Works (and two other sites in Tennessee), which Major General Leslie R. Groves proposed in December 1942, to produce plutonium for the bomb. In twenty-eight months, the plant was built and seventeen thousand people worked secretly and with limited knowledge of their purpose to produce the first atomic explosion, the Trinity bomb test in Alamogordo, New Mexico. Only one month later, Nagasaki was destroyed.

The Hanford Engineering Works was constructed on the Hanford Reservation, 560 square miles of desert in southeastern Washington State. The Yakima River, the Yakima Range, and the City of Richland border the site to the south, and Richland is the "company town" that houses workers and hosts, participating businesses, and laboratories. The city of Pasco is fifteen miles south, and the city of Yakima is about forty miles west of the reservation. The Columbia River flows through the northern part of the reservation and continues southward to form part of the border between Washington and Oregon until it empties into the Pacific Ocean. Of all the rivers flowing into the Pacific Ocean from the coast of North America, the Columbia River is the largest.

Players at the Hanford site have included government agencies, agency-sponsored scientific laboratories, and private industrial contractors. The Atomic Energy Commission (AEC) operated the site originally, but the Department of Energy (DOE) has been the controlling agency since its formal creation. The General Electric Corporation managed the Hanford site for the DOE, but Westinghouse Electric Corporation is the current manager. Exxon Nuclear and Rockwell International have also participated. There are also commercial nuclear dumps in the Hanford area that receive nuclear waste shipped from elsewhere. These were the subject of headlines in 1979 when Washington Governor Dixy Lee Ray closed one such dump because of improper waste handling.

At the Hanford works, eight reactors were built on the banks of the Hanford reach of the Columbia River between 1944 and 1955. These were single-pass reactors that pumped cooling river water into the reactor cores, stored the water in retention basins where it cooled for a few hours, then released it into the river. Single-pass reactors are illegal today because concentrations of sixty different radionuclides (radioactive species of atoms), heavy metals, and miscellaneous chemicals were released with the water. A ninth reactor that used recirculated cooling water, the N-reactor, was built in 1963 and operated from 1971 to 1988. Plutonium fuel from these reactors was reprocessed chemically at the Plutonium and Uranium Extraction plant (PUREX). The PUREX plant is also idle now.

Details of the Disaster

Production of plutonium at Hanford only began with preparation for Nagasaki. During the war and for twenty-five years afterward, its operations continued as the nation's arsenal of nuclear weapons was assembled. In 1950, the U.S. Public Health Service and Washington and Oregon states became concerned that contaminants might be entering the Columbia River from the Hanford Reservation. Data were not available from Hanford, where environmental records would remain cloaked in secrecy until 1986. Public health, the fishing industry, and the purity of the river were the driving concerns. Salmon, steelhead trout, and other fish thrive in water that is cold enough to limit disease-causing bacteria. Many game birds such as Canada geese, ducks, and pheasants also nest in the river's broad basin. After its discharge from the retention basins, the reactor water was still hot in both temperature and radioactivity; as a result, trout and salmon became smaller in size and disease-prone, and their population dropped. By 1959, the Hanford reactors were releasing about 13,300 curies of radionuclides per day; one curie equals the radioactivity of one gram of radium.

The health service's studies raised questions about the safety of well water (fed by the Columbia) for Richland and other towns, about the reproductive capabilities of the animal population along the river, and about the safety of these creatures as food for humans. Of course, release of contamination into a large body of moving water dilutes the contamination. But biologists (in studies done on the Hanford site in 1946 but not made public until 1986) found that radioactive materials ingested by fish were reconcentrated in tissues so that radioactive levels were six to thirty times higher than in the river water; the same studies repeated in 1947 showed that these reconcentration levels had climbed to 100,000 to 170,000 times the concentrations in water. These concentrations build up in organs rather than flesh, and warnings were issued to the local Native American population, who relish fish organs as delicacies, to avoid them. Ducks living on the river also had concentrated radioactive levels. In 1961, radiation surveys of shellfish near the river's mouth along the Washington and Oregon coasts (over three hundred miles away) were found to be heavily contaminated. Had reactor production ceased immediately after the highest level of production in 1945, a radioactive legacy would still be present in river-bottom sludge, seepage effluent, and the biological food chain.

In addition to undergound storage tanks holding high-level waste, some N-reactor fuel remains encapsulated in underwater storage basins near the Columbia River; these basins are now forty years old, and three to five percent of the fuel elements are known to be leaking.

To extract plutonium, PUREX and the other plants produced large volumes of radioactive liquid waste. The United States has produced 100,000 kilograms of plutonium for nuclear weaponry. To produce one kilogram of plutonium, the PUREX plant generated 340 gallons of high-level (highly radioactive) waste, 55,000 gallons of low- and mid-level waste, and over 2.5 million gallons of cooling water. When production was more urgent than concerns about disposal, the high-level waste was designated for "temporary" storage in a series of single-walled steel and concrete underground storage tanks. Soft (carbon) steel was used to construct the tanks because of the cost and lack of availability of stainless steel. Having set the precedent for "temporary," high-level waste continued to be stored in this fashion, and there are now 149 single-walled and 28 double-walled million-gallon tanks below ground on the Hanford Reservation. Some of these may also contain chemical mixtures that have unknown effects and may be explosive. Even some N-reactor fuel (approximately 6,100 cubic feet of solid waste) remains encapsulated in underwater storage basins near the Columbia River; these basins are now forty years old, and three to five percent of the fuel elements are known to be leaking.

By 1991, the single-shell tanks contained 157 million curies of radioactive material, equivalent to more than five times the amount of radioactive material released during the Chernobyl disaster. The double-walled tanks contained 111 million curies. The single-walled tanks are known to have leaked 750,000 gallons of waste; the exact volume is not known because the liquid placed in some of the tanks initially was not measured, and some records are missing. Monitoring systems were installed in some of the tanks; but these were primitive at best, read by untrained operators, and not maintained. Production of waste also continued faster than new tanks could be built, and operators had no choice but to place the liquid in leaking tanks.

The chemicals in these witches' brews may vary from tank to tank. During reprocessing, nitric acid was used to dissolve the spent fuel, and sodium hydroxide was added after the plutonium and uranium were removed to neutralize the acid so that it would not corrode pipes and storage tanks. Ferrocyanide and sodium titanate were added to cause cesium 137 and strontium 90 to precipitate from the liquid. These radioactive precipitants are stored in capsules at Hanford. These chemicals remain in storage with high-level waste that also contains unknown chemicals and a variety of radioactive species, all of which are heated by radioactive decay. One analogy equates a single-shell tank to "a railroad tank car full of radioactive liquid hot enough to boil itself for years on end and knock a Geiger counter off scale at a hundred paces." Hydrogen may accumulate in some tanks to explosive levels. In other tanks, a cesium-bearing crust has formed over the liquid and an oatmeal-like slurry has settled to the bottom; bubbles of hydrogen and nitrous oxide form in the slurry, raising the container top until the material "burps." Hydrogen is flammable at a threshold of five percent, and these bubbles contain thirty-five percent. Officials claim the chances of explosion are low because there are no ignition sources—and because such an explosion hasn't occurred yet.

The low- and mid-level waste was pumped into "cribs," an innocuous word for trenches or open-bottomed pits that were excavated into the soil and function like septic-tank drainage fields, allowing the contaminated liquid to percolate through the soil until it reaches the ground water. Surface ponds were used to hold the cooling water. An estimated two-hundred billion gallons of low-level wastes are known to have been dumped in Hanford's trenches and ponds. Pressure to keep up with production also excused improper disposal of over one-hundred million gallons of high-level waste into surface trenches. These wastes contain cesium 137, strontium 90, cobalt 60, technetium 99, and two plutonium isotopes, 239 and 240. Disposal of radioactive waste in trenches on federal land was banned by the Atomic Energy Commission in 1970.

Contaminants from the leaking tanks and the unlined excavations have entered the underlying groundwater aquifer and traveled to the Columbia River. So much liquid has entered the subsurface that the groundwater table under the Hanford Reservation has risen by seventy-five feet. Tumbleweeds growing in the desert have sucked water from the contaminated soil, and their stems are likewise contaminated. Radioactive chemicals were found to attract wild animals, which use them as a salt lick; in 1958, radioactive coyote and rabbit dung was found over a two-thousand-acre area.

Processing of the fuel rods in acid also generated radioactive iodine that was released into the air. These releases were greatest from 1944 to 1947 but continued until 1973. Humans who drank the milk from cows that grazed on grass on which the iodine 131 had settled experienced swollen glands, cancer, and especially thyroid conditions, since iodine concentrates in the thyroid gland in humans. A study showed that from 1944 to 1947, twelve-hundred children in the Hanford area received cumulative doses of 15 to 650 rads of airborne radioactive iodine. Five percent of the local population may have been exposed to 33 rads or more; one rad equals the amount of radiation from a dozen chest X rays. The same human population exposed to the airborne iodine also drank water and ate fish from the Columbia River containing radioactive phosphorus 32.

Impact

The history of operations at the Hanford reservation has resulted in several occurrences that can be described as "disasters," separately or cumulatively. With the exceptions of the blasts at Alamogordo and Nagasaki, the Hanford disaster has happened without a boom, blast, collapse, or roar. Dramatic headlines have not shouted it to the world; instead, small and often technical articles must be sought by diligent readers, or they must face mountains (sixty thousand pages by 1986) of previously classified environmental and scientific documents. Polls have found that most people are unaware of the nuclear waste hazard. The same people may think that nuclear power plants pose the greater hazard, when the weapons industry has produced many times over the volume of waste generated by peaceful uses of radioactivity. The harm done at Hanford has several causes: the urgency to produce the bomb, leakage of repositories, outright human bungling, faulty and immature technology, and scientific arrogance that stopped those aware of the real or potential hazard from admitting to it. For a time, production of plutonium was more important to decision makers than the future of the Columbia River basin and its residents.

Solutions to the many environmental hazards at Hanford have not been found absolutely, but the brain trust has been diverted to this search. The Pacific Northwest Laboratory (PNL), which is operated by the Battelle Memorial Institute, conducts many programs related to operations at the Hanford Reservation, including the Environmental and Molecular Sciences Laboratory, established recently to research waste management techniques and environmental remediation methods. Money has also been dedicated to environmental restoration, with $6 billion committed for the 1990-95 period with an estimated "ultimate" cost of $60 billion. Health risks associated with radioactivity including the cumulative nature of doses and the bioconcentration of radionuclides are better understood. Training programs and monitoring systems have advanced rapidly. And there is no shortage of ideas for solving the problems posed by the wastes themselves, although each potential route will require expensive characterization, study, field trials, and anticipation of the consequences.

The high-level liquid waste appears most threatening. Dilution with water is impossible; the volume of forty times the Great Lakes would be needed to dilute the relatively small quantities of strontium-90 and cesium-137 in storage at three radwaste sites including Hanford. The volume can be reduced by evaporators, but this leaves a tank in the ground filled with a hazardous solid. The multiple-level wastes in the double-shell tanks can be pumped to a plant and segregated into a large volume of low-level waste and a small amount of high-level material. It is proposed to mix the low-level liquid with a cementlike material and bury the resulting "grout" on site. The high-level waste may also be converted by vitrification to a glasslike substance that also can be buried. A potential danger using this method was identified in 1992, when it was found that humic acid occurring naturally in groundwater can leach radioactive substances from confinement in borosilicate glass, the most stable of the glasses. This sand trap in the theory is presently being researched; such leaching may take years, but because plutonium only loses half of its radioactivity in twenty-four thousand years, long holding times are essential.

Studies have also been made of the possibility of storing solidified waste in deep repositories in basalt (a fine-grained igneous rock) at Hanford, tuff (naturally compacted volcanic ash) in New Mexico, salt domes along the Gulf Coast, and bedded salt in Utah and New Mexico. Again, time, money, study, and anticipation of the unforeseen are required for research. For the contaminants in soil, field tests are now underway in which in situ vitrification is accomplished by inserting electrodes into waste-bearing soil and heating and fusing it until a glassy rock results. Gases driven out must be trapped, but the treated contaminants can be left in place. Bioremediation using microbes present naturally in soil may work in some situations, although this method is slow. Other experiments with molecular dynamics and creation of molecular clusters, in which a harmful molecule is cloaked in a cluster of safe molecules, are being done in the laboratory. Robotics have been proposed for both hazardous laboratory and field work. Clearly, the decontamination and decommissioning of radioactive waste at the Hanford Reservation represents a challenging frontier for scientists and engineers.

GILLIAN S. HOLMES

Where to Learn More

Brown, M. *Laying Waste: The Poisoning of America by Toxic Chemicals.* Washington Square Press, 1981.

Gillette, R. "Radiation Spill at Hanford: The Anatomy of an Accident." *Science,* Vol. 181, No. 4101, pp. 728-730.

League of Women Voters Education Fund. *The Nuclear Waste Primer: A Handbook for Citizens.* Nick Lyons Books, 1985.

Levi, B. G. "Hanford Seeks Short- and Long-Term Solutions to Its Legacy of Waste." *Physics Today,,* March 1992, pp. 17-21.

Lipschutz, R. D. *Radioactive Waste: Politics, Technology, and Risk,* a report of the Union of Concerned Scientists. Ballinger Publishing, 1980.

Long, Robert Emmet, ed. *The Problem of Waste Disposal,* The Reference Shelf, vol. 60, no. 5, H. W. Wilson, 1989.

"Of Humus, Glass Beakers and Radwastes." *Science News,* March 14, 1992, p. 173.

Panel on Hanford Wastes, Committee on Radioactive Waste Management, Commission on Natural Resources, National Research Council, National Academy of Sciences. *Radioactive Wastes at the Hanford Reservation: A Technical Review.* National Academy of Sciences, 1978.

Stenehjem, M. "Indecent Exposure." *Natural History,* September 1990, pp. 6-21.

"There Was Death in the Milk." *Time,* July 23, 1990, p. 28.

Walker, C. A., L. C. Gould, and E. J. Woodhouse, eds. *Too Hot to Handle? Social and Policy Issues in the Management of Radioactive Wastes.* Yale University Press, 1983.

Mercury poisoning

Minamata Bay, Japan

(1955 to present)

Industrial pollution has caused often fatal Minamata disease among local residents, leading to legal action against the firm responsible and local and federal governments.

Background

Since 1953, at least 2900 residents of the Minamata Bay area of the Japanese island of Kyushu have fallen victim to organic mercury poisoning. The organic mercury was contained in or converted from the industrial wastes of the Chisso Company, which had been founded early in the century. The firm's production at mid-century of huge quantities of acetaldehyde and vinyl chloride generated the mercury-laden wastes that were responsible for the disease outbreak. Although the firm had acknowledged damages from its industrial wastes since as far back as 1925, federal policies allowed it to continue dumping untreated wastes into the local waterways. Fish and shellfish absorbed the organic mercury, which in turn was ingested by larger animals and humans, where its chronic effect is extremely toxic or fatal.

Minamata is the name of a city, a river, and a bay in Kumamoto ken prefecture on the island of Kyushu, the southernmost large island in the Japanese archipelago. The city of Minamata has a population of approximately 37,000, and is surrounded by villages of fisherman and farmers. The Amakusa islands guard Minamata Bay from the Shiranui Sea, resulting in a protected body of water that was known for its abundance of fish.

The Sogi Electric Company was established in 1906, and built an industrial plant in Minamata a year later. In 1906, the company changed its name to Japan Nitrogen Fertilizer Company (Nippon Chisso Hiryo Kabushiki Kaisha). The firm has since changed its name to New Japan Nitrogen Company, but is still popularly known as the Chisso Company ("chisso" is the Japanese term for nitrogen).

In 1932, the Chisso plant began producing acetaldehyde (C_2H_4O), which is important in the manufacturing of plastics, pharmaceuticals, photographic chemicals, perfumes, and other products. Mercury also was used to make other products at Chisso, especially vinyl chloride. The plant became one of

Inorganic mercury discharged from a local chemical plant contaminated fish and shellfish eaten by the residents of Minamata, Japan, many of whom contracted mercury poisoning as a result. Here, a victim of Minamata disease, or mercury poisoning, is being bathed by a relative.

the largest production facilities for these compounds in the 1950s and 1960s. In Chisso's production of acetaldehyde, vinyl chloride, and other products, mercury was used as a catalyst. A catalyst serves as an aid to the production process, and usually does not become incorporated into the product. A catalyst sometimes can be recycled, but this is not always economical, and the mercury used by the Chisso plant was discharged with its other industrial wastes.

The mercury used in these production processes was in its inorganic form. Inorganic mercury is the fluid found in silver thermometers, and is classified as a heavy metal. Inorganic mercury has a long history of human use; in ancient times, it was known as quicksilver and used for medicinal purposes. Between five hundred and one thousand grams of inorganic mercury were contained in the wastewater discharged from the Chisso plant for every ton of acetaldehyde produced, and it is estimated that over two hundred tons of inorganic mercury were discharged into Minamata Bay between 1949 and 1953. The toxicological effects of inorganic mercury have been recognized in modern times. Inorganic mercury poisoning results in tremors, loose teeth, headaches, fatigue, and minor psychological changes. In laboratories and hospitals, great precautions are taken to ensure that inorganic mercury does not come into dermal (skin) contact and is not ingested.

Even more hazardous, however, was the organic mercury formed during the Chisso Company's production processes, or resulting from microbial action on the inorganic mercury it dumped into the bay. Current scientific evidence indicates that specific microbes found in lakes, rivers, or ocean sediments can convert inorganic mercury into organic mercury ("organic" indicates that the compound is based on carbon, the basis of living organisms). Water conditions—such as acidity, temperature, and the presence of other metals—appear to influence this reaction.

The quantities of organic mercury generated were small in relation to the volume of wastes being discharged into the bay, and it is not certain whether organic mercury directly affects fish or shellfish. Nonetheless, organic mercury accumulates as it goes up the food chain, so that as birds, larger animals, and humans eat the fish and shellfish, sufficient quantities accumulate to affect higher organisms over time. The chronic (long-term) effect of organic mercury is very toxic to humans. Organic mercury concentrates in the brain, liver, kidneys, and in the fetus of humans. It is detectable in hair, blood, and urine.

The procedures for analyzing the levels of organic mercury in fish are complex because the levels present are very low. In the United States, the Federal Drug Administration standards are for a maximum of 0.5 ppm (parts per million) for fish. Fish in Minamata Bay register levels of organic mercury of up to 50 ppm. Existing procedures can detect organic mercury in levels as low as 0.5 ppb (parts per billion), but few laboratories are equipped to perform these tests.

Beginning in 1925, long before the chronic effects of the organic mercury became evident, Chisso acknowledged the damage its wastes were doing to the fishing in Minamata Bay, and began providing token compensation to the local Fishermen's Cooperative. Nonetheless, Chisso made no effort to treat its wastewater since the government did not require it. The Japanese government was encouraging industrial growth throughout the country, and there was little concern for the potential effects of industrial pollution. Criticism and complaints stemming from previous outbreaks of poisoning due to industrial wastes were suppressed by the central government and the industries through intimidation and heavy-handed procedures.

Details of the Disaster

During 1956, the medical head of the Chisso Company hospital, the Kumamoto University School of Medicine, and the local health office recognized that some strange disease was breaking out in the villages around Minamata. The disease became widely known as Minamata disease, and later investigations established that the disease had taken its first official victim in 1953.

Dr. Hajime Hosokawa, medical director of the Chisso Hospital, began examining patients with symptoms of the disease in April 1956. He reported to the Minamata Public Health Department that an unidentified disease had afflicted a number of people in the area. He was concerned that it was a contagious disease, perhaps a form of polio or something similar. The disease affected birds and mammals as well as humans, and cats were the most notorious victims. Dr. Hosokawa discovered the actual cause of the disease in July 1959 by giving production waste effluents to a cat (No. 400) and watching it undergo the same symptoms as the humans and cats that were known to be ill. The Chisso Company responded by forbidding Dr. Hosokawa to continue his experiments or report his findings to the local authorities.

Dr. Hosokawa discovered the cause of the disease by giving production waste effluents to a cat and watching it undergo the same symptoms as the humans and cats that were ill. The Chisso Company responded by forbidding Dr. Hosokawa to continue his experiments or report his findings to local authorities.

By August 1959, concern over the disease prompted Minamata public health officials to turn to the Kumamoto University School of Medicine for assistance. The medical school sent a research team to investigate and submit a report directly to the city officials. The team suspected that the symptoms resulted from the ingestion of heavy metals, and studied the effects of magnesium, selenium, thallium, and other heavy metals. Finally, in 1962, they were certain that organic mercury caused the symptoms of the disease, and that the source of the organic mercury was the industrial wastewater from the Chisso plant.

The cause of Minamata disease was eating raw fish and shellfish that had absorbed organic mercury. The initial symptoms of poisoning included numbness of the lips and limbs, followed by constriction of vision. Muscular coordination then became unsteady, with the victim having exaggerated reflexes and moving like a drunken person. Loss of speech, hearing, and taste also occurred, along with emotional outbursts or depression. Advanced victims became stiff and bent, and registered high blood pressures and temperatures just prior to dying. Babies were born retarded, unable to learn to walk or speak, with no sight or hearing and with deformed limbs.

Impact

It is difficult to estimate the number of victims of the Minamata disease. There were twenty-nine hundred official victims recognized by the government as entitled to damage awards, but hundreds and perhaps thousands more claimed to be victims. The symptoms of the latter group of plaintiffs did not match the official profile of Minamata disease, and they have not been recognized. In addition, some victims of Minamata disease have avoided being identified due to the resulting stigma on their families.

Chisso was indicted and taken to court to establish that it had caused Minamata disease by discharging industrial wastes into Minamata Bay, and to make Chisso legally responsible for damages. The trial took place from July 1969 until March 1973. The plaintiffs, the victims of the Minamata disease victims, won their case. Chisso paid group damages and began negotiations to pay damages to individuals.

The plaintiffs also took the prefecture and federal governments to court to force them to recognize their part and to pay damages to the victims. In 1992, counter to the decision of a 1907 Kumamoto prefecture court that both the federal and prefectural governments were legally culpable, a federal district court determined that the federal government was not responsible for damages to the victims. To date, however, the legal recourses of the Minamata disease victims have not been exhausted.

Minamata Bay is being filled in to cover the mercury-laden industrial wastes, and there is a proposal to convert the area into an environmental park. Fishing has been banned from the bay, but contaminated fish and shellfish from just outside the bay are still caught, sold, and eaten.

Minamata residents carry a stigma that affects their ability to find jobs and spouses in other parts of Japan. The city and company are both suffering. Most local citizens sided with Chisso and resented the fishermen's actions against the company, believing that the lawsuits would financially hurt Chisso and therefore the standard of living in Minamata. This is exactly what has happened, but the blame for the tragedy lies not with the victims of the disease, but with the irresponsible policies of Chisso and of Japanese government officials over the course of the century.

JULIAN LARSON

Where to Learn More

D'Itri, Patricia A. and Frank M. *Mercury Contamination: A Human Tragedy*. John Wiley, 1977.

Ishimure, Michiko. *Paradise in the Sea of Sorrow*. Trans. by Livia Monnet. Yamaguchi, 1990.

Kuznetsov, D. A. "Minamata Disease: What Is a Keystone of Its Molecular Mechanisms? A Biochemical Theory on the Nature of Methyl Mercury Neurotoxicity." *International Journal of Neuroscience*, July 1990, pp. 1-51.

Mishima, Akio. *Bitter Sea: The Human Cost of Minamata Disease*. Trans. by Richard L. Gage and Susan B. Murata. Kosei, 1992.

Sanger, David E. "Japan and the Mercury Poisoned Sea: A Reckoning that Won't Go Away." *New York Times*, January 16, 1991, p. A3.

Schlesinger, Jacob M. "Japan Government Cleared of Blame in Minamata Industrial Pollution Case." *Wall Street Journal*, February 10, 1992, p. A10.

Smith, W. Eugene and Aileen Smith. *Minamata*. Holt, Rinehart and Winston, 1975.

Tsuchiya, Kenzaburo. "The Discovery of the Causal Agent of Minamata Disease." *American Journal of Industrial Medicine*, 1992, pp. 275-80.

Agent Orange contamination

Vietnam

(1961-1971)

Use of the defoliant Agent Orange during the Vietnam War caused debilitating medical problems among both the Vietnamese population and American military personnel.

Background

Agent Orange is a very effective herbicide used for clearing heavy vegetation, especially in forests, jungles and rights-of- way. Two herbicides—2,4-D and 2,4,5-T—combine in equal measure to make the Agent Orange formula. On a commercial level, Agent Orange was used in forestry control as early as the 1930s. Chemical manufacturers in the United States had discovered the defoliating properties of these herbicides and marketed the chemicals domestically and for export. New Brunswick, Canada, was the site of heavy Agent Orange spraying to control forests for industrial development during the 1950s and 1960s. In Malaysia in the 1950s, the British used compounds with the chemical mixture 2,4,5-T to clear communication routes.

The herbicides were considered for military use in the United States toward the end of World War II, during the action in the Pacific. In 1959, the Crops Division at Fort Detrick, Maryland, initiated the first large-scale military defoliation effort—the aerial application of Agent Orange to about four square miles of vegetation.The experiment proved highly successful; military personnel believed they had discovered an effective defoliation tool.

By 1960, the South Vietnamese government, aware of these early experiments, requested that the United States conduct trials of the herbicides for use against insurgent guerrilla forces. President Kennedy agreed to use of the chemicals, believing that the program would be carefully and selectively controlled. Kennedy was anxious that the initial program would not mean a denial of food. Secretary of State Dean Rusk was also concerned; he warned the United States would probably become the target of an intense "germ warfare" campaign initiated by Communist countries and echoed by some neutral countries. Rusk urged close control and supervision be retained in Washington.

The military, however, recognizing the limitations of fighting in foreign territory with troops that were not accustomed to jungle conditions, argued that clearing the communication lines and opening up areas of visibility would enhance their opportunities for success. Admiral Harry Felt, then head of operations in the Pacific, and the joint chiefs urged that the military plan and conduct future herbicide operations without having to obtain specific Washington approval for each strategic decision. Defense Secretary Robert McNamara was in full agreement with this plan, and Rusk was overruled. Spraying began in 1961.

In the early stages, fifteen varieties of chemicals were employed as herbicides, with different components and formulae; many contained some measure of 2,4-D or 2,4,5-T. The "rainbow of destruction" colors included blue, purple, pink and green. These early varieties were used in low quantities (a total of 281,000 gallons) and on a limited geographical area. With Project Ranch Hand—the Air Force team assigned to the spray missions—Agent Orange became the most widely produced and dispensed defoliant in Vietnam, and after 1962 the spray missions increased monthly.

Details of the Disaster

Military requirements for herbicide use were developed by the Army's Chemical Operations Division, J-3, Military Assistance Command, Vietnam (MACV). While the military promised to stay away from civilians or to resettle civilians and resupply food in any areas where herbicides destroyed the food, the promises were never kept. The use of herbicides for crop destruction peaked in 1965, when 45 percent of the total spraying was designed to destroy crops.

Initially, the aerial spraying took place near Saigon, but the geographical base eventually widened. During the 1967 expansion period of herbicide procurement, when requirements had become greater than the industries' ability to produce, the Air Force and joint chiefs of staff become actively involved in the herbicide program. All production for commercial use was diverted to the military, and the Department of Defense (DoD) was appointed to deal with problems of procurement and production. Commercial producers were encouraged to expand their facilities and build new plants, and the DoD made attractive offers to companies to manufacture herbicides. Working closely with the military, certain chemical companies even sent technical advisors to Vietnam to instruct personnel on the methods and techniques necessary for effective use of the herbicides.

During the peak of the spraying, approximately 129 sorties were flown per aircraft. Twenty-four UC-123B aircraft were used, averaging 39 sorties per day. In addition, trucks and helicopters went on spraying missions, backed up by such countries as Australia. C-123 cargo planes were also used. Helicopters flew without cargo doors so that frequent ground fire could be returned. Down below was the enemy, hiding and protected by the thick, broad leaves. In the case of dense jungle growth, one application of spray was made for the upper and another for the lower layers of vegetation.

The herbicidal effects of Agent Orange in Vietman were readily apparent. Above, an untreated mangrove forest; below, another forest as it appeared approximately two years after being sprayed with Agent Orange.

American troops were heavily exposed. Rotary blades would kick up gusts of spray, delivering a powerful dose onto the faces and bodies of the men inside the plane. Soldiers could inhale the fine misty spray and then splash in the sudden and unexpected deluge of an emergency dumping.

Readily absorbing the chemicals through their skin and lungs, hundreds of thousands of U.S. military personnel were exposed as they lived on the sprayed bases, slept near empty drums, drank water in areas where defoliation had occurred, and swam or washed in water that had been standing in anything from bomb craters to rivers. They ate food that had been brushed with spray. Empty herbicide drums were indiscriminately used and improperly stored. Volatile fumes from these drums caused damage to shade trees and to anyone near the fumes.

Those handling the herbicides in support of a particular project goal were directly exposed on a consistent basis. In addition to the Ranch Hands, at least three other groups received consistently high exposure:

1. The group considered secondary support personnel. This included army pilots who may have been involved in helicopter spraying, along with navy and even marine pilots.

2. Those who transported the herbicide from anywhere from Saigon to Bien Hoa and Da Nang in omnipresent fifty-five-gallon containers.

3. Specialized mechanics, electricians, and technical personnel assigned to work on various aircraft. Many in this group were not specifically assigned to Ranch Hand but had to work in aircraft that was repeatedly contaminated.

Agent Orange was used in Vietnam in undiluted form at the rate of three to four gallons per acre;13.8 pounds of the chemical 2,4,5-T were added to 12 pounds of 2,4-D per acre, a nearly fifty-fifty ratio. This concentration is 13.3 pounds per acre more than was recommended by the military's own manual and far more than was ever used domestically. Computer tapes now available show that some areas were sprayed as much as twenty-five times in just a few short months, dramatically increasing the exposure to anyone within those sprayed areas.

Agent Orange was effective. Evaluations show that the chemical killed and defoliated 90 to 95 percent of the treated vegetation. Thirty-six percent of all mangrove forest areas in South Vietnam were destroyed. Viet Cong tunnel openings, caves, and aboveground shelters were revealed to the aircraft after the destruction.

Overall, until 1971, a total of between seventeen and nineteen million gallons of the liquid acid was sprayed, covering at least 10 percent of the total land in Vietnam. (The Vietnamese government claims over 44 percent of the land was sprayed.) Sections of both Laos and Cambodia were also sprayed, secretly.

Agent Orange is contaminated by the deadly chemical dioxin, or TCDD. Dow Chemical (one of the Agent Orange manufacturers) knew that 2,4,5-T was contaminated by some toxic unidentified chemical since 1937. By 1965, having identified the contaminant as dioxin, and fearful of a congressional investigation and possible restrictive legislation on the manufacture of pesticides with high dioxin content, Dow called all the Vietnam War contractors together in order to inform the representatives about dioxin's toxicity. Dow admitted that chloracne, one sign of exposure, was an indication that the whole body was poisoned. Reports from the 1965 meeting identify dioxin as "one of the most toxic materials known, causing not only skin lesions but also liver damage [and] fatigue." Dow personnel were required to take extreme precautions in handling the materials.

Rather than share this information with the military, Dow and its fellow contractors simply recommended keeping the dioxin contamination low—a possibility if the manufacturing temperature is controlled. Repeat exposure at one part per million was acknowledged to be a significant health hazard.

In Vietnam, the dioxin toxicity varied from many parts per billion to parts per million, depending on each manufacturer's production methods. The contamination, up to 40,000 parts per billion, was considered "unacceptable" by the military after the war. The Environmental Protection Agency [EPA] evacuated Times Beach, Missouri, when tests revealed soil samples there with two parts per billion. The EPA eventually stated that one part per billion is dangerous to humans, although that figure has been revised downward over the years as more and more evidence reveals dioxin's toxicity at lower doses.

Impact

Returning Vietnam veterans began noticing a cluster of medical problems for which they had few answers. After a CBS television exposé of Agent Orange in 1978, the chemical companies, the Veterans Administration, and the military made the decision to deny any relationship between the chemicals and these growing health problems. Millions of dollars in potential compensation were at stake. Thus began the official Agent Orange coverup. Veterans were given the runaround.

However, the facts of dioxin's legacy mount. The federal Agency for Toxic Substances and Disease Registry Toxicological Profile for Dioxin produced enough evidence to "conclude that dioxin deserves our greatest respect." According to a congressional research brief, "TCDD is classified as class 6, super toxic in a range of 1 (practically non-toxic) to 6. TCDD may be the most toxic and potent teratogen known to man based on toxicological data in guinea pigs and rabbits." According to Dr. Barry Commoner, director of the Center for the Biology of Natural Systems at Washington University, the chemical is "so potent a killer that just three ounces of it placed in New York City's water supply could wipe out the populace."

Dioxin has been studied in some manner, mostly on animals, in laboratories since the 1940s. Studies on select humans known to be exposed are also adding to the data base. The most recent information indicates that it appears capable of interfering with a number of physiological systems.

Environmental Protection Agency scientists, along with a broad spectrum of researchers, call TCDD the most "carcinogenic" compound ever studied. During the late 1980s and early 1990s, its potential to cause rare forms of cancer in humans—soft tissue sarcoma and non-Hodgkin's lymphoma especially—has been confirmed.

The EPA considers dioxin in food one million times more toxic than cadmium or arsenic. Recent EPA studies indicate that low-level exposure leads to immune system damage and reproductive dysfunction.

The EPA considers dioxin in food one million times more toxic than cadmium or arsenic. Recent EPA studies indicate that low-level exposure leads to immune system damage (a degraded immune system leaves the body less able to defend itself against hostile forces in the natural environment) and reproductive dysfunction. It appears to be most damaging to young animals exposed in utero. It also appears to affect behavior and learning ability, suggesting that it is neurotoxic.

These recent studies confirm what Vietnam veterans and their counterparts in Vietnam have long known: dioxin affects reproduction via maternal circulation. Unusually high numbers of birth defective and developmentally delayed children are born to dioxin-exposed veterans and to exposed Vietnam mothers. A long-term, twenty-year study conducted in Vietnam identifies spontaneous abortions, stillbirths, and congenital defects. Recent studies on infants born to mothers eating contaminated fish found the infants had an impaired ability to learn, with more than a 10 percent decline in visual recognition memory.

Dioxin has the ability to play havoc with the body's chemical systems year after year, functioning like a steroid hormone. Newer findings conducted on wild life around the Great Lakes have shown that dioxin has the capacity to feminize male chicks and rats and masculinize female chicks and rats—an issue with ominous consequence for humans. Testicle size is reduced as is sperm count.

When the American Legion did its own studies on Vietnam veterans and reviewed the scientific literature, it concluded that at least thirty cancers and diseases are related to Agent Orange because of the dioxin contaminant. In 1993, the National Academy of Science's Institute of Medicine announced that dioxin is now linked to Hodgkin's disease and a skin-blistering condition knows as porphyria cutanea tarda. Veterans suffering from the above along with veterans who suffer two rare cancers and chloracne have finally been acknowledged by the Veterans Administration as acceptable for compensation. All other medically damaged veterans and their birth defective children have been ignored. In the meantime, thousands of exposed veterans have already died.

LIANE CLORFENE-CASTEN

Where to Learn More

Gough, M. "Agent Orange: Exposure and Policy." *American Journal of Public Health*, March 1991, pp. 289-90.

Husar, R. B. *Biological Basis for Risk Assessment of Dioxins and Related Compounds*. Cold Spring Harbor Laboratory Press, 1991.

———. *The Health Risks of Dioxin*. Hearing Before the Human Resources and Intergovernmental Relations Subcommittee on Government Operations, House of Representatives, One Hundred Second Congress, June 10, 1992. U.S. Government Printing Office, 1993.

Husar, R. B. and M. Gough. *Dioxin, Agent Orange: The Facts*. Plenum, 1986.

"A Review of the Scientific Literature: Human Health Effects Associated with Exposure to Herbicides and/or Their Associated Contaminants—Chlorinated Dioxins: Agent Orange and the Vietnam Veteran." Compiled by the Agent Orange Scientific Task Force, working with The American Legion, Vietnam Veterans of America, The National Veterans Legal Services Project. April 1990.

Schmidt, K. F. "Dioxin's Other Face: Portrait of an 'Environmental Hormone'." *Science News*, January 11, 1992, pp. 24-7.

Tschirley, F. "Dioxin." *Scientific American*, February 1986, pp. 29-35.

Zumwalt, Admiral Elmo R. Jr. USN (Ret.) "Report to the Secretary of Veterans Affairs, The Hon. Edward J. Derwinski. From the Special Assistant: Agent Orange Issues." First Report, May 5, 1990.

Dioxin contamination

Times Beach, Missouri

(1971)

A Missouri town had to be evacuated after the town was contaminated by the deadly chemical compound, dioxin.

Background

In 1971, chemical wastes containing dioxin were mixed with oil and sprayed along the streets of Times Beach to keep down the dust. The spraying was done by Russell M. Bliss, who was in the business of collecting and disposing of waste oils and chemicals from service stations and industrial plants; mixing with oil and spraying on streets was thought to be an expedient and efficient means of re-using waste. Much of the extremely toxic waste oil was sprayed on roads throughout the state.

Shortly after the spraying around Times Beach (a town of approximately twenty-eight hundred located about thirty miles southwest of St. Louis), horses on local farms and stables there and in other areas that had been sprayed began to sicken and die, as did pet dogs and cats. People also became ill for no apparent reason.

In an effort to determine if some toxin in the environment might be responsible, soil samples were sent to the Centers for Disease Control (CDC) in Atlanta for analysis. After three years of testing, the agency determined that small but dangerous levels of dioxin were present in the soil, representing a potentially serious hazard to the environment and local residents.

A decade after the spraying, Times Beach experienced the worst flooding in its history. On December 5, 1982, the Meramac River overran its banks and inundated the town, flooding homes and contaminating them and most everything else with dioxin. Since the compound binds tightly to soil and degrades very slowly, dioxin was still present in significantly high levels ten years after being spread around in the oil that was used in 1971 as a dust suppressant.

At one time, the CDC considered dioxin in the soil to be potentially hazardous to humans at levels of one part per billion and above (which is the equivalent on a time scale of about one second in thirty-two years). Levels found in Times Beach in some areas reached 100 to 300 parts per billion, and even higher in a few locations. By early 1993, federal and state officials had located some one-hundred other sites in Missouri where dioxin wastes had been dumped or sprayed, with levels of the compound reaching as high as 1,750 parts per billion in some areas.

After the CDC warned that Times Beach had become uninhabitable, most of the families were temporarily evacuated. The U. S. Environmental Protection Agency (EPA) eventually agreed to buy up the town for some $33 million and to relocate its inhabitants elsewhere.

Times Beach thus became the third town and neighborhood—along with Seveso, Italy, in 1976, and Love Canal, New York, in 1980—rendered uninhabitable and evacuated because of toxic chemical contamination.

Details of the Contamination

Dioxin (or tetrachlorodibenzo-*para*-dioxin [TCDD]) occurs mainly as an unintended contaminant of some herbicides (such as Agent Orange, and its major component, 2,4,5-T), and as a byproduct of certain chemical processes, such as bleaching processes for paper and consumer products. It has long been considered to be the most toxic chemical known.

In some species of laboratory animals, dioxin has been found to be toxic at every level tested. In other cases, scientists have been unable to find a level at which it does not produce adverse health effects, even when fed to animals at levels of a few parts per trillion, the average level at which it now occurs naturally in human fat tissue.

In humans, dioxin has been blamed for causing numerous disorders, including rare forms of cancer among Vietnam veterans exposed to Agent Orange, and birth defects in their children; miscarriages among pregnant women exposed to the dioxin-containing herbicides Silvex and 2,4,5-T; and various illnesses among chemical workers exposed to dioxin produced during the manufacture of agricultural and industrial compounds.

Indeed, the contamination of Times Beach seemed to take a serious toll on the health of the town's residents. Local officials claimed that virtually every household in Times Beach experienced health disorders, ranging from nosebleeds, depression, and chloracne (a severe skin disfigurement), to gene mutations, cancer, and heart disease. Almost all of the residents tested for dioxin contamination by the CDC showed abnormalities in their blood, liver, and kidney functions.

This was unsurprising, given the wide range of illnesses apparently caused earlier by the spraying of the contaminated oil in 1971 around Times Beach

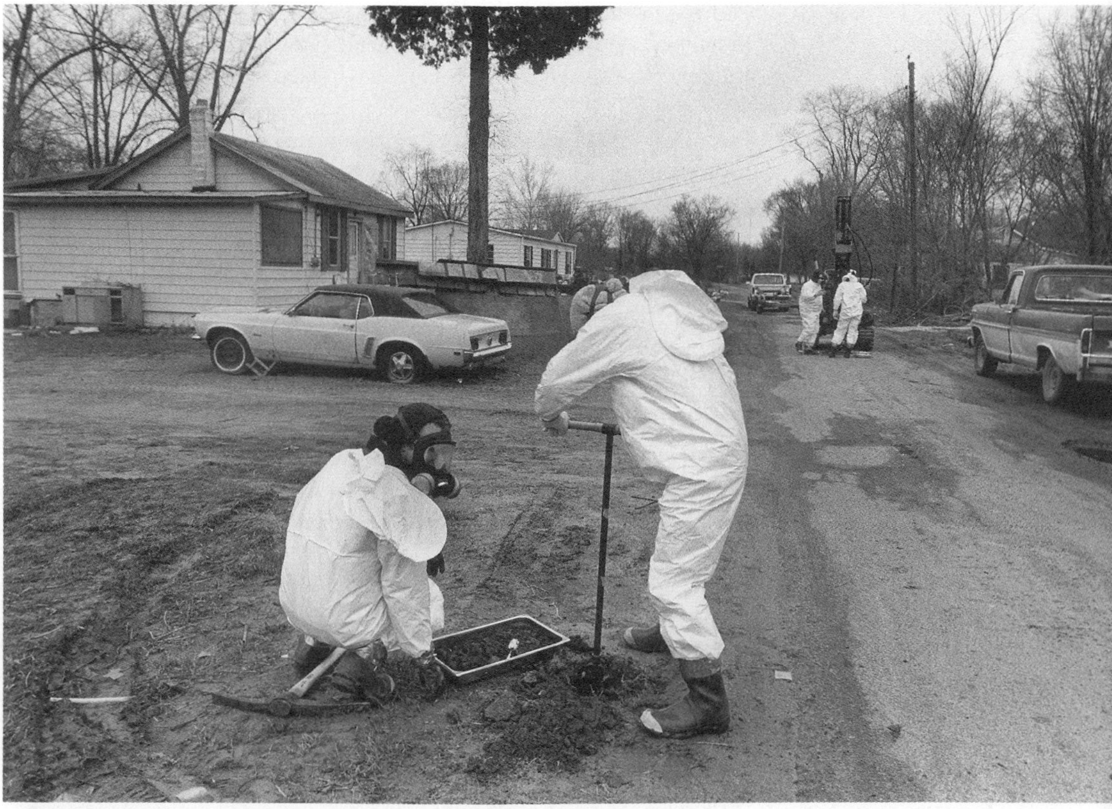

EPA technicians take deep soil samples from a lawn in Times Beach, Missouri, where chemical wastes containing dioxin were mixed with oil and sprayed on the streets to control dust.

and the Saint Louis area. For example, after such oil was used as a dust suppressant on a horse farm at Moscow Mills, Missouri, hundreds of sparrows and other birds that roosted in the barn were found dead on the ground. Within the next few weeks, hundreds more died, as well as cats, rodents, and horses. Even after the dioxin-laden soil had been removed to a depth of six inches to eight inches and replaced with clean river sand, the horses continued to sicken and die; an additional six inches of soil was removed, but to no avail.

Eventually, a total of forty-five horses, eleven cats, four of five dogs, and a large number of wild birds died from the spraying at the farm. Twenty-three other horses became ill, and there were twenty-six known miscarriages among pregnant mares, as well as six deformed foals born and several stillbirths. At least seven humans on the farm were also sickened for long periods, including six- and ten-year-old girls who had to be hospitalized. All suffered from headaches, nausea, diarrhea, fatigue, abdominal pain, blood in their urine, and other symptoms.

Impact

Perhaps the most significant impact of the tragedy at Times Beach was to dramatize and bring to nationwide attention the dangers of dioxin and other toxic chemicals so widespread in our environment and food chain.

For weeks, the Times Beach story dominated or was prominent in the national news. Dioxin became a household word nationwide. As a result, federal and local government officials and agencies came under strong pressure to put tighter restrictions on the release or disposal of dioxin into the air and water, and to ban herbicides and other chemicals and products containing dioxin.

In 1985, after years of pressure from citizens' groups and environmentalists, the EPA finally banned the widely used herbicides Silvex and 2,4,5-T. That same year, the agency set a water quality standard for dioxin, allowing no more than thirteen parts per quintillion, the equivalent of one leaf on all of the trees in North America. If enforced, such a standard would severely restrict the dumping into waterways of dioxin-contaminated wastes by chemical plants and pulp and paper mills.

In 1985 the EPA finally banned the widely used herbicides Silvex and 2,4,5-T. That same year, the agency set a water quality standard for dioxin, allowing no more than thirteen parts per quintillion, the equivalent of one leaf on all of the trees in North America.

In response to this greatly increased public concern over dioxin, some groups came to the defense of the compound, including several known for their longtime support of the industry and reluctance to recognize the dangers of toxic chemicals. In late June 1983, the American Medical Association reacted to the widespread publicity over Times Beach by approving a resolution calming fears about dioxin, and condemning the news media for making "dioxin the focus of a 'witch hunt,'" and engaging in "hysterical malreporting" on the compound, which "unnecessarily and ignorantly damaged... the lives and well-being of many inhabitants of regions contaminated with dioxin." A few days later, the Association backed away from the resolution, saying that it represented only the views of the Missouri delegation to the annual meeting of its house of delegates, which had passed the resolution.

The pulp and paper industry also became active in the campaign to persuade people that the compound was not as dangerous as previously estimated. Even some scientists and government officials lent their support to this effort to downgrade dioxin's dangers. In late 1989, Dr. Vernon N. Houk, then assistant surgeon general and director of the Center for Environmental Health and Injury at the CDC, supported the paper industry's position and supported a water standard that allowed for dioxin levels 544 times higher than the EPA had set. Meanwhile, the EPA stuck to its position that dioxin was one of the most hazardous substances known, saying unequivocally in a legal brief that the chemical was known to be "carcinogenic, teratogenic, mutagenic, and fetotoxic" and to cause a variety of other serious health problems.

Because of the continuing controversy over dioxin, the EPA has for several years been under strong pressure to downgrade its assessment of the compound's toxicity, and the agency has been engaged in intensive reassessment. But most experts and scientists who have studied dioxin say that there is no new evidence that would justify lowering restrictions, and many consider it even more dangerous than previously estimated. This view was most forcefully expressed on June 10, 1992, by Democratic Congressman Ted Weiss, chairman of the House Human Resources and Intergovernmental Relationals Subcommittee. While chairing hearings on the matter, Weiss stated: "Although this cancer-causing chemical is already considered the most toxic compound ever created, the latest scientific research indicates that it causes even more harm than previously believed."

The debate over dioxin's toxicity is given special urgency by the fact that the compound has so pervaded our society that virtually all Americans are being exposed to it on a regular basis, even nursing infants taking in contaminated mother's milk. It is believed that some 98 percent of people's intake of dioxin comes from milk and other dairy foods, fish, beef, and chicken, and that it takes about seven years for the body to remove one half of its burden of dioxin. Other major sources of exposure to dioxin are emissions from municipal and industrial incinerators (which produce the compound when plastics are burned); bleached white paper products (writing and typing paper, cardboard milk cartons, paper plates and towels, tampons, disposable diapers, coffee filters); and wood preservatives.

On April 16, 1986, Dr. Arnold Schecter, professor of preventive medicine at the State University of New York at Binghamton, reported that it was difficult to prove that dioxin was killing or otherwise harming people in the United States because the entire population, and that of other industrialized nations, was so contaminated with the compound. As he told the annual meeting of the American Chemical Society, "Our studies suggest that the fatty tissue of the average American contains about 7.2 parts per trillion of TCDD. This means that the general population of this country cannot be considered a control group in assessing the health effects on people known to have had unusually high exposure."

In March 1993, Dr. Linda Birnbaum of the EPA's Health Effects Laboratory, told a meeting of the Society of Toxicology, "Scientists from Europe and the United States agreed. . .that data indicate dioxin contamination levels in human blood are near the level expected to cause adverse health effects."

LEWIS G. REGENSTEIN

Where to Learn More

After Silent Spring. Natural Resources Defense Council, 1993.

Briggs, Shirley A., and the Rachel Carson Council. *Basic Guide to Pesticides*. Hemisphere Publishing, 1992.

Carson, Rachel. *Silent Spring*. Crest Books, 1962.

Dioxins. U. S. Environmental Protection Agency, 1980.

Gips, Terry. *Breaking the Pesticide Habit*. International Alliance for Sustainable Agriculture, 1987.

Harris, Art. "After the Deluge, a Tainted Town." *The Washington Post*, January 6, 1983, pp. A1, A6.

—. "Town Struggles with Toxic Legacy." *The Washington Post*, January 10, 1983, pp. A1, A8.

Health Effect of Toxic Pollution: A Report from the Surgeon General and a Brief Review of Selected Environmental Contamination Incidents with Potential for Health Effects. United States Library of Congress, August 1980, p. 235.

Nader, Ralph., et al. *Who's Poisoning America?* Sierra Club Books, 1981.

Pesticides in Children's Food. Environmental Working Group, 1993.

Regenstein, Lewis G. *Cleaning Up America the Poisoned*. Acropolis Books, 1993.

Regulating Pesticides in Food: The Delaney Paradox. National Research Council, 1987.

Reinhold, Robert. "Many Questions but Few Answers for People of Dioxin-Periled Town." *The New York Times*, January 10, 1983, pp. 1, 10.

———. "Missouri Governor Proposes Steps to Fight Dioxin Crisis." *The New York Times*, January 12, 1983.

———. "Missouri Now Fears 100 Sites Could be Tainted by Dioxin." *The New York Times*, January 18, 1983, pp. 1, 7.

Six Case Studies of Compensation for Toxic Substances Pollution. Committee on Environment and Public Works (U. S. Senate), June 1980, pp. 279-85.

"The 'River Rats' Want to Stay." *Time*, January 10, 1983, p. 21.

Whiteside, Thomas. *The Pendulum and the Toxic Cloud*. Yale University Press, 1979.

Methylmercury seed poisoning

Iraq

(1971-1972)

Human consumption of seed grain treated with fungicide and intended only for planting resulted in nervous system damage and death to many individuals.

Background

In late 1971 and early 1972, Iraq was the site of the largest outbreak of human poisoning from eating methylmercury-treated seed grain that was ground into flour to make bread. Both sexes and all ages were affected. The ingested quantities of methylmercury ranged from small amounts that produced no overt effects to lethal doses. Fetuses suffered the most damage. Among others the severity of the neurological and psychiatric symptoms was almost directly proportional to the amount of contaminated bread consumed. In contrast, in cities where bread is produced from government-inspected flour mills, not a single case of poisoning was reported.

Details of the Disaster

In October and November of 1971 a total of about seventy-three thousand tons of high yield Mexipac wheat seed grain and twenty-two thousand tons of barley seed grain was shipped to Iraq. Almost all of it had been treated with a methylmercury fungicide and was distributed by cooperatives to farmers for planting throughout the country. Because the grain was imported from Mexico, the poison warning printed on the sacks was in Spanish only. The message that the grain was intended for planting only was therefore not heeded. How much of this seed grain was used to prepare homemade bread is not known. The Iraqi government estimated that the treated grain was distributed to no more than five percent of the rural population, about two-hundred thousand people. Most of the fatalities occurred within three months after the end of the exposure although a few long-term illnesses resulted in fatalities as well.

By March 1973, up to forty-thousand persons were unofficially estimated to have been poisoned, some from every province in Iraq. These figures were not confirmed because news reporters were denied entry to the country and the movements of other foreigners were restricted. However, tourists reported that large numbers of people suffered brain damage, blindness, and paralysis. The total number of casualties will never be known precisely because hospitals were quickly overloaded and many victims could not gain access to them. In addition, most of the poisonings occurred in rural areas, and many were not reported to authorities. The government officially acknowledged that 6,530 persons were hospitalized and 459 died.

Other such incidents had occurred previously. The first in Iraq was documented in 1956. Among the two hundred persons afflicted, 70 died. Also in Iraq, in 1960 an estimated 1000 persons were affected in a similar incident and over 200 died. During the 1960s and 1970s other smaller but similar episodes of alkylmercury poisoning occurred on a more limited scale in the United States, West Pakistan, Guatemala, Ghana, and Mexico. The total known death toll was 42 with approximately 197 individuals less seriously affected.

These tragedies occurred because people ignored poisoning warnings and ate seed grain treated with chemicals to control diseases caused by seed-borne infections and to protect the germinating seedlings from soil-borne pathogens. Never intended to be consumed directly, the purpose was to increase the quality and quantity of agricultural crops. Without such seed treatment, over a few years crop production could be diminished substantially. Consequently, various organic and inorganic chemicals, including mercuric compounds, have been used as fungicides to treat seeds since the end of the nineteenth century. Their effectiveness was greatly increased when arylmercury (phenylmercury) and alkoxyalkylmercury (methoxyethylmercury) compounds were introduced around 1914.

They had a wider spectrum of fungicidal activity than nonmercury formulations and were used extensively until the mid 1940s, when the alkylmercury (methyl- and ethylmercury) compounds were introduced and very quickly became the seed dressing of choice because of their greater efficacy and lower cost. After that, alkylmercurials were applied to a wide variety of agricultural crops such as: barley, beans, corn, cotton, flax, millet, milo, oats, peanuts, tomatoes, peas, rice, rye, safflower, sorghum, soybeans, sugar beets, and wheat, as well as to protect fruit trees, rice, turf grasses, and golf courses. However, the introduction of the more effective and more highly toxic alkylmercurials has been responsible for nearly all of the severe poisoning incidents, a consequence that was not observed when either aryl- or alkoxymercurials were used as seed dressings. One notable exception occurred in Turkey between 1955 and 1959. After peasant families consumed seeds treated with hexachlorobenzene, an estimated three-thousand people became ill, and about three hundred died. This epidemiological poisoning pattern was very similar to that caused by the ingestion of alkylmercurials.

As these examples demonstrate, in developing countries hundreds of people died or became incapacitated either from direct consumption of grains treated with alkylmercury compounds or from eating the meat of animals that had also eaten treated seeds. Human consumption of treated seed grain, either direct or indirect, led to crippling illness, disfigurement, dementia, and death in numerous instances around the world. After alkylmercury seed dressings were banned in Sweden in 1966 and in the United States and Canada in 1970, they continued to be exported to nations where poor and uninformed farmers and their families swelled the ranks of the afflicted.

Among developed countries, Sweden initially recognized a problem in the 1960s, when wild birds were observed flopping helplessly on the ground before they died. Investigations indicated that birds and rodents could have consumed at least one of the eighty metric tons of alkylmercury fungicides applied to Swedish seeds between 1940 and 1966. The birds died after eating treated seeds left uncovered on the ground during planting, but this did not raise as great an alarm as the possibility that the poison could be transferred to human beings. As the practice of feeding excess treated seed grain to domestic birds and animals was widespread, fears were raised that the methylmercury could translocate up the food chain to human beings. This concern was substantiated as contaminated eggs, poultry, and meat were discovered in the market. Levels of alkylmercurials in Swedish eggs, for example, averaged twenty-nine parts per billion of mercury; and on a few farms they were as high as fifteen-hundred parts per billion.

In developing countries hundreds of people died or became incapacitated either from direct consumption of grains treated with alkylmercury compounds or from eating the meat of animals that had also eaten treated seeds.

After Austrian and Danish markets refused to sell Swedish eggs because of their high methylmercury levels, all alkylmercury seed treatments were banned in Sweden in 1966; but exports were still permitted even as the migrating wild birds returned and information was compiled on other seed-dressing related human tragedies. That farmers in Sweden had suffered toxic reactions when they planted treated seed had been demonstrated in incidents reported as far back as 1929, and at least twenty fatalities had occurred over three to four years before a list of cases was compiled in 1949. In Canada, while admitting that the farmers could not be monitored, legislation was also enacted against mixing treated seed grain with grain being sent to market.

Over the centuries the symptoms of inorganic mercury poisoning were well documented by the exposure of miners and industrial workers as mercury accumulated in their brains, kidneys, and livers. Loose teeth, tremors, and psychopathological symptoms were common at low exposure; but removal from the source would often enable the victims to recover. However, the effects of short-chain alkylmercurials, such as methylmercury, are more severe and constitute the greatest risk to human health because of their high toxicity and chemical stability, which prevents their degradation into the inorganic form, and slow elimination from the body. With a half-life of about seventy days, continued exposure elevates the levels. It also crosses the blood-brain and placental barriers, attacking the central nervous system and inducing teratogenic changes in the fetus. The neurological symptoms include loss of coordination in walking, slurred speech, constricted vision, loss of sensation (especially in fingers, toes, and lips), and loss of hearing. The effects of alkylmercury poisoning are generally considered to be irreversible. Severe poisoning can cause coma, blindness, and death.

The symptoms of alkylmercury poisoning were described in England after 1940 when four of sixteen workmen were poisoned while treating seed grain despite wearing dust masks, goggles, and gloves as they worked in a well-ventilated fungicide manufacturing plant. Exposure to the fumes for as little as two months brought reactions such as tunnel vision and a staggering gait, and a twenty-three-year-old workman exposed for four months became totally disabled. When Dr. Donald Hunter observed the four afflicted workmen, he issued a warning and advocated that such dangerous compounds should be outlawed; but he was ignored.

When the disabled workman died of pneumonia in 1954, an autopsy confirmed that some brain cells had been destroyed and the central nervous system damaged. This report subsequently enabled Japanese investigators to confirm that patients who died of what was called Minamata disease had, in fact, suffered from alkylmercury poisoning, in this case caused by eating contaminated fish. The limited number of cases of poisoning among seed treaters was attributed to the fact that they only worked two or three months out of the year and so could detoxify between seasons.

Impact

In the United States the sale of all alkylmercury seed dressings, except Panogen (methylmercury dicyandiamide) was banned in late 1970. Sale of Panogen was finally halted in May 1971, more than a year after the national evening television news reported on the poisoning of one family, the Hucklebys, who ate pork from a hog that was fed waste grain treated with this compound in Alamogordo, New Mexico. The other contaminated hogs were sold to a packing plant, causing concern over the possibility of broad dissemination in the population. Spread widely, the meat was not likely to evoke poisoning symptoms; but for the family eating the contaminated pork every day, the results were extreme. An older Huckleby son escaped because he was away from home at the time, and the older daughter made some recovery; but two younger children became blind and paralyzed. A baby born after the incident was blind and retarded, demonstrating the compound's impact on the fetus.

Other incidents of poisoning by eating treated seed grain directly were reported in California, Oregon, and Texas. After domestic sales of alkylmercury seed dressings were banned, in at least one case treated grain was obtained while it was being loaded for export. However, the shipment of treated grain to foreign countries generated very little interest in the American press; and the major epidemic in Iraq occurred after the sale of treated grain was outlawed in the United States.

All of the major epidemics of poisoning occurred in countries where food was in short supply and improved agricultural production had a high priority. In many episodes illiterate rural families either did not understand the words, recognize the poison symbols on the bags of grain, or believe government warnings that the seed grain was unsafe to eat. In addition, depending on the dose, the latency period of weeks to months between the ingestion of methylmercury-treated grain and the onset of overt symptoms gave the victims a false sense of security. Also, washing the seed removed the visible color and was falsely reassuring.

In some instances the families fed some of the grain to chickens and swine first. When they did not observe poisoning symptoms in the livestock and poultry after a few days, the farmers became convinced that the warnings were false and the grain was safe to eat. The process of poisoning is apt to be accelerated when treated seed grain is consumed directly because the doses are larger than when it is washed first or fed to animals. However, unlike exposure to methylmercury vapors, which may induce symptoms very quickly, usually at least three months elapses between exposure and the onset of symptoms, depending on the quantities consumed.

Since Iraq has often been at war since the 1970s, with the exception of a few follow-up scientific reports published between 1985 and 1989, hardly any new information relative to the long-term health effects on the thousands of victims has been published or released by the Iraqi government. However, in other parts of the world the relative advantages and disadvantages of the different kinds of seed dressings have been considered further and additional precautions recommended.

The major advantages of organomercurials are their low cost and effective control of a wide spectrum of seedborne diseases at low application rates. However, these reasons were not deemed sufficient to warrant the continued use of these highly toxic seed dressings in agriculture in several developed countries, considering not only their role in human poisonings from the 1950s to the 1970s but also the lack of sufficient safeguards to insure that such treated seed would not again be diverted for human consumption. After review by the World Health Organization (WHO), the Food and Agriculture Organization (FAO), and agricultural officials of several developed countries, these seed dressings were banned in the early and middle 1970s, forcing the pesticide industry to develop and market a number of more expensive, alternative nonmercury chemicals such as chlorobenzene (hexachlorobenzene, pentachloronitrobenzene, and quintozene), benzimidazole (benomyl and thiabendazole), and carbamodithiolate (thiram, disulfiram, maneb, and mancozeb) compounds, as well as other commercial fungicides such as captan, carboxin, drazoxolin, and methoxychlor.

Overall, most of these alternative chemicals are less toxic than the organomercurials, but they also exhibit a much more limited spectrum of effectiveness against pathogens. Therefore, they must be applied at higher concentrations than the organomercurials, further increasing their cost. This disparity in cost between the more effective and cheaper organomercury seed dressings and the less effective and more expensive nonmercury alternatives creates an economic incentive favoring the use of the more toxic organomercurials.

Therefore, until economical, effective, and broad-spectrum alternative fungicides become available, the WHO and FAO recognize that under certain conditions, the use of alkoxyalkylmercurials and arylmercurials is likely to continue under the safeguards set by individual countries. Accordingly, they have recommended that (1) the use of alkylmercury seed dressings be restricted to the treatment of nuclear stocks of cereal seed used for the first few generations of seed multiplication, and never be used to treat seed intended to be exported to produce food; (2) alkoxyalkylmercury, arylmercury, and hexachlorobenzene be used to dress cereal seed only if the need for such treatment has been assessed and the use of an alternative is precluded; (3) all dressed seed exported for food production be distinctly dyed and/or treated with a bitter, nausea-inducing substance to render the seeds unacceptable as food (the bags containing the dressed seed should also be adequately labeled in the appropriate language and with warning symbols, accompanied by documentation and an instruction leaflet on the handling and hazards of the dressed seed; and (4) appropriate government authorities should track all consignments of alkylmercury-treated seeds to ensure that they are being used, stored, and disposed of in a safe manner.

FRANK M. D'ITRI

Where to Learn More

Bakir, F., S. F. Damluji, A. Amin-Zaki, M. Murtad, A. Khalidi, N. Y. Al-Rawi, S. Tikriti and H.I. Dhahir. "Methylmercury Poisoning in Iraq." *Science*, vol. 181, 1973, pp. 230-241.

D'Itri, P. A., and F. M. D'Itri. *Mercury Contamination: A Human Tragedy*. Wiley, 1977, pp. 29-46.

Greenwood, M. R. "Methylmercury Poisoning in Iraq. An Epidemiological Study of the 1971-72 Outbreak." *Journal of Applied Toxicology*, vol. 5, no. 3, 1985, pp. 148-59.

Hunter, D., R. R. Bomford, and D. S. Russell. "Poisoning by Methylmercury Compounds." *Quarterly Journal of Medicine*, vol. 33, 1940, pp. 235-41.

Hunter, D., and D. S. Russell. "Focal Cerebral and Cerebellar Atrophy in a Human Subject due to Organic Mercury Compounds." *Journal of Neurological and Neurosurgical Psychiatry*, vol. 17, 1954, pp. 235-241.

WHO. *The Use of Mercury and Alternative Compounds as Seed Dressings*. Report of a Joint FAO/WHO Meeting, World Health Organization, Technical Report Series No. 555, 1974.

———. "Conference on Intoxication Due to Alkylmercury-Treated Seed." *Bulletin of the World Health Organization*, vol 53 (supplement), 1976, pp. 1-138.

Chemical plant explosion

Flixborough, England

(1974)

Sudden and devastating explosion at a British cyclohexane processing plant prompted a reevaluation across Europe of chemical plant safety.

Background

On June 1, 1974, a vapor cloud explosion destroyed the Nypro (UK) Limited Works at Flixborough, England. The powerful blast killed twenty-eight people and injured eighty-nine. In the tiny village of Flixborough, approximately two thousand feet away, 72 of the 79 existing houses were destroyed. Extensive property damage was reported for miles around; a survey indicated that 1,821 houses and 167 stores or factories had been damaged. There was no malfeasance involved in the disaster. Nypro was in complete compliance with the British safety standards of the time. As far as anyone knew, everything at the Nypro Works was normal right up to the moment when it virtually ceased to exist.

The Nypro Works was one of the most advanced chemical plants in the United Kingdom, a joint enterprise of the National Coal Board and Dutch State Mines. Its final product was caprolactam, which is used in the manufacture of nylon. Before the 1974 explosion, the Nypro Works had never suffered a mishap of any consequence. It coexisted with the village of Flixborough for several years without incident, and most residents were unaware of the nature of the plant. Most area residents interviewed after the disaster thought the plant was a fertilizer factory, which part of it had been some years before. The part of the plant that exploded in fact processed a highly flammable liquid under high pressure and temperature—a liquid that, if released, would flash into an explosive vapor.

Section 25A of the Nypro Works plant used an oxidation process in which cyclohexane was mixed with oxygen and a catalyst to form cyclohexanone and cyclohexanol, which subsequently were processed into caprolactam. For this process, the cyclohexane was heated to 311 degrees Fahrenheit, and pressurized to 125 pounds per square inch. The process occurred in a series of six reactor tanks. Each reactor was set a little lower than the one before it, and the cyclohexane flowed downwards from one reactor to another. Each tank, made of mild steel and coated on the interior with stainless steel, was approximately twelve feet in diameter by sixteen feet high.

Two months prior to the accident, the number five reactor developed cracks. The plant was shut down, and number five was drained and removed. The engineers decided to bypass it temporarily with a large pipe, though no plant engineer present was qualified to guarantee the safety of such a substitution. The men who made the pipe were experienced in plant installations, but they weren't chemical engineers or metallurgists. The one factor they calculated in designing the pipe was whether it would allow enough liquid cyclohexane to flow between the reactors. They neglected to consider how the pipe would behave under continual high temperature and pressure.

The part of the plant that exploded in fact processed a highly flammable liquid under high pressure and temperature—a liquid that, if released, would flash into an explosive vapor.

The pipe was built of stainless steel, and was supported by a scaffold. To accommodate the drop between the two reactors, it had a "dogleg" or crook. On either end of the pipe was a bellows, which provided a small amount of flexibility between the pipe and the reactor. The pipe was 20 inches in diameter, and weighed 2,400 pounds. When the engineers restarted the plant on April 1, the pipe seemed to function, although they did observe that under operating temperature and pressure it lifted itself off the supporting scaffold.

On May 29, the plant developed a cyclohexane leak, and was shut down for repairs. Other maintenance work was also undertaken. The plant was restarted early on the morning of June 1, and bringing it gradually up to operating temperature and pressure took a good portion of the day. When the afternoon shift came on duty at 3:00 P.M., everything seemed fine.

Details of the Explosion

The explosion occurred at 4:53 p.m., and no one knows exactly what caused it. The control room was destroyed, as were all the personnel, instruments, and information it contained. The bypass pipe was found bent in half, torn away from the two reactors it connected. At some point just prior to the explosion, the 20-inch bypass pipe tore loose from reactors 4 and 6. Released from its pressure, fifty tons of cyclohexane flashed into vapor and formed a gigantic cloud over the Nypro Works. When the cloud detonated, it formed an airburst explosion with a force estimated as equivalent to detonating between 15 and 45 tons of TNT. Given that the land surrounding the plant was completely flat, there was nothing to break up the concussion wave, which spread for miles. Parts of the plant were still burning three days later.

The British government, Nypro, and the various insurers wasted no time in investigating the disaster. A Court of Inquiry was formed, with a panel of four judges taking evidence for seventy days from experts and investigators. The main physical evidence was:

- The 20-inch diameter pipe was bent completely in half. Bits of wood and insulation were trapped in the folded steel.

Parts of the Nypro chemical plant continued to burn three days after the initial explosion—the force of which is estimated to have equaled detonating between 15 and 45 tons of TNT.

- An 8-inch diameter pipe nearby, also part of the cyclohexane circuit, had a fifty-inch-long split on the inside of a 90-degree elbow, with the edges splayed as if the contents had exploded out of the pipe. The pipe had been exposed to extremely high temperatures.

- A video taken between one and two minutes after the blast by a witness showed two separate flames: the still-expanding fireball of the main explosion and a vertical jet of flame shooting 250 feet in the air. This flame's origin was approximately the location of the 8-inch pipe.

- A high-pressure flange that connected the torn pipe elbow to a valve had two loose bolts. A section of gasket was missing in the area of the loose bolts. The flange and valve were located in a cast-iron box next to the elbow.

- At the behest of the Court of Inquiry, the Safety in Mines Research Establishment (SMRE) constructed an exact duplicate of the 20-inch pipe and subjected it to tests at high temperatures and pressures. Under conditions identical to those at the plant, the pipe was observed to "squirm" on its bellows. At a pressure very much higher than anything it should have experienced at the plant, one of the bellows blew out, causing a major leak. But nothing investigators could do to the pipe would cause it to jack-knife in the way the original pipe had.

Various theories were offered at the Inquiry to explain the disaster. One hypothesis was that boiling water trapped in the reactor tanks could have created high-pressure surges that caused the 20-inch pipe to fail. This hypothesis, though feasible, gained few adherents during the inquiry. The Court ultimately made the determination that the disaster was caused entirely by a spontaneous failure of the 20-inch pipe. This conclusion placed the blame squarely on those responsible for the construction of the pipe, all of whom were dead.

The Court's report satisfied most of those present, but caused a storm of controversy among metallurgists and chemical engineers who had studied the blast. These specialists claimed that the Court, in coming to its conclusion, ignored much or most of the available evidence.

A separate theory, which became known as the "8-inch hypothesis," held that although the 20-inch pipe was under stresses greater than it could safely endure, only an outside force could have caused it to buckle the way it did. Proponents of the 8-inch hypothesis claimed that the disaster began not with the 20-inch pipe, but with the 8-inch pipe, and actually occurred in three stages:

1. The gasket blew out at the two loose flange bolts, causing a high-pressure cyclohexane leak inside the cast-iron valve box. Hidden within the box, this leak could have escaped notice for some time. A static discharge or some other cause ignited the leaking vapor. The box directed the jet of flame onto the inside of the elbow in the 8-inch pipe.

2. After several minutes of this flame, the 8-inch pipe failed. The resulting explosion opened a 50-inch tear in the pipe, and threw a blast wave and debris toward the already-stressed 20-inch pipe. (These same investigators, using data from the SMRE tests, calculated that it would require a lateral thrust of only 4.5 PSI to buckle the 20-inch pipe.)

3. The 20-inch pipe jack-knifed and tore loose from the two reactors. The collapsing pipe trapped pieces of airborne debris in its folds as it fell to the concrete below. The cyclohexane flashed into vapor, and detonated moments later.

Impact

The Flixborough explosion ended the British chemical industry's record of having harmed very few people outside of plant boundaries in an accident. The disaster triggered a wave of recommendations for new strategies to deal with the dangers posed by large processing plants and their proximity to populated areas.

The implications of the Flixborough blast were not confined to the United Kingdom. After the disaster, internationally-accepted standards for health and safety were reevaluated. This movement gained momentum two years later with a plant accident in Seveso, Italy that unleashed several tons of caustic chemicals and several kilograms of highly-toxic dioxin onto the area. This accident caused approximately four hundred injuries and forced the sterilization of the surrounding land and the slaughter of contaminated animals. The European Community responded with the "Seveso Directive," parts of which were incorporated in Britain's Control of Industrial Major Accident Hazards (CIMAH) regulations. These regulations form an overall strategy for the design and operation of chemical plants, as well as the handling of unstable, flammable, and toxic substances. Levels of control over this strategy extend from the government ministries down to plant management.

The approach taken by the European governments in responding to the new safety concerns about chemical processing plants was not what might be expected in the United States. Rather than producing a number of federal regulations that establish guidelines for plant operation and materials handling, the European governments generally impose the obligation to operate safely. As one commentator observed, "In other countries, particularly the UK, it has been realized that it is impracticable to control a developing, high technology industry by detailed regulations. They soon get out of date, or circumstances arise which were not thought of by those who wrote them and they encourage the attitude that all you need to do to get a safe plant is follow the rules."

This "European" approach to regulation is based on two assumptions:

1. "It is up to the employers to decide what is a safe plant; those who create a hazard know more about it than anyone else."

2. These employers have ample incentive to protect their investments from loss and liability.

In all cases, of course, the European governmental agencies maintain the rights of inspection and intervention.

This approach to regulation is not without controversy. It is certainly true that not everyone can be trusted with such responsibility. On the other hand, industrial disasters and ecological damage in the former Soviet Union make clear that centralized control is also no guarantee of safety.

In the twenty years since Flixborough, advances have been made in both developing safer chemical processes (for example, by using chemicals that are less flammable or toxic), and in designing and locating plants to minimize the hazard posed to the general citizenry. No industrial processing plant will ever be 100% safe, but considerable efforts have been made to prevent a disaster comparable to the Flixborough explosion from happening in the future.

JOEL SIMON

Where to Learn More

Ball, J. G. "After the Flixborough Report: Do We Know the Real Truth?" *Process Engineering*, December 1975, pp. 39-46.

———. "Was This What Caused the Flixborough Explosion?" *Process Engineering*, January 1976, pp. 53-60.

Butler, Paul. "The Lessons To Be Learned from the Flixborough Enquiry." *The Engineer*, December 1975, pp. 20-24, 50.

Cullen, John et al. "Flixborough—Ten Years On." *The Chemical Engineer*, June 1984, pp. 25-33.

"The Flixborough Cyclohexane Disaster." Her Majesty's Stationery Office, 1975.

McLain, Lynton. "Nypro: Where Fail Safe Failed." *The Engineer*, June 1974, pp. 8-9.

Warner, Sir Frederick. "The Flixborough." *Chemical Engineering Progress*, September 1975, pp. 77-84.

Wood, Robin. "Flixborough: Technical Lessons Obscured by Doubts." *Process Engineering*, February 1976, p. 53.

Dioxin contamination

near Seveso, Italy

(1976)

Worker errors and an unanticipated chemical reaction led to the release of a highly poisonous cloud of dioxin and widespread contamination.

Background

At 12:37 P.M. on July 10, 1976, a runaway chemical reaction ruptured a pressure disk in a reaction vessel at the ICMESA chemical factory just north of Seveso, Italy, and released from one to fifteen pounds of the extremely toxic compound tetrachlorodibenzo*para*-dioxin, or dioxin, into the air. A light breeze blew the dioxin cloud southward over the surrounding neighborhoods, contaminating the area, killing many birds and small animals, and injuring many people in its path. No people died from the accident.

The ICMESA (Industrie Chimiche Meda Societa Anonima) chemical factory was owned by Givaudan Corporation, a subsidiary of the Swiss pharmaceutical conglomerate, Hoffmann-La Roche. The factory produced trichlorophenol, or TCP, a chemical used in deodorants and in the antibacterial soaps that doctors use before performing surgery. Trichlorophenol was produced at the ICMESA plant by mixing chemicals in a ten thousand-liter reaction vessel. The reaction vessel was heated with steam coils and had two sets of stirring blades to equalize the temperature inside the vessel. The chemicals were stirred at temperatures between 140 and 170 degrees Celsius for six to eight hours in the reaction vessel. After some of the chemicals were evaporated, the semirefined TCP remained in the reaction vessel, from which it was drained off into another vessel and further refined. Small amounts of dioxin, no more than ten parts per million, could be produced in the chemical reaction, depending on the temperature.

An important problem in the process of manufacturing chemicals is controlling temperatures, pressures, and reaction rates so that explosions do not occur and so that toxic byproducts are not produced. Dioxin, for example, is produced in significant amounts only at higher temperatures and would never have been produced in dangerous quantities at the ICMESA plant if the temperature in the reaction vessel had been properly controlled.

The ICMESA plant had many safety features to handle potential problems during the production of trichlorophenol. One of these safety features was the process of stirring the ingredients in the reactor vessel, which kept the temperature constant and helped to prevent runaway chemical reactions. A second safety feature was a reserve of three thousand liters of water used to cool the chemicals in the reaction vessel. A third safety feature was a rupture disk that would burst at a preset level to release excess pressure. Among other safety features was a level of steam heating that made it theoretically impossible to raise the temperature in the reactor vessel above 188 degrees Celsius.

Normally, in processing a batch of trichlorophenol, the workers followed a fourteen-hour timetable and a set of procedures. It took about one hour to feed the reaction vessel with its load of chemicals in the right temperature range. Then it took six to eight hours for the reactions to take place to produce the raw trichlorophenol. Next it took about three to four hours to evaporate some of the chemicals out of the reactor vessel. Finally it took fifteen minutes or so to cool the reaction mixture with the three thousand liters of water. At this point, the remaining liquid chemicals in the reaction vessel could be removed for further refining into trichlorophenol. Early in the morning of July 10, 1976, however, the chemical workers did not follow some important procedures.

Details of the Accident

Five problems led to the release of the dioxin cloud. The first problem was created by the chemical plant workers. They failed to follow company safety procedures when they cut short the evaporation process at 5:00 on the morning of July 10. Because the ingredients in the reaction vessel might overheat, produce unwanted byproducts, and explode, the workers were supposed to cool the contents of the reaction vessel by pouring water into it before they went off shift. This they failed to do. To make matters worse, fifteen minutes after the reaction was finished, the workers created the second problem: they turned off the stirring blades that would have kept the heat dispersed throughout the reaction vessel and would have prevented the liquid chemicals from forming in layers and releasing more heat. Since the plant had closed for the weekend and the workers' shift was ending at 6:00 A.M., they reasoned that, left to its own devices, the reactor and its contents would slowly cool without the addition of the water normally used to stop the reaction and bring down the temperature of the mix. This reasoning was obviously totally wrong.

The third problem was that the steam heat in the coils surrounding the upper part of the reaction vessel rose to a temperature of about 300 degrees Celsius since it had not been drenched with water. This rise in temperature by itself would not have been enough to do significant damage, but, with the other problems, it added to the temperature that allowed the dioxin to form.

The fourth problem was completely unanticipated. Over the period of seven and a half hours after the stirring stopped in the reaction vessel at 5:15 A.M.,

A view of the huge pit dug to contain the topsoil contaminated by the accidental release of dioxin near Seveso, Italy. When the topsoil was bulldozed for interment in the pit, clouds of dioxin-laden dust were raised and blown over previously uncontaminated areas.

the liquid chemicals in the vessel stratified, or formed layers. Although the process of stratification is well known, chemical researchers previously thought that the temperatures in the different layers in the reaction vessel would stay roughly the same. But that did not happen in the ICMESA reaction vessel. Instead, the liquid chemicals in the top layer of the vessel began a self-heating action, and over a period of seven and a half hours, they slowly heated up to 230 degrees Celsius.

At 230 degrees Celsius, long after the workers had gone home, the fifth problem occurred: a well-known runaway reaction took place that created still more heat and created the dioxin. The runaway reaction also raised the pressure in the vessel so high that it burst the pressure disk at the top of the reaction vessel and released the cloud of dioxin at 12:37 in the afternoon of Saturday, July 10, 1976.

At that time, many people in Seveso just to the south of the ICMESA plant were having lunch in their yards. The white cloud passed over them, and many went inside their houses to escape it. Some children went to play in the cloud, though they soon fled because of its bad odor. Officials of the ICMESA plant did nothing immediately. One chemist who was called in that Saturday afternoon told some Seveso residents close to the plant not to eat the fruit from their trees. Late Sunday afternoon, ICMESA managers visited the mayor of Seveso to tell him that the people should not eat the fruits from their trees because some herbicidal chemicals had been released. A few days later, on Thursday, the mayor gave public notice that people should not eat their homegrown fruits or vegetables. Fourteen days after the release of the dioxin, on July 24, 1976, the ICMESA plant authorities finally gave formal notice that a deadly chemical had been released and that evacuation of the area should be considered.

Impact

At first, the effect of the acrid cloud of dioxin on the inhabitants of Seveso was to create nausea, dizziness, headaches, and to burn like sores on the skin. Later in the day birds, chickens, and domestic animals such as cats and rabbits began to sicken and die. One man reported that when he tried to pick up his dying cat, its tail fell off. Some eighty-one domestic animals died in the contamination or were slaughtered as a health precaution.

The long-term medical effects of the dioxin contamination are uncertain. One hundred and eighty-seven confirmed cases of the skin condition chloracne were confirmed, but most of these cleared up after treatment. Thirty-five percent of the people in the northern part of the contaminated zone (i.e., the part closest to the ICMESA plant) were found to have enlarged livers, which was regarded as important, because dioxin is especially harmful to the liver. One health concern was dioxin's well-known effect of deforming the fetuses of animals. A herd of forty cows was being bred near the ICMESA plant, and there were ten spontaneous abortions among the thirteen cows that became pregnant. Of the three cows that carried their pregnancies to term, only one had a calf that lived very long after it was born. The fear of birth defects caused thirty-four women to obtain therapeutic abortions, with the permission of the Catholic church. One estimate said that another 120 women in the area may have had abortions without the permission of the church. The fear of birth defects has proven justified. Babies born with birth defects in the Seveso area in 1975, the year before the contamination, numbered three. In 1978, they totaled fifty-three.

On July 25, 1976, over two weeks after the dioxin contamination, the Seveso regional authorities finally ordered the evacuation of 739 residents from the 253 acres immediately adjacent to the ICMESA plant. Their evacuation took place from July 26 to August 2. Unfortunately, the evacuees took with them their cars, trucks, furniture, and bicycles that were certain to have been contaminated with dioxin. A fence nine feet high and four miles long was eventually built around the contaminated 253 acres. Another area of 665 acres with a population of five thousand people was sealed off from nonresidents, but the residents were allowed to stay and to come and go as they pleased.

The process of decontamination was only partially effective, and dioxin spread around the neighboring regions. After the evacuation, teams of decontamination workers began to remove the carcasses of dead animals, as well as trees, brush, other vegetation, and eight inches of the topsoil in many areas of the 253 acres of land adjacent to the ICMESA plant. But some decontamination measures only spread the dioxin around. As topsoil was bulldozed in the contaminated area, clouds of dust were raised and dioxin was blown away to yet other adjacent areas. As people tracked in and out of the contaminated area to remove their belongings, they spread the dioxin. Heavy and continuous rains in October of 1976 flooded the most polluted part of Seveso, washing dioxin and contaminated soil away. In Milan, thirteen miles south of Seveso, dioxin traces have been found in the mud. In the summer of 1977, over a hundred schools in the areas around Seveso were found to be contaminated with low levels of dioxin.

The owners of the ICMESA factory, the Hoffmann-La Roche pharmaceutical company, set aside over $134 million to pay for the dioxin contamination. Some evacuees were paid to have their houses completely renovated. Others received the equivalent of two years' salary. Because the legal system in Italy moves very slowly, it was almost seven years, that is, April of 1983, before a criminal case against five former employees of the ICMESA plant was begun. In an odd twist, one supervisor at the plant was murdered by terrorists before he could be prosecuted. Other legal cases against Hoffmann-La Roche are pending.

The fear of birth defects has proven justified. Babies born with birth defects in the Seveso area in 1975, the year before the contamination, numbered three. In 1978, they totaled fifty-three.

The accident at Seveso had repercussions throughout the chemical industry. After the dioxin contamination, plans to construct several chemical plants in Italy were halted and a plant that manufactured trichlorophenol in Coalite, England, was shut down. In addition, because the main cause of the Seveso dioxin contamination was worker error, authorities proposed that companies educate workers about mandatory procedures and the consequences of failing to follow them. One expert also advised that future trichlorophenol manufacturing plants should have total containment systems that would not vent any expanding gases to the outside environment.

A number of pollution regulations were developed in Europe after the Seveso accident. Companies have been required to disclose any dangerous processes used at their factories, and regulations have been developed to limit the transport of toxic wastes across European national boundaries. These latter regulations have made it very hard for Hoffmann-La Roche to dispose of the dioxin collected during its cleanup activities. The company was again embarassed by its handling of this problem when it was discovered that forty-one barrels of waste material contaminated with dioxin from Seveso had been located in an unused slaughterhouse close to the middle of Paris. The company had hired an Italian susidiary of a German engineering company to arrange for disposal, and it, in turn, subcontracted with a French individual who drove from Seveso to Paris in 1983 with the barrels. After the disclosure of this cross-country violation, the waste was later incinerated at a disposal plant in Switzerland.

The authorities in the Seveso area mishandled the accident. First, they were slow to react to the accident and took their time letting people know that there had been a problem. Without this delay, the residents of the Seveso area might have evacuated in time to escape harm. Furthermore, the cleanup process progressed slowly and sloppily. The contaminated zone was neither protected nor policed during this period, and many of the vacated houses were vandalized. After frequent delays and much conflicting information, evacuees became homesick and started pressuring the authorities to let them return to their land. The authorities partially relented in the fall of 1977, and some of the evacuees were allowed to return to their homes in the cleaned-up areas of the contaminated zone. The area immediately surrounding the factory was designated Zone A, and no one was allowed to return, but were settled later in Zones B and R. In 1989, a team of researchers published a study describing increased rates of cardiovascular deaths among Seveso residents. In 1993, that same team published another study stating that people living in contaminated areas around Seveso have an elevated risk for several types of cancer.

PATRICK MOORE

Where to Learn More

"Drug Giant Plagued by Dioxin's Poison." *Business Week*, May 2, 1983, p. 42.

Cardillo, Paolo, A. Girelli, and G. Ferraiolo. "The Seveso Case and the Safety Problem in the Production of 2,4,5-trichlorophenol." *Journal of Hazardous Materials*, August 1984, pp. 221-34.

"New Mechanism for Seveso Explosion Proposed." *Chemistry and Industry*, April 4, 1981, p. 197.

Dickson, David. "The Embarrassing Odyssey of Seveso's Dioxin." *Science*, June 24, 1983, pp. 1362-63.

"Fears Still Cloud Italy's Toxic Town." *Newsweek*, May 10, 1982, p. 14.

Sambeth, J. "What Really Happened at Seveso." *Chemical Engineering*, May 16, 1983, p. 44 ff.

Stevens, Jeffrey J. "What Happened at Seveso?" *Chemistry and Industry*, July 19, 1980, pp. 564-6.

Stone, Richard. "New Seveso Findings Point to Cancer." *Science*, September 10, 1993, p. 1383.

"The Deadly Cloud." *Time*, August 16, 1976, p. 39.

Whiteside, Thomas. "Contaminated." *New Yorker*, September 4, 1978, p. 34 ff.

Toxic vapor leak

Bhopal, India

(1984)

A toxic vapor leak resulted in massive loss of life and widespread injury.

Background

On the morning of December 3, 1984, a poisonous cloud of methyl isocyanate gas, an element in the pesticide produced by the Union Carbide plant in Bhopal, India, escaped and drifted over the city. Thousands of people were killed and many more badly injured as the gas attacked their nervous systems. Somehow, water was mixed into the wrong tank, causing an increase in pressure that blew open a valve. The ensuing explosion allowed fifty thousand pounds of deadly gas to escape, dealing unprecedented death and destruction to inhabitants. The horror at Bhopal prompted harsh public scrutiny of all chemical plants on a worldwide basis, and the courts began the long process of litigation over damage claims.

The Union Carbide site at Bhopal was first developed in 1969 as a mixing and packaging plant for pesticides imported from the United States. In 1980 it was expanded to manufacture the carbamate pesticides Sevin and Temik. One chemical used in large quantities in the production process was methyl isocyanate (MIC), a highly reactive, volatile, and toxic compound. Methyl isocyanate was stored on-site in large underground storage tanks. The Bhopal MIC storage tanks had capacities of approximately fifteen thousand gallons, over three times the size of MIC storage tanks at any other chemical manufacturing facility that stores MIC.

The toxicity of MIC is well established in the warnings issued by the chemical industry and regulatory agencies. The threshold limit value (TLV) of MIC established under United States Occupational Safety and Health Act (OSHA) standards is exposure to 0.02 parts per million of MIC in an eight-hour period. Exposure beyond this limit is considered deleterious to a worker's health. Methyl isocyanate causes irritation of the eyes, nose, and throat at 2 parts per million. Exposure at the 21 parts per million level is tortuous. The TLV of methyl isocyanate is one-tenth that of phosgene, the chemical used in World War I as the warfare agent mustard gas.

It is very important that stored MIC be protected from overheating, from escaping into the atmosphere, and from the introduction of contaminants, including water, with which it might react. Methyl isocyanate is also highly reactive with such common metals as iron, copper, tin, and zinc. During bulk storage MIC is often protected and cooled by a blanket of dry nitrogen. To prevent or slow reactivity, the temperature of bulk-stored MIC is maintained at 0 degrees Celsius, or below. Any contamination of MIC by a metal or by water, or any other substance with which it reacts, creates chemical reactions that have the potential to escalate into a runaway reaction. A runaway reaction, or a chemical chain reaction, occurs when the heat of reaction is not mitigated by surroundings that conduct the heat of reaction. Under these conditions, the heat of reaction rapidly builds to an explosive level.

The exact operative mechanisms of the Bhopal release are still subject to debate. In accidents and releases that occur in most modern processing plants, an extensive database of operator logs; electronic sensors and recording devices linked to computer control, monitoring, and data storage systems; and computer models of cause-and-effect scenarios are available to the investigator. At Bhopal, the array of gauging, monitoring, and recording systems typical of a modern processing plant were either not in place, out-of-service, or not set properly prior to the explosion. Thus, only a most-probable-cause scenario can be postulated.

Details of the Disaster

For the six weeks prior to the accident, the Bhopal plant's MIC production unit had been shut down due to an abundance of on-site stores of Sevin and Temik. As is usual during downtime, workers had been repairing, cleaning, and performing maintenance on the unit. A standard item of maintenance involved cleaning the filters in the four lines that carried MIC from the storage tank into the processing unit. These four lines were connected to the pipeline for the relief valve vent header (RVVH). This RVVH configuration was designed to carry toxic vapors from the MIC storage tanks to the vent gas scrubber (VGS) unit, should the pressure build up in the storage tanks.

The VGS would "scrub" any toxic release with caustic soda to neutralize the MIC and then shunt the scrubbed gas to a flare tower via a knockout pot (a device for removing liquid droplets prior to flaring). Any vapor release would be scrubbed in the VGS, passed through the knockout pot to collect any droplets that might have formed, and then burned in a flare tower.

To prevent water from traveling toward the MIC tank via the RVVH piping, a maintenance worker closed a valve to isolate the filter pipelines from the RVVH during the washing process. However, the worker did not insert a slip blind—a metal disc—into the valve to seal it, as is usually done when filters are washed. Also, two of the four bleeder valves that allowed water to flow out of the lines during the washing process were completely clogged; the flow rate from the other two indicated a partial blockage. This was noted, and the washing process was temporarily halted, but the filter washing procedure soon resumed.

A toxic vapor leak at this Union Carbide plant in Bhopal, India resulted in massive loss of life and widespread injury.

As noted in an inspection earlier in the year, many of the valves in the plant were leaky. The filter isolation valve was no exception. Thus, water pressure increased in the partially clogged filter washing system and water began flowing past the isolation valve and into the RVVH.

The piping between the RVVH and the MIC storage tank contained a number of closed valves which, under as-designed configurations, would prevent the entry of water into the storage tank via the RVVH. There was also a one-way rupture disc designed to break in the event of increased pressure from MIC vapor and allow the toxic vapor to flow into the RVVH. However, a jumper line had recently been installed between the RVVH and the process vent header (PVH). This header was designed as a route for venting gases released during movement of the MIC to the production unit. Any gases released from the tank would vent to the scrubber. The jumper line allowed gases from the storage tank to vent to the scrubber through the headers of either the PVH or the RVVH should one of the header units be down for repairs. The jumper line was installed instead of the more expensive solution of installing a backup line for the process vent headers.

The jumper line, which had been installed in May 1984, was opened on December 2 to allow repairs on the PVH. This open line allowed the water backing up the line from the filter washing process to enter the PVH instead of the RVVH. From the PVH, the water was able to enter tank 610 because of leaking valves in the piping connecting the PVH to the storage tank. It was known that the valves that were to seal the tank from the PVH were leaking because recent efforts to pressurize tank 610 with liquid nitrogen so that the MIC could be transferred to a Sevin production unit, when needed, had been unsuccessful. Although the pressurization efforts had been going on for a week, nothing was done to locate and correct the faulty valve, or valves.

There were several other conditions that contributed to the explosive tank failure and subsequent toxic release. In addition to the leaking water—which reacted with the MIC to create intense heat—tank 610 was 75 to 87 percent full, well above the 50 to 60 percent recommended in Union Carbide safety standards. Tank 619, which was supposed to be empty as a safety overflow, contained approximately twelve tons of MIC. The MIC in tank 610 was contaminated by chloroform at a level much higher than that recommended in Carbide's safety standards. Chloroform reacts both with MIC and with the stainless steel walls of the storage tank. A final precondition was the temperature of the MIC in the storage tank. The MIC was between fifteen and twenty degrees Celsius. Standard safe storage procedures specify a temperature of zero to five degrees Celsius, preferably lower.

When the operators on duty noted a sudden rise in the pressure in tank 610, from two to thirty to fifty-five pounds per square inch, there was nothing that could be done to prevent, or even slow down, the runaway reaction. The operators attempted to actuate the site safety systems, but the systems either were not working, could not handle the vapor pressure load, or would create additional problems if used.

The vent gas scrubber was on standby status, and could not be turned on in time to neutralize some of the escaping gases. The flare tower was not available to burn any of the release because the piping linking the MIC tank to the vent gas scrubber, and subsequently to the flare tower, had been removed. Also, the operators feared that moving the reacting MIC into the overflow tank—which had twelve tons of MIC in it—would only create additional dangers. A water spray for dousing the leaking toxic vapors rushing from a stack on the relief valve vent header did not work. The stack was 120 feet high, but the spray reached only 100 feet.

Approximately twenty-seven tons of MIC vapor, and fourteen tons of reaction products, were released into the atmosphere during the explosion failure of tank 610. The release of this searing cloud of toxic vapor occurred over a period of approximately ninety minutes and affected an area of fifteen square miles.

For an hour and a half, the toxic contents of tank 610 and associated reaction products spread into the slums downwind of the plant. The narrow streets and flimsy dwellings of this area became a charnel house of human beings and animals gasping for breath amid a heavy white vapor—twice as heavy as air—boiling from the plant and filling the low-lying areas. The gas was immediately deadly, killing persons up to five miles away from the site. A neighborhood where the greatest loss of life occurred was four miles from the plant.

The Indian Supreme Court's official estimate of those killed by the gas is three thousand. Other investigators put this figure in the twenty-five hundred to five thousand range, and medical personnel who directly treated the victims in Bhopal have stated that up to twelve thousand died as a direct result of the leak. The victims died immediately from either asphyxiation or suffocation caused by pulmonary edema (fluid in the lungs), or in subsequent days from causes directly attributable to exposure to the effects of the toxic vapor.

Those injured by the release of MIC and its reaction products suffered from symptoms that included acute respiratory distress, eye irritation, and problems of the circulatory, gastrointestinal, and central nervous systems. Long-term effects included chronic lesions of the eyes, permanent scarring of the lungs, and injuries to the liver, brain, heart, kidneys, and immune system. An epidemiological study of the yearly rate of spontaneous abortions and infant deaths in the years after the accident found the rate in Bhopal to be three to four times the regional rate.

An Indian government commission estimate of those injured by exposure to the vapor included 30,000 with permanent injuries, 20,000 with temporary injuries, and 150,000 with minor injuries. This figure is disputed by victims' rights organizations who say that this estimated is much too low.

Impact

The magnitude and implications of this event created a great sense of uneasiness in the industry and in the general public, leading to questions of the principles of plant operations and the nature of the chemical industry in general, and of the manner in which Union Carbide operated the plant in India. In addition to industry and site-specific safety design configuration, the operation readiness of the Bhopal plant was examined closely.

After Bhopal there was an inquiry into the causes of the release on the site, and the potential for accidents at other processing facilities using and storing MIC. These inquiries were concerned with such issues as tank construction and monitoring; site safety warning and containment systems; the potential for failures caused by operator error; identifying site configurations that undermine the safety inherent in the original design of the MIC storage tank and production unit; and storage system safety maintenance and emergency response procedures.

These are important safety-directed considerations. They are crucial to understanding and guarding against system failures capable of releasing toxic compounds into the communities in the vicinity of a plant. However, the management decisions that allowed the Union Carbide India plant to

The toxic fumes leaking from the plant were immediately deadly, killing persons up to five miles away. Eventually, thousands were evacuated from their homes.

operate in an unsafe manner and for a shantytown to develop around this plant without an appropriate measure of emergency planning, are important issues. Industry management and interest groups and regulatory agencies must address these issues to effect change in the policies and actions that allowed these preconditions for a disaster to exist.

The questions of criminal prosecution for negligence and fiscal compensation to the victims have been the subject of intense debate and litigation in the government chambers and court systems of India. Initially, Indian prosecutors brought charges of criminal negligence against the Indian and American management of Union Carbide. Some months after the accident, Union Carbide stated that an angry employee sabotaged the plant by deliberately allowing water to enter the MIC storage tank. However, no one associated with the Bhopal plant has been directly charged with this sabotage.

After several years of negotiations, Union Carbide agreed to a settlement of $470 million with the government of Rajiv Ghandi. The dropping of all criminal charges against Carbide officials initially charged was part of this settlement.

The amount was challenged on legal grounds, but was upheld as constitutional by the Indian Supreme Court in 1991. However, the legitimacy of this amount is not recognized by the present Indian government. The original claim against Union Carbide was for $3.3 billion, and the present government feels this is a more substantive settlement. They are also interested in pursuing criminal charges against Carbide management. Union Carbide India has paid, as ordered, $190 million in interim compensation, which the Indian government is to distribute among the survivors pending the final outcome of the settlement.

Presently, under the authority of the Bhopal Claims Act, the Indian government is giving survivors 200 rupees a month (approximately U.S. $10). These payments are to last as long as the Bhopal case remains unsettled in the courts and governing bureaus. It is likely that litigation and the debate over responsibility and compensation will continue for several more years.

LAMONT INGALLS

Where to Learn More

"Union Carbide's Bhopal Bill." *Business Week*, February 27, 1989, p. 40.

Everest, L. *Behind the Poison Cloud*. Banner Press, 1985.

"Bhopal Tragedy's Health Effects." *Journal of the American Medical Association*, December 5, 1990, p. 2781.

"Public Health Lessons from the Bhopal Chemical Disaster." *Journal of the American Medical Association*, December 5, 1990, p. 2795.

Kurzman, D. *A Killing Wind: Inside Union Carbide and the Bhopal Disaster*. McGraw-Hill, 1987.

Lepowski, W. "Lessons from Bhopal," *Chemical & Engineering News*, November 17, 1986, pp. 39-45.

———. "Indian Activists Press Bhopal Accident Issues in U.S. Visit." *Chemical & Engineering News*, May 13, 1991, pp. 22-3.

Toxic vapor leak

Institute, West Virginia

(1985)

An explosive failure of a temporary storage tank containing an aldicarb oxime mixture caused a toxic vapor leak at a chemical plant, injuring one hundred and thirty-five people.

Background

Immediately following the tragic methyl isocyanite gas leak at Bhopal, India, on December 3, 1984, which killed twenty-five hundred people and injured two hundred thousand, the Union Carbide Corporation (UCC) plant at Institute, West Virginia, was shut down. This facility was the only chemical plant in the United States that manufactured and used methyl isocyanate (MIC). The same substance made at Bhopal, the MIC was manufactured at Institute and then shipped to a plant in Georgia. At this plant, the MIC was reacted with aldicarb oxime, another toxic substance, to form aldicarb, a component of the pesticide Temik.

In the Bhopal incident, a cloud of MIC spread rapidly from a UCC facility into a heavily populated area, causing the largest industrial accident in history. After Bhopal, there was intense concern among UCC management and the residents of Institute that an MIC release could occur. MIC is highly corrosive, toxic, and unstable, and its manufacture and storage must take place under stringent conditions of isolation and control. From 1980 to 1984 there had been twenty-eight MIC leaks documented at the Institute plant. Although these leaks had been confined to the site, the safety of the plant was in doubt and it was closed down for a safety review.

During this period of closedown for inspection and analysis of plant process design, UCC installed several additional safety features. One of these was a computerized air-monitoring system designed to detect and identify any releases of hazardous vapors in the plant. The system could also predict the movements and the toxicity or explosive levels of a vapor release. The predictive modeling would enable UCC safety personnel to act using real-time information, making it possible for them to warn off-site emergency response teams and nearby residents if the vapor cloud appeared to be spreading beyond plant boundaries.

Other changes were also made to alleviate the possibility of a release of MIC, or associated toxic substances. Safety devices with primary and redundant systems to control MIC were installed, and the on-site storage of MIC was reduced.

Prior to the Bhopal incident, MIC was shipped unmixed off-site. These shipments were discontinued and were not resumed following the restart of the plant on May 4, 1985. Instead, plant processes were reconfigured to allow MIC to react with aldicarb oxime (AO) to produce aldicarb. The aldicarb was then shipped to the Georgia plant for additional processing. Although not as toxic as MIC, small amounts of AO were known to cause nausea, tremors, eye and lung irritation, and heart palpitations. Aldicarb oxime had also been found by the Environmental Protection Agency (EPA) to cause cancer in laboratory animals under certain conditions.

During the months of the shutdown, the Institute plant was subject to a close safety review by federal officials who audited the site for violations of Occupational Safety and Health Administration (OSHA) regulations. In February 1985, OSHA officials declared that a Bhopal-scale accident was a very remote possibility in the United States and that the Institute plant could safely return to operation.

Details of the Accident

The storage tank and site safety systems were ruptured at several points by excessive system vapor pressure. This buildup of pressure before the explosion was the product of a thermal runaway reaction. This type of chemical chain reaction occurs when more heat is released during the reaction than can be carried away by cooling or through absorption by the walls of the storage vessel.

The post-Bhopal process redesign called for installation of a storage tank and attendant support and safety systems. This configuration would be used for storing the AO prior to movement to the mixing vessel where the AO would react with MIC. This storage area was to be a temporary facility, to be used only until a permanent storage tank could be constructed.

The storage tank had a capacity of five thousand gallons and normally contained a mixture of AO and methylene chloride, a common solvent used to enhance the MIC-AO reactive process. The lower two-thirds of the storage tank was enclosed by a steam jacket. This jacket was used, as needed, to heat the mixture because methylene chloride becomes solid below seventy degrees Fahrenheit. The steam to the jacket was controlled by a motor-operated valve.

On August 1, 1985 an aldicarb oxime-methylene chloride mixture was pumped into the storage tank from a mixing tank. This mixture was excess from the aldicarb production process.

From August 1 to August 7, a faulty gasket on the motor-operated valve allowed steam to leak into the heating jacket. This heat caused the methylene chloride to vaporize. The vapor condensed in other parts of the system, including a spare storage tank, the crystallizer, and the "knockout pot," a vessel that collects the gas prior to venting to a vapor scrubber (a vapor scrubber treats gases and renders them less harmful to the environment). This vaporization of the methylene chloride caused the AO to concentrate in the bottom of the storage tank, in turn creating conditions which allowed a chemical heat reaction to proceed more rapidly than in the standard AO mixture.

On August 7, operators started pumps to transfer the contents of the storage tank to the mixing tank. After two and a half hours, the pumps stopped functioning. The operators assumed the tank was empty. In fact, there were still about five hundred gallons of the concentrated AO mixture in the tank. No attempt was made to verify that the storage vessel was empty. The tank level gauge was broken near the top and remained out of service. No other efforts were made to gauge the contents.

The toxic cloud caused immediate eye, throat, and lung irritation, and nausea. So many were impaired by the toxic cloud at the golf course than an emergency aid center was set up at one of the greens.

From August 7 to August 11 the concentrated AO mixture continued to be heated by the steam leaking into the jacket. Plant operators were not aware of this gradual heating because they had not programmed the electronic sensing system to display the temperature and pressure of the tank. At 9:25 A.M. on Sunday, August 11, a runaway thermal reaction began in the storage tank. The operators heard a deep rumbling noise in the tank. Within seconds, they noticed excess flow to the flare designed to burn escaping gases that had been treated in a caustic solution in the scrubber. This high pressure at the flare indicated that an automatic safety valve had opened in the scrubber.

Within thirty-five seconds, the rupture disk on the knock-out pot failed, sending the toxic vapor mixture directly into the atmosphere. This rupture was followed immediately by blowout of the gaskets at the cleaning access holes on top of the storage tank. This gasket failure also released toxic vapor into the atmosphere.

The control room was rapidly engulfed by a thick, toxic cloud that prevented the operators from seeing their controls. However, one of the operators was able to set off the toxic gas release warning. The six operators, trapped by the gas, huddled in the control room, sharing two breathing apparatuses.

Within seven minutes the release was complete. The vapor cloud continued drifting through the plant. Personnel monitoring the cloud through the computerized sensing system noted that the release did not appear to be heading off-site. However, the toxic cloud continued to move, drifting into nearby populated areas.

The toxic vapor cloud—composed of AO, methylene chloride, and various breakdown products, including sulfur compounds—spread into the vicinity of the plant. People in their homes and backyards, driving on nearby roads,

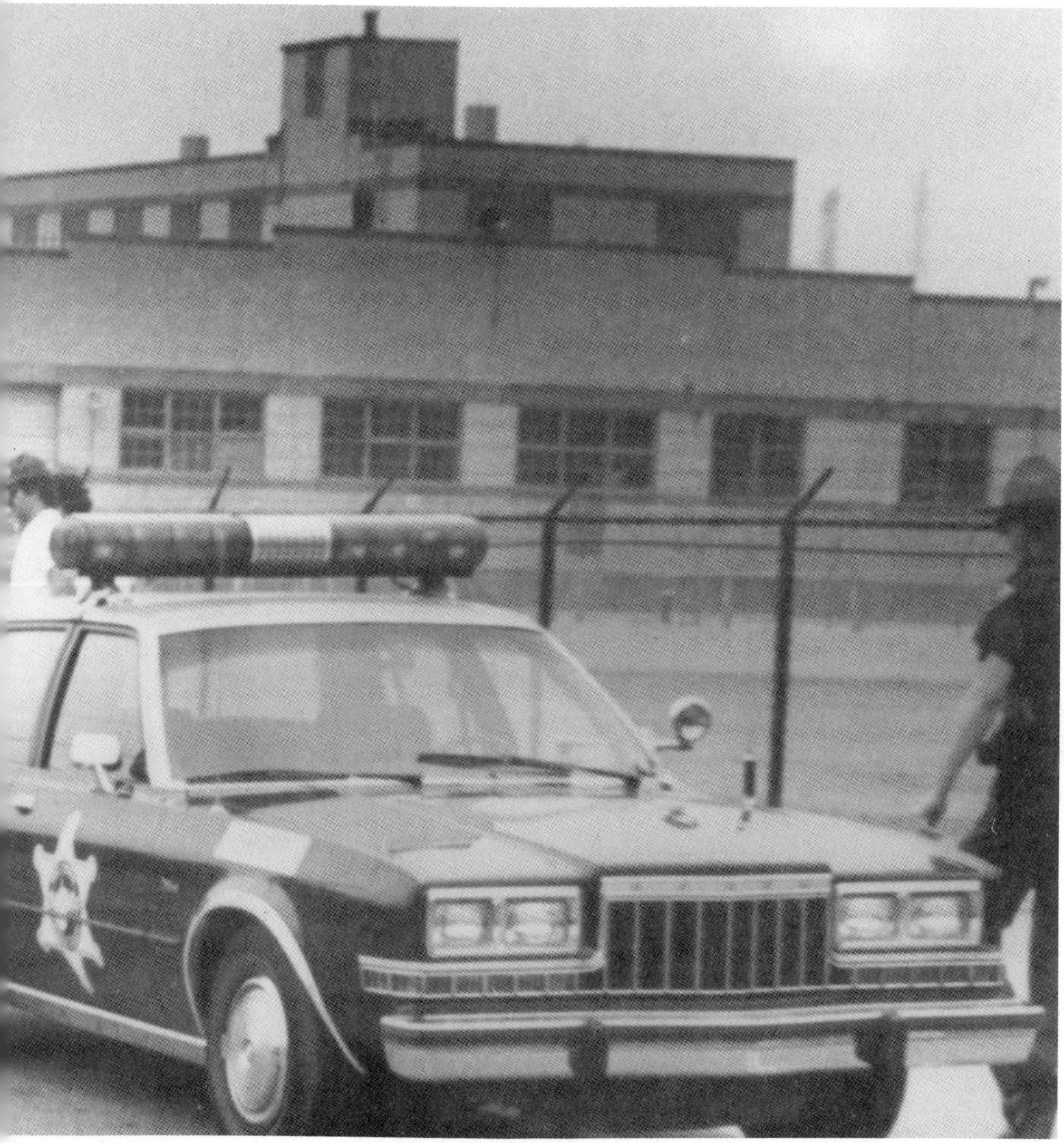

Following the accidental release of toxic vapors at its West Virgina plant, Union Carbide was cited for over 200 violations of OSHA regulations.

and playing golf at an adjacent golf course were affected. The toxic cloud caused immediate eye, throat, and lung irritation, and nausea. Altogether, one hundred and thirty-five people were seriously affected. So many were impaired by the toxic cloud at the golf course than an emergency aid center was set up at one of the greens. Of the one hundred and thirty-five people given emergency treatment, twenty-eight were admitted to area hospitals. Eight of those hospitalized were workers from the Institute plant.

A poll of area residents living in the vicinity of the plant found that most reported learning about the leak from seeing the vapor cloud or feeling the effects of the gas rather than from hearing the emergency warning sirens. However, UCC management said that the lag in warning time of twenty-one minutes initially reported on the computerized emergency response log was due to a computer problem. Union Carbide management contended that the actual time from detection to activation of the warning sirens was slightly over one minute.

Site safety staff operating the newly installed air-monitoring system reported that the system predicted the cloud of toxic vapor would not drift outside plant boundaries. A subsequent investigation disclosed that UCC had not purchased the software that would have enabled the system to predict the movements of vapors associated with the release of aldicarb oxime.

Problems with the release containment systems were also found. The piping and pressure gaskets were not designed to hold a flow of the magnitude caused by the runaway thermal reaction. Although earlier modeling indicated that design criteria were adequate, the phenomenon of two-phase release may have affected the system's vapor handling capacity. This type of release creates a mixture of vapor and liquid under pressure—like foam on a mug of beer. A two-phase release does not allow the proper release of pressure in systems designed for vapor-only pressure releases.

Another concern was the venting of toxic vapor directly to the atmosphere from the knockout pot and the storage tank when a rupture disk and gaskets failed. This design was in violation of site safety design and operations procedures. All systems in place to contain tank failure for toxic substances of this type (category four according to UCC: the same classification as MIC) should be vented into a closed system. Another safety system—a water containment sprinkler—did not have enough volume to suppress a release of this magnitude.

UCC's review of the operating history of this storage unit also disclosed several operator errors. These errors included not repairing a level gauge that would have shown that the tank still contained an aldicarb oxime-methylene chloride mixture; shutting off or ignoring high-pressure alarms; and not repairing a broken high-temperature alarm. The operators also did not request information on the tank from the unit's sensing computer, although the computer had been recording the increases in pressure and temperature since the steam line began leaking.

Impact

After the August 11 toxic leakage, OSHA auditors made a thorough inspection of the plant. Union Carbide was cited for two hundred and twenty-one violations of OSHA regulations, including failure to follow plant safety operating procedures when transferring the AO mixture and not having proper safety systems in place to prevent air contamination in the event of a leak. As punishment for the violations, OSHA imposed a $1.4 million fine, the largest fine the organization had handed down in its sixteen-year history. The vapor release also heightened the industry-wide need for review of safety backup systems and the design of vessels for storing volatile toxic substances. Also, the need for further research into determining the time-to-maximum-rate before thermal runaway for common chemical reactions was highlighted as a concern by industry groups.

Coming as it did a few months after the tragedy at Bhopal, India, this accident caused a general awareness in the chemical industry of safety and design topics related to toxic release containment. Among the public and legislators, there was also political action related to both the need for site safety control through legislation and the study and classification of toxic substances.

LAMONT INGALLS

Where to Learn More

"Union Carbide: New Accidents Revive Safety Issues." *Chemical and Engineering News*, August 19, 1985, p. 4.

"Carbide Springs Another Leak." *Chemistry and Industry*. August 19, 1985, p. 530.

"Beyond Bhopal: Toward a Fail-Safe Chemical Industry." *High Technology*, April 1985, pp. 55-61.

Diamond, Stuart. "Carbide Blames a Faulty Design for Toxic Leak." *New York Times*, August 13, 1985, p. 1.

———."Chemical Pipe Size Called Key Safety Factor." *New York Times*, August 14, 1985, p. 19.

———. "Carbide Asserts String of Errors Caused Gas Leak." *New York Times*, August 24, 1985, p. 1.

"OSHA Cites Union Carbide for Leak." *Professional Safety*, November 1985, p. 8.

Dams

- Austin Dam (1911)
- St. Francis Dam (1928)
- Malpasset Dam (1959)
- Vaiont Dam (1963)
- Baldwin Hills Dam (1963)
- Teton Dam (1976)
- Stava Dam (1985)

Austin Dam failure

Austin, Pennsylvania

(1911)

Failure of a poorly constructed concrete dam founded on unsuitable bedrock caused a destructive and fatal flood, followed by an explosion of gas lines and a devastating fire.

Background

The afternoon of September 30, 1911, was sunny and "perfect" in the towns of Austin (population 3,200) and Costello (approximately 400-500 residents) in the mountainous country of north central Pennsylvania. Many of the townspeople were strolling down Rutherberger Street in Austin and enjoying summer's last hurrah. The atmosphere changed in a matter of seconds when, after a loud blast, a torrent of water roared through the valley, channeled by its narrowness and rugged walls. The force ripped gas mains from beneath the streets; and as soon as the wall of water passed, an errant flame lit the gas and the fire jumped from gas pipe to gas pipe and house to building throughout the standing remains of the thirty-year-old town of Austin. Initial reports claimed that 1,000 people had perished and 2,000 were left homeless, although later information placed the death toll at between 50 and 149. The source of this grief was also the source of livelihood for Austin and the surrounding towns. The Bayless Pulp and Paper Mill owned the concrete dam, which it had constructed in 1909 to provide a water storage reservoir for its water-intensive pulp- and papermaking business. There had been a precursor to this disaster in January 1910 when, following intense winter rain and snowmelt, cracks had been observed in the dam. The cracks were repaired, but they were not recognized as indications of problems related to the foundation, design, and construction of the structure.

Industry in the United States of the early twentieth century was booming when the Bayless Pulp and Paper Mill decided to replace its small dam with a much larger one in 1909. Industry and its right to manipulate resources such as water were largely unregulated at this time, so when Bayless decided to enlarge its reservoir to expand its production, agency approvals were not needed. Engineers were engaged to design the new dam and observe its construction, but when cost restrictions and orders to finish the dam before the winter of 1909-10 were placed on them, the seeds for disaster had been planted.

The town of Austin is in the valley of Freeman Run, and the dam spanned this steep valley with a length of 530 feet at its base (across the floor of the valley) and 544 feet at its crest. The dam was 49 feet high at its crest, 42 feet high at the crest of its emergency spillway, 32 feet wide at its base, and about 2.5 feet wide at its crest. The reservoir was designed to contain two hundred million gallons of water that would fill one-half mile of the valley.

The dam had two major components to its design. The concrete gravity dam was made of sixteen thousand cubic yards of cyclopean concrete, which consisted of a cement matrix bonding together boulder-sized ("cyclopean") pieces of native quarried sandstone rock, as opposed to the gravel-sized rock used as aggregate in modern concrete construction. The boulders were placed to break joints, and the concrete was placed in stepped sections that were dovetailed together. The concrete was also reinforced with twisted steel rods that were each 1.25 inches in diameter, 25 feet long, and placed at 2.7 feet apart in the structure. Steel-reinforced pilasters or buttressing sidewalls were placed on either side of the spillway, and extra reinforcement was installed in the top or crest of the dam. The second major component of the dam was an earthfill embankment that was rolled into place against the concrete on the upstream or reservoir face of the concrete dam. This embankment was 27 feet high and sloped at three feet horizontally for every one foot in vertical height down from the point where its top met the concrete dam into the floor of the reservoir.

The foundation material consisted of the rock in the valley floor; thin layers of shale were nestled between horizontal layers of sandstone. About eight thousand cubic yards of soil, compacted and cemented alluvial gravel, and loose rock were excavated from the area of the base of the dam, and drill holes were advanced eight feet into bedrock to explore for fractures and fissures. The surface for the base of the dam was washed and grouted with cement. The dam was founded on a concrete footing that extended four feet into bedrock and was four feet wide, and the abutments (sides) were set twenty feet into bedrock. But it is not known whether studies were made of the permeability of the rock layers and of the need for a cutoff wall that would prevent seepage from traveling through the horizontal sandstone under the dam.

The dam was still under construction as the winter of 1909-10 approached. Temperatures were below freezing when some of the concrete was placed, and the final stages of construction were finished hurriedly. The dam was completed on about December 1, 1909, and a crack running from the crest of the dam vertically to the ground was visible when construction was finished. This crack also passed through the cyclopean sandstone boulders. By the end of the month, a second crack had appeared. Both cracks seemed to have resulted from contraction of the concrete; there was no measurable settlement of the dam, and there was no water in the reservoir yet. Within six weeks, the reservoir was filled, and the maximum load was placed on the fresh concrete.

On January 17, 1910, a warm spell brought heavy rains and caused rapid snowmelt, and four days later, flood water was pouring over the spillway.

An eyewitness to the Austin Dam disaster recalled: "The break ran across the front of the dam like the crack of a whip and the water gushed out with a force that carried a whole section . . . sixty feet away."

The next day, heavy seepage was seen emerging from the toe of the dam and cracks also appeared near the bottom of the downstream face of the dam from the ground as far as fifty feet downstream from the dam. Cracks also appeared near the bottom of the downstream face of the dam. Dynamite was used to blast two notches through the crest to lower the operating level of the reservoir. Superficial repairs were made. They were also hurried because of the time of the year. The two blasted notches were not filled in—and the reservoir was not allowed to rise again—until late September 1911. The ineffectiveness of these repairs would become all too obvious on September 30, 1911.

Details of the Failure

On that September afternoon, a hole opened between the dam and the west abutment and continued to spread. The force of the water shifted major portions of the dam; pieces overturned, and a large section of the center of the dam moved downstream as if on rollers. The released flood entered Austin as a solid wall like a locomotive. Because the valley walls were high and near-vertical, the water could not disperse or flatten out. An eyewitness saw leakage emerging from five or six cracks on the face of the dam as it failed. He said, " ... one after another six great masses of concrete came crashing through the valley just as though you were to knock down one domino after another. The break ran across the front of the dam like the crack of a whip and the water gushed out with a force that carried a whole section ... sixty feet away."

Someone rang a fire bell as a warning, and more people ran into the streets to face the crushing flood. Some escaped into the surrounding hills and later described the roar of the rushing water and the white foam on its crest. Homes were ripped from their foundations and thrown into the torrent. In the shops of the Buffalo & Susquehanna Railroad, large machines were overturned onto their operators. The largest gas main in the business district of Austin ruptured, and a fire started immediately. Sucked by the draft of the receding flood, the flames traveled from house to house. Other pipes broke and fed the fire. The stove wood prepared for shipment at the Standard Lumber Company was more kindling for the town. The Bayless Pulp and Paper Mill also was a victim of the flood's wreck and the fire's ravage. Austin's fire protection was limited, and what there was had been wiped out by the flood. The wall of water had left the injured and dead in the streets; and rescuers could not reach them because of the immediately subsequent inferno. State Senator Frank E. Baldwin was injured in the melee; among the dead were his wife, children, parents, and sister. Three miles down Freeman Run, the town of Costello was leveled; and eight miles below Costello, Wharton was heavily damaged but suffered no fatalities.

A lineman for the telephone company saw the flood strike the town and called for help from a home on the hillside. In an hour, a train with doctors, nurses, and food and medical supplies was bound for Austin. Assistance also came by automobile over the mountain roads. Fortunately, the hospital was sited on the hillside and escaped the devastation; a schoolhouse also survived, although all the town's churches were destroyed. The fire burned itself out by midnight. The rescue effort then began in earnest.

On October 1, the Pennsylvania State Water Commission and the District Attorney of Potter County began two separate inquiries into the dam's failure. Chalkley Hatton, the engineer who designed the dam and reviewed its condition after the 1910 failure, stood by his original design but claimed the foundation had been undermined by seepage in 1910 and not repaired with a concrete cutoff trench upstream of the dam and reinforcement of the downstream face as he had recommended to Bayless Pulp and Paper Company. He stated that this seepage condition had not been anticipated during the original design and construction, despite the foundation exploration. He also blamed the placement of concrete during freezing weather and the inadequacy of the time allowed for the concrete to cure before the reservoir was filled.

Reports after the failure state that the break lines between the failed masses of concrete correspond to work done on different days, supporting hurried construction as one of the contributors to failure. The faces of all but two of

Austin, Pennsylvania, following the failure of the Austin Dam in 1911. Flooding and the ensuing gas-main fire claimed between 50 and 149 lives.

the cracks were also discolored, indicating that the September failure had occurred on older break lines. Excavations were also made into the failed foundation; investigators found that rock had slid from about twelve feet upstream of the base of the dam, and the bottom of the slide was six feet below the concrete foundation. The hydrostatic pressure uplifting the dam was recalculated, based on these post-failure observations, and was found to be much higher than that estimated during design.

In fact, all technical aspects of Austin Dam were poor by both the standards of 1909 and those of today. The foundation conditions were not studied adequately, so the foundation treatment was ineffective. By definition, a gravity-type dam (such as the Austin Dam) uses mass to resist motion; and the mass, area of the base of the dam, foundation embedment, and the other methods of anchorage were insufficient. The construction failures were obvious and numerous and included use of weak, oversized aggregate placed in improperly cured concrete in freezing weather. When the first crack appeared, the dam had already failed, but this fact was ignored. When the second crack appeared, failure was restated emphatically and ignored again. When the January 1910 blatant failure occurred, it showed that the dam structure and the founding bedrock had failed, and these were again met with a more blatant ignorance. The blasting of holes in the dam represented an astounding, almost comical blunder that probably weakened whatever remained of the structure's integrity. And the owner/operator's disregard of the engineer's recommended repairs was the fatal seal. The reservoir should never have been filled after the cracks were observed in December 1909. It should have been abandoned completely after the events of January 1910.

Impact

The failure of Austin Dam was costly in fatalities, injuries, and loss of private property. This failure and other, similar failures of structures owned by industries and individuals certainly led state and federal authorities to promote legislation for the responsible design, operation, and maintenance of the hundreds of such dams nationwide. Alexander Rice McKim, inspector of dams and docks for the Conservation Commission of the State of New York visited Austin, Pennsylvania, and also inspected the dams in New York State. He noted, "I find the dams in this State generally in a very neglected condition, even when originally well constructed. The responsibility and the danger of impounding waters behind insufficient barriers is generally not recognized." We have the benefit of hindsight, but the failure of Austin Dam is particularly sad because every event in the dam's history compounded multiple errors at the same time as it offered an opportunity to stop, reconsider, act responsibly, and avert disaster.

GILLIAN S. HOLMES

Where to Learn More

"Austin Dam Failure." *Engineering Magazine*, November 1911, pp. 253-57.

Jansen, Robert B. "Austin Dam." *Dams and Public Safety: A Water Resources Technical Publication*, U.S. Department of the Interior, Water and Power Resources Service, 1980, pp. 119-20.

"Nearly 1,000 Dead in Austin, Pa., Devastated by Flood and Fire." *New York Times*, October 1, 1911, p. 1.

"Two inquiries Begun into Dam's Breaking." *New York Times*, October 2, 1911, p. 3.

"Austin Dam Was Weak." *New York Times*, December 8, 1911, p. 10.

"Failure of the Austin Dam." *Scientific American*, October 14, 1911, p. 331+.

St. Francis Dam failure

San Francisquito Canyon, California

(1928)

Failure of a dam constructed over a defective foundation killed hundreds and attested to the importance of thorough geological investigations of dam sites.

Background

The sudden and catastrophic collapse of the St. Francis Dam in California occurred just before midnight on March 12, 1928. Because the reservoir was filled to capacity, a tremendous volume of water (12 billion gallons) was released onto the canyon below. Many residents of the canyon were asleep in their homes when the wave of water struck, and approximately 450 lives were lost. Buildings and property located at the bottom of the canyon were swept away by the on-rush of water, and the property damage amounted to over $10 million, five times the construction cost of the dam. The collapse of the St. Francis Dam was California's second most destructive disaster, exceeded in loss of life and property only by the San Francisco earthquake of 1906.

The St. Francis Dam was constructed from April 1924 to May 1926 by the Los Angeles Bureau of Water Works and Supply. The dam was located in the San Francisquito Canyon, about forty-five miles northwest of Los Angeles. The dam was part of an ambitious and large-scale project to provide water to the growing city of Los Angeles; the reservoir created by the dam was to serve as a terminal water storage area near the lower end of the Los Angeles Aqueduct.

The St. Francis Dam was designed and built as a solid concrete, gravity type dam, whereby the massive weight (500 million pounds) of the concrete itself would hold back the water in the reservoir. The dam had a maximum height of 205 feet, and the length of its main section was 700 feet.

The consensus among engineers and geologists is that the St. Francis Dam failed because of adverse geological conditions at the site. Specifically, the failure was due to a defective foundation. Three distinct foundation conditions could have led to the disaster: (1) the slipping of the rock beneath the easterly side of the dam along weak geological planes; (2) the slumping of rocks on the westerly side of the dam as a result of water saturation; or (3) the seepage of water under pressure along a fault beneath the dam. Although the information available is not conclusive, evidence indicates that seepage of water beneath the dam was the probable cause of the disaster.

As early as 1926, when the reservoir began to be filled, water leaked through the fault zone beneath the dam. Workmen putting the final touches on the dam were ordered to pack oakum into the fault area to seal off such a leak. This remedial device was not entirely successful, and water continued to seep beneath the dam. A two-inch diameter pipe was then laid to the fault zone, and the seepage was collected and carried down to the dam-keeper's cottage for domestic use.

On the morning of March 28, 1928, significant leaks developed beneath the St. Francis Dam. The dam-keeper noted that one leak was passing muddy water and informed the Chief Engineer and General Manager of the Los Angeles Bureau of Water Works and Supply, William Mulholland. Muddy water appearing in such a leak generally indicates that the earthen foundation material that is packed beneath a dam is being washed away. If so, because of the high water pressure exerted by a full reservoir, it is only a matter of time until a blowout occurs, or the dam settles unevenly and fractures.

Although Chief Engineer Mulholland inspected the dam less than twelve hours before it collapsed, he believed that the dam-keeper was mistaken and the dam was not endangered. Therefore, no alarm or warning about a possible failure of the St. Francis Dam was raised.

Details of the Failure

The failure of the St. Francis Dam the night of March 28 showed that Chief Engineer Mulholland had been mistaken. The dam collapsed suddenly, and water began roaring down the canyon, attaining a maximum flow depth of about 125 feet. A massive powerhouse attached to the dam, built of concrete and sixty-five feet in height, was crushed like an eggshell from the force of the flood. Huge pieces of the concrete dam, some weighing 20 million pounds, were washed more than one-half mile down the canyon.

Even one hour after the dam collapsed, no warning had been given to the people in the path of the oncoming water. Word finally began to spread, via telephone operators, deputy sheriffs, highway patrolmen, and volunteers, that a flood was on the way. The heroic efforts of a few men and women saved many lives, and evacuations took place over the next three hours as the flood travelled past the San Francisquito Canyon toward the Pacific Ocean.

An estimated 450 persons were killed by the flood. Only 319 of the bodies could be recovered; the rest were either buried beneath deep mud or washed into the Pacific Ocean. The dam-keeper and his six-year-old son lived in a cottage located immediately below the dam, and were the first casualties of Mulholland's error of judgment. The dead included almost every member of the Ruiz family, pioneer settlers of the San Francisquito canyon. Only three of seventy residents of the canyon area survived, and they provided tragic accounts of the catastrophic flood. About nine hundred buildings were demolished, three hundred others were heavily damaged, and twenty-four thousand acres of agricultural land were destroyed.

Impact

The cause of the St. Francis Dam failure was investigated by a commission of engineers and scientists appointed by the governor of California. They concluded that failure was related to a fault beneath the dam—a fault that had been duly noted in the 1922 Seismological Society of America map of California. Such faults, formed by the shearing and ripping of land as two plates of the earth's crust grind past each other, are common to the state. The fault beneath the St. Francis Dam created sheared and fractured rock that the commission described as an "exceedingly fragile flaky material" inappropriate for a dam site.

The governor's commission determined that the initial foundation failure of the St. Francis Dam occurred near or at the fault due to the percolation of water into and through that section of the foundation. The infiltration of water resulted in a softening of the fractured rock beneath the dam. Either the fractured rock was blown out by water pressure, or the dam settled at the weakened portion of the foundation, or both. In any case, the complete collapse of large segments of the dam quickly followed.

The governor's commission held that the St. Francis Dam, if built upon a solid rock foundation, would have been secure and permanent. It was the geological characteristics of the site alone that led to the failure.

The governor's commission also held that the St. Francis Dam, if built upon a solid rock foundation, would have been secure and permanent. It was the geological characteristics of the site alone that led to the failure. The collapse of the St. Francis Dam, which has not been rebuilt, therefore illustrates the fact that a dam is only as strong as the rock that it rests upon. Thus, the failure of the dam underscored the importance of undertaking extensive geological investigations as part of the process of siting a dam.

William Mulholland of the Los Angeles Bureau of Water Works and Supply accepted full responsibility for the failure of the St. Francis Dam. "Don't blame anyone else, you just fasten it on me. If there is an error in human judgement, I was the human," Mulholland announced. His career had begun in 1876 as a ditch-tender for the primitive water system of Los Angeles. A self-educated engineer, Mulholland earned a formidable reputation as "Father of the Los Angeles Municipal Water System" through hard work and diligent study during twenty-five years of service. Mulholland built nineteen dams for the city of Los Angeles before the St. Francis Dam disaster led to his retirement and ended his career.

Los Angeles, which has a semi-arid climate, must import water for use in the hot and dry summer months. Because of the essential nature of the St. Francis Dam and reservoir project to Los Angeles, Mulholland may have either overlooked or ignored the defective foundation conditions. An additional determination of the governor's commission, therefore, was that the failure of the dam indicated the desirability of having such projects constructed and

The only portion of the St. Francis Dam left standing after its failure in 1928.

maintained under the supervision and control of state authorities. This determination reflected the fact that while the city of Los Angeles was the primary beneficiary of the project, the disaster struck canyon residents who had no control over the design, construction, or maintenance of the St. Francis Dam.

ROBERT W. DAY AND DEBORAH A. DAY

Where to Learn More

"Causes Leading to the Failure of the St. Francis Dam." *Committee Report for the State.* California State Printing Office, 1928.

Failure of the St. Francis Dam. Association of Engineering Geologists, Southern California Section, 1978.

Geiger, Charles W. "Why the St. Francis Dam Collapsed." *Scientific American*, June 1928, pp. 520-22.

Griswold, Wesley S. "Day the Dam Burst: St. Francis Dam, Calif." *Popular Science*, March 1964, pp. 88-91+.

"Engineering Verdict on the St. Francis Dam Disaster." *Literary Digest*, April 21, 1928, p. 19.

Longwell, Chester R. "Lessons from the St. Francis Dam." *Science*, July 13, 1928, pp. 36-7.

"274 Perish, 700 Missing, in Torrent Loosed by Bursting California Dam; Flood Engulfs Victims as They Sleep." *New York Times*, March 14, 1928, p. 1.

"Charges Weakness in California Dam." *New York Times*, March 16, 1928, p. 24.

"Dam Collapse Laid to Base." *New York Times*, April 11, 1928, p. 31.

Outland, Charles F. *Man-Made Disaster: The Story of the St. Francis Dam.* Arthur H. Clark, 1963.

Malpasset Dam failure

Fréjus, France

(1959)

Failure of the rock foundation caused the collapse of a thin-walled concrete arch dam.

Background

The 197-foot-high Malpasset Dam near Fréjus, France, collapsed on December 2, 1959. A huge volume of water released when the dam failed, flowed down through the Reyan Valley and killed 421 people in the village below. Many of the victims were already in bed for the evening. The village—founded in 49 B.C. by Julius Caesar—contained numerous priceless monuments, many of which were destroyed.

The purpose of the dam was to store the waters of the Reyan river and its tributaries. The dam was to provide flood control and supply water for drinking, irrigating crops, and fighting forest fires in southeastern France. The site chosen for the dam was declared to be absolutely watertight and ideally suited for an arch dam. A geologist reported, "A brief visit to the site on the Reyan established that an arch dam could easily be constructed regardless of the elevation adopted. Furthermore, there is no foundation problem. The basin is entirely waterproof; there are no limits to the elevation of the reservoir as far as the natural conditions are concerned." A geotechnical engineer, rather than a geologist, normally inspects the site by drilling to extract rock samples (*borings*) which are examined to determine the overall geology of the area as well as the load-bearing capacity of the rock.

Details of the Failure

The dam design was developed in 1951 by the firm of Coyne & Bellier, led by the renowned and innovative designer, André Coyne. Construction began on April 1, 1952 and took twenty months to complete. From 1955 to 1959, the level of water in the reservoir behind the dam slowly rose. An exceptionally large amount of rainfall in the fall of 1959 caused a further rise of the water to near capacity; seepage developed about nine hundred feet downstream from the dam. This seepage was dismissed as due either to deep infiltration of water or heavy rainfall. At 9:10 P.M. on December 2, 1959, the dam burst with such force that 300-ton concrete blocks from the dam were found nearly a mile downstream. The right bank and central section remained relatively intact, but the left portion of the dam was nearly obliterated.

The dam burst with such force that 300-ton concrete blocks from the dam were found nearly a mile downstream. The right bank and central section remained relatively intact, but the left portion of the dam was nearly obliterated.

The advantage of the arch has been well known since the time of the Roman Empire, when arches were used to build aqueducts. The forces acting upon the arch, due either to building loads or water pressure, subject the member to compression. Compression strengthens concrete, which means that an arch dam can be built with less concrete than other types of dams; using less concrete lowers construction costs.

Arch dams are specifically designed to transfer water pressure through the relatively thin walls to the rock foundation and abutments. The foundation and abutments must be able to resist this force without excess deformation or shear displacement of the rock. If the rock deforms or shears, the thin-walled arch can crack, causing a catastrophic failure.

The Malpasset Dam was designed as a thin-walled concrete arch type structure. At its base, the dam was 22.6 feet (seven meters) thick, while the top of the dam was five feet thick (1.5 meters). The walls of the dam were relatively thin, compared with their height. The ratio of dam thickness to dam height—only 11 percent—was the lowest such ratio of any other dam then in existence. The dam's walls arched, or curved, from one abutment (where the dam meets the side wall of a canyon) to the other, and it was this arched design that made the thin walls possible. In addition to the vertical curvature, the dam differently curved on the upstream and downstream faces.

At its right bank, the arch of the Malpasset dam intersected a massive rocky peak. At its left bank, the arch was founded on a concrete abutment protected by a wing wall, a construction made necessary by the local topography.

It was the left bank abutment that suffered a near complete collapse. Most of this side of the dam was sucked clean away by the force of the onrushing water. Concrete blocks connected to the bedrock were found intact—the strength of the joints between the blocks and bedrock were tested as adequate. It was later determined, therefore, that a fault or weakness in the bedrock was the cause of collapse.

The French Ministry of Agriculture ordered a commission of engineers to investigate and determine the cause of the failure. The commission concluded in 1960 that "the most probable cause of the disaster should be attributed either to the presence of a slip plane, or to the higher upstream fault described in the report." Slip planes and faults are similar and can be described as planes of weakness in the rock. They can be formed by the shearing and ripping of the earth's crust during earthquakes. The slip plane or fault could have caused a local increase in deformability of the foundation or actual shearing of the rock beneath the dam.

Initially one engineer, Jacques Dargeou, was charged with involuntary homicide by negligence. Dargeou demanded a new investigation which

Although engineers declared the site of the Malpasset Dam to be perfectly suited to the project, a shift in the dam's foundation caused its arched walls to collapse.

found him negligent in not performing adequate exploration and investigation of the dam's foundation; the investigation also pointed to the designer, Coyne (who died six months after the collapse), as being negligent in collecting and evaluating data on the foundation. After many legal actions, Dargeou was acquitted.

But the legal problems did not end there. In 1964, in accordance with French law, additional charges against four other engineers on the project. The four were eventually acquitted, but during the trial two new facts about the collapse were unearthed.

First, photographs, taken a month before the tragedy showed twenty-eight changes in the dam's shape and foundation from its originally built dimensions. Second, it was revealed that demolition for a nearby highway—as close as 260 feet (78 meters)—may have weakened the dam foundation. Each blast at the demolition site used nearly 2.5 tons of dynamite, eight times above the "safe" limit.

Impact

Initial geologic investigation did not discover the slip plane or fault within the rock foundation. Prior to the construction of the dam, subsurface exploration was performed to investigate the character of the rock. In addition, during construction the rock foundation was inspected prior to placing concrete, but this also did not disclose any adverse geologic conditions. One lesson learned from this failure is the importance of thoroughly investigating the foundation conditions prior to dam construction. Because the arch dam concentrates the force of the water pressure on a relatively small area, any undetected defects in the rock, such as a slip plane or fault, can substantially weaken the foundation.

The failure of the Malpasset Dam did substantially alter the design practices in the United States. Much greater arch dams, some up to 600 feet (183 meters) in height, had already been constructed and were performing adequately at the time of the Malpasset Dam failure. The collapse did, however, confirm that a concrete arch dam is only as strong as the rock that supports it. The French investigating commission concluded: "As a result of these investigations, the Commission can affirm that the catastrophe of Malpasset should not diminish the confidence of engineers in dams of the arch type, the safety of which is ensured as long as the entire supporting structure is capable of permanently carrying the loads transmitted by such a structure."

ROBERT W. DAY AND DEBORAH A. DAY

Where to Learn More

"Final Report of Failure of Malpasset Dam." *Civil Engineering Magazine*, p. 101.

"The Malpasset Dam Design." *Engineering*. 1959, pp. 642-43.

"Malpasset Acquittal, Court Declares Engineers Not Guilty in Dam Collapse." *Engineering News-Record*. 1966, p. 19.

Levy, Matthys and Mario Salvadori. *Why Buildings Fall Down*. W. W. Norton & Co., 1992, pp. 166-72.

Ministry of Agriculture. "Final Report of the Investigating Committee of the Malpasset Dam." Paris, 1960. [Trans. D. Ben-Yakov].

Vaiont Dam landslide disaster

near Longarone, Italy

(1963)

A huge mass of rock and mud rushed into the reservoir behind the Vaiont Dam, forcing a wall of water over the dam and onto villages in the valley below and causing thousands of deaths.

Background

On the night of Wednesday, October 9, 1963, the four thousand people of Longarone and several other villages in the steeply-sided Piave River Valley were preparing for bed when they heard a thunderlike roar. Within fifteen minutes, an air blast followed by a three-hundred-foot-high torrent of water, rock, mud, and gathering debris swept down the valley, wiping out three-fourths of the population, erasing all traces of many Renaissance-age buildings, and leaving a wasteland of mud. The stunned world first thought that the Vaiont Dam, the highest (869 feet) thin-arch dam in the world, had failed. But daylight showed that the dam had miraculously sustained only minor damage, and the rush of water had been created by a huge mass of rock and soil that had slid down the side of neighboring Mount Toc, filling the reservoir (the lake behind the dam) and forcing its water over the dam, like a child jumping into a bathtub and sloshing water over the side.

In the northeastern corner of Italy, where the Alps of Italy, Switzerland, and Yugoslavia meet, the villages in the picturesque river valleys had no electricity until the early 1960s, when a series of water collection and power systems were designed and built. The Piave River and its tributary, the Vaiont, flow toward the Adriatic Sea, and under this scheme of the Società Adriatica di Eletricitta (SADE), their waters were to be tapped for electricity. The manager of SADE's hydroelectric department, Carlo Semenza, supervised the Vaiont project. The dam site was a narrow, steeply walled gorge. The dam itself was designed as a double-curvature arch (curving both from top to bottom and side to side), 625 feet long, 11 feet thick at the crest (top), and 73 feet thick at the base.

Extensive analyses were performed on the proposed dam. A scale model was built and analyzed at the Istituto Sperimentale Modelli e Strutture (ISMES), a structural laboratory famous among engineers and located in Bergamo, Italy. Construction began in 1956 and was completed in 1960. A second scale model was built after the foundation had been excavated so that support provided under the dam and up the abutments (sides where the dam and rock meet) by the limestone bedrock could be analyzed more exactly. Other facilities included an overflow spillway for emergencies, two outlet tunnels through the rock wall of the gorge (also for use in releasing water), power tunnels for hydroelectric power, the underground powerplant at the downstream base of the dam, and an office and hotel high above the dam on opposite rock walls for the workers.

When the foundation for the dam was excavated, the rock was found to be fractured, weak, and pitted with solution cavities (holes dissolved in the limestone by water). To strengthen the foundation and abutments, tunnels were drilled into the rock, and grout (cement) was injected through the tunnels to fill these gaps. This is often done to dam foundations to improve the natural conditions. An unusual addition to the site was the drilling of a tunnel under part of the reservoir from upstream of the dam to over a mile further upstream. The purpose of this bypass tunnel was to allow water to flow to the dam if landslides should ever block the lower (near dam) part of the reservoir. This "unique and remarkable" tunnel was added in 1961. The previous year, a landslide had flowed into part of the reservoir, and the designers wanted to preserve the capability of the dam to release water for generating power and also for emergencies.

The reservoir area was also studied during development of the project, and there was some concern regarding the stability of Mount Toc. Major landslides were known to have occurred in the Vaiont valley over geologic time. The geology was described as "adverse," meaning that the materials and the natural shape or geometry of the area were a bad combination. Rock in the area had formed from thick sediments, including limestone, claystone, and marl. These were low or weak in shear strength and were badly affected by ground water, which caused natural joints and bedding planes to enlarge and weaken, creating solution cavities.

During construction of the dam, landslides and cracks appeared on the mountain slopes. These were repaired and stabilized with concrete, but instruments installed in the mountainside showed stresses and strains that indicated potential for movement. Slow creep of soil and rock down the face was also observed. Small earth tremors, like local earthquakes, were recorded from 1960 through 1963 in association with the observed slope creep. This movement was measured to be as much as twelve inches per week between 1960 and 1961, but slope movements decreased after this period until the autumn of 1963.

On November 4, 1960, the landslide that prompted design of the bypass tunnel occurred. When the bypass tunnel was drilled in 1961, there were no indications of significant slide planes (failure surfaces which landslides move down) in either small-diameter exploration holes or in the bypass tunnel. Cracks continued to appear on the mountain's slopes and to grow above the 1960 slide until they formed an outline of the massive slide that would eventually move so disastrously. The reservoir was not filled for two years

The Vaiont Dam, before a huge mass of rock and mud from a landslide rushed into its reservoir, forcing a wall of water over the dam and onto villages in the valley below.

because of concern over these observed changes. From 1960 to April 1963, the lake level was also kept low as a precaution. Creation of the lake, however, added new and weakening water forces. Heavy rains in August and September 1963 caused water pressure and the weight of the rock to increase, while the strength of the rock decreased. The rate of slope creep began to increase again on September 18, 1963.

Details of the Disaster

Intense rainfall occurred again September 28 through October 9. On Mount Toc, grazing animals moved away from the area when they apparently sensed danger. The mayors of Erto and Casso, villages on the shore of the reservoir and in the landslide area, cautioned fishermen that dangerous waves might be caused by landslides; Erto was actually evacuated. The level of the reservoir was lowered by ten feet between September 28 and October 8. On October 8, technicians monitoring instruments realized that the huge rock mass was moving and that it was five times larger than previously assumed. The reservoir operators began to lower the lake level further, but the heavy influx of rainwater counteracted this.

At 10:39 P.M. on October 9, 1963, a rock mass estimated at over 300 million cubic yards sped 1.2 miles down the mountainside at 100 feet per second. In Vienna, Austria, and Brussels, Belgium, the movement of the landslide registered as a strong earth tremor. It pushed ahead of it a wall of air that blasted water and rocks almost 800 feet higher than the lake surface. As the mass entered the reservoir, the air wall was followed by a wall of water that reared up out of the reservoir as it was displaced by the rock avalanche. It washed through the village of Casso, high on the slope above the dam; the water wall was 330 feet higher than the top of the dam when it went over. It demolished the workers' hotel, then surged to the other abutment, where it swirled over the office for the dam's operations. None of the technicians who worked at the site survived to describe its destruction.

At 10:43 P.M., the water roared through the town of Longarone, where the Vaiont River enters the Piave River one mile downstream from the dam. The water was 230 feet high when it passed over Longarone. The villages of Pirago, Rivalta, Villanova, and Fae were also in the path of the deadly wave, as were smaller hamlets like Codissago and Castellavazzo. By 10:55 P.M., the wall of water had become only a flood downstream, and the intense noise was replaced with a deadly calm. In Pirago, only a church tower and some electric power poles remained standing. Railway steel was twisted into spirals, massive steel doors at the dam were ripped, all communications lines were destroyed, and access to the valley was blocked by mud, rubble, and dead cattle.

The scene in the eerie daylight was of a mud-drenched holocaust. The dam, now obscene in its beauty, had survived almost untouched, with only a small notch cut out of the crest. The force of water and rock thrown against it is estimated to have been about four million tons, which is many times greater than the dam's design pressure. The blast of air and water opened and destroyed all of the internal workings of the dam and underground power plant. Behind the dam, two-thirds of its reservoir had been replaced by a new mountain that towered 574 feet above the water level. And downstream from the dam, only an occasional building remained in a quagmire of mud.

The dam had survived almost untouched, with only a small notch cut out of the crest. The force of water and rock thrown against it is estimated to have been about four million tons—many times greater than the dam's design pressure.

Rescuers had trouble reaching the scene. The river was twice its normal size, and entire villages were buried in a sea of mud. Bodies had been tossed into the treetops. Whole families had perished, and so few local people survived that identification of the bodies was impossible. Even the final death toll is vague; it isn't known whether all the bodies were recovered from the mud, because records of the number of people in each hamlet were lost. In at least one miracle, two children were found alive in the cellar under their buried home on October 12, three days after the event.

Official inquiries into the disaster were highly charged with emotions and political pressures. Two types of inquiry were mandated by the Italian government. A court of inquiry was established quickly but also under pressure from the Communist party of the Italian Parliament. The engineers through the Superior Council of Italy's Ministry of Public Works also studied the disaster. Italy's nationalized electrical industry, including SADE, was found to be at fault for allowing the site to be used, and the project's builders and engineers were tried in the courts as responsible individuals. Inquiries determined that the suitability of the location of the dam and reservoir was still being researched at the time of the disaster. The problems were several and were mostly related to the reservoir and general site location, not the dam itself.

The hazards of landslides, creep, and tremors due to stresses in the mountainside were well known, but actions were taken neither to stop the project nor to protect the population. Exploratory holes had been drilled in the reservoir and landslide area, but none was deep enough to intercept the slide plane of the massive landslide. In fact, the monstrous size of the rock mass had never been approximated or even guessed, despite the pattern of cracks visible on the mountainside, the data from instruments, the local geology, and a number of other indicators. The lack of an early warning system to protect the local population was also a major contributor to the death toll.

Some consideration was given initially to clearing the reservoir, but the Superior Council concluded that the task was impossible because of the mass of earth to be removed. Distance to haul terrain, the enormous expense, and the question of where to place the huge volume of earth complicated the issue. Vaiont Dam was eliminated as a power source within the area's hydroelectric network. The villages downstream were rebuilt, and new factories were constructed on the valley floor.

Impact

The most horrific impact of the landslide at Vaiont Dam was the loss of three thousand lives. The disaster was not foreseen and may have been unforeseeable, because this set of compounding conditions had never occurred before. Other dam projects have been considered failures because reservoirs filled with silt or were unable to hold water. Vaiont was the first case in which a landslide filled a reservoir so completely, let alone with such devastating consequences. That the dam itself survived is a triumph of engineering, although that fact is dwarfed by the immensity of the disaster.

In fairness to the Vaiont engineers, no landslide had previously been known to move so quickly over such a long distance. Also, it may not have been possible to envision movement of a mass of that scale, even though indications seem clear in hindsight; this topic is still debated by engineers. Today, considerable study is given to the suitability of a site for both the dam and reservoir; much like a bathtub, all sides must be strong and stable if it is to hold water successfully. The Vaiont disaster showed that the geology of the area must be studied thoroughly and under two conditions: as it exists before the dam is constructed and as it will behave after the reservoir is influenced by the weight, volume, and movement of water.

The other lesson from the Vaiont disaster was the need for protecting the public. The mayors of two of the villages above the dam had the foresight to warn and evacuate their citizens, but no word of caution was given by any person, official, or agency to the towns downstream. Early warning systems that allow enough time for evacuation are now required practice. Such systems allow changing and compounding circumstances to be considered when warnings are issued (for example, the rainfall and slope movement data from the Vaiont site), and they require actions from a series of agencies auch as police and fire departments, so that if one element of the early warning system fails, others are available. The need to warn and protect the public seems obvious, but events at the Vaiont Dam showed otherwise. It may be some consolation that lives elsewhere may have been saved as a result of this legacy from the Vaiont Dam in Italy.

GILLIAN S. HOLMES

Where to Learn More

Canning, John, ed. *Great Disasters: Catastrophes of the Twentieth Century*. Longmeadow Press, 1976, pp. 76-77.

"Vaiont Slide—Seven Years Later." *Civil Engineering*, June 1970, p. 86.

"Alps Spawn Notable Dams." *Engineering News-Record*, November 2, 1961, pp. 32-40.

"Vaiont Dam Survives Immense Overtopping." *Engineering News-Record*, October 17, 1963, pp. 22-3.

"Lessons from Vaiont." *Engineering News-Record*, October 24, 1963, p. 84.

"Vaiont a Total Loss?" *Engineering News-Record*, October 24, 1963, p. 21.

"Vaiont Dropped as Power Source." *Engineering News-Record*, October 31, 1963, p. 18.

"Before and After at Vaiont." *Engineering News-Record*, December 5, 1963.

Frank, Beryl. *Great Disasters of the World*. Galahad Books, 1981, pp. 160-63.

Jansen, Robert B. *Dams and Public Safety: A Water Resources Technical Publication.* U.S. Dept. of the Interior, Water and Power Resources Service, 1980, pp. 214-20.

Kiersch, George A. "Vaiont Reservoir Disaster: Geologic Causes of Tremendous Landslide Accompanied by Destructive Flood Wave," *Civil Engineering,* March 1964, pp. 32-39.

Legget, Robert F., and Paul F. Karrow. *Handbook of Geology in Civil Engineering.* McGraw-Hill, 1983.

"Giant Dam Falls in Italy; Hundreds Reported Dead as Water Engulfs Town." *New York Times,* October 10, 1963, p. 1.

"Alpine Dam Resists Slide—U.S. Army Aiding Rescuers." *New York Times,* October 11, 1963, p. 1.

Baldwin Hills Dam failure

Los Angeles, California

(1963)

Subsidence of land beneath an earthfill dam caused the dam to leak and, within hours, to fail, releasing water into the suburbs downstream and causing five deaths and massive property damage.

Background

Saturday afternoon in the Los Angeles suburb of Baldwin Hills was shattered on December 14, 1963, by police sirens and broadcast warnings for residents to evacuate their homes. Within five hours after the first ominous signs, the Baldwin Hills Dam, a twelve-year-old structure surrounded by the suburb, leaked water, then mud, then its full contents in a floodlike fury. The site had not been ideal because soft rock made up the foundation and the local network of faults, but designers had used a number of techniques to compensate for the poor features of the site. Unfortunately, neighboring withdrawal of oil and gas from deep below ground had caused the soft material to subside or settle, which slowly caused one corner of the reservoir to sink and opened cracks in the protective system of liners and underdrains through which the water, under pressure from its own weight, could leak and erode to disastrous proportions.

In California, considerable attention is devoted to water resources because demands of the large population often exceed the water supply. Water for the city of Los Angeles is carried by aqueducts from the Owens River and the Colorado River. After the water reaches Los Angeles, it is stored in a system of reservoirs that are owned, designed, constructed, and maintained by the Los Angeles Department of Water and Power (DWP). In 1947, construction was started on the Baldwin Hills Dam and Reservoir, a hilltop site on the highest hills in southwestern Los Angeles where the reservoir would be ideally located to serve the south and southwest parts of the city. In 1947, the area around the site was not developed and there were no homes on the hillsides. The only notable feature near the site was the Inglewood oil field to the south and west.

Explorations for the dam's construction and excavation of its foundation revealed a zone of minor faults of late Pleistocene or recent geologic ages that showed fifty-nine feet of historical movement and were thought to be part of the neighboring Newport-Inglewood fault system, which is known to generate earthquakes. The Inglewood fault itself was about five hundred feet from the west rim of the reservoir. Although earthquakes might not occur along these faults, they might act as planes of weakness for other kinds of movement. The gate tower at the dam was relocated so it would not be on top of this zone.

The rock forming the foundation for the dam and reservoir was also not the best material. It consisted of soft and highly erodible siltstone and sandstone. The dam itself was 232 feet high and 650 feet long. It was constructed of 850,000 cubic yards of compacted soil, based on a design by Ralph R. Proctor, a pioneer in soil compaction. The reservoir was formed on three sides by dikes of compacted earth (the dam made up the fourth side), and it contained about 300 million gallons of water. The site covered an area of about nineteen acres. The Baldwin Hills Dam and Reservoir were dedicated on April 11, 1951.

Because of the soft rock under the dam and the fault system found during construction, the DWP took extreme care in designing and building the facility. The design included a system of liners and drains that would keep water from reaching the soil in the dam and dikes and the erodible rock under the reservoir. This lining was constructed of layers like a sandwich. The uppermost layer (under the water) was three inches of asphalt pavement. Under this was a layer of compacted, impermeable (water-blocking) soil; this layer was ten feet thick in the reservoir but decreased to five feet thick at the tops of the dam and dike slopes. Below the compacted soil was a two-inch-thick layer of sand mixed with gunite (a bonding cement) that would keep the fine soil particles from moving into the next layer which was composed of four inches of pea gravel that would collect any seepage water that managed to penetrate the upper layers and would carry this water to pipes. The pipes were 4 inches in diameter and carried the water to the drainage inspection gallery so it could be seen and monitored. Finally, below the pea-gravel drain, there was a membrane of asphalt directly on the foundation materials. Like the asphalt above and the compacted soil liner, this membrane was relatively impermeable. In March 1957, after 6 years of operation, the reservoir was drained completely and the floor was cleaned and inspected; there were no cracks or other disturbances in the asphalt at that time.

The DWP also required an unusually rigid program of regular surveillance after operations started. The drainage inspection gallery and drain networks were inspected daily. Movements of the reservoir and neighboring area were monitored monthly by surveying. Instruments that had been installed during construction were monitored regularly, and various technical specialists also visited the site once each month to observe any changes that might affect the safety of the dam and the public.

In the first year of operation (1951-52), a crack was seen in the drainage inspection gallery near the fault zones that had been pointed out during construction. Other, slight movements were seen during the twelve-year existence of the reservoir but were considered normal settlement. In April

When the Baldwin Hills reservoir broke, a flood of water, mud, and debris erupted from the downstream face of the dam, down a nine-hundred-foot-long ravine, and into the suburb below.

1963, seepage from the reservoir began increasing slightly, and water from some drains increased then decreased then increased in an unusual pattern of fluctuation.

Although movements in the dam seemed small and normal, there were other shifts in the Baldwin Hills that were more dramatic. Before the reservoir was built, survey stations in the hills began to subside (become lower in elevation). After 1957, cracks developed in the earth southwest of the reservoir as a bowl-shaped area developed from 1950 to 1963. This bowl included the southwest corner of the reservoir, and the survey data showed that a point on the ground about 1/2 mile west of the reservoir had subsided about 9.7 feet in forty years. Because of the shape of this bowl, the southwest corner of the reservoir subsided 3 feet more than the northeast corner. Several of the greatest of these elevation drops were at points where the most oil was being extracted from the Inglewood oil fields. In 1957, the oil companies began injecting saltwater to prevent the ground from sinking further, and in some areas, it actually rose up again somewhat. The damage to the reservoir had been done, however. The subsidence "stretched" the weak rock in the foundation and caused it to separate in blocks along the preexisting fault lines. The blocks were like steps on a staircase down into the subsidence bowl; the gaps between them widened, gradually destroyed the sandwich of linings and drains in the reservoir, and allowed water to leak into the openings in the foundation.

Details of the Disaster

At 9:30 A.M. on Saturday, December 14, 1963, a caretaker at the Baldwin Hills Dam inspected the site and saw nothing unusual. At 11:15 A.M., he performed his next routine inspection and heard water running through the drainage inspection gallery. The noise was unusually loud, so he also checked the drains that had been designed to carry water out from under the reservoir's asphalt lining and noticed that they were flowing freely. The caretaker summoned Mr. Richard E. Hemborg, the operating system engineer, and the two opened the discharge lines to the large storm drain under La Brea Avenue that would lower the water in the reservoir and that would eventually release it safely in about twenty-four hours. The engineer also alerted local authorities to the unusual conditions.

The time was now 12:20 P.M. Police on motorcycles and in patrol cars began evacuating the residents in the suburbs below the dam. They knocked on doors, started sirens, and closed streets to incoming traffic. Back at the dam, workmen faced increasingly hazardous conditions as the leakage got worse. By 1:00 P.M., water was emerging from the downstream end of the east abutment and the water was muddy, indicating that piping, or internal erosion of the dam's soil, was occurring. The workers entered the inspection chamber under the dam, worked in the streets immediately downstream to clean inlets to the storm drain system, and dangled on ropes down the upstream side of the dam to try to find the source of the outflow and stop it. By this time, all of these actions were heroic.

At 2:20 P.M., radio and television stations were told of the impending disaster and they began broadcasting alerts. Helicopters flew over the endangered area, and the police used loudspeakers to warn residents to flee. The water level had lowered enough by this time to reveal, on the upstream face of the dam, a three-foot-wide break in the lining of the reservoir that was directly in line with the point on the downstream face where the seepage was emerging. The workmen again ventured perilously close to this void to try to plug it with sandbags, but the bags simply vanished into the abyss. The men retreated to safety as the breach grew wider. By 3:20 P.M., about sixteen hundred people from neighborhoods downstream had been evacuated.

At 3:38 P.M., an explosion of water, mud, and debris including the reservoir's asphalt lining erupted from the downstream face of the dam, flooded down a nine-hundred-foot-long ravine, and emerged at the top of Cloverdale Avenue. An eyewitness said, "[The] sloping earth wall was leaking muddy water from a top-to-bottom crack. It gushed in a ten-foot-wide brown stream from a fissure at the bottom. Next, there was a roar like a great cannon, then a rumbling and shuddering of the ground as the face of the slope erupted. A mighty jet of water and mud and fragments of earth shot out. The explosion point was at the bottom, opening a huge hole into which the upper wall portions fell."

By 4:55 P.M., all the water had emptied from the reservoir. Downstream, despite the remarkable evacuation, the toll was still high. Five people, several trapped in their cars, had drowned. Of the houses in the suburb, 41 immediately below the dam were completely obliterated and 986 others, as well as 100 apartment buildings, were damaged. Cloverdale Avenue and other streets as well as their underlying utilities had been gouged out of the ground. Curbs and sidewalks were gone, and swimming pools had been destroyed or filled with muck. The flood waters had been eight feet high when they reached the lower, flatter part of Cloverdale Avenue. In the reservoir, a crack through the floor of the reservoir pointed like an ominous finger to the seventy-five-foot-wide, V-shaped notch that had been cut through the dam from the top to 90 feet down. The water drained into the sewer and storm drain systems and into the Balboa Creek flood control channel. It left several feet of mud, automobiles, and debris from the dam, wrecked homes, and eroded streets.

Impact

The immediate and tragic impacts of the disaster were the loss of five lives and property damage estimated at $50 million, not including the cost to repair the dam itself. President Lyndon B. Johnson quickly named the Baldwin Hills a disaster area, and the U.S. Army Corps of Engineers was ordered to help. Mayor Samuel W. Yorty of the city of Los Angeles established a board of inquiry composed of seven legal and technical experts, including professors from the California Institute of Technology and the universities of Southern California and California at Los Angeles. Robert R. Wilson, chief geologist of the DWP, described a program he had directed of drilling borings and excavating trenches through the floor of the failed

Ariel photograph showing the extent of the break in the Baldwin Hills reservoir dam.

reservoir; they showed no displacement (change in level) at the surface but, at ten feet below the floor, one side of the crack was eight inches lower than the other. There had been no earthquakes or other earth movements measured by instruments to cause this change.

The city of Los Angeles and the DWP sued several oil companies that extracted petroleum, natural gas, and water from the Inglewood oil field. The city contended that the extraction had caused the land to subside. The designers and builders of the dam were not held accountable because they could not have anticipated that subsidence would occur and that it would have such a disastrous effect over time. The lawsuit was for $25.8 million, but the City accepted an out-of-court settlement of $3.9 million.

There were several positive aspects to the failure. The alertness of the caretaker and the engineer gained time for the evacuation, the police and other agencies responded quickly in evacuating the area, the failure occurred during daylight, and the capacity of the reservoir was relatively small. The authorities were also determined after the failure to investigate all of the facts and make the public aware of their diligence and findings. This gave the public the sense that the truth would be told regardless of politics, and it allowed engineers access to information so that technical lessons could be applied at other sites.

The Baldwin Hills disaster opened a wider area of questions about the siting of dams in populated areas and the legal liability for dam failure. California State Assembly speaker Jesse Unruh appointed a committee of ten members of the state legislature to explore these questions. Los Angeles County supervisor Kenneth Hahn requested that the Legislature consider, among other legislation, laws regarding location, design, construction, operation, and maintenance of dams and reservoirs; installation of electronic early warning systems and implementation of evacuation plans; and the power for the State to veto dams proposed where active faults are located.

GILLIAN S. HOLMES

Where to Learn More

Casagrande, Arthur, Stanley D. Wilson, and E. Douglas Schwantes, Jr. "The Baldwin Hills Reservoir Failure in Retrospect." *Performance of Earth and Earth-Supported Structures, June 11-14, 1972,* vol. 1, pt. 1. American Society of Civil Engineers, 1972, pp. 551-88.

"Subsidence Blamed in Earth Dam Failure." *Engineering News Record,* December 19, 1963, p. 50.

"Dam Failure in Spotlight." *Engineering News-Record,* January 2, 1964, p. 60.

"Multiple Probes Start on Dam Failure." *Engineering News-Record,* January 2, 1964, p. 15.

"Earth Shift May Have Wrecked Los Angeles Dam." *Engineering News-Record,* January 9, 1964, p. 15.

"Earth Shifts Wrecked Reservoir." *Engineering News-Record,* February 27, 1964, pp. 22-23.

Hamilton, Douglas H., and Richard L. Meehan. "Ground Rupture in the Baldwin Hills." *Science,* April 23, 1971, pp. 333-44.

Jansen, Robert B. *Dams and Public Safety: A Water Resources Technical Publication.* U.S. Department of the Interior, Water and Power Resources Service, 1980, pp. 121-25.

Jessup, Walter E. "Baldwin Hills Dam Failure." *Civil Engineering,* February 1964, pp. 62-64.

Leadabrand, Russ. "Day the Dam Broke." *American Forests,* February 1964, pp. 30-33.

Legget, Robert F., and Paul F. Karrow. *Handbook of Geology in Civil Engineering,* McGraw-Hill, 1983, pp. 25-27.

Leps, Thomas M. "Analysis of Failure of Baldwin Hills Reservoir," *Performance of Earth and Earth-Supported Structures, June 11-14, 1972,* vol. 1, pt. 1. American Society of Civil Engineers, 1972, pp. 507-50.

"Why?" *Newsweek,* December 30, 1963, p. 17.

"Baldwin Hills Injections: Oil and Earth Movements." *Science News,* May 1, 1971, p. 300.

"Reservoir Failure under Study." *Water Works and Water Engineering,* January 1964, p. 82.

Teton Dam collapse

Upper Snake River Valley, Idaho

(1976)

The collapse of a dam built on fractured, water-permeable volcanic rock inundated an agricultural valley and killed eleven people.

Background

On June 5, 1976, only six months after construction was completed, the Teton Dam in southeastern Idaho collapsed. The federal Bureau of Reclamation had built the $86 million dam and reservoir project across a canyon consisting of highly fractured volcanic rock. When filling of the reservoir was nearly complete, water seepage undermined the dam and unleashed billions of gallons of water on the valley below. Fortunately, rivulets appearing downstream from the dam warned of the disaster in advance. Even so, the Teton Dam failure claimed eleven lives and destroyed the communities and farms of the predominantly Mormon population of the upper Snake River Valley.

Although the Teton Dam was constructed relatively quickly in the 1970s, it was the product of over seventy years of development. Three federal agencies—the Geological Survey, the Army Corps of Engineers, and the Bureau of Reclamation—participated in choosing the site for Teton Dam. Feasibility studies were conducted as early as 1904, and detailed site investigations began in 1932. All of these studies centered on the Upper Teton River, far upstream from where the Teton Dam was eventually raised. In 1956, field examinations were made at the mouth of the Teton Canyon, but this site was deemed unsuitable. In 1961, representatives of the Teton County Wheat Growers Association proposed to the Bureau of Reclamation that the dam be built near the mouth of the North Fork Teton River. The dam was primarily intended to provide agricultural irrigation (along with flood control, electrical power, and reservoir-based recreation), and the association held that this upriver location would service the largest area.

The Bureau of Reclamation, however, decided to place the dam as far downstream in the Teton River Canyon as possible. The Bureau based this decision on two reasons unrelated to geological suitability: increased reservoir capacity, as the water would fill down the canyon; and reduced irrigation costs, as the water would extend closer to the agricultural valley below. These priorities would prove fatal to the project, which was authorized by Congress in 1964.

In geological terms, the area chosen for the dam essentially consisted of two-million year old rhyolite, a rock based on volcanic ash. While the rhyolite itself was strong, the rock was highly fractured; one-fourth to two-inch cracks were readily visible across the canyon. These interconnecting cracks easily allowed groundwater to flow through, especially in the right abutment (the right-hand side) of the canyon. In addition, large rock cavities were uncovered during excavations for the dam. Such cavities, if not structurally dangerous, may also have served as significant water conduits.

In 1969, the Bureau of Reclamation conducted a grouting program on the left abutment of the canyon to test the feasibility of the site. This program involved boring 23 holes into the abutment and injecting a grout mixture to fill in the cracks in the rock. Several of these holes took in significant volumes of grout; the grout injection into just two of the exploratory holes took 15,720 sacks of cement and 17,787 cubic feet of sand, exceeding estimates of the grout required for the entire test. In 1970, ten more holes were drilled to check the effectiveness of this grouting. The results indicated that it would be more economical to excavate and refill a portion of the abutments than to grout the entire site.

The Bureau's preparation of the Teton Dam site was therefore a major engineering project. To remove loose soils, a foundation trench was dug along the river bed and into the abutments. To remove the most fractured strata of rock, an additional "keytrench" was dug seventy feet into the abutments. An extensive network of grout holes was then drilled, with some of the bores extending down three hundred feet. This network extended thirty-two hundred feet across the canyon and up to one thousand feet into the abutments, and was to take in almost 600,000 cubic feet of grout. The Bureau of Reclamation, given this unprecedented trench and grouting program, believed it had made the site impermeable to water.

The multilayered dam, built up of approximately ten million cubic yards of earth and rising over three hundred feet above the riverbed, remained to be constructed. The core of the dam consisted of a silt-based mixture. This core was wider than normal because silt was both plentiful at the site and more permeable than the clay typically used in dams. Soil filters of sand and gravel were placed on both the upstream and downstream slopes of the core to prevent erosion under the high water pressures that would be generated by the reservoir. However, the sand and gravel mixture used for the soil filters did not, in relation to the silt core of the dam, meet the Bureau of Reclamation's own criteria to prevent erosion. Moreover, no soil filters were set beneath the core of the dam, or on the downstream side of the keytrenches.

The final design of Teton Dam was completed in early 1971. On March 17, 1972, excavation started on the project. By November 26, 1975, the dam was essentially complete. In October 1975, the reservoir had begun filling. It filled slowly and steadily until April 1976, when increasing flows accelerated the

Water seepage into the highly permeable volcanic rock below caused the Teton Dam (above) to collapse in June 1976, only six months after being completed.

rate of filling. On April 13, the Bureau's decision to capture the unusually high spring runoff of water caused the rate of filling, a specified maximum two feet per day, to be exceeded. A maximum rise of 4.3 feet per day occurred on May 18. This rapid filling would tend to separate and crack the soil within the dam.

Details of the Collapse

On June 3, 1976, clear water springs appeared on the dry riverbed about thirteen hundred feet and fifteen hundred feet downstream from the Teton Dam. These springs were flowing through the weaker right abutment of the canyon. On June 5, at approximately 8:30 A.M., more water began to flow from the downstream right abutment, but the springs were now muddy and much closer to the back slope of the dam. This muddy water indicated that the silt from the dam and the trenches was beginning to wash away. About 10:00 A.M., a leak emerged from the back slope of the dam proper, close to the leaks on the right abutment.

A whirlpool had formed in the reservoir by 11:00 A.M., signalling that the core of the dam had been breached. Efforts to bulldoze rock and soil into the expanding gap on the back slope of the dam were futile. The gap was expanding at such a dangerous rate that the bulldozers had to be abandoned, with the operators fleeing on foot. The bulldozers fell into the expanding sinkhole and shot out from underground farther downstream.

The community in the valley was well organized and generally received timely warning of the collapse of the dam. Even so, eleven lives were lost, over a thousand people were injured, and 25,000 more people were left homeless.

Between 11:15 and 11:50 A.M., the back slope of the dam continued to erode. By 11:50, a tunnel large enough to endanger the crest of the dam had developed. Minutes before noon, the crest collapsed in a cloud of brown dust and water began surging through the breach. The breach was quickly enlarged by the torrent of water, and half of the dam ultimately washed away. The reservoir emptied completely, releasing eighty billion gallons of water into the river valley.

As the predominantly Mormon community in the valley was closeknit and well organized, the population generally received timely warning of the collapse of the dam. Even so, the failure of the dam destroyed the communities of Rexburg and Sugar City, claiming eleven lives, injuring over a thousand people, and leaving 25,000 more homeless. The flood washed over 100,000 agricultural acres and killed up to 20,000 head of cattle. The power plant attached to the project was also destroyed. The $1 billion cleanup effort took several years to complete, and the dam has not been rebuilt.

Impact

An independent panel to review the cause of Teton Dam failure was formed by the Department of Interior and the Governor of Idaho. One year after the collapse of Teton Dam, the panel released the following conclusions: (1) The dam failed by internal erosion, particularly in the right keytrench; (2) the grouting did not adequately seal the fractured volcanic rock; (3) once water began to flow through the dam, the breach progressed rapidly; (4) the design of the dam did not adequately take into account the foundation conditions or the characteristics of the soil used for fill; and (5) construction essentially conformed to design specifications.

A separate governmental Failure Review Group released its own Report of Findings on the collapse of the dam. This group made four major recommendations toward preventing similar failures of large dams: (1) independent review boards should assess the design; (2) design decisions should be documented formally; (3) design engineers should remain involved with the project through the construction phase; and (4) monitoring instruments should be in place from construction onward to ensure the safety of the project. Most observers concluded from the findings of the two panels that the Bureau of Reclamation design engineers should accept responsibility for the disaster.

The *New York Times* quoted Gilbert G. Stamm, then commissioner of the Bureau of Reclamation, as declaring that "theoretically, what happened could not happen." Such a claim points to the limitations of human and institutional expertise. Soon thereafter, in a tacit acknowledgment of the Teton Dam catastrophe, the Bureau of Reclamation changed its name to Bureau of Water and Power Resources.

David J. Elton

Where to Learn More

Boffey, Philip M. "Teton Dam Collapse: Was It a Predictable Disaster?" *Science*, July 2, 1976, pp. 30-32.

———. "Teton Dam Verdict: Foul-up by the Engineers." *Science*, January 21, 1977, pp. 270-72.

"Teton Dam Failure." *Civil Engineering*, August 1977, pp. 56-61.

Cowan, Edward. "Teton Dam Disaster Laid to Design." *New York Times*, January 7, 1977, p. 10.

"Teton Haunts BuRec." *Engineering News-Record*, June 16, 1977, p. 13.

"Economic Consideration May Have Helped Doom Teton Dam to Failure." *Engineering News-Record*, February 3, 1977, p. 12.

"Final Teton Report Faults Designer." *Engineering News-Record*, February 14, 1980, p. 31.

Independent Panel to Review Cause of Teton Dam Failure. *Report to U.S. Department of the Interior and to the State of Idaho on Failure of Teton Dam*. GPO, 1976.

Penman, A. D. M. "Failure of Teton Dam." *Ground Engineering*, September 1977, p. 18.

Peterson, Frank Ross. *The Teton Dam Disaster: Tragedy or Triumph?* Utah State University Press, 1982.

"Teton: Eyewitness to Disaster." *Time*, June 21, 1976, pp. 56-59.

Stava Dam failure

Stava, Italy

(1985)

Dam failure in Italy killed 269 people and drew attention to national policy issues.

Background

On July 19, 1985, near the resort community of Stava, Italy, a pair of mining dams failed. The flood wave that was unleashed washed away everything in its path as it traversed the Stava River valley, heavily impacting the villages of Stava and Tesoro. The failure occurred during the lunch hour, when tourists and visitors returned to Stava for their mid-day meal. The flood killed 269 people and buried hotels, businesses, homes and roads under tons of mud. Survivors estimated the flood wave at approximately 100 feet high and 150 feet wide.

Located 40 feet above and less than a mile upstream from the village of Stava, the two dams, one placed directly above the other, were part of a fluorspar mining operation. Also called fluorite, fluorspar is a translucent mineral used to manufacture opalescent glass. The dams were necessary to dispose of the wash water and mine tailings (residues) from the process whereby fluorspar was extracted from the mined material.

Mine tailings were separated into clays and sands. The heavier sandy material was evenly deposited on the upstream embankment faces while the finer clay material was deposited in the upper basin of water. These sediments settled out first in the upper embankment and then in the lower embankment as the wash water passed through. The filtered water was then discharged to an old river bed that flowed into the Rio Porcellini ("rio" is the Italian term for "river"). The embankments had grown over the years. At the time of failure, the basins measured 150 x 300 feet and rose to 165 feet from the toe of the lower to the crest of the upper embankment.

The dams were built in 1961. The mining complex was originally owned and operated by two Italian chemical companies, Edison and Montecatini, that later merged to form Montedison. The mine and tailings dam operation changed ownership several times. In 1979, the complex was sold to two units of the Italian state energy company, Enil. In 1981, the property was sold to the firm Prealpi Mineraria, the owner at the time of the failure.

The location for the dams was selected without sufficient consideration of the site's geological aspects. The site, named "Pozzoli," was at an elevation of 4,425 feet above sea level on the west side of the Stava River. The dams were built on top of glacial deposits. These sand, gravel, and cobble deposits, formed by the movement and meltwater of glaciers, are common to the region.

The damsite was near the Rio Porcellini, which exists as a surface stream uphill of the dams and resurfaces downstream from the dams as a spring. Investigations at the failure site revealed calcareous deposits directly beneath the dam site. Calcareous deposits are made up of calcium and/or magnesium carbonates (dolomites or limestones) that are easily eroded by water action. Thus, the site contained many subsurface caves, sinkholes, and springs. Some surface water courses and springs at the site were filled and partially covered. After the failure, several of these springs reappeared just as they existed before construction began in 1961.

In 1984, the year before the failure, the townspeople in Stava and Tesoro complained to the mine owners and local authorities that the Rio Stava water was increasingly dirty. This indicated that the silts in the wash water were not completely settling out in the dams before the water was discharged into the old river bed. The villagers protested on ecological grounds when mine owners applied for a permit to again enlarge the embankments.

Details of the Failure

At 12:23 P.M. on July 19, 1985, the residents and tourists in Stava felt a rumbling in the ground, which was followed by a white cloud that some mistook for smoke. A few believed they had experienced an earthquake (earthquakes are common in the area). Several minutes later, everything was engulfed by a surging mass of water, white-gray mud, and debris. Without warning, the Prealpi Mineraria's two dams had failed, leading to widespread destruction in Stava and, to a lesser degree, in the downstream village of Tesoro. The flood reached Tesoro, four kilometers downstream from the dam, in five to six minutes. Approximately 200,000 cubic meters of mud and debris were deposited in the Stava Valley.

Although the dam failure occurred in a matter of minutes, several events during the previous six-month period suggested that serious problems were occurring. In January 1985, an earth slide occurred in the lower portion of the upper dam, where decant pipes passed beneath the dam. The slide appeared to be the consequence of water freezing in and/or around the service pipe. Water continued seeping from this area until March, 1985.

In June 1985, the decant pipe beneath the lower dam broke, and the water and liquid mud in the lower basin was released through the broken pipe, creating a crater above the break point. The failed decant pipe was plugged and a

A huge chasm at Stava, Italy, cut by the rush of water following the collapse of two mining dams above the town.

bypass pipe installed. During this repair period, the water level in the upper basin was lowered as much as possible to excavate embankment material for the repairs.

On July 15, both ponds were filled and on a modified basis returned to service. At 11:30 A.M. and again at 12:15 P.M. on July 19, the day of the disaster, power failures occurred in an electrical line running across the Rio Porcellini depression, about 660 feet (200 meters) downhill from the lower dam. The failures were not understood at the time. At 12:23 P.M., the dams gave way.

Impact

The failure of the Stava tailings dams was the worst of its type in European history. Initial speculation about the cause of the failure focussed on damaging runoff from recent heavy rains, liquefaction of the embankment, and/or piping and subsurface erosion, all compounded by poor operation and maintenance and lack of regulatory oversight. Liquefaction occurs when water pore pressure increases to the point that the liquid acts a lubricant, and any additional stress causes movement in the earth material. Piping is a subsurface erosion that occurs when a very small leak in dam face slowly enlarges to a critical size, at which point it enlarges rapidly. This erosion can be accelerated by sudden increases in reservoir water depth and any actions causing the embankment to settle, compact, or shift.

The enormous loss of life and property involved in the accident led to a complete judicial investigation under the direction of the District Attorney. A group of experts was enlisted to establish whether or not the failure of the dams could have been foreseen and prevented. The lack of before-failure data about the dams made their task very difficult. A large part of the history of the dams had to be reconstructed on the basis of assumptions about embankment composition, material placement quantities and characteristics, and general mining operations.

The construction of the lower embankment was based on rough outlines drawn up by a Montecatini engineer who had studied at the Colorado School of Mines. The engineer subsequently wrote a paper that delineated specific steps that were necessary to improve the embankment design, but such modifications were never undertaken. Moreover, the upper basin was added and then expanded without adequate attention to engineering ramifications and construction norms. Thus, the original Montecatini engineer was cleared early in the court investigation of any responsibility for the failure.

The investigating team made several important determinations about the tailings dams operation:

1. The dams were constructed without any prior geologic, hydrogeologic, geotechnical or hydrologic investigations. This seemed to be common practice during the early 1960s.

2. Neither the formulation of a comprehensive plan nor a systematic stability analysis was ever undertaken. Construction and operation were makeshift.

3. The dams' reserve stability was always very low. While the factor of safety (FS) started out above 1.0, the FS was estimated to have fallen by the middle of the 1970s to between 0.75 and 1.0, well below acceptable minimum standards.

4. The dams were never instrumented to monitor stress, pore water pressure, or earth movement.

5. No tests were ever performed to ascertain the quality or characteristics of the sediments in the ponds, despite the use of the "upstream construction method" that hinges on those sediment characteristics.

6. Sufficient sediment-filtering could only be achieved by using both ponds in series and at very high water levels. This brought water into contact with the dam crests over large portions of their lengths, which is extremely hazardous to an upstream method dam.

7. No measures were taken to effectively divert surface runoff from the ponds or to drain springs.

Investigators determined the disaster could have been avoided, since the dam collapsed "because it was projected, built, and operated in such a way that it did not offer the kind of safety civilized society expects from works that put the lives of human beings at risk."

Overall, then, the tailings dams were haphazardly engineered, badly constructed and improperly maintained. The experts concluded that a high failure potential must have existed in these two dams as soon as they reached certain dimensions (which were not specified), and that the structures would be prone to failure given any trigger mechanism. The potential trigger mechanisms, perhaps acting in combination, included: movement in the embankment zones; excess pore water pressure from refilling ponds too quickly; water contact with the dam crests; the decant pipe failure; and the excavation of slope material. Some experts believed that excess pore pressure in vulnerable parts of the foundation was the most likely trigger mechanism, and could alone explain the explosive power and extent of the failure.

The investigators determined that the disaster could have been avoided, observing caustically that the dam collapsed "because it was projected, built, and operated in such a way that it did not offer the kind of safety civilized society expects from works that put the lives of human beings at risk." Italy's lax environmental and dam safety laws, which created conflicting inspection and enforcement responsibilities among various levels of government, contributed to the disaster. The tragedy at Stava thus revealed the need for national policy and regulatory reform.

The Italian court found all current and past owners of the mining operation, as well as local politicians and safety officials, negligent and guilty in the failure. Some were sent to prison, including ten of the mining operation's owners and employees. All are liable in the court-assigned damage payments.

NEAL J. GRUBER

Where to Learn More

"More Than Rain at Fault in Failure." *Engineering News-Record*, July 25, 1985, p. 10.

"Dam Collapse Analyzed." *Engineering News-Record*, July 24, 1986, p. 11.

"Basin Owners Convicted." *Engineering News-Record*, August 11, 1988, pp. 17-18.

Hallenstein, D. and D. MacKenzie. "Design Faults Revealed in Stava Dam." *New Scientist*, July 25, 1985, p. 18.

Manson, Tycho. "A Valley of Death." *Maclean's*, July 29, 1985, p. 27.

Palmer, Jay D. "The Mountainside Exploded." *Time*, July 29, 1985, p. 28.

"Actions and Omissions." *Time*, August 5, 1985, p. 41.

Van Zyl, D. J. A. and S. G. Vick. "Hydraulic Fill Structures." *Proceedings from a specialty conference at Colorado State University. Geotechnical Special Publication No. 21*, American Society of Civil Engineers, 1988.

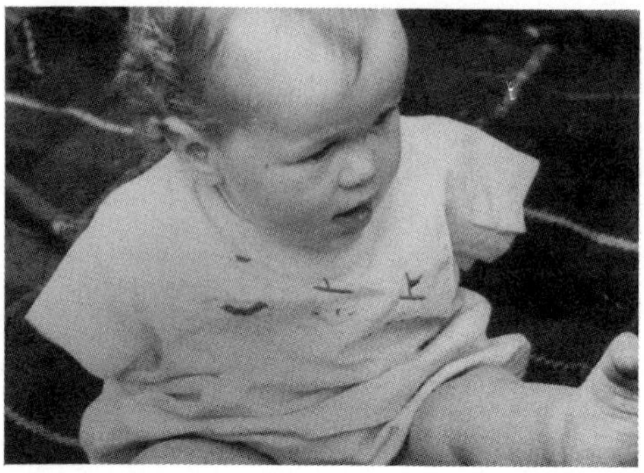

Medical Disasters

- DES (1940s-1970s)
- Thalidomide (1950s-1960s)
- Silicone-gel implants (1960s-1990s)
- Dalkon Shield (1970s-1980s)

Diethylstilbestrol (DES)

(1940s-1970s)

Diethylstilbestrol, a synthetic estrogen, can cause cancer and reproductive abnormalities in individuals whose mothers used the drug.

Background

In 1971, an article in the *New England Journal of Medicine* reported on seven cases of a rare form of vaginal cancer—adenocarcinoma—in girls and young women. There had been other reports of such cases, all of which were startling both because of the unexpectedly high incidence of cases and, even more surprising, because the cases reported were adolescent girls and women no older than their early twenties. Prior to this "epidemic," the few cases of adenocarcinoma reported in the medical literature had all occurred in postmenopausal women over the age of fifty. However, the *New England Journal* article also reported, for the first time, that a common link had been found in its seven cases: the mothers of these young women had all taken diethylstilbestrol (DES) during their pregnancies.

Estrogen was the name given to the biologically active substance first extracted from women's ovaries in 1912. By the late 1920s, Edward Doisy, an American, had prepared the first nearly pure estrogen. Doisy later won the Nobel Prize for medicine.

DES was the first of the synthetic estrogens. It was discovered in 1938 by Sir E. Charles Dodds, a British biochemist. The drug behaves physiologically like naturally occurring estrogens but is quite different in its chemical structure. In fact, it has a close chemical relationship to some of the cancer-causing substances found in coal tar. But DES was easy to make, inexpensive, and convenient—it could be administered orally, which natural estrogens could not.

The public and medical community's intense interest in estrogens was fueled because estrogen deficiencies were recognized as causing abnormal conditions. In 1934, George and Olive Smith, a husband-and-wife team of respected Harvard researchers, had reported that at least four weeks before the occurrence of late complications of pregnancy, such as fetal death or premature delivery, estrogen levels in the urine of pregnant women declined sharply.

Diethylstilbestrol was not officially approved by the U.S. Food and Drug Administration (FDA) until 1940, when it became one of the first major drugs to be approved under the revised Food, Drug, and Cosmetics Act of 1938. This version of the act (which itself was hastily passed in response to the elixir sulfanilamide disaster of 1937) required that drugs be proved safe before they could be marketed.

It was the claim that DES might prevent complications occurring late in pregnancy that launched the wholesale prescription of DES as a prophylactic measure, that is, as a drug that could prevent complications of pregnancy. In many cases, the drug was prescribed for women who had no symptoms or other indication that they required treatment.

However, long-term safety issues were not well appreciated at the time. There was, for example, no requirement for long-term toxicity studies in animals. Furthermore, there was no requirement at all, not even a weak one, that drugs be shown to be effective until the Kefauver-Harris Act was added to the act in 1962.

But the real surge in DES use did not begin until the late 1940s, when the Smiths published papers in the *American Journal of Obstetrics and Gynecology*. In particular, a landmark paper published in 1948 by Olive Smith claimed that DES could increase the chance for a successful pregnancy in women who had had previous miscarriages or who had high blood pressure or diabetes, and that it might also prevent complications that occur late in pregnancy. It was this last claim that launched the wholesale prescription of DES as a prophylactic measure, that is, as a drug that could prevent complications of pregnancy. In many cases, the drug was prescribed for women who had no symptoms or other indication that they required treatment.

At the same time that DES was becoming a popular drug among humans, animal breeders also discovered it. They realized that the estrogenic effects of DES made animals grow to marketable weight faster and on less feed. In 1947, the U.S. Department of Agriculture (USDA) approved implantable DES pellets, to be inserted under the neck skin of chickens.

Details of the Disaster

The DES disaster is unusual because of the long interval—more than a generation—between the cause and its consequences. It can reasonably be said that the tragedy occurred because no one imagined that a drug could have such long-delayed side effects. Perhaps not surprisingly, laws and regulations devised with more conventional hazards in mind failed to function effectively either to prevent or contain the disaster.

The DES mystique began to fade beginning in 1953, with the publication of two independent studies that had been conducted to see if DES was actually effective. Both these studies were controlled, double-blind trials in which neither patient nor physician knew if the prescribed pill was real or a placebo—a "dummy" pill. Both tests showed that DES did not reduce miscarriages, premature births, or other late-pregnancy complications. In fact, DES slightly increased the number of miscarriages. The larger of the two studies, begun in 1950 by William Dieckmann of the University of Chicago, involved some two thousand women and produced evidence that DES-treated women had smaller babies and twice as many miscarriages.

The Dieckmann study was a turning point, but by no means the death knell, for DES. Sales of the drug peaked in 1953 and started gradually to decline, primarily because of unfavorable reports that appeared regularly in medical journals. But the drug continued to be legally marketed and widely prescribed. At the time, there was no law against marketing ineffective drugs, nor were drug companies required to publicize, or even disclose, that a drug was ineffective. As far as its use in pregnancy was concerned, no further developments occurred until 1971.

Meanwhile, DES continued to be used in animals, in the forms of both feed additives and implantable pellets. In 1950, three years after the USDA approved the use of implantable DES pellets in chickens, the agency was sued by mink ranchers, whose minks had become sterile after eating DES residues left in the heads and necks of treated chickens. Nevertheless, in 1954, the USDA extended the use of DES to cattle, sheep, and hogs, with the proviso that treatment be stopped forty-eight hours prior to slaughter so that no drug residues would occur in the edible meat.

However, residues of DES continued to be found in edible animal tissues. Since DES was known to cause cancer in animals, increasing concerns about food safety spurred the passage of the Delaney amendment to the Food, Drug, and Cosmetic Act in 1958. This amendment flatly prohibited the use of any additives in human food that caused cancer in either animals or humans. The Delaney amendment applied to additives that were put directly in foods, as well as those that entered human food indirectly as a result of the diet of an edible animal. However, the FDA argued that the Delaney amendment could not be applied retroactively to products whose use had already been approved—so-called grandfathered products.

In 1959, using improved analytic techniques, dangerous levels of DES were detected in chickens and the use of DES in poultry was banned. However, use of the drug in cattle continued since residues were not detected in beef. The use of DES grew steadily. By 1970, seventy-five percent of the beef cattle in the U.S. were fed DES.

In November 1971, seven months after Arthur Herbst and his colleagues at Massachusetts General Hospital published their paper linking DES with vaginal cancer, the FDA issued a ruling requiring drug companies to warn against the use of DES during pregnancy. Not long after, it became public knowledge that DES was being dispensed by university health clinics as a morning-after contraceptive, which a woman who had unprotected sexual intercourse would take to induce an immediate miscarriage. DES had never been approved for this purpose.

Also in 1971, in the wake of the Herbst report, information was released indicating that residues of DES in beef had been found as far back as 1966. A new method of detection was also introduced that revealed the existence of

DES residues in meat that the established methods had declared safe. Since no amount of a cancer-causing agent, however small, is considered permissible, the inescapable conclusion was that there was no way to avoid DES residues in meat short of banning DES entirely. However, the controversy and the battle continued through 1979, when DES use in food-producing animals was finally banned.

Today DES is banned worldwide as a growth hormone in cattle and poultry, but other naturally occurring hormones are used. By some estimates, U.S. farmers save between $2 billion and $4 billion every year by using hormones. Although there is still some controversy surrounding the use of any hormones for these purposes, prevailing expert opinion is that the use of such products does not produce hormone levels in edible tissues that are distinguishable from normal levels in the animals. In Europe, however, cattle hormones of any kind are illegal.

In humans, DES has been studied extensively since 1971. The results of these studies are pessimistic. In the late 1970s, it became clear that in addition to their higher incidence of vaginal cancer, DES daughters were subject to a host of anatomical and functional irregularities of the reproductive tract: vaginal adenosis, T-shaped uteri, cockscomb cervixes, hooded cervixes, cervical ridges, and vaginal or uterine cysts. When these young women reached childbearing age, they encountered high rates of infertility, ectopic (tubal) pregnancies, and premature births.

Damage has not been limited to DES daughters from the original study. Sons of DES mothers have a higher-than-average rate of infertility, testicular cancer, and other noncancerous anatomical abnormalities. The DES mothers themselves, decades after taking the pills, may be especially prone to vaginal adenocarcinoma, breast cancer, and cancer of the uterus.

Impact

Today, the federally approved labeling for DES carries a boxed warning stating that use of estrogens increases the risk of cancer of the endometrium (the lining of the uterus), and that estrogens should not be used during pregnancy. The label also carries an unboxed warning that DES should not be used as a morning-after contraceptive. Despite such cautions and the consequent risk of birth defects if the method fails, DES is widely used for that purpose. The principal approved uses of DES in women are for relief of symptoms of menopause, and for treatment of a few specific conditions.

Diethylstilbestrol was first marketed in 1941 by Eli Lilly and Company, a pharmaceutical firm that manufactured and sold about seventy-five percent of the DES consumed in the United States. The remaining twenty-five percent of the market went to some twenty other drug companies, including American Pharmaceutical Company, Rexall Drug & Chemical Company, E. R. Squibb & Sons, and Upjohn Company. Between 1947 and 1971, an estimated six million pregnant women took DES.

Numerous lawsuits have been brought against the drug companies. As of 1992, more than one thousand DES cases had been filed nationwide, with Lilly named in six hundred of them. About 10 percent of the cases involve so-called third-generation claims—that is, claims on behalf of the grandchildren of women who took DES. Among other impacts, the DES experience has changed the practice of American law, allowing plaintiffs to sue all manufacturers of DES when they are unable to discover which brand their mothers took.

In October 1991, a New York state jury handed down a $12.2 million verdict against Eli Lilly in a case involving a woman whose cancer and infertility were linked to her mother's DES use. It was considered the largest award in a DES case, and also marked the first time a DES spouse was awarded damages. The plaintiff's husband won $550,000 in addition to the settlement awarded to his wife. The jury was considering whether the company should also pay punitive damages when the two sides settled the case out of court; terms of the final settlement were kept confidential. The case was expected to trigger drug companies to settle hundreds of other suits over the antimiscarriage drug.

From forty to sixty percent of DES daughters have some form of benign abnormality in their reproductive organs, according to the National Cancer Institute. This puts them at higher risk for premature or protracted labor, which can result in problems at birth. There is an "indirect" link between the drug and third-generation babies being born with serious handicaps caused by uterine and cervical abnormalities in the DES daughters. While researchers say there is no evidence that DES-induced defects are passed down genetically to the third generation, a growing number of DES daughters, advocates, and medical experts are calling for more research into possible direct and indirect third-generation effects.

ELLEN WILLIAMS

Where to Learn More

Apfel, R. J., and S. M. Fisher. *To Do No Harm: DES and the Dilemmas of Modern Medicine.* Yale University Press, 1986.

Dieckmann, W. J., M. E. Davis, L. M. Rynkiewicz, and R. E. Pottinger. "Does the Administration of Diethylstilbestrol during Pregnancy have Therapeutic Value?" *American Journal of Obstetrics and Gynecology*, November 1953, p. 1062.

Epstein, S. S. *The Politics of Cancer.* Anchor Press, 1979.

Fenichell, S,, and L. Charfoos. *Daughters at Risk: A Personal DES History.* Doubleday, 1981.

Herbst, A. L., H. Ulfelder, and D. C. Poskanzer. "Adenocarcinoma of the Vagina: Association of Maternal Stilbestrol Therapy with Tumor Appearance in Young Women." *New England Journal of Medicine*, April 22, 1971, pp. 878-881.

McGrory, M. "The Lingering Pain of DES," *Washington Post*, April 23, 1992.

Meyers, R. *D.E.S., The Bitter Pill.* Seaview/Putnam, 1983.

Orenberg, C. L. *DES: The Complete Story*. St. Martin's Press, 1981.

Schell, O. *Modern Meat: Antibiotics, Hormones, and the Pharmaceutical Farm*. Random House, 1984.

Seaman, B., and G. Seaman. *Women and the Crisis in Sex Hormones*. Rawson Associates Publishers, 1977.

Smith, O. W. "Diethylstilbestrol in the Prevention and Treatment of Complications of Pregnancy." *American Journal of Obstetrics and Gynecology*, November 1948, p. 821.

Thalidomide

(1950s-1960s)

Widespread use of an improperly tested drug led to the birth of thousands of malformed babies, half of whom died within weeks of birth.

Background

In the early 1960s, the over-the-counter (nonprescription) pharmaceutical Thalidomide, marketed as a sedative and antinausea drug in more than forty countries, was found to have caused ghastly birth defects in thousands of babies whose mothers had taken the drug to prevent morning sickness. Some of the babies born to mothers who had taken Thalidomide were born without arms or without legs. Others were born blind and deaf or with heart defects or intestinal abnormalities. Some of the babies were nothing more than a trunk with an eyeless, earless head mounted on it. Some were mentally retarded; most, perhaps more cruelly, were of normal intelligence. It is thought that all together more than ten thousand babies around the world were affected by Thalidomide. This horrible tragedy led directly to the strengthening of the laws regulating the development and sales of prescription and nonprescription drugs on four continents and to vastly improved testing of new drugs in the United States.

Historically, new pharmaceuticals have been developed by trial and error, the so-called suck-it-and-see method. Organic chemists synthesize new chemical compounds, which are then evaluated by pharmacologists (scientists who study the effects of drugs) for beneficial effects on lab animals. If a compound is found to have a desirable effect in the laboratory animals, defined as a beneficial change in the animal's biochemistry, the study is then expanded to human volunteers and finally to the general populace. Even when performed correctly by dedicated scientists under favorable conditions, this method does not always result in the development of safe and effective drugs. When the process is compromised, either by incompetent scientists or by unscrupulous executives intent on rushing a new drug to market, disaster can result.

Thalidomide was originally developed in 1954 by Chemie Grünenthal, a West German pharmaceuticals company. At the time, Grünenthal derived the majority of its revenues by marketing, under license, pharmaceuticals developed in other countries. Because the profit margins on marketing other companies' pharmaceuticals were low, Grünenthal's scientists (many of whom had no formal training in pharmacology) were under intense pressure to develop new drugs, which could then be marketed in West Germany without having to pay royalties and which could be licensed to pharmaceutical companies in other nations. Hence, it is not surprising that after submitting their new drug Thalidomide to limited animal studies (none of which later proved to be reproducible by other laboratories) and cursory clinical trials on human beings (none of which were done using standard double-blind testing in which neither the doctor nor the patient knows if the patient is receiving the drug or a harmless placebo), Grünenthal rushed the drug to the market.

By May 1961 Thalidomide was the most successful product in the history of the Grünenthal corporation. It accounted for more than half of Grünenthal's gross revenues and was by far the most popular over-the-counter sedative on the market in West Germany. Grünenthal marketed Thalidomide as a cure for everything from nervous exhaustion to the flu to, tragically, morning sickness. Their greatest selling point was Thalidomide's safety. It was, they claimed, completely nontoxic and without side effects.

The success of Thalidomide in West Germany did not escape the notice of pharmaceutical companies throughout the rest of the world. Licenses to manufacture Thalidomide were obtained by pharmaceutical companies in Britain, Japan, France, Italy, Australia, and in the world's largest market for over-the-counter remedies, the United States of America.

Details of the Disaster

Thalidomide had been on the market for more than six years and had caused nearly four thousand babies, mostly in West Germany, to be born with ghastly birth defects before anyone began to suspect that Thalidomide might be the cause of the sudden surge in birth defects that had occurred in very nation in the world where Thalidomide was being marketed. Ironically, Thalidomide was first identified as the cause of birth defects by Dr. William McBride, an Australian obstetrician. Dr. McBride had the misfortune to deliver three babies in the first six months of 1961 with nearly identical birth defects: the three babies were born with no radius bone (the shorter of the two bones in the forearm) in either arm and with a bowel atresia (their lower intestine had no opening to the anus). All three babies died shortly after birth.

Dr. McBride researched the medical annals for information on bowel atresias and on birth defects in general. What he read led him to believe that the abnormalities he had seen in his practice were caused by an adverse drug reaction. Upon examining the hospital records of the three women who had given birth to the malformed babies, he found that all three women had been given Thalidomide during their pregnancies and that it was the only drug any of them had taken. This seemed to him to be compelling evidence that Thalidomide was the cause of the birth defects and that it should be withdrawn from use until further studies could be conducted. Dr. McBride immediately notified the local distributor of Thalidomide that the drug was the cause of birth defects and submitted a paper to *The Lancet*, one of the world's most prestigious medical journals, detailing his theory about Thalidomide. He was ignored by the drug companies. *The Lancet* refused to publish his paper. It would be another six months before the medical establishment would admit that Dr. McBride had been right. Thousands more babies would die and thousands more babies would be born without arms and legs before Thalidomide would be withdrawn from the market.

Except for the courage of a single doctor working for the United States Food and Drug Administration (FDA), the Thalidomide disaster could have been a hundred times worse than it actually was. Shortly before Dr. McBride delivered his first Thalidomide baby, the American licensee of Thalidomide, Richardson-Merrel, makers of Vicks VapoRub, submitted their application to the FDA to market Thalidomide over the counter in the United States. Richardson-Merrel had big plans for Thalidomide and, in fact, had already distributed more than two million Thalidomide pills to more than twelve hundred doctors who had subsequently administered the drug to approximately twenty thousand patients—all before Richardson-Merrel submitted its application for FDA approval of Thalidomide for use in human beings.

Throughout 1961 Richardson-Merrel would have its application to market Thalidomide in the United States rejected on six different occasions by Dr. Frances Oldham Kelsey of the FDA. Dr. Kelsey almost single-handedly faced down a multi-billion-dollar company that had exerted strong pressure on her and her superiors to approve Thalidomide for use in the United States, at one point threatening Dr. Kelsey with a libel suit. Richardson-Merrel withdrew its application for FDA approval of Thalidomide only after it had been shown to cause birth defects in West Germany and had been withdrawn from the market there. For her courage, President John F. Kennedy awarded Dr. Kelsey the Distinguished Federal Civilian Service Medal, the highest award given to civilians in government service.

Biological Effects of Thalidomide

The first question to be answered in any study seeking to evaluate the effect of a drug on fetuses is whether the drug can cross the placental barrier. The placenta is the blood-rich organ through which fetuses draw nutrients and oxygen from their mothers. The placenta protects the fetus from substances present in the mother's blood by limiting contact—there is no actual blood-to-blood contact between a mother and her fetus. Many harmful substances cannot cross the placenta from mother to fetus. It would seem an elementary precaution that any drug that was to be marketed as a cure for pregnancy-induced morning sickness would be evaluated for its ability to cross the placental barrier. Yet neither Chemie Grünenthal, Distillers Company (Biochemicals)—the British distributor of Thalidomide—nor Richardson-Merrel had ever produced such a study. It was not until after Thalidomide was withdrawn from the market as a suspected teratogenic (a substance that causes fetal malformations) that Grünenthal performed this vital experiment.

Some of the babies born to mothers who had taken Thalidomide were born without arms or without legs; others with heart or intestinal defects. It is estimated that more than 10,000 babies worldwide were affected by the drug.

Using a radioactive-tagged sample of Thalidomide, Grünenthal scientists tested the drug in pregnant mice and found that it did cross the placental barrier.

To this day it is not certain how Thalidomide causes damage to developing fetuses once it has crossed the placental barrier. One theory is that the drug

Defects to the eyes and ears occur if the mother takes Thalidomide between the twentieth and twenty-fifth day after conception, arms are affected between the twenty-sixth and thirtieth day of gestation, and legs between the thirty-first and thirty-fifth day.

prevents the fetal blood vessels from forming properly. Blood clots form rather than blood vessels, resulting in tissue death in the areas affected. Depending on when during the fetal development cycle the drug is ingested, the effects can be wildly different. Defects to the eyes and ears occur if the mother takes Thalidomide between the twentieth and twenty-fifth day after conception, arms are affected between the twenty-sixth and thirtieth day of gestation, and legs between the thirty-first and thirty-fifth day. It is unfortunate that morning sickness, for which Thalidomide was often prescribed, is most common during the first months of pregnancy. Thalidomide apparently had no effect on fetuses if the mother took the drug after the third month of pregnancy.

Another side effect of Thalidomide, its inhibition of tumor necrosis factor (TNF), may prove useful in the treatment of AIDS. TNF, a natural substance that normally defends the body against cancer and infections, promotes the production of HIV viruses. Some AIDS patients who have taken Thalidomide have gained weight and no longer experience fevers. Although researches do not consider Thalidomide a potential cure for AIDS, they believe the drug may prove effective in slowing the disease's progress by suppressing the production of HIV viruses in infected cells.

Impact

One reason the United States escaped the Thalidomide disaster so lightly, with fewer than a dozen known cases of Thalidomide-deformed births, was that it already had an agency, the FDA, to regulate new pharmaceuticals. Unfortunately, the reason the United States had an FDA was that it had already experienced a major over-the-counter drug disaster. The FDA was founded in 1927 after an untested sore-throat medication produced by the S. E. Massengill Company of Tennessee killed more than one hundred people, including many children. After the Thalidomide disaster, the Food, Drug and Cosmetics Act of 1938, which the FDA is charged with enforcing, was modified by the passage of the Kefauver-Harris Act of 1962.

The Kefauver-Harris Act greatly strengthened the FDA. The act had two main components: it required that pharmaceutical companies prove that drugs were both safe and effective (prior to Thalidomide, pharmaceutical companies only had to show a drug was safe), and it widely expanded the FDA's powers to regulate the development and testing of new drugs.

Prior to Thalidomide, many pharmaceutical companies used their clinical trials more as a marketing scheme than as a genuine scientific study of a drug's safety and effectiveness. In the case of Thalidomide, Richardson-Merrel had given away, apparently to any doctor who would take them, more than two million free samples of Thalidomide prior to ever submitting the drug to the FDA for approval. No records were kept of the patients who took the drug, any side effects the patients might have experienced, or even the numbers and types of patients who were taking the drug. There was no requirement that the doctors taking part in the study inform their patients that they were prescribing an experimental drug. The pharmaceutical companies were not even required to report all of their data to the FDA and had been known to hold back damaging information—in 1963 three executives of Richardson-Merrel were fined $80,000 and placed on probation for falsifying and withholding test data on an anticholesterol drug (which was also withdrawn from the market). After Thalidomide, the pharmaceutical companies were required to show that the doctors testing their new drugs were qualified and that the tests being conducted were safe, controlled, and reproducible, and that no person was given a drug without being informed that the drug was experimental. The pharmaceutical companies were required to report all of the results of all of their tests, not just the ones that showed their new drugs in a favorable light.

After Thalidomide, all drug trials in the United States were required to study the effects of the drug on fetuses. It was found in the case of Thalidomide that the drug did not lead to deformities in rat fetuses, but it did produce nearly identical deformities to those found in human beings in rabbit fetuses. For this reason, every new drug tested in the United States is now evaluated for its ability to cross the placental barrier and is then tested in multiple animal species prior to its study in human beings.

Conflict of interest rules were passed forbidding FDA employees from accepting money from pharmaceutical companies—in the early 1960s, the head of the FDA's antibiotics division had accepted more than a quarter of a million dollars from pharmaceutical companies for authoring promotional articles.

But perhaps the most far-reaching result of the Thalidomide disaster was the example set by Dr. Frances Kelsey. Her moral courage under extreme pressure has been an example to which other FDA doctors and scientists have been able to turn ever since. It is one of the reasons why the United States has the safest pharmaceutical industry in the world.

JEFFERY R. RAINES

Where to Learn More

"Drug Scare Spurs Review of Safeguards." *Business Week*, August 4, 1962, pp. 70-71.

"Tough Drug Law Coming." *Business Week*, August 11, 1962, pp. 37-38.

Ferrara, J. L. M., and H. J. Deeg. "Graft-versus-Host Disease." *New England Journal of Medicine*, April 1991, pp. 667-74.

Gorman, C. "An Old New Drug For AIDS." *Time*, July 12, 1993, p. 49.

Knightley, P., H. Evans, E. Potter, and M. Wallace. *Suffer the Children: The Story of Thalidomide*. Viking Press, New York, 1979.

Lambert, E. C. *Modern Medical Mistakes*. Indiana University Press, Bloomington, 1978.

Lear, J. "The Unfinished Story of Thalidomide." *Saturday Review*, September 1, 1962, pp. 35-43.

Melville, A., and C. Johnson. *Cured to Death: The Effects of Prescription Drugs*. Stein and Day, New York, 1982.

"Abortion & the Law." *Time*, August 3, 1962, p. 30.

"The Case for Thalidomide." *Time*, April 27, 1992, p. 23.

Will, J. "The Feminine Conscience of the FDA: Dr. Frances Oldham Kelsey." *Saturday Review*, September 1, 1962, pp. 41-43.

Silicone-gel implants

(1960s-1990s)

Silicone-gel breast implants leak or rupture, causing inflammation, fever, arthritis, scleroderma (a hardening of the skin), and other health problems.

Background

For over thirty years, silicone-gel implants have been used for cosmetic reasons to restore the size of a woman's breasts after a mastectomy (breast removal), to augment the size of a woman's breasts in order to enhance her self-esteem, and to correct problems resulting from birth defects or medical traumas. Over one million American women have received silicone-gel implants, and up until January of 1992, from 120,000 to 150,000 women a year underwent silicone-gel implant surgery. Occasional case studies of health problems associated with the implants have been published in medical journals over a period of many years. These studies have never associated any deaths with the use of silicone-gel implants, but they have connected the implants with a variety of health problems.

Back in the early 1960s, when silicone-gel implants were first sold, they were not regulated by the federal government. But in 1976, the U.S. Food and Drug Administration (FDA) was made responsible for the safety of implants. Because of new regulations developed in 1988 by the FDA, manufacturers of the implants were required to supply more safety information about their products beginning in July 1991. As a result of the new requirements, concerns about the effectiveness of silicone-gel implants gained more publicity. The implants were found to have several problems: (1) they deteriorate in time and must be replaced; (2) the flexible sacs containing the gel very often leak at least tiny amounts of the gel; and (3) the sacs rupture, though sources vary on the rates of rupture. Conservative sources say the rupture rate is less than one percent, but government sources say the rate may be four to six percent.

Silicone's scientific name is polydimethylsiloxane, and depending on the length of the molecules in the polydimethylsiloxane compound, silicone can appear in liquid, gel, or solid states. Silicone has been used since World War II (1940-1945) in a wide variety of medical products, including needles, syringes, joint prostheses (e.g., an artificial hip joint), heart bypass technology (e.g., synthetic tubes used in the heart), the treatment of diabetes, hemodialysis (i.e., blood purification treatments), and test tube linings.

Silicone-gel implants are made in a variety of sizes, and have two main components: an outer sac, which is often made up of silastic, an elastic silicone rubber; and the silicone gel itself, which may vary in viscosity according to how firm the patient wants the implants to be. The silastic envelope is assumed to be biologically inert; that is, it will not react with the human immune system and create inflammation, pain, or other adverse reactions. Silicone gel, however, will react with the body's chemical defense system and cause adverse reactions.

Silicone-gel implants have been manufactured by the Bristol-Myers Squibb Company, Dow-Corning, Minnesota Mining and Manufacturing Company, Mentor Corporation, and other firms. Dow scientists discovered in 1974 that the silicone-gel implants created a reaction by the body's autoimmune system. But Dow did not make their results public for at least fifteen years. In 1992, an investigation by the United States attorney general discovered that prior to 1987, Dow-Corning employees had falsified data about the oven temperatures used in the process of manufacturing the implants. Until March of 1992, Dow made the large majority of silicone-gel implants used in the United States.

Details of the Health Problems

Many health problems have been associated with silicone-gel implants, including arthritis; inflammations; swollen joints; rashes; autoimmune system problems, which arise when the body tries to reject foreign objects; and scleroderma, a disease that causes the skin to become tight and leathery. The two cases that follow illustrate some of the reactions to silicone-gel implants.

In the first case, a thirty-two-year-old woman augmented the size of her breasts with silicone-gel implants in 1977. The day after the uneventful operation, she came down with a fever of 104 degrees Fahrenheit (40 degrees Centigrade) and drenching sweats. Four days after her operation, in the absence of any stimulus, she began to experience pain in her fingers and toes. Eight days after her operation, the joints in her hands, knees, and ankles began to swell. Ten days after her operation, she began to feel a generalized abdominal pain, and her doctors discovered that both of her kidneys were enlarged. During this time, the incisions on her breasts were healing more or less normally, and thorough physical examinations as well as tests of her blood, urine, and other fluids showed no other cause for the pain or swelling. Nonetheless, it became harder and harder for her to breathe, her kidneys began to deteriorate, and she became delirious.

Since her doctors could find no other explanation for her problems, they decided to remove her breast implants on the eleventh day after her operation. Her condition improved sharply within twenty-four hours. The silicone-gel implants that were removed looked normal. At the same time

Available in a variety of sizes, implants comprise an outer sac of silastic filled with silicone gel. While the outer sac is typically harmless, the inner gel works against the body's chemical defense system causing multiple adverse reactions.

that the doctors removed her implants, they took urine, blood, and sputum samples, and collected fluid from both breast pockets to try to find the cause of the woman's symptoms. All of these samples showed the presence of silicone compounds. Six days after the removal of her implants, most of the woman's symptoms subsided and she was released from the hospital. Three weeks after her discharge, further samples of blood and breast fluid showed no silicone present. At the end of three months, she had healed completely. After considerable testing, the woman's doctors could find no explanation for her rapid deterioration in health. Since her symptoms began shortly after the surgery for her silicone-gel implants, and since the symptoms subsided sharply after the implants were removed, her doctors could only conclude that the implants caused the incident.

The second case involved a forty-six-year-old woman who had silicone-gel breast implants in 1976 for cosmetic reasons, and had them replaced twice. Her second operation occurred in September 1987, when her doctors discovered that one implant had either ruptured or leaked silicone gel. For a while after her operation, she showed no ill effects. However, in November 1987 her face, hands, legs, and feet began to swell intermittently. She went to her doctor several times for treatment, which temporarily improved her symptoms. She continued to experience tightness in her skin in these areas and began to develop firm, discolored patches of skin on the front part of her lower legs. In July 1988, a skin biopsy showed that she had scleroderma, which, if left untreated, could begin to affect her internal organs and eventually lead to death. Another examination in August 1988 showed that, aside from the swelling and discolored patches of skin, her health was normal.

The woman's scleroderma symptoms, however, worsened, and the discolored patches of skin spread up her legs to her thighs. Her legs became so stiff that she could not continue with her running exercises or aerobics classes. Because of negative publicity in the mass media at the time about silicone-gel implants, the woman decided to have her silicone implants removed and replaced with saline (i.e., salt-water) implants. Two weeks after removing the silicone implants, she felt much better and her symptoms had improved significantly. One year later, she was completely recovered and she returned to her running and aerobics classes.

These are only two of a number of case studies that associate silicone-gel implants with health problems. Yet, despite these and other case studies, there are far more examples of silicone-gel implants that have been used without any complications. In fact, over 90 percent of women who have had silicone-gel implants are satisfied with them. A number of physicians have published studies of the effects of silicone-gel implants on women's health, and they have concluded that the implants are safe. Although negative reactions to the implants clearly occur, it is equally clear that such reactions are infrequent.

Impact

Three significant changes have resulted from the controversy over the safety of the silicone-gel implants: (1) the manufacture of the implants has slowed to a trickle; (2) several thousand lawsuits have been filed; and (3) the FDA has undertaken a long-range, large-scale study to see how safe silicone-gel implants really are.

In 1988, the Food and Drug Administration told manufacturers of silicone-gel implants that they had to present detailed safety information about their products to the FDA by July 9, 1991. Although the safety information showed that there was no clear link between silicone-gel implants and autoimmune disorders or other systemic diseases, the lack of long-term safety data about the implants caused Dr. David Kessler, the commissioner of the FDA, to declare a forty-five-day moratorium on the sale of the silicone-gel implants, beginning January 6, 1992.

Three significant changes have resulted from the controversy: (1) the manufacture of the implants has slowed to a trickle; (2) several thousand lawsuits have been filed; and (3) the FDA has undertaken a long-range, large-scale study to see how safe silicone-gel implants really are.

The publicity surrounding Kessler's moratorium only exacerbated the legal and financial problems of the silicone-gel manufacturers that had begun some time before. In 1984, a San Francisco, California, jury found against Dow-Corning and awarded a Nevada woman $1.5 million in damages. The case was settled out of court for an undisclosed sum after Dow-Corning appealed the decision. In December of 1991, another San Francisco jury issued a $7.3 million verdict against Dow-Corning. At least part of the reason for the large judgment was the evidence showing that Dow-Corning had not warned doctors that they had linked ruptured silicone-gel implants with autoimmune diseases. After Dr. Kessler's moratorium in January 1992, the volume of lawsuits increased substantially. By February 1993, an estimated three thousand lawsuits related to silicone-gel implants had been filed in state and federal courts. Late in January 1993, a Texas state court awarded $25 million to a claimant against the Bristol-Myers Squibb company. Bristol-Myers Squibb had closed its silicone-gel implant business in September 1991. Dow Corning followed suit in March 1992.

In September 1993, Dow-Corning announced plans to create a $4.75 billion liability fund to compensate those injured by silicone-gel implants. Each victim could receive between $200,000 and $2 million; removal of the implants would also be covered. The fund would be paid into by those being sued—insurance companies, doctors, and manufacturers. The plan awaits approval by a federal judge.

That same month, a settlement was reached to end the manufacture of silicone-gel implants by United States companies by 1995. By mid-1993, only one such company existed: The Mentor Corporation.

Despite the moratorium on the sale of the implants, the FDA will allow silicone-gel implant surgery for women who need surgery to replace a device already in place, and for women who need implants after a mastectomy.

Finally, one of the most significant outcomes of the implant controversy is that the FDA will conduct a long term, large-scale, scientifically controlled study on a limited number of women to determine—at long last—the safety and effectiveness of silicone-gel implants.

PATRICK MOORE

Where to Learn More

Angell, Marcia. "Breast Implants—Protection or Paternalism?" *New England Journal of Medicine,* June 18, 1992, pp. 1695-96.

Burton, Thomas M. "Breast Implants Raise More Safety Issues." *Wall Street Journal,* February 4, 1993, pp. B1, B8.

Fisher, Jack C. "The Silicone Controversy—When Will Science Prevail?" *New England Journal of Medicine,* June 18, 1992, pp. 1696-98.

Kessler, David A. "The Basis of the FDA's Decision on Breast Implants." *New England Journal of Medicine,* June 18, 1992, pp. 1713-15.

Sahn, Eleanor E., et al. "Scleroderma following Augmentation Mammoplasty." *Archives of Dermatology,* September 1990, pp. 1198-1202.

Smart, Tim. "Breast Implants: What Did the Industry Know, and When?" *Business Week,* June 10, 1991, pp. 94-98.

Uretsky, Barry F., et al. "Augmentation Mammoplasty Associated with a Severe Systemic Illness." *Annals of Plastic Surgery,* November 1979, pp. 445-47.

Dalkon Shield

(1970s-1980s)

An intrauterine birth control device is fatal for some women and causes spontaneous abortions, internal injuries, infertility, and birth defects.

Background

The Dalkon Shield, a birth control device implanted by a doctor in a woman's uterus, did not work as advertised by its manufacturer and, beginning in 1971, caused at least twenty deaths and thousands of cases of internal injuries. The A. H. Robins Company, which manufactured the Dalkon Shield, knew about potential problems with the shield all along, but marketed it anyway and promoted it heavily. After users' problems with the device were made known to Robins, the company kept the product on the market and only removed it in the United States in late June of 1974 after widespread bad publicity. In April, 1975, Robins suspended the sales of the Dalkon Shield outside of the United States. In October 1984, after almost ten years of more litigation, Robins finally began to urge doctors to remove the shields still in use.

The Dalkon Shield is a piece of flat, flexible plastic, shaped like a crab and measuring about three quarters of an inch across. To keep a woman's body from spontaneously expelling it, the shield has five prongs in each side, which keep it in place in the uterus. The prongs also make the shield difficult to insert and remove, and they can puncture the lining of the uterus. The shield has a plastic string called a tailstring tied to its bottom portion, which helps a doctor to remove it and allows a woman to check that the shield is properly in place. The upper end of the tailstring, which is threaded through a small opening in the bottom part of the shield and knotted, stays in the woman's uterus. The remainder of the string extends through the cervix, the lower necklike end of the uterus, and into the vagina.

The construction of the tailstring is an important factor in causing infections in the uterus. The string is made of tiny nylon strands encased in a sheath. The sheath, however, is not sealed shut at either end, which allows moisture and bacteria to travel up the sheath from the vagina into the uterus in a process called "wicking." In wicking, a fluid is attracted—even vertically—

through fibrous materials because of the surface tension of the fluid. In the case of wicking in the tailstring of the Dalkon Shield, bacteria traveled from the vagina to the uterus and caused dangerous pelvic inflammatory infections.

In June 1970, A. H. Robins Company purchased the rights to the Dalkon Shield from its inventors, Hugh Davis and Irwin Lerner. Despite a lack of any data on the long-term safety of the device, Robins began to sell and promote the Dalkon Shield vigorously in January 1971. Eventually, it was estimated that 3.6 million women had the shield inserted in their bodies.

Hugh Davis, a doctor, and Irwin Lerner, an engineer, invented the Dalkon Shield. After its invention, Davis tested the shield for twelve months on 640 women. He then published his results in the February 1970 issue of the *American Journal of Obstetrics and Gynecology*. In his article, Davis claimed an annual pregnancy rate of 1.1 percent, which was substantially better than any other intrauterine device. But Davis's study had important flaws. First, he had too few experimental subjects. Second, only eight of his patients had the device inserted in the first month of the study. Third, he prescribed contraceptive foam to be used simultaneously during the first few months after insertion of the shield. Fourth, he only waited three days after the end of the twelve-month period before compiling the data that he used in his article. That meant that he did not count any pregnancies that he discovered in his 640 patients after the twelve-month period had elapsed. Fifth, he did not disclose in the article that he had a financial stake in the success of the shield.

In June 1970, A. H. Robins Company, a Fortune 500 pharmaceuticals corporation based in Richmond, Virginia, purchased the rights to the Dalkon Shield from its inventors, Davis and Lerner. In the days before it purchased the shield, Robins found out that Davis had more complete data showing that the pregnancy rate from the shield was a minimum of 5.3 percent, and it also found out that Davis had no data on the longterm safety of the device. A few days after Robins bought the rights to the shield, Lerner warned Robins that the tailstring might be a problem. About two weeks after Robins purchased the shield, a Robins executive wrote to thirty-nine other Robins executives that the shield had a "wicking" tendency that should be looked into carefully.

Robins did nothing about the safety problems and began to sell and promote the Dalkon Shield vigorously in January 1971. Robins told its sales staff of the 1.1 percent pregnancy rate, and omitted any mention of the 5.3 percent pregnancy rate that Davis had subsequently discovered. Robins also advertised the shield in medical journals as being "trouble free" and "safe and effective." Eventually, it was estimated that 3.6 million women had the shield inserted in their bodies. The stage was set.

Details of the Health Problems

Even the process of inserting a Dalkon Shield was painful, causing many women to faint, to cry out in agony, and to bleed vaginally for days. Once the shield was inserted into her uterus, a woman could experience pelvic infections, infertility, perforations of the uterus, scarred internal reproductive organs, spontaneous abortions caused by infections, and premature or stillborn babies. Over five percent of the women who used the shield to avoid pregnancy became pregnant anyway. Several specific cases will illustrate the physical and emotional damage caused by the shield.

Opponents of the Dalkon Shield protest against the A.H. Robins Co.—producers of the Shield.

Pam Van Duyn had the shield inserted in October 1977. Within a few hours she began to run a high fever. The first doctor she visited thought she had food poisoning and gave her penicillin and other antibiotics. After spending five days in agony, Pam saw a second doctor, a gynecologist, who diagnosed her as having acute pelvic inflammatory disease (PID), a runaway bacterial infection affecting a woman's ovaries, uterus, and other pelvic organs. If unchecked, PID can be fatal. It can also cause scarring to the fallopian tubes, which carry the egg down to the uterus. The scarring can make women infertile. While examining Pam, the doctor found a large mass the size of an orange attached to her left ovary. A while later the doctor discovered the exact kind of bacteria infecting Pam and treated her with a special antibiotic, which controlled the infection. After nine days in the hospital, Pam went home, but she had to rest in bed for several weeks and felt weak for months after that. The large mass, however, did not go away, and Pam was left with a low-grade, chronic infection that flared up from time to time. In 1983, Pam had her left ovary and fallopian tube removed, and in 1984 she had surgery to remove the scarring in her right fallopian tube. The effects of the Dalkon Shield, however, were not restricted to Pam's body. Her physical injuries, and her diminished chances of ever having the children that she wanted, caused Pam great emotional pain, and she and her husband separated, though they reunited later.

Peggy Mample had a Dalkon Shield inserted in the autumn of 1971. Immediately her menstrual periods became very heavy and painful. In March 1972 she became pregnant despite the presence of the shield. She decided to leave the shield in her uterus to minimize the chances of a miscarriage. Seven months later, Peggy had an inflammatory reaction to the Dalkon Shield, which forced her to give birth to her daughter, Melissa, two months prematurely. Melissa was born with dislocated hips and hyaline membrane disease, which makes it hard for infants to breathe, and which, in turn, can lead to retardation if the infant's brain is deprived of oxygen. Melissa gained strength and was released from the hospital. Peggy found out a year later that Melissa did have brain damage, a form of cerebral palsy that paralyzed the muscles of her legs. Melissa could not walk far, and only then with the aid of leg braces and a walker. Though Melissa's mental abilities were fine, she had to get around in a wheelchair or by dragging herself across the floor, pulling her paralyzed legs behind her. Peggy Mample's husband had trouble accepting his daughter's disabilities, and he and Peggy were divorced. Peggy later remarried and has two healthy sons.

Pam Van Duyn and Peggy Mample are just two of many cases of the physical and emotional suffering caused by the Dalkon Shield.

Impact

After some women experienced problems with the Dalkon shield, they began to sue A. H. Robins. Many of the early cases were settled out of court for relatively small sums—under $10,000 for example. But in February 1975, a Kansas woman won $85,000 in damages against Robins. That broke the dam. By the middle of 1975, over five hundred legal cases demanding close to half a billion dollars in damages were pending against A. H. Robins around the United States. Later, some women were to win well over a million dollars each in their cases against the company, and one Kansas woman in 1985 won $8.9 million in damages. In litigating later suits, it was discovered that lawyers for Robins had destroyed valuable evidence at the request of their company's chief legal counsel and with the knowledge of the company's president. The evidence, some of which had been secretly photocopied and kept by a Robins lawyer with a guilty conscience, showed that Robins knew all along about the potential damage that the Dalkon Shield could inflict upon women.

By the mid 1980s, it became clear to the management of A. H. Robins that more large financial judgments would soon be assessed against the company. On August 21, 1985, the company, which was controlled by E. Claiborne Robins, Sr. and E. Claiborne Robins, Jr., filed for bankruptcy. The Robins family hoped that they could cope with their losses and ultimately emerge with their company still in their control. During the long bankruptcy process, however, Robins became a target for takeover by other companies. On December 15, 1989, Robins officially merged with American Home Products. For their family-owned share of the Robins stock, E. Claiborne Robins father and son received $385 million, which was tax free because the transaction was an exchange of stock. Shortly after their company was taken over, both Robinses were told that their services would no longer be needed by American Home Products.

During the bankruptcy process, which lasted over four years, over 325,000 claims against Robins were filed. Of these, the federal courts allowed 195,000. Ultimately, a sum totaling $2.7 billion was paid into a trust to settle the claims of women who used the Dalkon Shield. Those claims are being processed as of this printing and will probably continue to be processed into the twenty-first century.

PATRICK MOORE

Where to Learn More

Alexander, Charles P. "Robins Runs for Shelter." *Time,* September 2, 1985, p. 23.

Chakravarty, Subrata N. "Tunnel Vision." *Forbes,* May 21, 1984, pp. 214, 218.

Culliton, Barbara J., and Debra S. Knopman. "Dalkon Shield Affair: A Bad Lesson in Science and Decision-Making." *Science,* September 6, 1974, pp. 839-41.

Mintz, Morton. *At Any Cost: Corporate Greed, Women, and the Dalkon Shield.* Pantheon Books, 1985.

Perry, Susan, and Jim Dawson. *Nightmare: Women and the Dalkon Shield.* Macmillan, 1985.

Sobol, Richard B. *Bending the Law: The Story of the Dalkon Shield Bankruptcy.* University of Chicago Press, 1991.

Nuclear Plants

- Kyshtym (1957)
- Windscale (1957)
- SL-1 (1961)
- Browns Ferry (1975)
- Three Mile Island (1979)
- Tsuruga (1981)
- Ginna (1982)
- Chernobyl (1986)

Kyshtym power plant explosion

Sverdlovsk, Russia

(1957)

A failure in the cooling system caused an explosion at a nuclear waste storage site.

Background

On September 29, 1957, at a nuclear power plant in the Sverdlovsk region of the former Soviet Union, a mixture of sodium nitrate and sodium acetate used in the processing of nuclear wastes exploded. The blast carried seventy to eighty metric tons of wastes into the atmosphere. Fallout from the blast caused more than eight thousand deaths and contaminated hundreds of square miles of land around and downwind of the explosion. Information about the accident was not released by the Soviet government until thirty-two years later, in 1989.

The first Soviet nuclear reactors were constructed in the years following World War II. Before the end of the war, the Russian government knew about American efforts to build an atomic bomb and concluded that it had to have its own nuclear weapons program. The Soviets were not, however, as concerned about environmental and human safety as Americans typically have been. The Soviet nuclear program is replete with stories of accidental and intentional releases of hazardous materials at reactor sites.

The three main sites of nuclear weapons development in the former Soviet Union were at Chelyabinsk, in Sverdlovsk along the eastern slopes of the southern Ural Mountains, and at Tomsk and Dodonoro, located north of the Mongolian border. As of the early 1990s, only two of the sites have ever been visited by Western scientists.

The nuclear facility constructed at Kyshtym, near Chelyabinsk, was one of the secret communities called *atom grads* whose purpose it was to produce, process, and store the nuclear materials needed for atomic weapons. Construction of the Kyshtym plant began in 1945 on the site of an older munitions factory. Soviet authorities have now revealed that construction work at Kyshtym was carried out by about seventy thousand inmates taken from twelve labor camps in various parts of the Soviet Union.

Construction of the facilities was planned and managed by the production association Mayak, a name that is often associated with the specific reactors built at the site. The plant included a reactor for the production of plutonium and a reprocessing plant. The first reactor at Kyshtym was one of the graphite-moderated, light-water reactors so favored by Soviet scientists. The reactor, called Anotchka, or A-Reactor, was completed and put into operation in just eighteen months. A sphere of plutonium ten centimeters in diameter produced at Kyshtym became the fuel for the first Soviet atomic bomb exploded on August 29, 1949. In the meanwhile, the town around the nuclear facility continued to grow rapidly and, by 1947, had a population of several thousand.

Over the next six years, five more nuclear reactors were built at Kyshtym. In addition, a fuel reprocessing plant was opened at the facility in December 1948. The purpose of a reprocessing plant is to take spent fuel from nuclear reactors, treat it chemically, and recover uranium and plutonium that can then be reused in a reactor. For the first thirty years of its existence, the Kyshtym plant was used exclusively for the reprocessing of fuel from military reactors.

Any nuclear reactor requires huge amounts of water for its operation. For that reason, nuclear plants are usually built on lakes, rivers, or the ocean. In most cases, wastewater from the nuclear reactor is treated before being returned to the body of water. This process allows the water to cool and lose some of the radioactivity it picks up in the plant. At Kyshtym, all six nuclear reactors were built on the shores of Lake Kyzyltash. They used an "open-cycle" cooling system in which water was drawn directly from the lake, cycled through the reactor, and then returned directly to the lake. The thermal plume produced by this system was so apparent that it could be photographed from space by the United States' satellite *Landsat*.

Soviet concerns for safety did not improve over time. After the A-reactor at Kyshtym was completed, it discharged its wastewater and radioactive wastes into the Techa River (although it still used Lake Kyzyltash for its cooling water). By 1951, radiation carried by the waters of the Techa was detectable in the Arctic Ocean, more than a thousand miles away.

Eventually, plant officials decided to place the Techa River off limits to the general public because of the high levels of radioactivity it contained. They built a wire fence along the riverbanks to keep people from fishing or swimming in the river. Then, they changed their disposal system and began to dump their wastes in Karachay Lake, a body of water with no outlet. Their hope was that radioactive wastes would cause no harm if they remained trapped in the lake.

That plan did not work very well. In the first place, the amount of radiation reached astronomical levels. At its peak, the lake contained 120 million curies of iodine 131 and strontium 90. By comparison, the 1986 explosion at the Chernobyl nuclear plant released only 1.2 million curies of these isotopes. In 1967, an unexpected disaster occurred. A hot dry summer caused unusual amounts of evaporation from the lake. By fall, the shore was coated with radioactive dust that had precipitated out of the lake. When heavy winds began to blow, they carried dangerous levels of radiation over an area of eighteen hundred square kilometers (seven hundred square miles).

Details of the Disaster

Exactly what happened at Kyshtym ten years earlier in 1957 was not known until 1989, when the Soviet government opened its files to Western experts. Prior to that time, the disaster had been the subject of extensive speculation by scientists both inside and outside of the Soviet Union. In 1976, for example, the United States Central Intelligence Agency released a dozen documents that it had collected about incidents at Kyshtym. The documents described a number of possible events, including an explosion in a nuclear power plant, an aboveground test of a nuclear weapon, and "an accidental atomic explosion." Working with such scant evidence, scientists argued back and forth during the 1980s trying to piece together exactly what happened at Kyshtym.

About twenty million curies of radiation, about a quarter the level released at Chernobyl, spewed into the atmosphere as a result of the explosion. The blast was not a nuclear one, as most Western observers had thought. Instead, it resulted instead from a chemical reaction like those that take place in conventional bombs.

As a result of the new Soviet policy of *glasnost*, we now have a good idea of what transpired on September 29, 1957. As an initial step in reprocessing, fuel rods were treated with nitric acid to dissolve uranium, plutonium, and other elements in the fuel. Chemical methods were then used to separate the desired products (uranium and plutonium) from unwanted, "waste," materials. The waste products, like the fuel itself, were highly radioactive.

After removal of uranium and plutonium from the nitric acid solution, the remaining materials were treated with sodium acetate and sodium nitrate and transferred to large storage tanks. They were kept in these storage tanks for about a year while they became less radioactive. Then, after a second attempt to remove uranium and plutonium, the wastes were dumped into Lake Karachay.

In 1953, plant operators added another stage to the cooling-off process. They built a rectangular concrete basin underground with walls 1.5 meters (4.8 feet) thick. The basin was large enough to hold twenty stainless steel tanks, each with a volume of 300 cubic meters (10,000 cubic feet). The plan was to fill the tanks with radioactive wastes and store them underwater until they had cooled off.

The September 29 accident occurred because the storage system never worked as designed. Over time the steel tanks began to crack and break open. The water inside the tanks evaporated and the solid components began to precipitate out. A highly dangerous mixture of sodium nitrate, sodium acetate, and a combination of radioactive isotopes began to form in the tanks. A revised cooling method designed to deal with this problem was developed, but it too failed.

Finally, on September 29, heat produced by the radioactive materials had raised the temperature in one tank to 350 degrees Celsius (660 degrees

Fahrenheit). This heat was sufficient to ignite the nitrate-acetate combination that had also precipitated out inside the barrel. An explosion with the force of seventy to one hundred tons of TNT blew seventy to eighty tons of radioactive waste in the air. About twenty million curies of radiation, about a quarter the level released at Chernobyl, spewed into the atmosphere. The explosion was not a nuclear blast, as most Western observers had thought. Instead, it resulted instead from a chemical reaction like those that take place in conventional bombs. The effects of the explosion were complicated, however, by the intensely radioactive materials stored in the tanks that were carried into the atmosphere by the blast.

Impact

The earliest reports of the Kyshtym disaster carried descriptions of widespread devastation of the environment as well as extensive human death and injury. One Soviet émigré told of a trip through the Chelyabinsk area in which he saw "no villages, no towns, no people, no cultivated land." His driver explained that no one was allowed to stop for a distance of twenty to thirty kilometers around the area.

Another report claimed that refugee camps built to house people who were moved out of the Kyshtym area had grown into large villages. But so many deaths occurred, the report went on, that "still faster [than the villages] grew the small hamlet's cemetery, accepting the refugees who died from the radiation." In yet another report, an observer described survivors of the Kyshtym accident that he visited in a nearby hospital. "We could see," he said, "the skin of their face, hands, and other exposed parts of the body to be sloughing off." This pattern is characteristic of the radiation sickness that develops when a person is exposed to high levels of radiation.

In their original 1989 announcement, Soviet officials denied reports like these. They claimed that no one had died in the accident and that the rate of malignant tumors and blood disorders was no greater than normal among residents of the region.

A later report provided different information, however. In 1992, a study conducted by the Institute of Biophysics at the former Soviet Health Ministry in Chelyabinsk found that 8,015 people had died over the preceding thirty-two years as a result of the accident. In addition, the number of cases of leukemia was significantly greater than the number expected for the population being studied.

Although many victims were killed or injured from radioactive fallout, others were and still are at risk from the water supply. Radioactive particles from the explosion eventually contaminated lakes and rivers on which about 270,000 people depended. About 1,000 of those people were evacuated ten days after the blast, but it was more than a year before more massive evacuations were begun. In the meanwhile, unknown thousands of residents had been eating food grown on contaminated land and drinking water from contaminated reservoirs. The government eventually plowed under or removed from agricultural use 1,340 square kilometers (515 square miles) of land. More than 85 percent of this land was eventually restored to farm use by 1978.

Contamination remains a problem at Kyshtym, 36 years after the accident. Significant levels of radiation have been detected in groundwater up to three kilometers (two miles) from the lake. Lethal doses of radiation also continue to exist along the shores of the lake itself.

The accident at Kyshtym apparently had little short-term effect on work going on there. All six reactors continued to operate into the late 1980s. The A-reactor was finally decommissioned in 1987, while four other reactors were shut down between 1987 and 1990. Only a single heavy water-moderated reactor continues in operation.

At one time, officials had plans to build three breeder reactors at the Kyshtym complex. Two were started, and their concrete footings can still be seen there. But they were never completed, and the third plant was never started. There seems little likelihood that any of the plants will ever be built.

DAVID E. NEWTON

Where to Learn More

Cochoran, Thomas B., and Robert S. Norris. "A First Look at the Soviet Bomb Complex." *Bulletin of the Atomic Scientists*, May 1991, pp. 25-31.

Nikipelov, B. V., and Y. G. Droozkhko. "Explosion in the South Ural Mountains." *Pirorda*, May 1990, pp. 48-49.

Norman, Colin. "Soviet Radwaste Spill Confirmed." *Science*, April 16, 1982, p. 274, and responses in July 16, 1982, pp. 198ff.

Oberg, James E. *Uncovering Soviet Disasters*. Random House, 1988, Chapter 13.

Perea, Judith. "Soviet Plutonium Plant Killed Thousands." *New Scientist*, June 20, 1992, p. 10.

Trabalka, John R., L. Dean Eyman, and Stanley I. Awerback, "Analysis of the 1957-1958 Soviet Nuclear Accident." *Science*, July 18, 1980, pp. 345-53.

Windscale reactor complex fire

England

(1957)

World's first major commercial nuclear power plant disaster at Great Britain's facility for making plutonium.

Background

On Thursday evening, October 10, 1957, fire broke out in Unit 1 at the Windscale nuclear reactor complex on the coast of Cumberland in northwest England. The fire burned for more than twelve hours, releasing large quantities of radiation to the surrounding countryside. When safety officers found dangerous levels of radiation in milk produced by cows in the area, the milk was dumped into the ocean. A report of the accident released shortly after it occurred omitted some important details that were not discovered until thirty years later.

The immediate postwar years were difficult times for British nuclear scientists. On the one hand, they were under considerable pressure from government officials to build a nuclear weapon. The need to keep abreast of the United States and the Soviet Union in weapons development seemed obvious to the British. On the other hand, the British had no access to weapons technology information from the Americans, alongside whom they had fought in World War II. The United States Atomic Energy Act of 1946 prohibited the export of such information from the United States to any foreign nation, including its longtime ally, Great Britain.

The response by British nuclear scientists was remarkable. In 1947, they began construction of a pair of nuclear reactors for the production of plutonium at Windscale. Plutonium is one of the two materials that can be used to cause an explosion in a nuclear weapon. By the end of the 1940s, both Windscale units were in operation. Their enormous four hundred-foot-high exhaust stacks were familiar sights in the neighborhood, although residents of the area were never told what was going on at the plants.

The Windscale reactors were of a somewhat uncommon design. They consisted of huge honeycombed graphite blocks, fifty feet high and twenty-five feet deep. Running horizontally through the blocks were twenty thousand channels, each containing a long, thin fuel rod filled with uranium. When the reactor was in operation, neutrons bombarded the uranium, producing heat through fission (splitting) reactions and plutonium as a by-product. The plutonium was eventually extracted from the fuel rods at a chemical processing plant also located at the Windscale complex.

The graphite blocks were cooled by air. Huge fans at one end of each block blew a blast of air down each fuel channel, carrying away heat from the fission reaction and out the exhaust stacks. Large rectangular blocks at the top of each stack trapped radioactive materials carried on the heated air.

The British nuclear effort was extraordinarily effective. On October 3, 1952, their first atomic bomb was detonated at Monte Bello Island off the coast of Australia. The decision was also made to expand the Windscale complex with the construction of two more reactors for the manufacture of plutonium. The two new plants were constructed at Calder Hall, adjacent to the older plants. They had an added feature, however, in that the heat they produced was used to generate electricity rather than being vented into the atmosphere. With considerable ceremony, Queen Elizabeth II opened the Calder Hall facility in October 1956. On that occasion, the British were enormously proud of their accomplishments in nuclear science, as they had every right to be.

Details of the Disaster

That glow of success was to be replaced almost exactly a year later, however, by a much different and more sinister kind of glow: the fire at Windscale's Unit 1. The accident came about as the result of a somewhat unusual physical phenomenon known as the Wigner effect. The phenomenon is named for the Hungarian-American physicist, Eugene Wigner, who first predicted its occurrence.

The Wigner effect occurs in graphite reactors that have been in use for some period of time. Some of the energy released by fission in the reactor moves carbon atoms in the graphite block out of position. The longer the reactor is in use, therefore, the more distorted the block becomes. If nothing is done about this phenomenon, the reactor may eventually release its stored up energy all at once, causing a possibly dangerous overheating of the graphite block.

The solution to this problem is relatively simple. If the reactor is simply allowed to heat up to about 200 degrees Celsius, carbon atoms in the graphite block become more mobile and return to their normal positions, and the stored-up energy in the block is released safely.

Although American scientists were familiar with the Wigner effect, their British counterparts at Windscale apparently were not. The latter learned about the phenomenon by accident in September 1952 when they observed the release of Wigner energy after the reactor had spontaneously overheated. After that near-accident, operators devised an empirical, rule-of-thumb procedure for dealing with Wigner energy. They allowed the reactor to overheat about once every twenty thousand megawatt days of operation.

The tall stacks in the center of this photo are the exhaust stacks for the two original Windscale units, including Unit 1, in which the 1957 accident occurred. The dome at right is part of a later nuclear reactor built at the complex.

The policy appeared to work well until production pressures increased the length of time between Wigner heatings, first to once every thirty thousand megawatt days and then to once every forty thousand megawatt days. As these periods between energy release increased, so did the potential risk of a reactor overheating accident. Still, in October 1957, operators had performed the operation successfully and without incident on eight separate occasions.

The ninth operation, begun on Monday, October 7, 1957, was to have a different result. Historians believe that plant operators may have been under unusual pressure during this occasion. To perform the Wigner release, the reactor has to be shut down for a few days. But the British government was pushing forward aggressively on its weapons program. A bomb test originally scheduled for June 1958 had recently been moved forward by seven months, to November 1957. The pressure to generate plutonium was intense. Experts believe that such pressures may have caused Windscale operators to delay the Wigner release too long or to be careless in carrying out the procedure.

In any case, the October 7 release was to involve the same complex mixture of mechanical failures and human errors characteristic of most nuclear reactor accidents. The operation began normally enough on Monday as cooling fans were turned down, allowing the reactor to heat up of its own accord. By Tuesday morning, everything seemed to be going normally, with Wigner heat being released in the reactor as expected. Operators turned the cooling fans back on, expecting to return the reactor to normal operation while the release of Wigner energy continued.

By midday Tuesday, however, some strange happenings were observed. The temperature of the reactor appeared to be falling although it should have remained constant during the loss of Wigner energy. Operators decided to resume heating of the reactor although such an action had never been attempted in a similar operation before. The reactor had already reached its maximum permissible temperature of 250 degrees Celsius during the earlier heating.

Nonetheless, cooling fans were again turned off, and temperatures again began to rise, this time reaching 300 degrees Celsius within a matter of hours. By Wednesday evening, operators were no longer sure what was happening in the graphite block. Wigner energy was still being released, but temperatures in the block had now reached a dangerous 350 degrees Celsius and, in one region, more than 400 degrees Celsius.

Part of the problem at this point was a technical one. Thermocouples—which use metal wires, not liquid, to measure temperature—were built into the block at various locations to monitor temperatures in the reactor. However, later studies showed that their placement was such that regions of the block could give readings as much as 40 percent higher than readings given by the nearest thermocouple. Thus, a 400-degree Celsius reading observed by an operator might not reflect a temperature at least 100 degrees Celsius higher in a full channel only a few meters away.

As temperatures in the block continued to rise, operators decided to turn the cooling fans back on. The fans had little effect, however, and more bad news appeared on Thursday morning. Monitoring instruments in the caps on top of the exhaust stack indicated that high levels of radiation were escaping from the reactor.

Such readings would normally indicate that one or more of the fuel rods in the block had broken apart. Normally, operators would be able to find out if that kind of accident had, in fact, occurred. Scanners placed at various positions in the block had been designed to monitor the integrity of all fuel rods. Heat released in the block over the preceding two days, however, had damaged the scanners so that they could no longer be used to trace possible fuel breakdowns in the reactor.

By Thursday evening, October 10, operators were sufficiently concerned about circumstances to take an extraordinary step. They decided to remove external sections of the block and look directly into the heart of the reactor.

The procedure was an enormously dangerous, last-ditch effort to find out what was going on in the machine.

To their horror, they saw a glow of white heat. Some of the fuel rods had obviously burst open, uranium metal was glowing and the surrounding graphite was on fire. A major disaster was clearly at hand.

Workers tried a brute force approach to solving the problem by forcing the damaged fuel rods out of the reactor. The rods had become so warped, however, that they would not budge. As a secondary approach, workers then attempted to push out rods surrounding the damaged section, an effort that was successful. They had been able at least to contain the damaged area of the block.

The fire continued to be a problem, however. Unless it could be put out, it would certainly spread throughout the graphite block and engulf the whole reactor. Two alternatives were available. First, efforts were made to smother the flames with huge volumes of carbon dioxide pumped in from nearby tanks. When this failed, operators decided to try flooding the reactor with water.

This step was a risky one. There was a chance that the water would turn to steam and blow the reactor apart. Fortunately, that did not occur. When the first stream of water hit the graphite block at 8:55 A.M. on Friday, it hissed and turned to steam—but no explosion took place. In fact, two hours later, operators were convinced that the fire had been extinguished. As a safety precaution, however, they continued the flow of water for another twenty-eight hours.

Operators knew when they poured water on the fire in Unit 1 at Windscale that they were taking a serious step. Graphite absorbs water and does not dry out well. In putting out the fire at the reactor, they were also condemning it to its death. Unit 1 was never returned to operation.

Meanwhile, evidence of the accident began to appear in the area around the plant. Monitoring teams from the United Kingdom's Atomic Energy Authority (AEA) detected four radioactive isotopes, cesium 137, strontium 89, strontium 90, and iodine 131 in the surrounding countryside. Of these, only the last was present in concentrations high enough to be a matter of concern. Experts knew that iodine 131 collects on grass, which is eaten by cows. The isotope then becomes part of the milk the cows produce. As a cautionary step, therefore, authorities destroyed all milk from an area of two hundred square miles adjacent to the damaged plant.

Impact

Operators knew when they poured water on the fire in Unit 1 at Windscale that they were taking a serious step. Graphite absorbs water and does not dry out well. In putting out the fire at the reactor, they were also condemning it to its death. Unit 1 was never returned to operation.

At first, plans were made to continue the operation of Unit 2. Those plans were soon changed, however, when the decision was made to close down the undamaged reactor in order to see if it could provide information about the accident at Unit 1. Although these studies were made, Unit 2 was never reopened either.

British nuclear industry officials claim that they learned a great deal as a result of the disaster. In 1988, John Collier, chairman of the AEA, said, "A new safety culture was born out of the Windscale fire which highlighted faults that have been corrected, making nuclear power operations safe and effective. We have learned important lessons from the 1957 fire, and these have been taken up by the industry worldwide."

For many years, the health effects of the Windscale accident were thought to be minimal. A 1983 report by the U.K. National Radiological Protection Board estimated that thirteen people died as a result of thyroid cancer caused by iodine 131 released during the accident. The Cumbria Area Health Authority reported that deaths from leukemia were "not significantly altered from the national rate."

But other health authorities were not so sure. Historian Margaret Gowing has pointed out a possible factor to which little attention has been given, polonium 210. This isotope, used as a trigger in atomic bombs, was also being produced in Windscale's Unit 1 reactor. Its release during the accident was little noted or studied, according to Gowing. But there is evidence to support the view that the isotope may have been responsible for up to 150 percent the number of deaths of leukemia beyond the normal rate in the Copeland and South Lakeland areas downwind of Windscale. A 1983 article in the journal *New Scientist* suggested that the polonium factor may have made the Windscale accident "the worst environmental disaster that Western Europe has known [to date] this century."

DAVID E. NEWTON

Where to Learn More

Dickson, David. "Doctored Report Revives Debate on 1957 Mishap." *Science*, February 5, 1988, pp. 556-57.

Herbert, Roy. "The Day the Reactor Caught Fire." *New Scientist*, October 14, 1982, pp. 84-87.

Howe, Hartley. "World's First Atomic Alarm." *Popular Science*, October 1958, pp. 92ff.

Pearce, Fred. "Penney's Windscale Thoughts." *New Scientist*, January 7, 1988, pp. 34-35.

Urquhart, John. "Polonium: Windscale's Most Lethal Legacy." *New Scientist*, March 31, 1983, pp. 873-75.

Williams, Gurney, III. "Radioactive Accidents." *Science Digest*, August 1971, pp. 10-14.

SL-1 experimental reactor explosion

Idaho

(1961)

One of the United States' first experimental nuclear reactors exploded as a result of a combination of mechanical failures and operator error.

Background

An experimental reactor, known as the Experimental Stationary Low-Power Nuclear Reactor (SL-1), located near Idaho Falls, Idaho, exploded on the afternoon of January 3, 1961. The accident apparently resulted from a combination of mechanical failure and human error. Three technicians working on the reactor were killed, and lethal levels of radiation contaminated the site for months afterward.

As World War II came to a conclusion, the United States Atomic Energy Commission (AEC) turned its attention from the exclusive production of nuclear weapons to peaceful applications of nuclear energy. One of the locations it selected for a future research facility was the area around Arco, Idaho.

Arco, a tiny town of 780, had little in and of itself to attract the AEC's attention. What did interest the commission was the four hundred thousand acres of wasteland surrounding the little town. The absence of plants, animals, and humans seemed to make the region an ideal place to conduct nuclear experiments. Should an accident occur, there was little danger of damaging the environment or human life. The largest city of any size, Idaho Falls, was an apparently safe sixty-five miles to the east. A possible fringe benefit of the site was the porous lava beds that covered much of the area around Arco. Nuclear engineers thought the beds might provide a place to dump radioactive waste resulting from their experiments.

By 1951, the first experimental reactor, called Experimental Breeder Reactor No. 1, or EBR-1, began operation at the Arco site, now called the National Reactor Testing Station (NRTS). On December 20, 1951, the reactor was turned on, and it provided enough electricity to light four 200-watt bulbs. The next day, the reactor was taken to full power. It began supplying electricity to the reactor building itself and the equipment associated with it.

The success of EBR-1 was an exciting step forward for AEC engineers. But the plant was in operation only four years before experiencing a meltdown of its core. The reactor was shut down in 1955 and never returned to service.

Other construction was going on at NRTS, however, as engineers studied a number of possible reactor designs. One such experimental reactor was the Stationary Low-Power Reactor No. 1, or SL-1. The reactor was designed and built by scientists from the Argonne National Laboratory, outside Chicago. Those scientists had earned a reputation as the most reliable and dependable workers in the field. SL-1 was first put into operation in August 1958. Responsibility for its construction and operation was shared by four different entities: the Atomic Energy Commission, the U.S. Army, Phillips Petroleum Company, and Combustion Engineering.

The purpose of SL-1 was to test the feasibility of a small nuclear reactor for supplying electricity in remote areas such as the Arctic. Under full operation, the plant eventually produced three thousand kilowatts of power and seemed to be a completely successful experiment.

Details of the Accident

The first indication of a problem at SL-1 was the sounding of an automatic alarm at the NRTS Fire Station and Security Headquarters on January 3, 1961. Six AEC fire fighters, an assistant fire chief, and a security officer immediately set off on the eight-mile drive to the reactor. They arrived nine minutes later.

No sign of fire or other accident was visible to the arriving safety workers. They worked their way from building to building in the reactor complex, checking for smoke or fire, but with no results. Only when they approached the reactor itself did workers realize what had happened, for the radiation levels were greater than the limits of their meters (25 roentgens per hour). The needles of their Geiger counters jumped off scale, indicating a radiation level of more than 25 roentgens per hour. The rescue crew turned back at this point, having no instruments with which to measure the true level of radiation in the reactor. A second crew carrying meters that read to 200 roentgens arrived and found their instruments reading off scale also.

Finally, a third crew reached the site. The crew consisted of a firefighter and a health physicist from the Materials Testing Reactor, operated by Phillips, about ten miles away. The meters carried by these workers read to a maximum of 500 roentgens, far beyond the level one would expect to encounter in an accident. Again, the needles went off scale. The roentgen is used as a unit of exposure dose, and a single dose of 500 roentgens is fatal for half of all exposed individuals.

Rescue workers now faced a difficult decision. Radiation within the reactor was obviously at lethal levels. But a search of other buildings had failed to turn up the three-man maintenance crew (two soldiers and a sailor) that was supposed to be working on the reactor. It seemed likely that they were inside the reactor building. But were they still alive?

At 10:35 P.M., four officials from Combustion Engineering and the Phillips health physicist began going into the reactor room to search for the three military workers. They each stayed no more than three minutes at a time in order to reduce their total exposure to radiation.

Eventually, the rescue team found one worker dead on the floor of the reactor building and the second barely alive. He was removed but died on his way to the hospital. The third man was eventually found pinned to the ceiling of the reactor room by one of the control rods that had penetrated his body from groin to shoulder. His body was removed six days later through the heroic efforts of a rescue team and a remotely controlled crane.

In all nuclear reactors, a liquid or gas flows through the core of the reactor and heats up. Its purpose is to take away the heat generated by fission in the reactor core, so it is called a coolant. The heat is transferred from the reactor to a steam generator, where it boils water to steam and functions from then on like any power station powered by coal. SL-1, like most nuclear reactors, used water as a coolant. The SL-1 reactor vessel which surrounded the reactor was 4.5 feet in diameter and 14.5 feet high and resembled a silo. The reactor contained long, vertical fuel rods that held pellets of uranium, and similarly shaped control rods spaced between the fuel rods which absorb neutrons and thus kept the fission reaction going steadily by being raised or lowered.

Although a well-designed reactor in many respects, SL-1 appeared to have at least two fatal flaws. The first involved the core's control rods. When fully immersed in the core, the rods absorbed enough neutrons to bring the chain reaction to a halt. Fully withdrawn—as they never would have been—the chain reaction would have gone out of control and melted down the core. The margin for error was a small one. The instruction manual for the reactor warned workers to remove the rods by "not more than four inches," but it did not explain why this was imperative.

On January 3, however, workers were faced with two problems. First, the reactor had been shut down for eleven days for routine maintenance. The control rods had been disconnected from the automatic chain drive to which they were usually attached. Workers had to pull the rods up manually far enough to reattach them to the drive.

That task would not have been extraordinarily difficult except for the second problem. The channels holding the control rods and the rods themselves had become slightly warped. The rods did not slide smoothly in and out of the housings. Authorities hypothesize that a worker must have had to use unusual force to pull up on one of the control rods. As the rod finally broke loose and began to move, the theory goes, it slid more than four inches out of the core. For a fraction of a second, a flood of excess neutrons bombarded the fuel rods, the core temperature shot up a few thousand degrees, cooling water turned to steam, and the reactor lid blew off. The whole accident took less than a second to occur.

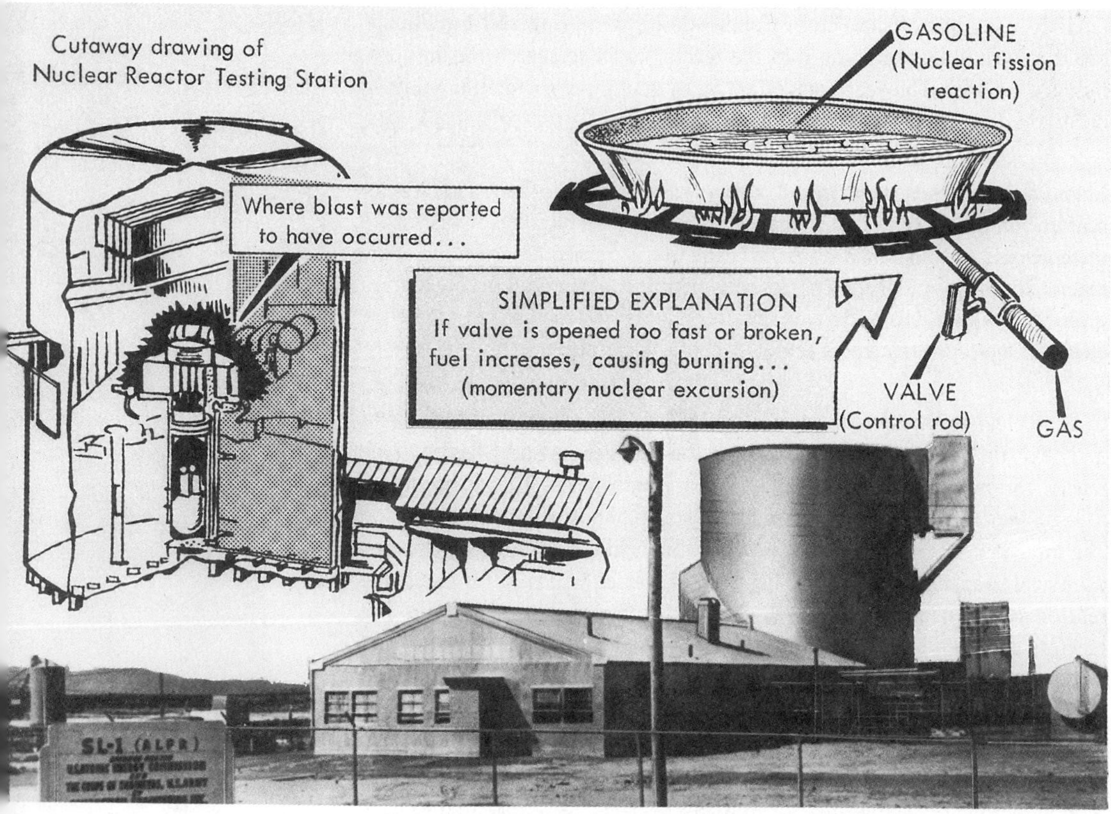

The SL-1 nuclear reactor near Arco, Idaho. Poorly designed control rods, which allowed operators little room for error when inserting and removing from the reactor core, probably contributed to the accident.

A second design element may have contributed to this sequence of events. SL-1 contained enough uranium to keep the reactor running without refueling for three years. As designed, the control rods alone were not adequate to regulate the reaction. The solution developed for this problem was to attach thin strips of boron to the aluminum surface of the fuel rods. Boron soaks up neutrons much as does the cadmium used in control rods. Then, the boron strips provided an extra way of controlling neutrons traveling through the core.

The unforeseen difficulty, however, was the ease with which the boron strips tore loose from the fuel elements. When scientists were eventually able to study the bottom of the reactor, they found pieces, strips, and flakes of boron that had come loose from the fuel rods. For some time—but no one knows exactly how long—the extra measure of safety supposedly offered by the boron strips had actually been disappearing. The "four inch caution" suggested in the instruction manual, based on the presence of the boron, was obviously not adequate.

Impact

Many individuals made mistakes in the design and operation of SL-1 that ultimately resulted in the January 3 explosion. But one early decision was obviously a wise one: to locate the reactor test site in a remote area. Other than the three military workers who were killed in the reactor blast, there were no apparent deaths or injuries. No report has been made of long-term health effects among the brave workers who entered the reactor in the first hours after the accident, fourteen of whom received radiation doses of more than five roentgens, and six of more than twenty roentgens. Radiation levels were monitored downwind from the accident, and iodine-131 was detected in air, vegetation, and milk. However, this contamination was reported not to have exceeded the maximum permissible levels for continuous exposure to the off-site population.

The major impact of the accident was probably its effect on people's attitudes about the commercial applications of nuclear power. Since the late 1940s, scientists and government officials had been painting an optimistic picture of the role of nuclear power in the world of the future. The public was told that it could look forward to atomic-powered ships and airplanes and a virtually limitless supply of safe, clean electricity from nuclear power plants. And for at least a decade, the nuclear power industry appeared to be keeping that promise. Although relatively minor problems were reported from time to time, more and more commercial plants were coming on line.

The SL-1 disaster, with the first loss of human life in a reactor accident, was like a dash of cold water on the high hopes and glowing promises of the industry. Official reports on the accident found plenty of blame to go around. For example, neither the Washington nor the local office of the AEC was aware of design errors in the reactor or the record of control rod problems that had been accumulating.

The SL-1 disaster, with the first loss of human life in a reactor accident, was like a dash of cold water on the high hopes and glowing promises of the industry. Official reports on the accident found plenty of blame to go around. For example, neither the Washington nor the local office of the AEC was aware of design errors in the reactor or the record of control rod problems that had been accumulating.

The commission did close down the only other reactor with a design like SL-1, located in Greenland, after the January 3 accident. It also reaffirmed its decision o locate nuclear power plants only in nonurban areas. Atomic Energy Commission scientists eventually developed a power-to-distance ratio that quantified and watered down this policy. According to that ratio, a 1.5 million kilowatt plant would have to be built at least 17.7 miles from a large urban area.

Combustion Engineering also came in for its share of the blame. Investigators pointed out that the company already knew enough about the reactor's problems to have closed it down long before the accident. The company did not disagree with this conclusion.

Correcting the design problems that led to the SL-1 accident was actually relatively simple. First, the practice of overloading the core with uranium so that it could operate longer between refueling was discontinued. This change made it unnecessary to use the boron strips that had caused so much trouble in the SL-1.

Second, the control rod problem was also resolved fairly easily. Investigators found that a single rod, the one in the center of the core, was critical. Its removal alone, regardless of the position of the other rods, was capable of causing an accident. Had the three military workers had to deal with a sticky rod in any one of the other positions, the incident might not have occurred.

The solution adopted for reactors of the SL-1 type, then, was to install automatic stops that physically prevented the removal of the control rod by more than some safe maximum distance. Had that simple precaution been in use on January 3, 1961, the SL-1 disaster might never have taken place.

DAVID E. NEWTON

Where to Learn More

Armagnac, Alden P. "The Atomic Accident That Couldn't Happen." *Popular Science*, September 1961, p. 52.

Lapp, Ralph E. "A Small Atomic Accident." *Harper's*, June 1961, pp. 47-51.

"Hot Stuff." *Newsweek*, February 27, 1961, pp. 64-65.

Norton, Boyd. "Atomic Roulette." *Chemtech*, July 1980, pp. 412-15.

Browns Ferry power plant fire

Fire in the electrical system of a nuclear power plant resulted in the most serious commercial nuclear power accident up to that time.

Decatur, Alabama

(1975)

Background

On March 22, 1975, workers at the Browns Ferry nuclear power facility were testing parts of the plant's electrical system. In so doing, they used a candle to check for air leaks between electrical cables and their housing. One of the candles set fire to insulation in the cables and the sealant packed around them, and this fire spread rapidly to Reactor 1 at the power plant, which had to be closed down. Reactor 2 was also shut down to avoid possible further damage. The fire was eventually extinguished about seven hours after it began. Direct damage to the plant was estimated at $6.7 million, while the cost of keeping both reactors out of service for eighteen months amounted to about $10 million per month. Overall, the total cost of the accident was probably between $100 million and $200 million.

Construction on the Browns Ferry nuclear power plant was started in the late 1960s. The original design called for three reactors to be constructed along the Tennessee River, about ten miles northwest of Decatur, Alabama. The complex was owned, built, and operated by the Tennessee Valley Authority (TVA).

All three reactors were to be boiling water reactors with a generating capacity of about eleven hundred megawatts each. In a boiling water reactor, the coolant water is allowed to boil, forming steam that is then pumped to a turbine generator. Unit 1 came on-line in 1974 and Unit 2 in 1975. Construction was still underway on Unit 3 when the electrical fire occurred. Its opening, originally scheduled for 1976, was eventually delayed by more than a year because of the accident.

The accident took place in a building known as the cable-spreading room (CSR) located directly beneath the plant's main control room. The CSR was 11 feet high, 35 feet wide, and about 180 feet long, the CSR was filled with thousands of cables and wire that supplied electricity to every part of the plant. Pumps, motors, valves, monitoring instruments, control rods, and nearly every other device in the plant was operated by this electrical system. Cables led from the control room to the reactors and to every other part of the complex.

On March 22, 1975, electricians were involved in a fairly routine operation in the CSR: sealing the space around cables where they passed in and out of the room. Designers were well aware of the potential fire danger posed by such openings. If an electrical fire were to get started in an electrical cable, they knew, it could spread rapidly down the cable, though the opening, and into the next building.

As a matter of fact, fires in electrical cables have long been common in nuclear power plants. Such fires have caused serious damage in reactors in England, Finland, Sweden, Switzerland, and at nuclear power plants—San Onofre, California; Quad City, Illinois; Salem, New Jersey; Indian Point and Nine-Mile Point, New York; Beaver Valley and Peach Bottom, Pennsylvania; and Oconee, South Carolina.

One way to limit the spread of electrical fires is to seal the space around cables with a spongy material such as polyurethane foam that will hold the cables tightly in place. Since polyurethane is flammable, however, the seal must also be covered with a flame-retardant material.

After a seal is made, electricians test its integrity by placing a lighted candle near the cable. If the candle flame does not flicker, the seal is tight. If the flame does flicker, the seal must be leaking.

The potential risk in this operation is obvious. It requires that a flame be placed near a material (such as polyurethane) that is flammable. If the flame retardant has not been applied properly, a fire is likely to occur. In spite of the apparent danger of this testing operation, it is generally regarded as standard procedure by electricians.

Nonetheless, fires during the testing of electrical cables were apparently not uncommon at Browns Ferry. During construction of the CSR, a number of small fires had been started. But in each case workers had been able to put them out easily, with no danger to the plant.

Details of the Accident

March 22, 1975, was a different matter. At 12:15 P.M., a section of polyurethane sealant was set on fire by a lighted candle held by an inspector. At first workers tried to put out the fire with portable fire extinguishers. They were successful in limiting the spread of the fire in the cable-spreading room (CSR). They did not realize, however, that the fire was continuing to spread along the insulation on cables hidden from their view. As cables lost their insulation, electrical wires came into contact with each other and shorted out. Within minutes, power to all parts of the plant began to fail. By 12:30, the fire had spread along cables into the Unit 1 reactor building.

Efforts to control the fire were slowed by a number of design problems and human errors. For example, at 1:10 P.M., workers attempted to call the nearest public fire department in Athens, Alabama, only to discover that an incorrect telephone number was listed on the emergency card. When the Athens firefighters did arrive twenty minutes later, they were not admitted to the plant for fifteen minutes. When they did gain access to the plant, managers did not allow them to use water to put out flames even though the fire captain specifically recommended that approach twice. When water finally was used on the fire at 7:00 P.M., nearly seven hours after it began, the fire was brought under control in forty-five minutes.

Fire was only one of the problems confronting plant managers, however. As electrical circuits failed, more and more automatic functions also failed. For example, indicator lights in the control room behaved erratically, glowing more brightly than normal or fading out completely. Operators could not be sure about the status of operating systems in either reactor building.

At 12:51 P.M., thirty-six minutes after the fires started, operators decided to shut down the reactor in Unit 1 manually. Nine minutes later, the reactor in Unit 2 was also taken out of operation. But these actions did not bring an end to the dangers facing the plant. Even when fission reactions are stopped in a reactor, the core remains very hot for days, and cooling water must still be circulated through it to keep the core from melting down.

Upon shutting the reactors down, workers were faced with the delicate task of cooling the reactors in both Units 1 and 2 manually. Plant managers eventually called in more than 150 off-duty workers to operate pumps, valves, and other systems by hand. At first, the water level in the Unit 1 reactor fell from its normal position of two hundred inches above the core to only forty-six inches. Eventually, however, enough water was pumped in to restore it to its two-hundred-inch level.

During most of the day on March 22, conditions in both reactors remained a matter of concern, as internal pressures continued to rise. Finally, at 6:30 P.M., the reactor in Unit 2 was brought under control and declared to be in a safe, shut-down condition. Reactor 1 remained a problem, however, as its internal pressure continued to climb throughout the evening and into the next morning. Finally, at 4:10 A.M., it also was brought under control and declared safe.

Impact

The significance of the Browns Ferry fire depends to some extent on whom one asks. The nuclear power industry has argued that the incident proves that nuclear power plant safeguards work. No lives were lost, no injuries sustained, no permanent damage done to the reactors, and no dangerous amounts of radiation released to the atmosphere. In fact, later studies showed that the release of radiation was about the same as that produced when a nuclear power plant is shut down for refueling.

One nuclear power journal observed that the event was "nothing like the near catastrophe so readily suggested by the opponents of nuclear energy." Yet a near catastrophe is just the way some observers described the fire. A story in *U.S. News & World Report*, for instance, claimed that "one strained pumping system was all that stood between control of the core and a possible meltdown."

Both construction and design errors were apparent. In some places, flame retardant was used on only 30 cm of cable adjacent to a seal, when manufacturer's specifications had suggested using it on 180 to 240 cm. Also, the wiring for both primary and backup systems were often laid in the same channel, guaranteeing that the loss of one system would mean the loss of all backup systems as well.

One thing on which nearly everyone could agree was that the flame of an electrician's candle revealed a great deal more about the Browns Ferry plant than initially seemed to be the case, as both construction and design errors became apparent. In some places, for example, flame retardant was used on only 30 centimeters of cable adjacent to a seal, when manufacturer's specifications had suggested using it on 1.8 to 2.4 meters (180 to 240 cm) of material. In some locations, the retardant apparently had not been used at all.

Investigations of the accident also revealed design errors. For example, it is a cardinal rule in nuclear power plant design not to have the wiring for backup systems placed too close to the wiring for the primary system. If a fire destroys the cables for a primary system, it should not also destroy cables for the secondary and tertiary systems. Yet at Browns Ferry, the wiring for both primary and backup systems were often laid in the same channel, guaranteeing that the loss of one system would mean the loss of all backup systems as well. After their study of the accident, fire experts produced a list of nearly two dozen major problems at the Browns Ferry facility. Included in their list were such items as the absence of water-type fire extinguishers, the lack of breathing apparatus in either the cable-spreading room or reactor buildings, the failure to provide an adequate in-house fire-fighting program, the failure of plant personnel to recognize a fire alarm, the improper operation of the carbon dioxide discharge system, the absence of emergency light systems, and an inadequate fire detection system.

Both the United States Congress and the Nuclear Regulatory Commission (NRC) conducted formal investigations of the Browns Ferry fire. The NRC report accused the Tennessee Valley Authority (TVA) of a number of violations of government rules at the plant. Among the most important of these was the failure to separate primary and backup safety systems from each other. The report also claimed that the carbon dioxide fire-fighting system in the CSR could not be operated manually because metal plates had been installed in front of the alarm stations. In addition, the NRC accused the TVA of creating emergency response procedures that were poorly understood by workers. Finally, the NRC criticized plant managers for not reporting and following up on earlier foam-related fires.

The TVA responded to the NRC report with denials and accusations of its own. It claimed that its wiring systems were sufficiently isolated from each other and, in any case, some of the designs for such systems had been approved by the NRC itself. Furthermore, operation of the carbon dioxide fire-fighting system was not obstructed by metal plates on the alarm stations, which had been installed only to prevent the alarms' accidental activation. The TVA also complained that NRC inspectors had refused to examine training records that had been made available to them and that would have validated the plant's emergency training procedures.

In the end, the NRC decided not to assess penalties of any kind against the TVA because of the accident. A congressional committee investigating the accident expressed some amazement at this decision, indicating that it raised "serious questions regarding NRC compliance with its own rules." In any case, both reactors were eventually repaired and, eighteen months later, restored to operation. In 1977, they were joined by their sister reactor, Unit 3, at the Browns Ferry complex.

A 1982 review of Browns Ferry by the Institute of Nuclear Power Operations (INPO) gave the plant a clean bill of health. The INPO report said Browns Ferry had "an excellent equipment monitoring and diagnostic testing program" and "an impressive industrial safety education and awareness program."

DAVID E. NEWTON

Where to Learn More

Kaplan, Gadi. "The Browns Ferry Incident." *IEEE Spectrum*, October 1976, pp. 55-61.

Miller, James Nathan. "The Burning Question of Browns Ferry." *Reader's Digest*, April 1976, pp. 103-8.

"Fire Investigation Report." *Nuclear News*, September 1975, pp. 46-50.

"TVA Replies to Charges." *Nuclear News*, October 1975, pp. 37-38.

"No Penalties, Says NRC." *Nuclear News*, November 1975, pp. 46-48.

"Browns Ferry, Sequoyah Reviewed Again." *Nuclear News*, February 1983, p. 55.

Pryor, Andrew J. "Browns Ferry Revisited." *Fire Journal*, May 1977, pp. 85ff.

Rippon, Simon. "Browns Ferry Fire." *Nuclear Engineering International*, May 1975, pp. 461-62.

"Fire at a Nuclear Plant: Official Report on a Near-Disaster." *U.S. News & World Report*, February 16, 1976, p. 52.

Three Mile Island accident

Middletown, Pennsylvania

(1979)

The worst accident in the history of American commercial nuclear power generation caused a partial meltdown of the reactor core.

Background

On March 28, 1979, a partial meltdown occurred at Unit 2 of the Three Mile Island nuclear power plant (TMI-2) in Middletown, near Harrisburg, Pennsylvania. A valve in a pipe carrying cooling water that was supposed to be closed accidentally remained stuck in the "open" position, allowing cooling water to flow out of the reactor. Heat built up within the reactor core, causing some fuel elements to melt and releasing radioactive gas and water to the surrounding environment. A number of operator mistakes contributed to the accident. No one was killed or injured in the accident, but expert opinion remains divided about possible future health problems that might result from the accident. The accident became a critical factor in the dramatic retreat from nuclear power in the United States that has occurred since 1979.

The two units that make up the Three Mile Island nuclear power plant sit on a small island in the middle of the Susquehanna River in south-central Pennsylvania. Unit 1 was put into operation in September 1974. Construction on Unit 2 began in 1968 and was completed ten years later. Unit 2 was formally declared open on December 30, 1978.

Both units at Three Mile Island were built by Babcock & Wilcox, a giant construction company that specializes in nuclear power plant components. The Three Mile Island complex is owned by a consortium of power companies consisting of Pennsylvania Electric Company and Jersey Central Power and Light Company, each of which holds twenty-five percent of its stock, and Metropolitan Edison, which holds the remaining fifty percent of its stock.

The Three Mile Island units are pressurized-water reactors (PWR). Heat is generated in the core of such reactors when neutrons bombard uranium atoms in fuel rods, causing fission (splitting the atoms and releasing energy) to occur. The TMI-2 plant had 36,816 fuel rods, each filled with hundreds of pellets of enriched uranium metal about the size and shape of a small checker.

Heat produced by the fission reaction is removed by cooling water that is pumped through the reactor core. The cooling water is then pumped out of the core to a heat exchanger, where it is used to boil water in a secondary system. Steam produced in the secondary system is used to drive turbines that turn electric generators. At their peak capacity, the Three Mile Island plants produced eight hundred and eighty megawatts of electricity, enough to meet the needs of 346,000 residents of surrounding Berks, Lebanon, and York Counties. Pressurized-water reactors of this design are among the most common of all reactors in use in the United States today.

Like all nuclear power plants, the Three Mile Island reactors contained a complex safety system with many backup components. The core itself is enclosed in a steel casing almost nine inches thick. The core and cooling system, in turn, are enclosed within a large containment dome, one hundred and ninety feet high and one hundred and forty feet in diameter. Walls of the dome are four feet thick and made of reinforced concrete. The purpose of the containment dome is to capture any gases, radiation, or other materials released during a leak or accident in the core or cooling system.

As carefully designed as TMI-2 was, it experienced a number of problems almost from the day it was licensed in February of 1978. During the test period, valves opened and closed when they weren't supposed to, they remained stuck in the wrong position, seals broke, recording instruments failed to work properly, and other "glitches" developed.

The situation did not improve even after the plant officially went on-line in December 1978. It had operated only two weeks when two safety valves failed, causing the plant to shut down for two more weeks. During February, additional problems with valves, seals, pumps, and instruments developed. The plant seemed jinxed by one mishap after another.

The existence of problems such as these was not entirely a surprise. More than a year earlier, a nuclear safety expert for the Tennessee Valley Authority, Carl Michelson, had discovered safety problems with Babcock & Wilson reactors. Michelson reported his findings to the Nuclear Regulatory Commission (NRC), whose job it is to monitor the safety of nuclear power plants in the United States. For whatever reason, the NRC did not act on Michelson's report and did not request any modifications in Babcock & Wilson reactors.

Details of the Failure

The accident at TMI-2 that began at 4:00 A.M. lasted for more than five hours. During that time, more than forty distinct events occurred, some caused by mechanical error and some caused by human error. The series of events began with a routine maintenance operation, the changing of water in a system of pipes. During the operation, air accidentally got into a pipe, cutting off the flow of cooling water to the reactor core.

Under normal circumstances, monitoring instruments would have detected this problem and automatically switched over to a backup cooling system.

The Three Mile Island complex, containing two reactors and four cooling towers. Unit 2 was the site of a major nuclear accident in 1979.

However, the pumps used in the backup system were also undergoing routine maintenance and so could not deliver water to the pipes. To make the situation even more complicated and more dangerous, tags attached to the pumps indicating that they were undergoing maintenance hung down in such a way that they hid the warning lights. Even though the system correctly indicated a problem existed, operators couldn't see the indicator lights.

Without its cooling water, the reactor core began to heat up rapidly. Its temperature eventually reached more than five thousand degrees Fahrenheit, and fuel rods began to melt. As they melted, they burned through the steel casing of the core and reached the floor of the containment vessel itself. Had the molten fuel melted through that barrier, a total meltdown could have occurred, perhaps resulting in an explosion that would have breached the containment barrier and released huge levels of radioactive material into the air.

During the meltdown, other components of the plant continued to fail. Water in the cooling system started to boil, and an emergency relief valve opened automatically to allow it to escape into the containment dome. But the valve then failed to close, and even more water was lost from the cooling system. All the while, operators were either reading gauges that were not working properly or reading working gauges but then taking the wrong actions.

Under the circumstances the meltdown of the fuel rods was to be expected. What was unexpected was the development of a huge bubble of steam and hydrogen gas that formed inside the containment dome. The steam came, of course, from cooling water boiled out of the core. But where did the hydrogen come from?

Scientists had not realized before the TMI-2 accident that cooling water would not only boil, but would also decompose to oxygen and hydrogen by radiation and heat released in the meltdown. Hydrogen is a special source of concern because it is explosive. It would have been possible—although it did not happen in this case—for the hydrogen to explode, releasing radioactive materials over hundreds of square miles surrounding the plant.

As it was, radioactive material did escape from the reactor in other ways. Some contaminated coolant water was automatically pumped out of the containment dome into a nearby holding building. From there, it was eventually dumped into the Susquehanna River. Radioactive gases from the core also escaped though vents in the containment dome. The level of radiation was so high in these gases that it damaged the instruments designed to monitor them.

The immediate reaction to the TMI-2 incident by scientists and politicians appeared to be uncertainty and confusion. Two hours after the backup pumps first failed, the plant's chief engineer was told that a "minor snafu" had occurred at the plant. An hour and a half later, the situation had escalated into a "general emergency" for Unit 2. And by 8:15 A.M., plant managers had set up a direct telephone line to NRC offices in Washington, D.C. In their early press conferences, plant officials said that less than one percent of the core's fuel rods were damaged and that the internal temperature of the reactor was only two thousand degrees Fahrenheit. Later they had to revise those figures to at least fifty percent loss of fuel rods and a temperature of five thousand degrees Fahrenheit.

Everyone involved seemed to be concerned about causing unnecessary panic. Pennsylvania Governor Richard Thornburgh first considered calling for a general evacuation of the area, but then decided more modest precautions were sufficient. He advised people to stay in their homes with their windows closed.

Radiation measurements taken the day of the accident showed levels ranging from thirty millirems immediately above the containment dome to five millirems three miles downwind from the plant. The problem is that experts still disagree about the health effects of low-level radiation. A person receives about twenty millirems of radiation from a typical chest x-ray, so the levels measured at TMI-2 seem reasonably safe. But some authorities believe that each additional dose of radiation, no matter how small, increases a person's chance of contracting cancer later in life. Exactly how much of the threat the TMI-2 accident was to long-term human health is still, therefore, a matter of some conjecture.

Cleanup efforts on the damaged reactor were begun almost immediately. The most important source of concern was the steam-hydrogen bubble in the containment dome. Engineers vented part of the gas into the atmosphere through filters that removed its radioactivity. They also were able to transfer some of the steam-hydrogen mixture to an outside building where the hydrogen was combined chemically with oxygen to make water.

Cleanup crews were able to remove twenty thousand gallons of radioactive water that had flowed into the auxiliary building by transferring it to holding tanks. Finally, a variety of methods were used to cool down the hot reactor core itself and prevent any further meltdown.

The long-term cleanup of TMI-2 was a slow, dangerous, and complex process that took eleven years. Eventually one hundred and fifty metric tons of damaged fuel rods and other reactor components were shipped to the Idaho National Engineering Laboratory for storage and analysis. The total cost of the cleanup process was $973 million.

There are no plans to repair TMI-2. Safety experts will continue to observe and monitor the plant until early into the next century. At that point, it will be decommissioned along with its companion reactor, TMI-1. It will have been a very expensive structure whose total lifetime cost was close to $2 billion without ever producing any significant amount of electricity.

Impact

The TMI-2 accident has had both short-term, local, and long-term national and international effects. In the immediate area surrounding the Three Mile Island complex, residents became—and have remained—concerned about the possible effects of the accident on their health. Given the nature of short-term radiation, it is possible that increased rates of cancer or birth defects may not show up for years.

A reflection of these concerns is the more than twenty-two hundred lawsuits that have been filed by area residents because of the accident. To date, two hundred and eighty claims have been settled at a total cost of about $14 million. Operations at TMI-1 were significantly affected by the events at the neighboring reactor. More than one hundred modifications were made in the older plant at a cost of $95 million. To complete the necessary renovations, the plant was shut down for six years. During that time, an improved and expanded training program for TMI-1 personnel was developed and implemented. Much of the training at the plant is now done in an $18 million full-scale replica of the TMI-1 control room.

The TMI-2 accident became the subject of a number of investigations and studies by both governmental bodies and industry groups. Probably the two

The top portion of the TMI-1 reactor, which is still in operation. The similarly designed TMI-2 reactor was the site of the 1979 accident.

most important of these were an NRC study, whose results became generally known as the Rogovin Report, and a study commissioned by President Jimmy Carter, eventually known as the Kemeny Report. As a result of these studies, the NRC issued new regulations in 1980 requiring modifications of existing plans and new safety features to be included in future plants.

The nuclear power industry also became more interested in safety issues after the TMI-2 accident. Companies established the Institute of Nuclear Power Operations (INPO) whose purpose it is to review and evaluate nuclear power plant performance and to recommend ways of improving performance.

Probably the greatest single result of the TMI-2 accident was its impact on the future of nuclear power in the United States. Orders for nuclear power plants had already begun to decline in the year preceding the TMI-2 event. But the accident seemed to mark a real turning point in the public's attitude about nuclear power. Activists around the nation finally had an almost-worst-case scenario for what might happen in a nuclear accident. They began to speak out against plants under development and to work to defeat plans for new plants.

Those efforts have been largely successful. Since 1979, no new nuclear plants have been ordered and fifty-nine planned reactors have been canceled. Seventeen new plants have been opened, but all were in final stages of planning or construction at the time of the accident.

DAVID E. NEWTON

Where to Learn More

Adams, John. A. "TMI plus 5: Part I—A Slow Comeback." *IEEE Spectrum*, April 1984, pp. 27-33.

Burnham, David. "Nuclear Experts Reportedly Knew of Flaw in Some Reactors in 1977." *New York Times*, May 25, 1979, p. 18.

Eisenhunt, Darrell. G. "TMI plus 5: Part II—NRC as Referee." *IEEE Spectrum*, April 1984, 33-9.

Fischetti, Mark. A. "TMI plus 5: Part III—Band-aids and Better." *IEEE Spectrum*, April 1984, pp. 39-43.

Ford, Daniel. "Three Mile Island, 2—the Paper Trail." *New Yorker*, April 13, 1981, 46-7.

Franklin, Ben A. "Files Show Many Prior Problems at Three Mile Island." *New York Times*, April 19, 1979, p. 18.

Gray, Mike, and Ira Rosen. *The Warning: Accident at Three Mile Island*. Norton, 1982,

Levine, Joe, Mary Hickey and Denise Laffan. "Cleansing the Atom." *Time*, March 1989, pp. 18-24.

Matthews, Tom, et al. "Nuclear Accident: Three Mile Island Plant, Pa." *Newsweek*, April 9, 1979, pp. 24-30+.

Moss, Thomas H., and David L. Sills, eds. *The Three Mile Island Nuclear Accident: Lessons and Implications*. New York Academy of Sciences, 1981.

Rogovin, Mitchell and George T. Frampton, Jr. *Three Mile Island: A Report to the Commissioners and the Public,* vols. 1 and 2. Nuclear Regulatory Commission Special Inquiry Group, January 1980.

U.S. President's Commission on the Accident at Three Mile Island. *Report of the President's Commission on the Accident at Three Mile Island: The Need for Change; The Legacy of Three Mile Island,* John G. Kemeny, chairman. Government Printing Office, 1979.

Radioactive waste spill

Tsuruga, Japan

(1981)

Due to human error, nearly four thousand gallons of radioactive wastes escaped from a nuclear power plant in Japan.

Background

On March 8, 1981, a worker at the Tsuruga Nuclear Power Plant in Fukui prefecture in Japan was cleaning out a pipe in the building where radioactive wastes were treated. He neglected to close a valve in the pipe, allowing the holding tank to overflow. Radioactive wastes covered the floor of the waste treatment plant, leaked into an adjacent building, seeped into the ground, and eventually worked their way into the Sea of Japan. Officials at the plant attempted to hide the accident from public notice, but it was discovered six weeks later during a routine study of seaweed in the area.

Japan has amazed the world by its economic development since the end of World War II. The one Achilles' heel of the "Japanese miracle," however, has been that nation's lack of natural resources, particularly energy. The Japanese do not have sufficient reserves of coal, oil, or natural gas to maintain their recent pace of economic development. As recently as 1978, the nation imported 90 percent of the energy resources it used, of which 75 percent was being petroleum.

The OPEC (Organization of Petroleum Exporting Countries) oil embargo of 1973 made it abundantly clear to the Japanese how vulnerable they were to the rest of the world for their energy supplies. The government decided to place a renewed emphasis on the development of nuclear power to establish energy independence for the nation.

Nuclear power in Japan, however, has had a troubled history. The nation that experienced the devastation of the world's first two atomic bombs has always had a fundamental concern about the use of nuclear energy, even for peaceful uses. When the Japanese government finally decided to permit and to sponsor nuclear research in 1955, it adopted a law that contained many safeguards for citizens and guaranteed that nuclear research would never by used for military purposes.

Progress on the development of commercial nuclear power was rapid after the 1955 Basic Law on Atomic Energy. The first research reactor was opened in 1957 at Tokaimura in Ibaraki Prefecture, and the first commercial plant came on-line eight years later.

The government's original goal was to open more than thirty nuclear power plants by the year 1990, reducing the nation's dependence on foreign oil from 75 percent to 50 percent. That objective was overly optimistic, but progress was rapid nonetheless. By 1986, the Japanese had opened thirty-one commercial reactors that were producing almost twenty-four megawatts of electricity. Atomic energy had become the largest single source of power in the nation, accounting for 26 percent of the nation's energy supplies.

The nuclear power plant at Tsuruga, the nation's second oldest, was opened in 1970. It was constructed by General Electric (GE) based on designs provided by Ebasco Services of New York State. After the March 8 accident, GE pointed out that the accident took place in a part of the treatment plant built by the Japanese, not by the American company. The plant was operated by Japan Atomic Power Company (JAPC), the company in charge of all Japanese reactors. JAPC's activities were, in turn, monitored by two government agencies, the Agency of Natural Resources and Energy (ANRE) and the Ministry of International Trade and Technology.

Details of the Accident

One of the routine operations that takes place in a nuclear waste treatment plant is the flushing out of pipes from time to time. On March 8, 1981, a worker at the Tsuruga plant was performing this operation. He neglected to notice, however, that a valve in one of the intake pipes had failed to close. As a result, radioactive wastes continued flowing into a holding tank until it began to overflow.

A later investigation by ANRE found that the indicator light for the valve was not working properly. The light indicated that the valve was closed when, in fact, it was still open. The ANRE report concluded, however, that the operator "should have known that [the] indicator light was malfunctioning."

What happened next was not revealed to the public until April 18. Amazingly, plant managers themselves did not know about the spill until two days after it had occurred. The story that eventually came out was somewhat confused, and it changed with each new report.

Apparently, radioactive water overflowing the holding tank covered the floor of the waste treatment building, which was connected to the reactor building by pipes, and then began to run into the laundry room next door. From there, it drained through cracks in the laundry room floor into the ground beneath. Eventually it diffused into a sewer line and worked its way into the nearby Bay of Urazoko, an inlet of the Sea of Japan.

When plant officials learned of the accident, they ordered workers to clean up the radioactive wastes by hand, using mops and buckets. The job was an enormous one, taking more than two weeks and requiring the hiring of about fifty additional workers from outside the plant. According to later estimates, anywhere from fifteen to forty tons of radioactive wastes had escaped from the holding tank, the worst spill to occur in Japan. Authorities estimated that no more than one cubic meter of radioactive wastes ever reached the bay, although such estimates are highly problematic.

The decision was made not to announce the leak to the outside world. It was discovered accidentally when local health officials were making a routine check of seaweed in Urazoko Bay. They found levels of radiation at least ten times greater than normal. The levels of some isotopes were even higher. The reading for cobalt 60, for example, was five thousand times greater than any previously recorded in the area. When workers traced the source of radiation, they found that it came from an outlet pipe for the sewer running underneath the nuclear power plant.

Confronted with this evidence, plant officials acknowledged on April 18 that the accident had taken place the previous month. They claimed that the accident posed no danger either to workers in the plant or to nearby residents. Those involved in the clean up, officials claimed, had been exposed to no more than 35 millirems of radiation, less than the level permitted under health regulations in both Japan and the United States. However, since plant executives admitted that the workers' radiation-exposure badges were altered, there was no firm evidence to support this claim. A rem is a measuring unit that stands for Roentgen Equivalent Man and measures the biological produced in the human body by radiation. The average annual radiation dose received by a person in the U. S. is about 180 millirems. Most of this dose is due to natural radiation and medical and dental x-rays.

Impact

The immediate result of the Tsuruga accident was to open up the plant's past history of safety violations. In the months following the March 8 spill, JAPC officials gradually revealed details of more than thirty accidents that had occurred in the plant's recent history. The most recent before the March 8 incident had taken place during repair of a cracked pipe between January 24 and 28, 1981. During that event, more than forty workers may have been exposed to dangerous levels of radiation. On January 10, 1975, a spill similar to the March 8 accident had released thirteen tons of radioactive material to the surrounding environment. On that occasion, thirty-seven workers had been exposed to radiation during the cleanup process.

Officials of JAPC continually tried to reassure the general public that the Tsuruga accident was a minor event of no great significance. It was, they said, "nowhere near as serious as American's Three Mile Island" and they promised that there was never any "real damage" as a result of the spill. However, local residents were not easily reassured. "How do we know what is the effect of 15 tons in the bay?" asked one townsperson. The plant itself

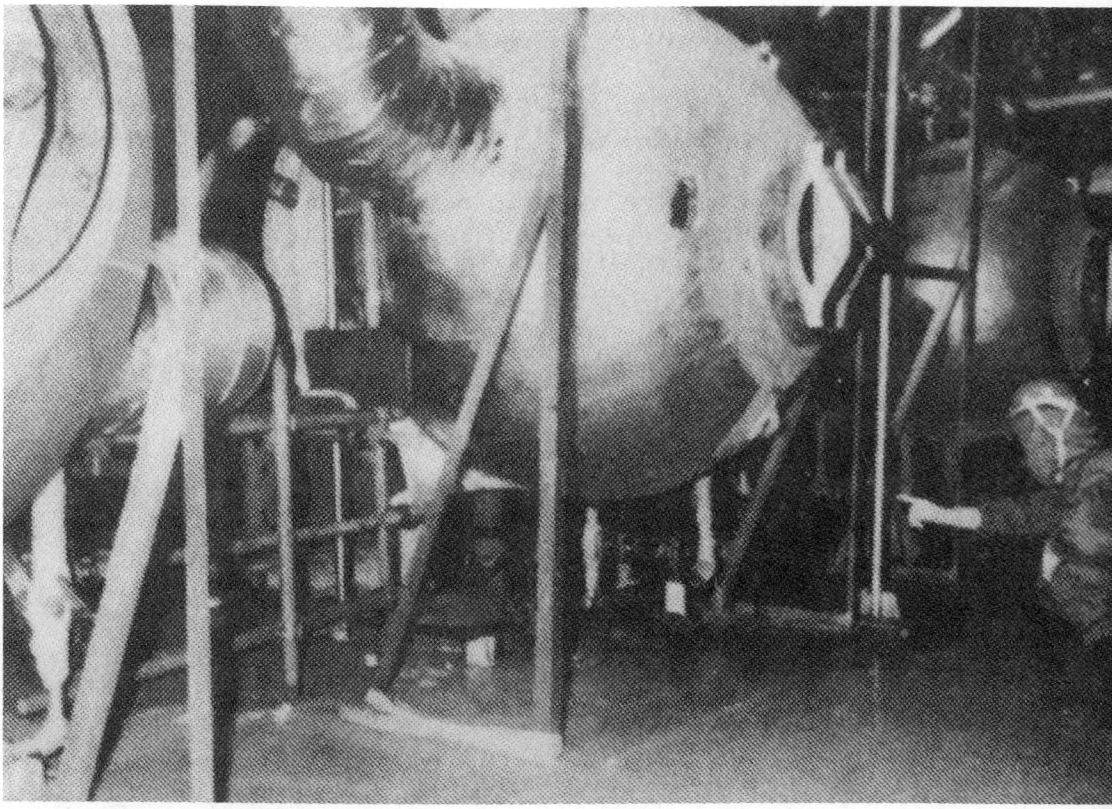

The valve that leaked radioactive waste at the nuclear power plant in Tsuruga, Japan.

was closed down for repairs. Those repairs ultimately cost about $10 million and kept the plant out of operation for six months.

National reaction to the company's belated April 18 announcement was also swift. The nation's second largest political party, the Socialists, called for an end to Japan's nuclear power program and a shutdown of all existing plants. And a grass-roots anti-nuclear movement soon enrolled more than 45,000 members throughout the nation.

In response to these actions, JAPC's chairman, Tomiichiro Shirasawa, and president, Shunichi Suzuki, announced their resignations on May 13. They accepted full responsibility for the accidents and coverups at Tsuruga and called for a return to confidence in the role of nuclear power in Japan's future.

The long-term effects of the Tsuruga accident have been relatively modest. In fact, the March 8 incident was minor, compared with the disasters at Three Mile Island and Chernobyl. And the role of nuclear power in Japan's energy equation has remained fairly constant. Japan's official policy is still one of weaning itself from dependence on foreign oil. As a result, the development

The immediate result of the Tsuruga accident was to open up the plant's past history of safety violations. In the months following the March 8 spill, JAPC officials gradually revealed details of more than thirty accidents that had occurred in the plant's recent history.

of nuclear power has continued at about the pace the nation set for itself two decades ago.

DAVID E. NEWTON

Where to Learn More

Firth, Suzanne. "Chilling Reminder of Things Past." *MacLean's*, May 4, 1981, pp. 32-33.

"Nuclear Energy Policy." *Kodansha Encyclopedia of Japan*. Kodansha International, 1983.

"Nuclear Power Plants." *Kodansha Encyclopedia of Japan* supplement. Kodansha International, 1986.

"Japan's Three-Mile Island." *Newsweek*, May 4, 1981, p. 42.

"Japanese Concede Errors in Nuclear Plant Mishap." *New York Times*, April 4, 1981, p. A6.

"Nuclear Contamination Found in Japanese Soil." *New York Times*, April 19, 1981, p. 7.

"Earlier Mishap Is Revealed at Japanese Nuclear Plant." *New York Times*, April 21, 1981, p. 5.

"Japan Says Nuclear Mishap Exposed 56 to Radiation." *New York Times*, April 22, 1981, p. 8.

"45 Workers Are Reported Exposed to Nuclear Radiation." *New York Times*, April 26, 1981, p. 6.

"Nuclear Executives in Japan Resign over Recent Mishaps." *New York Times*, May 14, 1981, p. 11.

"Repercussions Continue at Japan's Three Mile Island and at the Original." *Power*, July 1981, pp. 124-26.

Smith, R. Jeffrey. "Japanese Agitated by Nuclear Plant Spill." *Science*, June 5, 1981, p. 1124.

Stokes, Henry Scott. "For the Japanese, Sudden Misgivings about Nuclear Power." *New York Times*, May 16, 1981, p. 3.

Ginna power plant radioactive release

Failure of a pipe in a nuclear power plant caused shutdown of the reactor and release of radioactive gas to the atmosphere.

Ontario, New York

(1982)

Background

On January 25, 1982, a small pipe in the steam generating system of the Robert E. Ginna nuclear power plant in Ontario, New York, ruptured. Hot water from the primary cooling system leaked into the secondary cooling system. In dealing with the problem, operators were forced to release cooling water contaminated with radioactivity into an emergency relief tank, which overflowed and dumped about seventeen hundred gallons of the water onto the floor of the reactor building. At the same time, a small amount of radioactive steam was released through emergency safety valves in the top of the containment dome.

The Robert E. Ginna nuclear reactor was opened in 1970. It was built at Ontario, New York, about fifteen miles east of Rochester for the Rochester Gas & Electric Company (RG&E) by Westinghouse Electric Corporation, one of three United States companies making pressurized water reactors (PWR). The Ginna plant was originally known as Brockwood. At full power, it is able to generate 470 megawatts of electricity. In a PWR, heat produced by nuclear fission reactions in uranium fuel rods is absorbed and carried away by cooling water maintained under high pressure. The high pressure prevents the cooling water from boiling although it reaches temperatures of more than three hundred degrees Celsius.

The heated cooling water from the primary system is then pumped into a secondary steam generating system. The secondary system consists of thousands of long, narrow tubes through which hot water from the primary system passes. The tubes are surrounded by cold water that is heated by the primary water until it boils and changes to steam. The steam is then used to turn turbines that operate electrical generators. The Ginna plant has two steam generating systems, known as A and B.

One of the critical design problems in a PWR system is the construction of the tubes in the secondary system. These tubes must be thin enough to allow the efficient transfer of heat from the hot primary water to the cold secondary water. But they must also be strong enough to withstand mechanical vibration, turbulence, and corrosion by the surrounding water. The standard design calls for tubes about a millimeter thick made of a nickel alloy called Inconel.

Damage to secondary tubes has long been recognized as a problem in PWRs. Over time, the tubes have a tendency to become dented, pitted, or corroded. There are instances on record in which tubes actually break up, permitting primary water to escape into the secondary system.

Three different techniques are used for dealing with the problem of damaged tubes. When the tubes are easily accessible, they can simply be removed from the secondary system. Another approach is plugging, in which a tube is sealed up and essentially taken out of service. A third approach is sleeving, in which a tube of a smaller diameter than the damaged tube is inserted inside it. During a routine test of the secondary system at Ginna in May 1981, 4 tubes out of 3,260 in the B steam generator were plugged, 3 were pulled, and 15 sleeved. At the time, a total of 113 tubes had been plugged to date, 21 sleeved, and 7 pulled.

Details of the Accident

The first indication of a problem at the Ginna plant came at 9:25 A.M. on January 25, 1982. Five separate warnings went off at the same time. They indicated that pressure in the primary cooling system was dropping rapidly, by more than one thousand pounds per square inch in less than five minutes. Three minutes later, the reactor's automatic safety system was triggered. Control rods dropped into the core and shut down the nuclear fission chain reactions in the fuel rods. At the same time, the emergency cooling system turned on and cold water began flowing into the core.

At 9:40 A.M., a warning light indicated the cause of the problem. One of the 3,260 tubes in the B steam generating system had ruptured. Hot water from the primary cooling system had begun to flow into the cold water of the secondary system. As the primary water escaped into the secondary system, some of it changed into steam, which was vented automatically into the atmosphere. The automatic venting lasted only about two minutes and released a minimal amount of gas to the atmosphere. Operators declared an "alert," the lowest level of safety warning, at 9:40 A.M.

Operators were now faced with some immediate difficult decisions. They had to reduce pressure in the primary system to prevent more hot water from escaping into the secondary system. At the same time, they had to maintain pressure at a level high enough in the primary system to keep water from boiling and forming a steam cap above the core.

The Ginna nuclear reactor, which was the site of a 1982 accident that shut down the reactor and caused the release of radioactive material. Despite the accident, the reactor was back in operation within six months.

Their approach was to slowly equalize pressures in the primary and secondary systems by opening and closing a pressure relief valve in the primary system by hand. They did so successfully at 10:07 and 10:08 A.M., but during a third attempt at 10:09, the valve failed to close. Water flowed out of the primary system and pressure again started to drop rapidly. At 10:44 A.M., operators declared the next higher level of safety warning, a "site emergency."

The failure of the valve in the pressurizer to close created two new emergencies. First, the pressurizer tank started to overflow, dumping cooling water into a relief tank beneath the reactor. Eventually the tank became full and burst open (as it was designed to do), releasing radioactive water onto the reactor building floor and, from there, into the bay beneath the building. Workers quickly recognized this problem, however, and shut off a master valve that stopped the loss of cooling water from the pressurizer. A total of 1,690 gallons of contaminated water was eventually lost from the reactor.

At the same time, some of the pressurized water in the reactor began to boil, producing a bubble of steam above the core. The danger was that, as the steam bubble grew, it might force cooling water down until the fuel rods were exposed. If that were to happen, a meltdown like the one at Three Mile Island might occur.

Although the valve that stuck open at Ginna was the same kind of valve that caused the Three Mile Island problem, the Ginna failure was much less serious. One reason was that equipment had been improved as a result of lessons learned at Three Mile Island. More important, however, was the behavior of plant operators at Ginna. They knew what to do as the emergency developed, and they followed the correct steps to keep the reactor under control.

Operators attempted to deal with this threat in two ways. First, they gradually added more water to the primary cooling system, attempting to cool and compress the steam bubble. Second, they flooded the reactor core itself with cold water, trying to cool its metal structures as quickly as possible. The procedure worked, and by 11:15 A.M. the steam bubble had disappeared.

As workers were dealing with the problems in steam generating system B, they also began to shut down system A. At 11:29 A.M., they started venting steam from system A and pumping cooling water into that system. By 12:30 P.M., the temperature was dropping by two degrees per hour. Two days later, at 4:30 P.M. on Wednesday, January 27, the reactor was declared to be in a cold shutdown condition. The emergency was over.

Follow-up studies of the Ginna incident focused on two major issues. First, what was it that caused the tubes in the B steam generating system to rupture? Second, what damage, if any, had been done to the reactor structure by flooding it with cold water during emergency procedures?

The first question was answered during the Nuclear Regulatory Commission's (NRC) study of the accident. Investigators found a number of pieces of scrap metal in both steam generating systems. These included some long pieces and fragments of tubing from the steam generating systems, a piece of wire, and some strips of steel. Although it was impossible to know for sure, the NRC suggested that some of the debris may have resulted from sloppy workmanship. During routine repairs, it said, workers may have allowed bits and pieces of scrap metal—and, perhaps, even parts of their tools—to fall into the steam generating systems. When the reactor was operating, then, these materials may have been tossed around by cooling water, colliding with tubes and eventually causing one of them to break open.

The question of thermal shock to the reactor vessel required longer study. It is not possible in most cases to determine visually if metal has been damaged as a result of exposure to rapid temperature changes. By mid-May, however, the NRC had conducted detailed tests and determined that the reactor vessel had not been damaged. It gave RG&E permission to restart the reactor, and on May 27, the plant was once again restored to full power.

Impact

The problem at Ginna was inevitably compared to the far more serious incident at Three Mile Island of 1979. News reports recalled the earlier disaster and described the Ginna event as "one more setback for the nuclear power industry." And, in some ways, the two accidents were similar. For example, the valve that stuck open at Ginna was the same kind of valve that had caused the Three Mile Island problem.

Yet, in many ways, the two incidents were very different. As a spokesman for RG&E said, "This accident [at Ginna] didn't come within a country mile of Three Mile Island." One reason for the difference was that equipment had been modified and improved as a result of lessons learned at Three Mile Island. More important, however, was the behavior of plant operators at Ginna. They knew what to do as the emergency developed, and they followed the correct steps to keep the reactor under control. As a result, the accident was brought under control fairly quickly, and the plant was back in operation less than six months later. In comparison, the accident at Three Mile Island so badly damaged the facility that it will never again reopen.

As with all accidents at nuclear power plants, a great deal was learned about the operation and handling of reactors. A study by the Institute of Nuclear Power Operations (INPO), for example, praised Ginna's operators for their performance during the emergency. But they noted a number of improvements that could still be made in plant manuals and training. They found that actions other than those taken by operators would have resolved the crisis even more quickly.

Perhaps the most serious long-term effect of the Ginna incident has been the increased concern it raised over tubes in the steam generating system of PWRs like the one that failed. Every PWR contains thousands of such tubes, and engineers have long realized that they may be the weakest point in the whole plant. Procedures to deal with long-term corrosion and damage, described above, have long been known and used. But the lesson of Ginna is that tubes can also fail quickly as the result of violent impact or other still-unknown factors. As one report on the Ginna accident observed, tube failures had become "an expensive and complicated problem for utilities to solve."

DAVID E. NEWTON

Where to Learn More

Blake, E. Michael. "Ginna—A Special Case." *Nuclear News*, December 1982, pp. 52-53.

"Pipe Failure Closes New York Nuclear Plant." *Chemical & Engineering News*, February 1, 1982, p. 5.

Marbach, William D. "Again a Nuclear Mishap." *Newsweek*, February 8, 1982, pp. 63-65.

Marshall, Eliot. "Corrosion May Not Be Prime Culprit at Ginna." *Science*, March 5, 1982, p. 1215.

———. "NRC Reports on Ginna Nuclear Accident." *Science*, April 30, 1982, pp. 498-99.

———. "Reactor Mishap Raises Broad Questions." *Science*, February 12, 1982, pp. 877-78.

"Ginna Tube Burst Investigated." *Nuclear Engineering International*, March 1982, pp. 10-11.

"NRC Holds Off Restart in Wake of Tube Break." *Nuclear News*, March 1982, pp. 45-47.

"The NRC Task Force on Ginna Praised the Operators . . . ." *Nuclear News,* May 1982, p. 21.

"Unit Restarted: Generators To Be Inspected in September." *Nuclear News,* July 1982, p. 43.

"Hindsight on Ginna." *Nuclear News,* February 1983, p. 50.

Peterson, Ivar. "Ginna Atomic Plant Scrapped by Steel?" *Science News,* April 24, 1982, p. 277.

———. "The Ginna Problem: Isolated or Generic?" *Science News,* January 30, 1982, p. 68.

———. "Nuclear Shutdowns: Tubular Woes." *Science News,* February 13, 1982, p. 105.

Chernobyl accident

Ukraine

(1986)

The world's worst nuclear power plant disaster occurred when human errors led to a dangerous heat and steam buildup, triggering two enormous explosions.

Background

On April 26, 1986, two mammoth explosions blew apart Unit 4 of the Chernobyl nuclear power plant in the Ukrainian republic of the former USSR. The plant is located about seventy miles north of Kiev, the capital of the Ukraine. At least thirty-one workers and emergency personnel were killed immediately or died soon after the accident as a result of radiation sickness. Some two hundred thousand residents of the area were evacuated, and clouds of radioactive material were carried over most of northern Europe. Total fallout from the accident eventually reached a level ten times that of the atomic bomb dropped on Hiroshima, Japan, during World War II.

The Chernobyl complex consists of four reactors constructed between 1977 and 1983. By 1986, the four units were operating nearly at capacity and generating four million kilowatts of electricity. All four reactors at Chernobyl are graphite-moderated, water-cooled boiling water reactors known as RBMKs. At the time of the accident, twenty-one reactors of this design were in operation in the Soviet Union. Energy supplied by nuclear power plants accounted for about fifteen percent of all the electricity generated in the nation.

The RBMK design is uncommon, and plants of this kind are found almost exclusively in the countries that make up the former Soviet Union. They were built because they have the ability to perform two quite different functions at the same time: generate electricity and produce plutonium. The decision to build these dual-purpose plants goes back to the 1950s. The Soviet government committed itself to a weapons-development program that would keep it on a par with the United States nuclear arsenal. In order to accomplish that objective, weapons developers needed a constant and dependable supply of plutonium, the raw material needed for atomic and hydrogen bombs. The RBMK design was selected, therefore, because it could meet both domestic needs for electricity and military demands for weapons-quality plutonium.

The core of the RBMK reactor consists of a huge pile of graphite blocks weighing about two thousand tons. The purpose of the graphite is to slow down neutrons produced by fission in the fuel rods, a function performed by coolant water in most other reactors. Imbedded in the block are about seventeen hundred fuel rods, each containing hundreds of disc-shaped pellets of uranium. The reactor also contains control rods, which dampen the nuclear reaction when lowered into the block.

The RBMK is cooled by water that is pumped through the graphite block, from bottom to top. The cooling water travels through channels in the block that also hold the fuel rods. As the water passes over the fuel rods, it picks up heat and begins to boil. Steam formed by the boiling water collects at the top of the reactor, where it is removed and delivered to the steam turbine that generates electricity.

The RBMK reactor contains a number of design characteristics that make it inherently risky to operate as a commercial power source. Western scientists had been warning Moscow for a decade before the Chernobyl accident about these dangers. But the Soviets continued to have confidence in the RBMK. This confidence was driven not only by the practical reality of military demands, but also by one important advantage of the RBMKs: They can be refueled without the extended downtime needed to refuel Western reactors. Ironically, it was this very advantage that led to the disaster of April 26, 1986.

One problem with the RBMK design is that it makes no provision for a containment shell that is part of all reactors used in the United States. A containment shell is a last-ditch feature whose purpose it is to retain gases and radioactive materials released during an accident in the reactor core. The Three Mile Island accident in 1979 would have been immeasurably more serious if the products of the core meltdown had escaped to the atmosphere.

Another problem with the RBMK design is that a loss of cooling water increases the rate of fission (or nuclear splitting)—and hence heat production—in the core. The process is just the reverse of that in water-moderated reactors (such as those used in the United States), where loss of cooling water causes a decrease of power production in the core. This design flaw creates a peculiar situation that the reactor is most likely to go out of control when it is operating at lowest power.

Details of the Accident

At 1:24 A.M. on Saturday, April 26, Unit 4 of the Chernobyl Nuclear Power Plant was rocked by two enormous explosions. The roof was blown off the plant and radioactive gasses and materials were sent more than eleven hundred meters (thirty-six hundred feet) into the atmosphere. Two workers were killed instantly and another dozen received levels of radiation that would cause their death within the next two weeks.

As with most nuclear accidents on Soviet soil, government officials made no public announcement about the event. Not until monitoring instruments in

Unit 4 of the Chernobyl nuclear complex, which was the site of world's worst nuclear accident to date.

Sweden detected a dramatic increase in wind-borne radiation did the Soviets finally acknowledge the disaster. The news was reported on April 28 in a short five-sentence report by Tass, the official Soviet news agency.

But the political situation in the Soviet Union in 1986 was not what it had been a decade earlier. On May 14, First Secretary Mikhail Gorbachev, acting within his new policy of *glasnost*, took to national television and provided the world with a detailed description of all that was known at the time about the explosions. Still, it would be many months before all the details of the accident were unraveled.

The key event leading to the disaster was apparently a decision by the plant crew to carry out an unauthorized experiment. Operators wanted to know what would happen if there was a power outage and steam stopped flowing to the turbines. Would the kinetic energy of the spinning turbine blades be sufficient to maintain the cooling pumps until the emergency diesel generators turned on, they wondered. The way to find out, the crew reasoned, was to carry out a controlled test of this situation. What followed the crew's decision to do this experiment was a series of six major mistakes by workers. Any one error by itself would not have been fatal, but the combination of all six was to prove disastrous.

Perhaps the most important of these six mistakes was the crew's decision to disable the reactor's emergency coolant system. At the outset of the test, the reactor began to lose power. Because the test could only be continued if the reactor remained in operation, the crew disabled the coolant system. (As mentioned earlier, the reactor's design meant that the loss of any coolant water would increase its fission rate—its power level.) However, the reactor continued to lose power, so the crew removed all the control rods from the core. This move dramatically increased the reaction rate of the reactor. Almost instantly, the fission rate was cause for concern.

As power in the core began to increase, the crew attempted to reinsert the control rods manually. But the channels into which the rods were supposed to fit had deformed because of heat in the core. The rods did not drop properly, and power release in the reactor went out of control.

As steam vented from the reactors, water levels dropped dramatically in the core. Loss of water, in turn, increased power output from fission reactions. In less than a second, power output from the core increased a hundredfold. As temperatures increased to more than five thousand degrees Celsius, parts of the core melted. Molten metal reacted with the remaining cooling water, producing hydrogen gas and even more steam, until finally the top of the reactor was blown off—the first of the two explosions.

Details of the second explosion are less clear. Some authorities believe that it was largely a chemical and physical phenomenon, like the first explosion. It would take no more than a few seconds for the hydrogen gas initially produced to be ignited by heat released during the meltdown.

Other experts believe, however, that the second explosion may have been a pure nuclear reaction. Unlike most reactors in use today, an RBMK can at least theoretically explode like a nuclear (atomic) bomb. Some scientists believe that parts of the molten core may have achieved critical mass during meltdown. If that had occurred, then a true bomb-like explosion could have occurred, accounting for the second explosion.

Impact

The Chernobyl accident has had both short-term and long-term effects on the local area, on the world as a whole, and on the further development of the nuclear power industry. In addition to the thirty-one people who died immediately or within two weeks of the accident, another 299 were injured. About 135,000 residents were evacuated from the area within eighteen miles of the damaged plant, and later another 200,000 from other areas. But these numbers do not begin to reflect the magnitude of the damage to human health over the next generation and beyond.

Plants and animals in the immediate area of Chernobyl and downwind of the plant were heavily contaminated by fallout. Crops could not be harvested and most farm animals were destroyed to prevent their use as food. Almost a decade after the accident, levels of radiation are still so high in some areas that no native food can be grown or consumed. People survive on food that is shipped in from safe areas.

The chief contaminants remaining in the soil are cesium 137 and strontium 90. These two radioactive isotopes have half lives of thirty and twenty-eight years, respectively. That means that they will constitute a hazard for many more decades if they are not removed. Decontamination and removal of contaminated topsoil has not been progressing very rapidly, however. The result is that many residents of the area still face a constant and serious health risk from radioactive isotopes in the environment.

The fallout that covered the western Soviet Union spread to parts of Europe, causing concern about food supplies there also. On May 7, 1986, for example, Canadian customs officials announced that vegetables arriving from Italy were contaminated with radioactive iodine 31. At about the same time, member states of the European Community banned all fresh meat produced in Eastern Europe. As far away as Lapland, reindeer meat was so contaminated with radiation that it was declared unfit for human consumption.

And what of the reactor itself? Soviet engineers employed a wide array of techniques to put out the flames in the reactor, to cool it down, and to cover up the damaged facility. One of the first steps, for example, was to pump liquid nitrogen into the core to cool it down and to put out fires. Next, thousands of tons of sand, clay, lead, and boron were dumped on top of the plant from helicopters. These materials absorbed neutrons and helped put out fires. In the surrounding area, dikes were built to contain contaminated water, and several inches of soil were removed and transferred to a storage area.

The damaged reactor at Chernobyl is visible in the center of this photo, taken about two weeks after the 1986 accident. Eventually, workers built a steel-and-concrete tomb to isolate the reactor, but by 1992 the shell had cracked and begun to leak radioactive material into the environment.

Eventually, workers began to build a huge sarcophagus over the damaged plant. The steel-and-concrete tomb was designed to isolate the ruins of Unit 4 for the hundreds of years during which it would continue to emit dangerously high levels of radiation. The first shell did not last nearly that long, however. By 1992, it had cracked and had begun to leak radioactive material into the environment. The Ukrainian government found it necessary to make plans for a second shell, stronger than the first, to install on top of the original sarcophagus.

In a horrible disaster like the one at Chernobyl, scientists always hope to learn new information that can be used to increase safety in the future. At first, officials of the former Soviet Union were reluctant to consider changes in reactor design of their RBMK plants since they felt that operator errors, and not plant problems, had caused the accident. Eventually, however, they adopted a number of modifications that would make the reactors safer. For example, control rods were designed so that they could not be removed completely from the reactor core.

Scientists outside the Soviet Union, however, did not benefit significantly from events that occurred during the accident. Most non-Soviets had long been leery of the RBMK design and had rejected the use of such plants in their own countries. As one American nuclear expert said, "Most of the lessons from Chernobyl have been learned already and applied in the United States."

The final legacy of the Chernobyl accident is probably its impact on opinion worldwide about nuclear power. That impact varied from country to country. In the United States, where memories of the 1979 Three Mile Island accident had not yet faded, Chernobyl merely confirmed the fears that many had about the use of nuclear power. In France, Japan, Belgium, and other nations that depend heavily on nuclear power, the Chernobyl accident had hardly any impact. And in a few nations, such as Great Britain, the accident for the first time opened up a heated debate about the role of nuclear power as a source of energy.

DAVID E. NEWTON

Where to Learn More

Barringer, F. "Four Years Later, Soviets Reveal Wider Scope to Chernobyl Horror." *New York Times*, April 28, 1990, p. A1.

———. "Chernobyl: Five Years Later, the Danger Persists." *New York Times Magazine*, April 14, 1991, p. 28.

Cooke, S. "Human Failures Led to Chernobyl." *Engineering News Record*, August 28, 1986, pp. 10-11.

Diamond, S. "Moscow Now Sees Chernobyl's Peril Lasting for Year: Big Area Stricken." *New York Times*, August 22, 1986, p. A1.

———. "Design Flaws, Known to Moscow, Called Major Factor at Chernobyl." *New York Times*, August 26, 1986, p. A1.

Flavin, C. *Reassessing Nuclear Power: The Fallout From Chernobyl*. Worldwatch Institute, 1987.

Jackson, J. O. "Nuclear Time Bombs." *Time*, December 7, 1992, pp. 44-45.

Medvedev, G. *The Truth about Chernobyl*, trans. Evelyn Rossiter. Basic Books, 1991.

Moore, T., and D. Dietrich. "Chernobyl and Its Legacy." *EPRI Journal*, June 1987, pp. 4-21.

Reinhardt, W. "Soviet Reactor Was Flowed." *Engineering News Record*, May 22, 1986, p. 16

Seneviratne, G. "Soviets Alter Nuclear Designs." *Engineering News Record*, September 4, 1986, p. 10.

Shaherhak, T. *Chernobyl: A Documentary Story*, trans. Tan Press. St. Martin's Press, 1989.

Sweet, W. "Chernobyl: What Really Happened." *Technology Review*, July 1989, pp. 42-52.

Toman, B. "Disaster's Impact on Health Won't Be Known for Years." *Wall Street Journal*, April 23, 1987, p. 28.

U.S. Council for Energy Awareness. *Chernobyl Briefing Book*. March 26, 1991.

U.S. Department of Energy. *Health and Environmental Consequences of the Chernobyl Nuclear Power Plant Accident*. 1987.

U.S. Nuclear Regulatory Commission. *Report on the Accident at the Chernobyl Nuclear Power Station*. The Commission, 1987.

Van Der Pligt, J. "Chernobyl: Four Years Later: Attitudes, Risk Management and Communication." *Journal of Environmental Psychology*, June 1, 1990, pp. 91-99.

Ships

- *Titanic* (1912)
- Liberty ships (mid-1940s)
- *Andrea Doria-Stockholm collision* (1956)
- Amoco *Cadiz* (1978)
- Exxon *Valdez* (1989)

Sinking of the Titanic

off Newfoundland

(1912)

A state-of-the-art ocean liner sank after hitting an iceberg.

Background

On the night of April 14, 1912, the White Star Line's *Titanic,* on her maiden voyage from Southampton to New York, struck an iceberg off Newfoundland. It sank just two hours and forty minutes after the collision at 11:40 P.M., with a loss of over fifteen hundred lives. The *Titanic,* the largest and most luxurious ship afloat, was also believed to be the safest—"practically unsinkable." Heralded as a symbol of the triumph of technology, after the sinking it became a symbol of human presumption.

The *Titanic* was the conception, in 1907, of Lord Pirie, head of Harland & Wolff, a shipbuilding firm known to build the sturdiest and best ships in the British Isles, and J. Bruce Ismay, chairman of the White Star Line. In 1902, the White Star Line had been swallowed up by the monster trust International Mercantile Marine, which made the American financier J. Pierpont Morgan the principal owner of the *Titanic.* The venture was a marriage of American money and British technology. However, the *Titanic* was thought of as being British: it was registered as a British ship, was manned by British officers, and would revert to the British Navy in time of war.

The beginning of the century was a time of great rivalry among steamship companies. Since Pirie and Ismay did not think they could compete with the crack Cunard liners in speed, they decided to surpass them in size and grandeur. Also, in view of the growing immigrant traffic, they decided to greatly increase accommodations for steerage or low-fare passengers. The *Titanic* was built in Belfast alongside its sister ship the *Olympic,* which the *Titanic* exceeded in gross tonnage but not in length. The *Titanic* was launched on May 31, 1911.

The *Titanic* was eight hundred and eighty-two feet in length, ninety-two feet in width, and weighed 46,328 gross tons; its nine steel decks rose as high as an eleven-story building. The stability afforded by its vast size was deemed to be one of many safeguards against its foundering. It also embodied a greater proportionate mass of steel in its structure than had been used in previous ships. It was built with a double bottom, both skins of which were heavier and thicker than they had been in ships built before. The outer skin of the bottom was a full inch thick.

The huge hull was divided by fifteen transverse bulkheads, or upright partitions, extending the width of the ship, into sixteen "watertight" compartments, any two of which might flood without affecting the safety of the ship. The six compartments that contained boilers had their own pumping equipment. The doors between watertight compartments could be closed all at once by a switch on the bridge, or individually by crewmen. The bulkheads rose from the double bottom to five decks above forward and aft, and to four decks above amidships (the middle).

The *Titanic* was not only the biggest ship afloat but the last word in comfort and elegance. It had the first shipboard swimming pool, a Turkish bath with gilded cooling room, a gymnasium, and a squash court; a loading crane and compartment for automobiles; and a hospital with a modern operating room. The dining rooms, staterooms, and common rooms were furnished in various styles and periods; there was a Parisian café, for instance. The first-class cabins were especially opulent, some with coal-burning fireplaces in the sitting rooms and full-size four-poster beds in the bedrooms.

There were few detractors of this venture, probably because the *Titanic* embodied the spirit of overconfidence of the times. Some dissenters, though, pointed out that there wasn't a dock in America big enough for the *Titanic* or the *Olympic*. Others said that because these ships were the acme of luxury, they would represent an increase of expense rather than an advance in economic transportation. Still others said that the ships by their great size would concentrate too much wealth and too many lives in a single vessel, and also might surpass the underwriters' ability to insure them.

Details of the Sinking

The evening of April 14, the fifth day of the *Titanic's* maiden voyage, the sea was exceptionally calm and the sky was starry but moonless. The *Titanic* had received a number of warnings of ice in the region, and although Captain E. J. Smith is known to have seen at least four of these, he did not alter his speed of about twenty-one knots. There are several probable reasons for this. That visibility seemed to be good and that the ship was believed to be practically unsinkable must have contributed to his overconfidence. Also, ice fields were a hazard that he dealt with all the time, and robust captaincy was a matter of pride among sea captains, especially those with the White Star Line. Finally, he was probably trying to make good speed on the maiden voyage.

Yet April is one of the worst months for icebergs—the seaward tips of glaciers that break off, most of them coming from the west coast of Greenland. About a thousand of these reach the shipping lanes each year. On the night in question, the pair of lookouts—who were working without

The Titanic *as she set out on her voyage from Southampton.*

binoculars, which were supposed to be standard equipment on the White Star Line—did not see the iceberg until it was only a quarter mile away. The bow was swung swiftly to port (left), but it was too late. The underwater shelf of the ice tore through the plating on the starboard (right) bow, perhaps aft to amidships. Six watertight compartments were thought to have been breached. The bow started to sink, and later more compartments filled with water, which soon sloshed over the tops of the transverse bulkheads.

The forty-six thousand ton *Titanic* was traveling at twenty-one knots. So insignificant was the energy absorbed in the collision in proportion to the total energy of the vessel that the impact was scarcely felt. The gravity of the situation was comprehended only gradually, and many passengers had to be awakened. The collision occurred at 11:40 P.M., and the order to ready the lifeboats was given at 12:20 A.M.

There was great confusion in the loading of the lifeboats. There were only sixteen lifeboats and four emergency rafts—enough for roughly half of the passengers. Although the *Titanic* had complied with British Board of Trade rules regarding lifeboats, these were based on the tonnage of the ship rather than on the passenger capacity. Also, a reduction in the number of boats was allowed for ships deemed to have satisfactory watertight subdivision, as was the *Titanic.*

Since there had been no boat drill, the crew and passengers did not know to which lifeboats they should go. Furthermore, the officers in charge of loading the boats were afraid that if they were fully loaded, either the boats would buckle as they were being lowered or the davits (cranes) holding the boats over the side would break (neither of which would have happened, as boats and davits had been tested). Thus they sent the boats down only partly loaded, with instructions to come alongside the cargo ports to pick up more passengers. However, the cargo ports were never opened, so many boats went away only partly filled. There was room enough in the boats for 1,178 persons; since about seven hundred and eleven were saved in boats, about four hundred and sixty-seven needlessly lost their lives in the loading mishap.

As the *Titanic's* bow sank lower and lower, those remaining on the ship climbed to the stern, from which some ended up jumping into the 28 degrees Fahrenheit water in their lifebelts. As those in lifeboats watched spellbound, the immense stern reared up almost to the perpendicular and remained still for a few moments before sliding slowly, at 2:20 A.M., into the sea. Many survivors reported hearing sounds like thunder, or thick detonations, or a kind of "death rattle" before the stern went down.

Although some passengers asserted that the great ship broke in two, the explanation generally accepted to explain the thunderous sound was that as the stern rose, the boilers crashed down through the bulkheads. From what we know now about the position of the bow and the stern on the ocean floor—they are facing in opposite directions and are about 1,970 feet apart—it seems likely that the ship did break in two at or near the surface. As the bow sank and the stern rose, the pressure on the keel probably increased until it snapped.

The other startling discovery made by the 1985 Franco-American expedition and the 1986 American expedition led by Dr. Robert D. Ballard of the Woods Hole Oceanographic Institute concerned the gash in the hull. These expeditions located and photographed the wreck in thirteen thousand feet of water a few miles from where it had been thought to be. Dr. Ballard observed many buckled plates below the waterline, but no gash. It is possible, however, that the gash remains hidden deeper down where the bow sank with great force into the sediment.

It is likely that mishaps in wireless communication contributed to further unnecessary loss of life on the *Titanic*. On the *Californian*, a ship stopped for the night in ice fields not more than twenty miles away, the wireless operator had stopped working only fifteen or twenty minutes before the operator from the *Titanic* tried to get through with a distress call. At the time, wireless operators were employees of the Marconi company and did not follow around-the-clock shipboard watches. The next closest ship, which the *Titanic* did succeed in reaching, was the *Carpathia*, about fifty-eight miles away. The

Contemporary drawing of how the Titanic *may have struck the iceberg.*

Carpathia picked up the first lifeboat at 4:10 A.M. If the *Titanic* had had lifeboats for everyone aboard, there might have been no loss of life.

In 1993, a team of architects and engineers released a report in which they argued that the tragedy was caused not so much by the collision with the iceberg as by the structural weakness of the ship's steel plates. Low-grade steel such as that used on the *Titanic* is subject to brittle fracture—breaking rather than bending in cold temperatures. If a better grade of steel had been used, the ship might have withstood the collision or, at the very least, sunk more slowly, thus allowing more passengers to be saved. Interestingly, the team also suggested that the roar heard by the passengers as the ship sank may have been the steel plates fracturing, not the boilers crashing through the bulkheads.

Impact

A U.S. Senate inquiry into the *Titanic* tragedy began the day after the *Carpathia* landed in New York with the survivors. It was headed by William Alden Smith. A British investigation by the Board of Trade, with Lord Mersey as wreck commissioner, followed in May through July 1912. An international

At 13,000 feet below the surface, the Titanic's *deck plays host to the Alvin submarine of the 1986 Woods Hole Oceanographic Institute expedition. Researchers from the Institute were unable to locate the gash in the ship's hull caused by the iceberg.*

Safety of Life at Sea (SOLAS) conference met in London from November 1913 to January 1914. Many new rules and regulations resulted from these and later forums.

New regulations required that the number of lifeboats be based on the number of passengers rather than on the tonnage of the ship. There were new rules about lifeboat drills being held soon after a ship sails, with crew members and passengers assigned a particular place in a boat, about the training of the seamen assigned to the boats, and about the permanent equipment of the boats.

Shipping lanes were moved farther south, away from the ice. Ships approaching ice fields were required to slow down or alter their course. (If a ship is traveling at half speed, the reactive blow of a collision is reduced by one fourth, in that the energy of a moving mass increases as the square of the velocity.) The International Ice Patrol, based in Groton, Connecticut, was set up to monitor iceberg movement. It uses aircraft with side-looking radar to spot icebergs, and also makes computerized predictions about their whereabouts. To date, no further loss of life has occurred as a result of collision with an iceberg in the North Atlantic shipping lanes monitored by this control.

New regulations were also brought to bear on wireless operations. Thenceforth, a ship's wireless had to be manned day and night, and to have an auxiliary source of power. Also, it was the interference of amateur wireless operators in the *Titanic* tragedy, mostly in the period after the sinking and before the survivors had been brought to shore, that led to the licensing of all operators in the United States. This licensing was the beginning of the Federal Communications Commission.

From a strictly technological point of view, the main feature of the *Titanic* tragedy was the speed at which it sank. It was thought that the last word in modern shipbuilding should not have succumbed so quickly. Although new regulations after the tragedy about bulkhead subdivision led to the reconstruction of many passenger ships, it is questionable how well many ships built today—even with high-grade steel construction—could withstand the kind of blow suffered by the *Titanic.*

The *Titanic* disaster had other kinds of effects that were much less tangible. After a century of steady technological growth, there was widespread disillusionment. The tragedy was likened to Greek tragedies in which heroes were destroyed by their *hubris*, and to medieval morality plays. Ministers preached that God was punishing people for their excessive faith in material progress. Most profound, perhaps, was the sense of a loss of innocence.

SALLY VAN DUYNE

Where to Learn More

Ballard, Robert D. and Rick Archbold. *The Discovery of the Titanic.* Madison Press Books, 1987.

Broad, William J. "New Idea on Titanic Sinking Faults Steel as Main Culprit," *New York Times*, September 16, 1993.

Davie, Michael. *Titanic: The Death and Life of a Dream.* Alfred A. Knopf, 1987.

"Wreck of the Titanic." *Engineering Magazine*, June 1912, pp. 446-49.

"Scientific Aftermath of the Titanic Disaster," *Literary Digest,* May 25, 1912, pp. 1096-97.

"Tragedy of the Titanic—A Complete Story," *New York Times*, April 28, 1912, pp. 1-8.

Wade, Wyn Craig. *The Titanic: End of a Dream.* Rawson, Wade, 1979.

Liberty ship breakups

(mid-1940s)

Fracturing of all-welded cargo ships used in World War II put the use of welding rather than riveting in shipbuilding under question.

Background

During World War II, the United States produced 2,710 Liberty ships, a type of all-welded cargo ship designed primarily for war use. There had been great fear that the war might be lost due to the Allies having an insufficient number of ships, and the Liberty ship has often been credited with saving the world from disaster. Some of these vessels, however, suffered sudden fractures (unexplained at the time) that were accompanied by a sound like an explosion. In the years 1943 through 1945, about seven of the ships actually broke in two or sustained such severe fractures that they had to be abandoned. Hundreds more developed less severe fractures that could be repaired.

The United States Maritime Commission had been established in 1936 to expand the fleet of merchant vessels. On March 14, 1941, two new shipyards contracted with the commission to produce Liberty ships, modified from a British design in the interest of speedier construction. On February 7, 1942, after the U.S. entered the war, the War Shipping Administration was established by President Franklin D. Roosevelt to control all American shipping and to commandeer the vessels of the Maritime Commission for the duration of the war.

The Liberty ships were 441 feet 6 inches in overall length and weighed 7,176 gross tons. Since their top speed was only eleven knots, in wartime they usually traveled in convoy. The vessels had seven watertight bulkheads, all extending to the upper deck, which divided the hull into five cargo holds, some equipped with tanks. The deck space was equally important in carrying cargo. Each cargo hold had a rectangular hatch on deck that was twenty to thirty-five feet long and twenty feet wide (many of the deck fractures started at the corners of these hatches). The Liberties were intended to carry a crew of forty-five men, but this number was increased by 36 gunners to a total crew of 81 after the laws that prohibited American merchant ships from being armed and entering combat zones were suspended.

The first Liberty ship, the *Patrick Henry*, was launched on September 27, 1941, a day that was proclaimed by President Roosevelt as the "Liberty Fleet Day." The *Patrick Henry* had spent one hundred fifty days under construction on the slipway at the Bethlehem-Fairfield Shipyard in Baltimore, and took another ninety-five days to outfit, a total of two hundred forty-five days. After building twenty ships, this yard could complete a ship in one hundred twenty days; and after fifty ships, in fifty-eight days. Eighteen shipyards were ultimately involved in building Liberty ships, and almost unbelievable records were achieved and then surpassed. In November 1942, at the Permanente Metals Corporation in Richmond, California, the *Robert E. Peary* spent less than five days on the slipway and was outfitted in another three days.

In September 1942, American shipyards reached their goal of three ships a day: of the ninety-three ships delivered that month, sixty-seven were Liberty ships. A sobering statistic of the losses suffered by Allied shipping shows how imperative this rapid ship-building was. In 1942, eleven million tons of merchant ships were constructed for the Allies, of which eight million tons' worth were built in the United States. Axis destruction was so great, however, that the Allies suffered a net loss of one million tons in that year.

A vast amount of prefabrication work was done before a Liberty ship was actually constructed on the slipway. This was especially true at the yards controlled by Henry J. Kaiser, the principal figure in mass production shipbuilding in the United States. The Kaiser yards were laid out on quite revolutionary principles—they were assembly plants for more than 30,000 components produced en masse in thousands of factories in more than thirty-two states. Few of the workers at these yards had ever seen a shipyard before, and few of the executives had ever dealt with ship construction. Nonetheless, there was no indication that ships built at the Kaiser yards were more likely than others to fracture.

The speed with which the Liberties had to be built made the use of welding necessary. Although welding was still relatively new, by World War II it had largely replaced riveting. Welding saved in time, manpower, training for operators, and equipment. There were further advantages: the difficulty of making riveted structures watertight was eliminated, as was the weight of plate flaps and rivet heads. Nevertheless, welded ships had not yet been subjected to the unusual stresses and heavy loads of wartime conditions. The use of welding in the shipbuilding effort thus had its critics, especially after the fractures began to appear.

Details of Two Breakups

The Liberty ship *J.L.M. Curry* was completed at the Alabama Drydock and Shipbuilding Company in Mobile on May 15, 1942. On March 7, 1943, the vessel was traveling at five knots in very high seas in the Greenland Sea off

The bow of the Liberty ship Robert E. Peary *at the Kaiser shipbuilding yards. Despite construction problems, the speed with which the Liberty ships could be built (the* Peary *had been built in less than 5 days) made them invaluable to the Allied war effort.*

Iceland. The air temperatures ranged from fourteen to thirty degrees Fahrenheit. At 1:20 P.M., the ship came down from a very high wave and split in four places. The upper deck plating, side shell, and bulwark fractured on the port side after ends, and on the starboard side forward ends, of the Nos. 3 and 4 hatches. The cracks, which occurred not along the welds but in the plating, extended down through the shell plating to below the decks. The cracks opened and closed with the movements of the ship. The vessel was abandoned at 11:15 A.M. on March 8, and was sunk by shells from an Allied vessel.

It should be noted that some Liberty ship breakups occurred in calm-water conditions. On January 16, 1943, a ship docked in Portland, Oregon suddenly broke in half, though the water was calm and the temperature mild.

The *John P. Gaines*, completed on July 8, 1943 at the Oregon Ship Building Corporation in Portland, met a similar fate. On November 23 of that year, the vessel was making its way from Dutch Harbor in the Aleutian Islands to Seattle, on a sea with long ground swells. At 10:00 P.M., in air temperatures of forty to fifty degrees Fahrenheit, loud noises were heard, the source of which could not be identified in the dark, but was later believed to be the fracturing of the No. 3 hatch cover. At about 2:40 P.M. on November 24, off Shumagin Island, a large swell hit the vessel and washed over the port bow, propagating the fracture. As it passed over or between swells, the vessel began breaking apart. Ultimately, the vessel's forward end completely broke away. All passengers and crew were by this point in the aft end, and U.S. Army Transports were able to pick up most of the men. Nonetheless, eleven lives were lost, including six soldiers who were in one lifeboat. The bow of the *John P. Gaines* presumably sunk; the stern later drifted ashore at Big Koniuji Island, Alaska.

It should be noted that some Liberty ship breakups occurred in calm-water conditions. On January 16, 1943, a ship docked in Portland, Oregon suddenly broke in half, though the water was calm and the temperature mild.

Impact

In April of 1943, the Secretary of the U.S. Navy established a Board of Investigation to Inquire into the Design and Methods of Construction of Welded Steel Merchant Vessels. The committee issued several interim reports before its final report in 1946. The investigation included technical and statistical analyses of the casualties, research on designs and materials used in fabrication, strength studies of each type of vessel, and examination of convoy routes, with attention to sea and weather conditions.

A number of risk factors became apparent early on, such that improvements could be incorporated immediately. A large proportion of the fractures on Liberty ships were known to have begun at the corners of the hatches. (It was later determined that 25 percent of all fractures and 52 percent of all serious fractures on Liberty ships originated at these corners.) Ultimately, 2,047 Liberty ships were fitted with hatch corner reinforcements. Various "crack arresters"—one method involved riveting seam straps over the deck—were also installed. Since the risk was greater for vessels in cold weather and heavy seas, the reinforced vessels were assigned to low temperature areas. Through such measures, the number of casualties in all-welded steel merchant ships (of which the Liberties were one type) fell from a maximum of 140 per month in March 1944 to 20 per month by January 1946.

In its final report, the Board of Investigation concluded that a satisfactory all-welded ship structure could be produced. Proper design and high-quality workmanship (which could not always be secured during the war effort) were essential. The investigation found that porosity or inclusions in welds could lead to fractures; extensive use now is made of automatic welding and of gammaradiography for the inspection of welds. The investigation also recognized the hazard posed by steel that is brittle when notched. The steel that had been furnished to shipyards complied with the strength requirements of the time, but impact tests of steel taken from fractured vessels showed that in many cases it was "notch-brittle" or "notch-sensitive"—that is, its ability to absorb energy after becoming notched, especially at low temperatures, was low. This finding led to new specifications for the notch-toughness of steel to be used for shipbuilding.

In June 1943, Britain established its own committee to investigate fracturing in welded steel ships. An important outcome was the complete vindication of the use of welding in shipbuilding. This investigation, like its American counterpart, concluded that with sound design, good workmanship, and ductile steel, welded ships were fully reliable.

SALLY VAN DUYNE

Where to Learn More

Bunker, John. *Liberty Ships: The Ugly Ducklings of World War II.* Naval Institute Press, 1972.

"Calm Sea, Broken Ships: Tankers Break Up in Calm Weather." *Business Week,* June 22, 1946, p. 54.

Final Report of a Board of Investigation Convened by Order of the Secretary of the Navy to Inquire into the Design and Methods of Construction of Welded Steel Merchant Vessels, 15 July 1946. Government Printing Office, 1947.

Investigation of Plate Fractures on Welded Ships, Interim Report. House Report 1685, 76th Congress, 2nd Session, June 20, 1944.

Sawyer, L. A., and W. H. Mitchell. *The Liberty Ships: The History of the "Emergency" Type Cargo Ships Constructed in the United States During World War II.* Cornell, 1970.

Andrea Doria-Stockholm collision

off Massachusetts

(1956)

The Andrea Doria *sank after colliding with another liner, the* Stockholm, *off Massachusetts.*

Background

On July 25, 1956, the liners *Andrea Doria* and *Stockholm* collided near the Nantucket Lightship off Massachusetts. The *Andrea Doria* sank, with the loss of forty-three passengers and crew, and was the first big liner to be lost in peacetime since the *Titanic* went down in 1912. On the *Stockholm* three crew members were never seen again, and several later died from their injuries. The collision was caused primarily by misinterpretation of radar signals, and the sinking was a result of faulty ballasting.

The *Andrea Doria* was the flagship of the Italian Line. Built by Ansaldo of Sestri, near Genoa, the vessel was launched in 1951, the first passenger ship to come off the ways in Italy after World War II. Genoa was its home port. The *Doria* was seven hundred feet in length, weighed 29,083 tons, had a service speed of twenty-three knots, and was known for its graceful lines. On what was to be its last voyage, the ship left Genoa on July 17, 1956, and was due in New York City on July 26.

The *Stockholm*, owned by the Swedish-American Line, was built in Sweden by Götaverken of Göthenberg and launched in 1943. It was five hundred and twenty-five feet overall and 12,644 tons, and had a service speed of nineteen knots. On July 25, 1956, the ship was on its first day out, beginning its homeward voyage from New York City to Sweden.

Both ships were heading toward the Nantucket Lightship, a vessel anchored fifty miles off Nantucket, to guard the shoals off that island. The Nantucket Lightship is an important focal point for the sea lanes of the world. Although the shortest great circle—a circle outlined on the Earth's surface, such as the Equator, by a plane passing through the Earth's center—track for ships traveling between New York and European ports would be to the north of the vessel, they must pass to the south of it before turning north. Certain shipping lines had agreed to having their eastbound ships pass in a track twenty miles south of the light vessel and their westbound ships pass close to it. However, neither the Italian Line nor the Swedish-American Line was a party to this agreement.

Details of the Collision

On the afternoon of July 25, 1956, the *Andrea Doria*, eight days out of Genoa and due to dock in New York the next morning, encountered fog when it was still about one hundred and fifty miles from the Nantucket Lightship. Captain Piero Calamai observed the normal precautions for these conditions: closing the watertight doors, sounding regular prolonged blasts on the siren, having an officer keep a constant watch on the radar screen for any sign of approaching ships, and reducing the ship's speed. Only the last of these can be criticized, as the *Doria* had reduced its speed only five percent, from twenty-two knots to twenty-one knots, and the "moderate speed" called for in the collision rules is defined as a speed that allows the ship to be stopped within its visibility distance. By 8:00 A.M., the Doria's visibility was only half a mile, and stopping her would take considerably more distance than that. However, it was the custom among sea captains bound by schedules to make only token reductions in speed in bad weather, while keeping a close watch on their radar.

At 10:20 A.M., the *Doria* passed one mile to the south of the light vessel. At about 10:45, Second Officer Franchini saw a "blip" on the radar screen that represented a ship that appeared to be not quite due ahead, but fine, or slightly, to starboard. Captain Calamai checked the radar screen himself and judged that the rapidly approaching ship was to starboard—that is, to the right—of his heading marker. Captain Calamai felt quite certain that the two ships were on a starboard-to-starboard path, but thought that the distance between would be uncomfortably small. He gave the order to alter course to four degrees to port (left). In so doing, he was going against rule eighteen of the collision rules, which states that "when two power-driven vessels are meeting end on, or nearly end on, so as to involve risk of collision, each shall alter her course to starboard, so that each may pass on the port side of the other." He had two justifications for going against this well-known rule: he believed that the two ships were clearly on a starboard-to-starboard course, and he thought that an alteration of course to starboard might force his ship into the shoal water to the north of the light vessel.

When the two ships were less than five miles apart, the *Doria* still could not see the *Stockholm*; the *Stockholm* could see the masthead lights, but not the sidelights denoting port and starboard, on the *Doria.* When the ships were about two miles apart—and at their combined speeds of forty knots would close in on each other in three minutes—the *Doria* could discern the first navigation lights on the *Stockholm,* and could see that they were not green as expected in a starboard-to-starboard passing, but red. The red lights meant that the *Stockholm* was crossing the Doria's bow to make a port-to-port passing. At that moment Captain Calamai gave the fatal order, "Hard-a-port," which was consistent with his plan for a starboard-to-starboard

The Andrea Doria *listing heavily before sinking. The ship's severe list (eighteen degrees, then twenty-two degrees) rendered the port-side lifeboats useless and caused the starboard-side lifeboats to swing far out over the water, making them hard to access.*

crossing. There was nothing the crew could do as they watched the bow of the *Stockholm* plow into their starboard side.

The weather encountered by the *Stockholm* as it traveled east toward the Nantucket Lightship was very different from that encountered by the *Andrea Doria.* Somewhat before 11:00 P.M., as the *Stockholm* traveled through a clear moonlit and starlit night, Third Officer Carstens-Johannsen first saw the sign of an approaching ship at the twelve-mile radius on the radar screen. He recorded the position of the approaching ship as two degrees to port of his heading marker. Another plot of the ship a short time later also showed the ship to be slightly to port. As there was no sign of fog, Carstens, as he was called, saw no reason to call Captain Nordensen up to the bridge. But he did post a lookout on the port wing of the bridge to spot the approaching vessel, at which time he planned to alter his course to starboard to make a safe port-to-port passing. When the bridge lookout suddenly called out "Lights to port," Carstens immediately ordered a change in course of twenty degrees to starboard. Then he went out onto the bridge and saw to his horror that an immense liner was about to cross his bow, presenting the green lights of her starboard side. He shouted "Hard-a-starboard!" to the helmsman. As Captain Calamai had tried at the last moment to increase the distance in a starboard-to-starboard passing, Third Officer Carstens-Johannsen tried to increase the distance in a port-to-port passing. Then Carstens pulled the telegraphs to "Stop" and "Full Astern."

The stem of the *Stockholm*, stiffened for travel through the ice in northern oceans, plowed into the side of the *Andrea Doria* just below the bridge down to her starboard fuel tanks with a momentum of thirty million foot-pounds. Yet only one main compartment of the *Andrea Doria* was damaged, and it would have been able to withstand the collision had it not been for the severe list that developed soon afterward. Since the *Doria* was near the end of its journey, the starboard fuel tanks contained only air. When the starboard tanks were rent in the collision, water rushed to fill them, making the starboard side much heavier, and causing a list to starboard of eighteen degrees and soon after, twenty-two degrees. A narrow tunnelway that led to a pump room containing valves for flooding the tanks with sea water for stability ran between the fuel tanks on the port and starboard sides. Since this tunnel was so far inboard, a watertight door had not been fitted to it. The tunnel flooded quickly after the collision, and it was no longer possible to get to the valves to flood the port tanks to make them as heavy as the starboard tanks.

When the port side lifeboats were readied, it was found that they were fitted with davits—the cranes that project over sides of a ship—that could work only if the vessel's list was less than fifteen degrees. That meant that the port lifeboats were useless; the starboard-side lifeboats could not accommodate everyone and were hard to climb into because, with the list, they swung far out over the water.

The stem of the Stockholm, *stiffened for travel through the ice in northern oceans, plowed into the side of the* Andrea Doria *just below the bridge down to her starboard fuel tanks, with a momentum of thirty million foot-pounds.*

A number of nearby ships rushed to the rescue, working against time as the *Doria*'s list gradually increased and it was more and more likely to capsize. Of the 1,706 passengers and crew on the *Doria*, 1,663 survivors were taken off; most of the dead were killed on impact. The surviving passengers were all off by 4:00 A.M. on June 26; at 5:30, when the list reached forty degrees, the officers also left. At 10:09 A.M., eleven hours after the collision, the *Andrea Doria* rolled over onto its side.

Impact

In the legal proceedings that followed, each line claimed that the other line was totally to blame and that its own officers were entirely blameless. The Swedish-American Line sued the Italian Line for four million dollars and the Italian Line sued the Swedish-American Line for 1.8 million dollars. The Swedish-American Line's case was by far the stronger. Most important was the immediate eighteen degree list of the *Andrea Doria*, which put the responsibility for the capsizing and sinking onto the owners of the *Doria*, as did the *Doria*'s turn to port to make a starboard-to-starboard crossing, thus violating the Rules of the Road. Also at issue were the speed of the *Andrea*

Doria in fog, the failure of its crew to plot the bearings of the approaching *Stockholm*, and the disappearance of the *Doria*'s ship's log.

The elements of the Italian Line's case were Third Officer Carstens-Johannsen's delay before making a starboard turn, his youth and inexperience, and the fact that the *Stockholm* was far to the north of the recommended track for eastbound vessels.

After the hearings had gone on for about three and a half months, the two lines came to an out-of-court settlement, dropping their claims against one another. And so the court never did apportion blame for the collision. It seems likely, however, that the two ships were not on safe parallel courses in which they would pass either port-to-port or starboard-to-starboard, but were converging at a very small angle. A lesson relearned in the disaster is that in order to conform to the collision rules, officers of the watch should treat fine bearings as head-on collisions, and should make large alterations clearly discernable to the approaching ship in ample time.

There have been considerable changes in both officers' understanding of radar and in the technology of radar since the *Doria-Stockholm* collision. Captain Calamai probably did not understand that a four degree change in course would not make his intent to make a starboard-to-starboard passing clear to the *Stockholm*. However, today, largely as a result of this disaster, officers on ships are required to take courses in radar. In addition, by international convention, automatic radar plotting aids have had to be fitted in all large ships since the 1980s. Apparently, no plot was made on the *Andrea Doria* of the *Stockholm's* bearings, and although one was believed to have been made on the *Stockholm* of the Doria's bearings, it could not be produced, presumably because it had been rubbed out.

The true motion radars in use today allow the officer of the watch to see the actual course and speed of another ship, instead of the bearing relative to a heading marker. The latter system could result in errors because a ship can get a few degrees off course at any time, which may explain why the *Andrea Doria* thought the *Stockholm* was fine to starboard and the *Stockholm* thought the *Andrea Doria* was fine to port.

Another lesson learned from the collision involved the ballast tanks. The designers of the *Andrea Doria* had followed the practice common at the time of obtaining sufficient stability to satisfy the seven degree limit on listing by stipulating that the oil tanks when empty be flooded with sea water. This was an unreasonable burden to lay on the engineers, who did not want to contaminate their fuel tanks with sea water. For in order to clean the tanks, they then would have to pump the contaminated oil and water onto barges for removal ashore. The lessons learned here were the advisability of having separate tanks for ballast water needed for stability, and of having the capability of remote control of any valves needed for counterflooding of the ballast tanks. The 1960 Safety of Life at Sea (SOLAS) convention criticized the practice of using oil tanks for ballast, and recommended that full stability information be issued to the staff and that diagrams of ballasting arrangements be displayed on board.

A final effect of the *Andrea Doria-Stockholm* collision was the introduction of a traffic separation scheme for ships passing near the Nantucket Lightship, which was introduced in 1977. It required that eastbound ships keep to a lane south of the light vessel, and that westbound ships keep to a lane north of the light vessel. There is a separation zone three miles wide, with the light vessel in the middle, between the two lanes.

SALLY VAN DUYNE

Where to Learn More

Barnaby, K. C. *Some Ship Disasters and Their Causes.* A. S. Barnes, 1970.

Hoffer, William. *Saved: The Story of the Andrea Doria—The Greatest Sea Rescue in History*. Summit Books, 1979.

Marriott, John. *Disaster at Sea.* Ian Allan, 1987.

Moscow, Alvin. *Collision Course.* Grosset & Dunlap, 1981.

Amoco Cadiz oil spill

off France

(1978)

An oil tanker ran onto rocks off the coast of France after its steering system failed

Background

On the morning of March 16, 1978, the Amoco *Cadiz,* a very large crude carrier (VLCC), lost its steering in gale force winds off the rocky coast of Brittany, in France. The vessel ended up impaled on the rocks and eventually emptied 220,000 tons of oil to pollute the surrounding area. The disaster—the worst oil tanker pollution disaster that the world had known—was attributed to faulty design and improper maintenance of the steering system.

The Amoco *Cadiz* was nearing the end of a six week trip from the Persian Gulf, where the vessel had been laden with crude oil, to Rotterdam, via the southern tip of Africa. The registered owner of the Amoco *Cadiz* was the Amoco Transport Company, whose parent company, Standard Oil of Indiana, had decided to acquire four VLCCs in the late 1960s, after the Six-Day War closed the Suez Canal. Enormous tankers seemed to be the most economic way to carry oil around the tip of Africa.

The VLCCs were ordered from Astilleros Españoles, considered to be the best shipbuilders in Spain, where boats can be built for less money than in the United States. Since its delivery in 1974, the *Cadiz* had been on charter to Shell International Petroleum Company for $28,000 a day. The oil carried by the *Cadiz* belonged to Shell. Complicating matters further, the *Cadiz* carried a forty-man Italian crew and, for tax purposes, flew the Liberian flag.

The Amoco *Cadiz* was 1,095 feet long and 167 feet wide, and had a top theoretical service speed of 16.1 knots. Unladen, the ship weighed about 45,000 tons, but when laden carried an additional 4,000 tons of bunker oil and, in fifteen cargo compartments, 220,000 tons of crude oil. Built up on the stern was a six-story accommodation block with living quarters, engine room, and bridge.

Details of the Disaster

As the Amoco *Cadiz* was turning east to enter the English Channel on March 16, 1978, it was traveling through gale force winds, rain squalls, and very high seas. The fact that the wind was in the process of veering from the southwest to the northwest made the waves become steeper—about twenty to twenty-seven feet tall—and more jagged. At about 9:45 A.M., a massive body of water smashed over the tanker's starboard (right) quarter. Since a wave that is twice as high as the average wave is statistically likely to occur every three hours, it is speculated that the wave may have been forty to fifty-four feet.

Whether or not it was a three-hour wave, it was big enough to create overpressures in the ship's steering gear. The *Cadiz* was suddenly unmaneuverable, with the rudder hard to port (left), in the middle of the world's most crowded area of ship traffic. At this point, the vessel was fifteen miles from the French coast and seven and a half miles north of the island of Ouessant, with the northwest wind blowing it toward shore.

The steering gear on the Amoco *Cadiz* was a four-ram, hydraulic system, like those on most large modern ships. It was made by Manises, a subsidiary of Astilleros Españoles, the builder of the ship. The basic working of the gear is as follows: Turning the wheel on the bridge works to activate twin high-pressure pumps in the steering-gear room that send hydraulic oil through pipes into a distribution block, from which it is shunted into cylinders. Two pairs of cylinders face each other. Each cylinder encases a ram, a rolled steel piston as thick as a man's waist, which slides back and forth in its cylinder in response to surges of hydraulic oil. When the ship's course is straight ahead, a horizontal bar lies across the pivot of the rudder, but when a turn is being made the hydraulic rams on one side extend, pushing the bar away and causing the ship to turn. When the pistons on one side are under pressure, those opposite are under suction.

Hydraulic systems are able to generate the enormous power required to move the rudder against the inertia of the tanker's more than a quarter million tons of deadweight that wants to move straight ahead. The hydraulic energies harnessed are accompanied by great potential dangers in that such high pressures are involved. In order to hold back pressures of ten to twenty tons per square inch, the piping, flanges, and connections must be overbuilt and beyond reproach, and maintenance routines must be followed meticulously. This "perfect machine," as *Cadiz*'s steering gear appeared to be on the blueprint, also needs to work in antiseptically clean conditions.

When Captain Pasquale Bardari went down to the steering-gear room at 10:05 A.M., he found a fan-shaped fountain of oil spurting from the distribution block. It was determined later that the O-shaped flange that held the pipe fast against the distribution block had come away from its seat because five of the six studs that held it in place had sheared off.

The engineers on board were working to shut down valves to stop the oil from sprinkling around, and to purge the system of air by opening and

When the Amoco Cadiz *broke in two after running aground off the coast of France, it spilled 220,000 tons of crude oil into the ocean, creating a 100-mile long slick and fouling 245 miles of coastline.*

closing purging cocks. However, the rudder was slamming back and forth wildly, forcing oil out of the system and air into it. The rudder on the *Cadiz* was a steel slab about forty feet high and weighing about 160 tons. Unlike the steam steering systems on older, smaller vessels, in which a friction brake could be cranked down to immobilize the rudder while the steering was being fixed, on VLCCs the pressures were too great for this type of makeshift rigging. Eventually, as the mixture of oil and air in the *Cadiz*'s steering gear became churned up into a steamy froth, the effort to bleed the system of air had to be abandoned.

Over the course of the long afternoon, two attempts by a sea-going salvage tug to tow the immense tanker were unsuccessful, and at about 9:00 P.M. the *Cadiz* was impaled on a pinnacle of rock off the fishing village of Portsall. Oil leaked out, creating dense fumes and the possibility of explosion. At 1:45 A.M. on March 17, all of the crew except for Captain Bardari and one other, who were still hopeful that the arrival of a second tug would make a tow possible, were removed by helicopter. A few hours later the ship started breaking in two. At 5:10 A.M., the last two crew members were taken off in another helicopter.

The thick black mass of oil that spilled from the *Cadiz* polluted 130 miles of coastline, or 245 miles of shoreline when the inlets are taken into account. Oil slicks extended 100 miles out to sea, and oil sank six to ten feet into the sediment. As would be expected, the fishing and tourist industries were severely affected. It is estimated that twenty-two thousand birds died after being covered with oil. The manpower employed in the cleanup—military, civil defense, volunteers—was on average 5,500, but at peak times 8,000, per day. The oil had to be scraped out of the crevices in the rocks by hand. By some time in May 1978, two months later, most surface signs of the spill were no longer evident.

Impact

There were two major public investigations into the Amoco *Cadiz* grounding. The first began two months after the incident, and was held by the Marine Board of Investigation of the Republic of Liberia in London, with Sir Gordon Wilmer presiding. The hearing, which lasted six weeks, had no power to award damage payments.

The Liberian inquiry demonstrated that the Manises steering gear had defects in design and construction that made it too weak to withstand the pressures to which it might be submitted. The failure happened at a flange where a high-pressure hydraulic pipe from the port pump joined the distribution block. The flange came away from the wall of the block when an overloaded stud gave way, and then four more studs gave way. The pipe freed itself and began spurting oil.

The studs were found to be too few and too weak. Instead of the eight one-inch studs required by British and American standards, there were six one-half-inch studs. Similarly, the flanges were twenty millimeters thick, whereas American standards required that they be thirty millimeters. The flanges were also flattened on the sides, instead of being round, which causes uneven stresses. A further shortcoming of the Manises gear was that the pressure at which the relief valves would operate was substantially higher than the working pressures in the system.

The Liberian inquiry also found that the steering gear, once broken, could not be fixed because there was no effective backup system. The Manises system had two independent pumps, with separate leads to the rams, but only one distribution block. Thus, the system had neither the overbuilt gear nor the redundancy of equipment that guarantees safety.

The second proceeding, which took place in Chicago with U.S. District judge Frank McGarr presiding, was the result of the multitude of lawsuits that were filed after the grounding. It was far longer and more complex than the first investigation. Judge McGarr split the trial, which began in September 1978 and lasted more than six years, into two segments: liability and damages.

As was found in the earlier proceeding, Judge McGarr found defects in the design of the Manises steering gear. An additional defect uncovered in this investigation was the use of cast iron for the bushings that grasped the steel rams. Since cast iron is harder than steel, these bushings scratched the rams as they slid in and out, letting oil leak out and air and dirt leak in.

The most significant new discovery in the second legal proceeding concerned Amoco's grossly negligent maintenance of the hydraulic steering system and refusal to make repairs.

But the most significant new discovery in the second proceeding concerned Amoco's grossly negligent maintenance of the Manises system and refusal to make repairs. In the discovery portion of the liability trial, it came out that the oil had never been changed, it had never been sampled for contaminants (such as the minute metal fragments from the scored rams), the filters had never been cleaned, and the crews had not been given training in the maintenance of high-pressure hydraulic systems.

Furthermore, *Cadiz* and its three sister ships were found to have had a long history of steering-gear problems. For instance, during February 1978 *Cadiz's* steering system was known to be leaking twelve liters of oil per day (one liter per week is considered normal), but maintenance was being deferred until the end of its $28,000 per day charter. Eventually, Standard Oil of Indiana—Amoco's parent company—was found guilty of negligence and ordered to pay $85 million in damages.

As a result of the Amoco *Cadiz* spill, the International Maritime Organization (IMO) amended its requirements for steering gear, and these were written into the international Safety of Life At Sea (SOLAS) convention. One requirement was for a switching system to shut down the damaged side of a hydraulic motor and activate the other side. The IMO also called for tankers to be better equipped for emergency towing, and helped to develop a new traffic separation scheme to move vessels carrying dangerous cargo farther out to sea in the English Channel. Also in response to the *Cadiz* disaster, the French set up a regional operations center, the Centre Régional Opérationel de Surveillance et de Sauvetage, on the westernmost point of land in France to function like an air traffic control center, and identify and follow every ship passing into the English Channel.

SALLY VAN DUYNE

Where to Learn More

Carter, L. J. "Amoco *Cadiz* Incident Points up the Elusive Goal of Tanker Safety." *Science* May 5, 1978, pp. 514-16.

Chelminski, R. *Superwreck: Amoco* Cadiz: *The Shipwreck That Had to Happen.* William Morrow, 1987.

Marriott, J. *Disaster at Sea.* Ian Allan, 1987.

"What Happened to Amoco *Cadiz?*" *New Scientist* March 30, 1978, p. 837.

Exxon Valdez oil spill

Prince William Sound, Alaska

(1989)

An oil tanker ran aground in Prince William Sound, Alaska, resulting in a catastrophic oil spill.

Background

In the early morning of March 24, 1989, the tanker Exxon *Valdez* ran aground on Bligh Reef in Alaska, spilling 10.8 million gallons of crude oil into Prince William Sound. It was the worst oil spill in U.S. history to date. More than fifteen hundred miles of shoreline were polluted, and many thousands of birds and sea otters were killed. Although the immediate cause of the grounding was human error, the accident can be seen in a broader perspective as a technological failure because oil is so integral to the functioning of our technology-based society, and because technology played an important role in attempts to prevent and minimize the damage of the spill.

The Exxon *Valdez* oil spill and a few other spills have been widely reported, but there have been many other major spills of which the public has been generally unaware. Between 1978 and 1990, there were nineteen hundred oil spills of more than ten thousand gallons worldwide. Of these, twenty-six were of more than ten million gallons, with the *Valdez* spill being toward the low end of this group. The Exxon *Valdez* spill became so widely known not only because it was the largest spill in the United States and in North American waters, but also because it did so much damage in the enclosed waters of Prince William Sound, where the oil could not disperse and break down as easily as it could have on the high seas. Before the spill, Prince William Sound—with its snow-capped mountains, many islands, and abundant wildlife—was a place of rare pristine beauty.

The oil carried by the *Valdez* had traveled eight hundred miles by the Trans-Alaska Pipeline (TAP) from Prudhoe Bay to the port of Valdez. In 1970, Alyeska, a consortium of oil companies that included Exxon, had applied to Congress for a permit to build a pipeline to transport the 9.6 billion barrels of oil that lay below the frozen ground of Alaska's North Slope. In 1973, Middle Eastern countries temporarily cut off oil supplies to the West, and that same year Congress voted to authorize the pipeline. Construction of the TAP began in 1974 and was completed in 1977. Two million gallons of oil now travel through the pipeline every day, and seventy to seventy-five tankers leave the port of Valdez every month.

To appease environmentalists and the citizens of Alaska who opposed the building of the Trans-Alaska Pipeline, fearing that a catastrophic oil spill would be inevitable, the federal government and the oil industry made promises, many of which were not upheld. For example, promises in the early 1970s by the federal government to require double bottoms on tankers, and by oil industry members to build them, were a key factor in Congress's subsequent approval of the TAP route. But when the oil industry encountered hard times in the 1980s, double bottoms were for the most part forgotten. Then also, when the permit was issued, oil industry and federal officials assured the public that ice conditions in Prince William Sound would be carefully monitored with state-of-the-art equipment to minimize the chances that a tanker would collide with an iceberg. This monitoring never occurred.

Safety standards also declined over the years. In 1977 the average tanker going out of Valdez carried about forty crew members, but by 1989 the same size tankers carried only about twenty in crew, with twelve- to fourteen-hour workdays and exhaustion being the norm. In the first years of the pipeline, Alyeska had a spill-readiness team that worked around the clock running drills and maintaining equipment, but by 1982, when public watchfulness had slackened, these team members had been assigned to other duties as well, and the team and equipment could not be as quickly assembled. Finally, as a result of Coast Guard cutbacks in the 1980s, fewer people were assigned to the Valdez Coast Guard station, and tankers were no longer inspected when they left the terminal. The radar at the Valdez station was replaced by a less powerful unit, which meant that tankers could no longer be inspected when they left the terminal and could no longer be tracked as far as Bligh Reef. Unfortunately, the tanker crews had not been informed about the limits of the new radar. If the old radar had still been in use, the Coast Guard might have been able to warn the Exxon *Valdez* to turn sooner to avoid Bligh Reef.

Details of the Spill

The Exxon *Valdez* left the dock at Valdez at 9:12 P.M. on March 23, 1989, to embark on its five-day run to Long Beach, California. There had been a report of a heavy flow of Columbia Glacier ice into Prince William Sound, and this ice soon showed up on the ship's radar. A couple of hours into the voyage, the *Valdez* notified the Coast Guard station at Valdez of its intention to depart from the lane for outbound vessels in the traffic separation scheme and to travel out in the lane for inbound vessels in an attempt to avoid the ice. However, the *Valdez* went farther out of its way than it had reported, crossing first the two-thousand-yard zone separating traffic, then the fifteen-hundred-yard inbound lane, which also appeared to be full of ice, and continuing straight without making a right turn. Bligh Reef lay six miles ahead; as there appeared to be a gap of only nine-tenths of a mile between the

The first major challenge in cleaning up the area was mobilizing equipment and materials. Containment booms, such as that shown here, took fourteen hours to arrive at the spill.

edge of the ice and Bligh Reef, a well-timed turn would be necessary. Since six-tenths of a mile was needed to make the turn and the tanker was two-tenths of a mile long, the turn would have to start before the gap between the ice and the reef. Before leaving the deck to go below to his cabin, Captain Joseph Hazelwood, who had nineteen years experience with Exxon Shipping, told Third Mate Gregory Cousins to make a right turn when the vessel was across from Busby Island light and to skirt the edge of the ice, but failed to specify an exact course.

Perhaps because Cousins was looking at the radar screen, the tanker slipped past the light without starting to turn. At 12:04 A.M. on March 24, 1989, the Exxon *Valdez* grounded on Bligh Reef, tearing open its bottom and spewing oil into the surrounding waters. Although there is significant evidence that Hazelwood had been drinking before boarding the *Valdez*, and may have been drinking during the voyage, a jury subsequently acquitted him of all charges except negligent discharge of oil.

The management and cleanup of the spill were taken over from Alyeska by Exxon, which in turn was monitored by the Department of Environmental

Protection representing the state and the Coast Guard representing the federal government, as well as by special interest groups. The response to the spill, by all accounts, was slow and confused.

Mobilizing the necessary equipment and materials was the first major challenge. It took fourteen hours for the first barge to be loaded in Valdez and travel to the spill with containment booms, which are intended to float on the water and keep the oil from spreading. A day and a half into the spill, only a small line of booms were visible from the air. In addition to being too late and too few, there were further problems with the booms. Some sank because of leakage or punctures in their buoyancy compartments, and some intended for use in protected harbors were mistakenly deployed in open seas. However, about five days into the spill, after the effort to contain the great mass of the spill had failed, the booms were used successfully to protect a number of fish hatcheries.

Another fairly successful aspect of the cleanup was the lightering operation, or transferral of the remaining oil from the Exxon *Valdez* into another tanker. By early Saturday morning, more than twenty-four hours into the spill, the Exxon *Baton Rouge* was tied up alongside, and about four-fifths of the cargo of oil was recovered. Even so, by day four, the prevailing currents had carried what had spilled forty miles past islands and shoreline to the southwest; by day seven, ninety miles; by day eleven, one hundred forty miles; and by day fifty-six, four hundred seventy miles.

Because the long-term effects of two other remedies—chemical dispersants and bioremediation—are unknown, these were used only minimally. Dispersants don't remove oil from the water; rather, they break oil into tiny droplets that descend from the surface into the water column and, through rapid release of hydrocarbons from the oil, cause a severe toxic hit on marine organisms for about six hours. In general, it is considered better not to use dispersants close to shore where marine life is densest, but there exists a diversity of opinion here. Three days into the spill the Coast Guard authorized Exxon to use dispersants, and by that time gale force winds were blowing up and churning the oil into a mousse that could no longer be treated with dispersants. Even if authorization had been secured earlier, there was not nearly enough dispersant available to treat a spill of this magnitude. The skimming equipment was also useless in rough seas because mousse doesn't stick to the absorbent belts used to suck up the oil. By Monday morning, a blizzard was producing twenty-five-foot waves and less than one percent of the spilled oil had been recovered.

In the other controversial process, bioremediation, fertilizer is sprayed onto beaches to stimulate growth of naturally occurring bacteria that degrade oil and change it to a harmless substance. Only seventy miles of shoreline were treated with this fertilizer, however, because of the uncertain long-term effects of the bacteria on other life forms.

After the failure to contain and recover the spilled oil, the operation devolved into one of cleaning shorelines. At the peak of operations, more than eleven

More than one hundred endangered bald eagles and thirty-six thousand seabirds were found dead. Many more are thought to have died, but were never found.

thousand people, fourteen thousand vessels, and eighty-five aircraft were involved in the cleanup. The rocks along shorelines were washed with hot pressurized absorbent towels, or simply with absorbent towels. Many of the workers were involved with the wildlife—collecting dead animals so that the ingested oil wouldn't get into the food chain, or bringing sick ones in for treatment. More than one hundred endangered bald eagles, thirty-six thousand seabirds, and seven thousand sea otters were found dead. Many more are thought to have died but were never found.

Oil causes external damage to birds through loss of water-repellency and consequent loss of feathered insulation. The ingestion of oil lowers organisms' resistance to disease, reduces reproductive success, inhibits normal growth and development, and interrupts normal biochemical processes and behavioral patterns.

About 627 birds and 200 otters were saved in the rescue effort, which seemed to be mostly of symbolic and public relations value. Each otter saved is estimated to have cost Exxon between $40,000 and $89,000. From a strictly biological viewpoint, the rescue effort can't be seen as having had much

The Oil Pollution Act of 1990 requires all oil-carrying vessels operating in U.S. waters to be equipped with double hulls by the year 2015. Such a hull on the Valdez *would have cut the spill by sixty percent.*

value. Scientists do not know how long it will take for populations to restore themselves but estimate that for many species five to twenty years will be necessary.

When Exxon finished the bulk of its cleanup in September 1989, some of the shorelines appeared to be clean, but a layer of tar still existed not far below the shore surface. In some places, the oil had penetrated four feet and threatened to penetrate deeper.

Impact

As a result of this catastrophic spill, Congress passed the Oil Pollution Act of 1990, requiring all oil-carrying vessels operating in United States waters to be equipped with double hulls by the year 2015. A Coast Guard study performed after the spill has shown that a double hull on the *Valdez* could have cut the amount of spilled oil by sixty percent. Another Coast Guard report had shown that the cost of a double-hulled tanker, spread over its probable twenty-year lifetime, would be as low as three cents per barrel of oil carried.

Double-hulled tankers are believed to be advantageous in a low-energy grounding, in which the outer hull might be punctured, but the inner, nestled six feet or more inside the outer, remains intact. In a high-energy grounding, however, a rock or reef might rip both hulls apart. Japanese shipbuilders are championing an intermediate-deck design that might be better in high-energy groundings. In this design, a deck divides the tanker into upper and lower compartments. The lower section is below the ocean surface. If the hull is punctured, water should run in instead of oil running out, except for a small amount of oil released as the hull rolls.

In another response to the spill, the petroleum industry contributed $7 million to create the Petroleum Industry Response Organization, which established five regional centers along the coasts of the United States for fighting oil spills. These are staffed twenty-four hours a day and increase the availability of equipment for removing and cleaning up oil spills.

The Alaska state legislature also acted to require all tanker captains leaving port to take a breath alcohol test no more than one hour before boarding. In addition, tankers now must keep two pilots on the bridge until after they have passed Bligh Reef, and tankers must be accompanied by two tugboats while still in Prince William Sound. Tanker pilots are no longer allowed to change lanes in the traffic separation scheme.

The litigation that followed revolved around two questions: who had been at fault and whether there was negligence regarding the attempted cleanup. The state sued Exxon, Alyeska, and all the other owner companies, and Exxon countersued, charging that the state should pay much of the cleanup cost because it had interfered with the use of dispersants. The most significant settlement has required that Exxon set up a $900 million restitution fund.

SALLY VAN DUYNE

Where to Learn More

Cahill, Richard A. *Disasters at Sea: Titanic to Exxon Valdez.* London: Century, 1990.

Davidson, Art. *In the Wake of the Exxon Valdez: The Devastating Impact of the Alaska Oil Spill.* Sierra Club Books, 1990.

Keeble, John. *Out of the Channel: The Exxon Valdez Oil Spill in Prince William Sound.* Harper-Collins, 1991.

National Research Council. Committee on Tank Vessel Design. *Tanker Spills: Prevention by Design.* National Academy Press, 1991.

Skerrett, P. J. "Designing Better Tankers." *Technology Review,* September 1992, p. 8.

U.S. Coast Guard, American Petroleum Institute, U.S. Environmental Protection Agency (sponsors). *Proceedings of the 1991 International Oil Spill Conference,* March 4-7, 1991, San Diego Calif. American Petroleum Institute.

Spaceships

- Nike missiles (1958)
- Soviet rocket fire (1960)
- *Apollo 1* (1967)
- *Soyuz 1* (1967)
- *Apollo 13* (1970)
- *Soyuz 11* (1971)
- *Skylab* meteoroid shield (1973)
- *Challenger* (1986)

Nike missile explosions

Leonardo, New Jersey

(1958)

Explosion of seven ground-to-air Nike missiles during installation of new safety and arming mechanisms resulted in ten casualties.

Background

On May 22, 1958, at a missile launching site in Leonardo, New Jersey, a U. S. Nike ground-to-air missile exploded while civilian and military personnel were performing authorized modifications. The explosion set off six other missiles within the battery, killing ten workers and injuring two. An eighth missile was fired into a hill close by, and the accident scattered debris over a three-mile radius. There was no evidence of carelessness or incompetence, and the Army investigating board concluded that the probable cause of the explosion was the rupturing of a detonating cap. The Army settled all local damage claims, and discontinued the procedure that may have brought about the explosion.

The Nike missile, as part of the anti-aircraft system established in the 1950s to protect American cities from aerial attack, was a product of the Cold War between the United States and the Soviet Union. Designed by a service-industry team composed of the Army, Western Electric Company, Bell Telephone Laboratories, and the Douglas Aircraft Company, the Nike was the end result of ten years of guided-missile research and development.

The Nike, named after the goddess of victory in Greek mythology, was the Army's first anti-aircraft guided missile designed to intercept and destroy enemy targets regardless of evasive action. Development of the Nike had been completed by 1945, and firing tests were initiated in the fall of 1946. The first Nike missile battery was established at Fort Meade, Maryland in December 1953. By 1958, the Army had set up Nike installations to protect 23 zones across the country, embracing every major city in the nation. It described this system as an inner ring of defense for highly-populated, industrial, and strategic areas.

The Nike missile system consisted of two parts: the expendable missile, and the ground-based launching and control equipment. Also called the Nike

Ajax because of its Ajax-type upper stage, the Nike was a liquid-fueled supersonic anti-aircraft missile with two sets of fins for guidance and steering. Approximately 20 feet long and one foot in diameter, the Nike weighed more than one ton and normally was fired from an almost vertical position. Each missile contained several hundred pounds of TNT in three warheads with shrapnel projectiles, and was fueled by a highly explosive mixture of high-grade kerosene and fuming nitric acid.

Given its essentially defensive purpose, the Nike missile was launched from fixed emplacements or batteries. Each Nike battery was ready to fire at a moment's notice, and remained operative 24 hours a day. A battery consisted of three launching sections, each capable of launching four missiles without reloading. Because the emplacements had to be located within the zones they were defending, the batteries were almost always near major cities. The land requirement for each battery was modest, and public-owned property was often available for use; nonetheless, the defensive network did require that some batteries be established within privately-owned areas.

When it was installing these missile sites, the Army took great pains to persuade local populations and officials that they should not be concerned about safety issues. In one press release, the Army stated that a Nike missile site was "as safe as a gas station, as important to the security as the police or fire departments." This press release specified that: "Safety precautions are taken for storage of explosives and volatile fuels. Assembled missiles are stored underground. Fueling areas are surrounded by high earthen revetments." Thus was the Manhattan/New Jersey public assured that the 526th AAA Missile Battalion was perfectly safe and virtually accident-proof.

Details of the Explosion

On May 22, 1958, three separate activities were taking place at the 526th Missile Battalion in Leonardo, New Jersey. First, battery personnel were checking missiles in preparation for shifting to a higher state of alert. Second, an ordnance repair team composed of civilian personnel was repairing a launcher. Third, a team of ordnance civilians and enlisted men was performing authorized modifications to a group of Nike missiles. Although the first two activities did not contribute to the accident that occurred, they compounded its severity.

The modifications underway entailed the installation of new safety and arming or trigger devices on the Nike missiles. These modifications, which had been performed without incident on hundreds of missiles at other sites, involved removing the warhead and replacing the old trigger and brackets. By mid-day on May 22, the work team at the 526th Battalion had converted three Nike missiles. The procedure had become so routine that the missile battery control panel glowed with a "no danger" yellow light. Shortly after 1:00 P.M., the team began what was to be its fourth and final missile conversion of the day. At some point during this routine process, a mishap made the rocket explode. In theory, the arming device could not set off the missile until it was high in the air, but the warhead had been detonated.

One explosion led to another, and seven Nike missiles were completely detonated. While the eighth missile did not fully explode, its booster was ignited by a flying red-hot pellet, blasting the missile into the side of a nearby hill.

Although Army safety regulations specified that no more than two Nike missiles should be above ground at any given time, the overlapping activities meant that eight Nikes were out of storage and exposed. Consequently, one explosion led to another, and seven Nikes were completely detonated. While the eighth Nike missile did not fully explode, its booster was ignited by a flying red-hot pellet, blasting the missile into the side of a nearby hill. In 1955, a Nike missile accidentally had been fired from the launching pad at Fort Meade, Maryland, but no harm or injuries had resulted. The explosion at the 526th Missile Battalion, killing ten and injuring two, was the first fatal accident involving the Nike missile. The blast showered fragments over a three-mile radius, and was heard up to 15 miles away.

Impact

Understandably, both the Army and public officials were alarmed that such an unexpected event had occurred. This accident potentially meant that the other batteries and the corresponding populations were at risk. The 526th Missile Battalion itself, located on the New Jersey coast, was only 25 miles south of midtown Manhattan.

The Army immediately established a three-person Board of Inquiry and sent ordnance experts to the scene to investigate. The 526th Battalion was deactivated, and all missile conversion work at other Nike batteries was suspended. The Army, in cooperation with local officials, also instituted a plan to reassure the American public about safety concerns, and thereby retain support for the Nike program. State Department and Department of Defense officials also worried about the impact of negative publicity in Europe, particularly in Germany, where the United States was building Nike batteries.

On July 3, 1958, the board presented its findings to the general public through a press release. This press release indicated that: "After careful consideration of all the evidence, the Board concluded that the most probable cause of the explosion was the rupturing of a detonating cap."

The press release did not offer any details, or indicate how such a rupturing might have occurred. Nor did it acknowledge that, contrary to the Army's own safety regulations, eight Nikes had been above ground at the same time. The press release did state that "there was no evidence of gross carelessness, smoking, inattention to the operations, or any possible cause such as sabotage." The Board also reassured the public that "[c]orrective actions have been taken" to prevent a recurrence of the accident, and (somewhat confusingly) that "The procedure which may have caused the accident has been corrected and eliminated."

Ironically, the fatal explosion in New Jersey occurred just as the Nike Ajax became obsolete, and was being replaced by a superior version of the missile, the Nike Hercules. Whereas Nike Ajax had a speed of 1,200 miles per hour and a range of 20 to 30 miles, the new Nike Hercules would top 2,000 miles per hour and range more than 50 miles. However, the primary motivation for switching to the Nike Hercules was that this new missile could carry a small nuclear warhead. The United States had progressed in reducing the size, weight, and cost of nuclear munitions, and now intended to arm its domestic anti-aircraft network with a missile that could destroy not just one bomber, but an entire formation of enemy aircraft.

As the nuclear-tipped Nike Hercules shared various parts as well as the same guidance and control systems with the Nike Ajax, many public officials speculated about the possibility of a domestic nuclear disaster should a similar accident occur with a Nike Hercules missile. The Army, however, pursued its plans to use nuclear-tipped Nike Hercules in the domestic defense system, and launched a skilled publicity campaign to allay fears and retain public support. The emplacement of thousands of nuclear warheads in the suburbs of America's greatest cities thus began in 1958, and was completed by 1963. This network of emplacements, comprised at its height of 143 batteries, began to be scaled back in 1966. By 1972, all the Nike Hercules batteries had been deactivated except for a few in southern Florida.

Fortunately, there was never an incident relating the nuclear missiles comparable to the Nike Ajax accident at the 526th Missile Battalion. It could be said that the fatal explosion of the conventionally-armed Nike Ajax missiles in New Jersey alerted the Army to the fallibility of its systems and procedures, instigating reforms that may have prevented the occurrence of a domestic nuclear disaster.

LEONARD C. BRUNO

Where to Learn More

"Nike Ajax Probe Nears End." *Aviation Week*, June 2, 1958, p. 26.

Bruce-Briggs, B. *The Shield of Faith: A Chronicle of Strategic Defense from Zeppelins to Star Wars*. Simon and Schuster, 1988.

Department of Defense, Office of Public Information. News Release No. 648-58, Exhibit P. July 3, 1958.

"The Impossible." *Newsweek*, June 2, 1958, p. 18.

Raymond, Jack. "Army Stops Work on Nike Triggers, Blast Inquiry On." *New York Times*, May 24, 1958, p. 1.

"Death in the Neighborhood." *Time*, June 2, 1958, p. 16.

Rocket fire

Former Soviet Union

(1960)

Unexpected ignition of the fourth stage section of a rocket undergoing repairs after a failed launch attempt killed a Soviet Field Marshal.

Background

On October 24, 1960, Field Marshal Mitrofan Nedelin and scores of Soviet space engineers and technicians were killed when a four-stage launch vehicle undergoing repairs on the launch pad burst into flames. The high death toll resulted from Nedelin's decision to attempt repairs immediately after a failed launch attempt, bypassing the basic safety measures of draining the rocket's fuel tanks and deactivating its electrical systems. When the launch vehicle's fourth stage ignited its engines, the space probe was turned into a mammoth fire bomb. The Soviet Union largely withheld information about the disaster; two decades passed before details become available to the West.

During the early years of the space age, when the Cold War between the United States and the former Soviet Union spurred the development of rival space programs, the Soviets hoped to set a record for sending the first space probe (unmanned satellite) to Mars and Venus. Such space travel, in additional to the technological requirements, is subject to calendar-based constraints, since the logistics of an interplanetary voyage are controlled by orbital paths. In the case of Mars, a "launch window" opens up for a few days every 26 months, allowing a satellite to be launched from Earth with a minimum expenditure of propulsion power and fuel. A similar window opens to Venus every 18 months.

During the September 1960 visit of Soviet Premier Nikita S. Khrushchev to the United Nations headquarters in New York, Khrushchev hinted and bragged that such an interplanetary launch was about to take place. There were even rumors that he had brought along souvenir models of the Soviet interplanetary probes to give away during his visit. No such launch occurred during Khrushchev's trip to the United Nations, however, and his return home was followed by the report that Soviet Field Marshal Nedelin had been killed in an airplane crash. At this time, there were unsubstantiated reports that a manned launch attempt had somehow ended disastrously. Among the stories circulating was a December report from an Italian news agency claiming that a space-vehicle explosion had killed not only Nedelin, but one hundred Soviet scientists, engineers, and workmen as well.

Allegations that such a disaster had occurred did not surface again until the 1965 publication of *The Penkovskiy Papers*, the memoirs of a Soviet official who was executed for passing state secrets to the West. Penkovskiy provided few details about the accident, but did indicate that Nedelin and over three hundred people were killed when they approached the rocket pad fifteen or twenty minutes after the vehicle had failed to launch. The next mention of the disaster also appeared in Kruschev's posthumously published memoirs. Published in 1974, these papers essentially duplicated Penkovskiy's account. In 1976, the Soviet exile Zhores Medvedev confirmed the accounts of Penkovskiy and Kruschevby describing the event as "an irreparable catastrophe."

It was not until 1982, however, that the Soviet émigré S. Tiktin provided what many now consider the most plausible and reliable account of the event. Tiktin was responding in writing to the speculations of James E. Oberg, an American expert on the Soviet space program, about the disaster.

Details of the Fire

The Soviets were using the newly developed SS-7 ICBM (inter-continental ballistic missile) for the space probe that Khrushchev bragged about during his September 1960 trip to New York. Bugs had not entirely been eliminated from the system, however, and the politically charged project was subject to several delays. The launch was finally scheduled for October 24, 1960, under the supervision of Field Marshal Nedelin. When the countdown for the launch expired and the rockets failed to ignite, Nedelin must have felt compelled to go to any lengths to avoid yet another major delay.

Once the launch had clearly failed, Nedelin emerged from the protective bunker and indicated that the necessary repairs to the fuel-feed system would be made then and there, without performing the time-consuming safety precaution of draining the fuel tanks. This safety principle had been in force ever since the Germans began their liquid-fuel rocket experiments during World War II, and only someone of Nedelin's stature in the Soviet bureaucracy could have initiated such an order.

A similar gamble had paid off in July 1960, during the launch of an unmanned Vostok spacecraft. In this case, workers were able to weld a leak shut in the booster (first stage) rocket while the launch vehicle was still on the pad and without draining out all of its dangerous propellants. Nedelin may have had this earlier, successful gamble in mind when he ordered his crew to proceed immediately with repairs. Furthermore, as Nedelin undoubtedly recognized, his willingness to oversee the repairs personally would have a reassuring effect.

Once the launch had failed, Nedelin emerged from the protective bunker and indicated that the necessary repairs would be made then and there, without performing the time-consuming safety precaution of draining the fuel tanks. However, the fourth stage launch mechanism did not receive the shut-off command. When it ignited on schedule, the vehicle was still on the launch pad, not high above the earth.

Tiktin's report of the disaster explains that Nedelin and his crews set to work without realizing that the launch vehicle's fourth stage had for some reason not received the stand-down (temporary shut-off) command. According to the electronic information processed by the fourth stage, the launch had indeed occurred as planned. Thus, as the work supervised by Nedelin proceeded, the timer that controlled the fourth stage was ticking away, just as if the rockets of the first and second stages had completed their work and then fallen away from the upward-streaking rocket. Within a short period of time, this timer indicated that the small third stage was firing and placing the rocket into a low parking orbit. Following this stage, the fourth stage would ignite its 2,000 pounds of fuel to escape from the earth's gravitational pull and shoot on toward Mars.

When the time came for the fourth stage to fire its rockets, it was in fact sitting on top of an inert but fully fueled launch vehicle. The fourth stage was 120 feet above the launch pad, and only inches from the top of the third stage's fuel tank. When the fourth stage fired its engines, it took only seconds for the flames to burn through the thin metal top of the third stage's fuel tank and detonate the 20 tons of kerosene contained there. The million pounds of kerosene and liquid oxygen held in the two large, lower stages would then have ignited immediately.

Tiktin described the ensuing explosion as a roaring, monstrous fire rather than a destructive concussion. Those individuals unfortunate enough to be on the pad at the time of the explosion were instantly immolated, while those inside the bunker were unharmed. Remarkably, some of the bus drivers who had brought the crew to the pad and were sitting at the wheel 100 or 200 yards away from the explosion received only slight burns from scattering kerosene. Bus windows were not even shattered.

Impact

The only official Soviet reference to the accident was the brief announcement that Nedelin had died on October 25, 1960 in a plane crash. Nearly twenty years later, an official biography of Nedelin made no mention of an airplane crash as the cause of his death, stating only that Nedelin died "in the performance of his official duties." Following Tiktin's 1982 account of the accident, the Soviet government neither confirmed nor denied the story.

Tiktin's attempt to explain how such a seasoned space professional as Nedelin could violate the most basic principle of rocket safety places the blame on politics. Although Nedelin would have been fully aware of the risk he was taking, he responded to the enormous political pressure upon him to produce a successful launch to Mars. He was simultaneously constrained by the inflexible "launch window" inherent in every interplanetary attempt. Evidently, Nedelin felt he had no choice but to gamble.

This disaster cost the Soviet Union not only the experience and leadership of Mitrofan Nedelin, but a substantial portion of its skilled launch vehicle technicians. The loss of this body of expertise represented a major set-back to the Soviet's manned space program. If the explosion had not occurred, the epoch-making space flight manned by Yuri Gagarin might have taken place even earlier than April 12, 1961. But perhaps the ultimate lesson of this catastrophe, which took the lives of anywhere from 40 to 300 individuals, is the often unforgiving nature of space technology—especially when politics takes precedence over safety.

LEONARD C. BRUNO

Where to Learn More

Medvedev, Zhores. "Two Decades of Dissidence." *New Scientist*, November 4, 1976, pp. 264-267.

Newkirk, Dennis. *Almanac of Soviet Manned Space Flight*. Gulf Publishing, 1990.

Oberg, James E. *Red Star in Orbit*. Random House, 1981.

———. *Uncovering Soviet Disasters: Exploring the Limits of Glasnost*. Random House, 1988.

Penkovskiy, Oleg. *The Penkovskiy Papers*. Avon, 1965.

Talbott, Strobe (ed). *Khrushchev Remembers: The Last Testament*. Little, Brown, 1974.

Tiktin, S. "U. S. S. R. and Space." *Posev*, October 1982, pp. 46-50.

Apollo 1 capsule fire

Cape Canaveral, Florida

(1967)

Deaths of three astronauts during a ground test due to a flash fire evidently set off by a wiring malfunction exposed need for higher design, manufacturing, and safety standards at NASA.

Background

On January 27, 1967, at the NASA (National Aeronautics and Space Administration) base in Cape Canaveral, three astronauts were killed during a routine ground test of the Apollo command module. Although the exact cause of the fire remained undetermined, it probably began with an electrical arc caused by poor wiring design and installation. Once some combustible materials ignited, the fire was fed by the pure oxygen under pressure in the module. The astronauts had no equipment to suppress the fire, and were unable to open the six-bolt escape hatch in the short time that elapsed before the fire blazed.

With the successful completion of NASA's two-man Gemini program, AS-204 was to be the first manned Apollo mission. The United States was in its race with the Soviet Union to land men on the Moon, and it was with the three-man Apollo capsules that the Americans planned to accomplish this feat. In August 1966, the AS-204 crew was named. It consisted of two space veterans, Virgil I. "Gus" Grissom and Edward H. White, and one new astronaut, Roger B. Chaffee, who had not yet flown in space. NASA planned for the crew to make a December 1966 "shakedown" flight of up to two weeks in Earth's orbit. This flight would test the Apollo's new command and service modules (CSM) built by North American Aviation, which had not been deployed on a manned mission. Following a series of mechanical problems, however, the flight date was postponed until February 1967.

In preparation for this flight, the astronauts were to complete a four-phase ground test of the CSM's systems. One phase of this process required that the astronauts enter the command module, which was perched on a *Saturn IB* launch vehicle and emptied of fuel; seal and pressurize the module; and then begin a "plugs out" test to see if the spacecraft could run on its internal power system.

Apollo 1 *crew members Virgil I. (Gus) Grissom, Edward H. White, and Roger B. Chaffee answer questions at a 1966 press conference in California. All three perished during a routine ground test of the Apollo command module.*

The command module was designed to function in space, and therefore to be subject to greater internal than external pressure. To simulate this condition as closely as possible, NASA's engineers planned to pressurize the cabin with pure oxygen at approximately 16 pounds per square inch, rather than the normal 5 pounds per square inch of air. Although pure oxygen itself will not ignite, it can, when pressurized, rapidly feed an existing fire. No one, however, considered a cabin fire likely. There was no fire extinguisher in the capsule, and, since the launch vehicle was not fueled, the fire crews were on standby rather than maximum alert. Furthermore, no one considered excessive the 90 seconds it took to open the new six-bolt escape hatch; most experts were convinced that the alternate quick-release hatches were accident-prone and dangerous.

NASA's apparent complacency about safety precautions stemmed from the fact that, in six years of manned space flights, no astronaut had died in the course of a mission. The "plugs out" test, conducted with an unfueled launch vehicle, seemed both routine and low-risk.

Details of the Fire

On January 27, 1967, "Gus" Grissom, Edward White, and Roger Chaffee undertook the "plugs out" test on the new Apollo command and service modules. By 1:00 P.M. on that Friday afternoon, the three-man crew had crawled through the open hatch and assumed their flight positions. After two hours of tests conducted with the hatch open, the capsule was sealed and the cabin pressurized to 16.2 pounds per square inch of pure oxygen. The crew then went through a practice countdown and ran simulation tests for more than three hours, regularly interrupted by minor problems.

Shortly after 6:00 P.M., fifteen minutes before lift-off was to be simulated, the spacecraft switched to internal power. NASA engineers called a hold to check on some problem, and the astronauts waited yet again. The astronauts had been in their couches some five and one-half hours; it had been a long day. Suddenly, at 6:31 P.M., telemetry from the spacecraft indicated that a major short had occurred somewhere in the nearly twelve miles of electrical wiring packed into the command module. Less than ten seconds later, Roger Chaffee made an almost casual report: "Fire, I smell fire."

The spacecraft was not equipped with internal cameras, but a camera was focused on its porthole. At the first report of fire, all the camera operator could see was a sudden bright glow. The operator then saw flames flickering across the porthole and Edward White's hands reaching above his head to get at the bolts securing the hatch. He saw a lot of movement, and then another pair of arms struggling with the hatch. Soon, dark smoke completely obscured the scene. The last sound from the astronauts was a now frantic cry from Roger Chaffee: "We've got a bad fire—let's get out . . . we're burning up!" Seconds after this last transmission, the tremendous pressure inside the cabin split the capsule open, and a blaze of flame gushed out. Only 18 seconds had passed from the first call of fire to this explosion.

Help was close at hand, but the control personnel in the White Room were momentarily held back by the explosion. Due to the thick smoke, it then took five men working in shifts five and a half minutes to remove three separate hatches: the boost protective cover that shielded the command module during launch, the ablative hatch, and the inner hatch. Inside the capsule, once the smoke thinned, they found the three dead men. Chaffee was still strapped onto his couch, and the bodies of White and Grissom were lying close together below the hatch. White's handprint was outlined in ash on the hatch.

Official autopsies would later identify the cause of the deaths as asphyxiation, observing that although each astronaut had suffered serious burns, these were not fatal. However, the heat during those few terrible seconds had been so intense—the holes burned in aluminum tubing indicate temperatures of at least 760 degrees Celsius—that the astronaut's suits had melted and fused with the molten nylon and Velcro inside the capsule, forming a synthetic liquid that solidified as it cooled. Doctors arrived fourteen minutes after the first alarm of fire, but it took them seven hours to remove the bodies; those of White and Grissom had been welded to the capsule floor.

Impact

NASA's entire multi-billion dollar effort to put a man on the Moon virtually came to a standstill as a special Board of Inquiry sought explanations for the accident. On February 22, 1967, almost a month after the disaster, a seven-man review board issued an interim report stating that although no definite cause of the fire could be established, the most likely origin was an electrical malfunction. A fourteen volume report of some three thousand pages came out in early April; this report specified that the fire was probably caused by an electrical arc that occurred in the vicinity of the environmental control equipment under Grissom's couch.

The full report was highly critical of the conditions that enabled the fatal accident. It indicated that the some of the wiring unaffected by the fire revealed "numerous examples of poor installation, design, and workmanship." Moreover, the capsule was loaded with highly combustible materials such as Velcro and the nylon netting used to prevent loose objects from floating around in the zero gravity of space. In its conclusion, the report implied that disaster may have been inevitable given the substandard manufacturing procedures and the lack of safety measures.

The critique of NASA at least implicit in this report was not lost on Congress, and a House space subcommittee opened hearings on April 7, 1967 to assess the space program and the Apollo accident. Most observers decided that NASA had been pushing the Apollo program too hard and too fast; there simply had not been enough time to thoroughly test all its systems. The Apollo spacecraft was still an unproven and evolving craft, having undergone 623 changes between August 1966, when it was delivered by North American Aviation to NASA, and the fatal accident in January 1967. By the end of the Congressional hearings, the prevailing view was that making limited adjustments to the Apollo capsule and allowing the space program to continue on its timetable would be ill-advised. An exhaustive review of the entire spacecraft was necessary, however long it took.

NASA was therefore obligated to step back and reassess the full scope of its systems and procedures, particularly in terms of safety standards. NASA

The flash fire that spread through the Apollo 1 *spacecraft, killing all three occupants, was probably caused by an electrical arc that occurred in the vicinity of the environmental control equipment under Grissom's couch.*

took good advantage of this enforced break from its rivalry with the Soviet space program, and the outcome of a year and a half's worth of reevaluation was a completely redesigned Apollo command module. Some fifteen hundred modifications were made to the command module, resulting in a considerably more secure and fireproof vehicle.

First, NASA installed high-quality wiring. Flameproof coatings were applied over all wire connections, plastic switches were replaced by metal ones, and soldering became more meticulous. Almost all flammable materials inside the module were removed. A new, fire-resistant material known as Beta cloth was developed for spacesuits. Instead of igniting at 500 degrees Celsius, as did the old suits made of Nomex, the Beta cloth suits could withstand temperatures of more than 800 degrees Celsius.

A considerable debate over the use of pure oxygen resulted in a compromise in the favor of safety. When they were in space, the crew would breathe pure oxygen at five pounds per square-inch. For ground testing and launching, however, the cabin would be filled with a mixture of oxygen and nitrogen at sea-level pressure. To prevent this nitrogen from causing "bends" (nitrogen narcosis or poisoning), the astronauts would breathe only through their spacesuits, which contained pure oxygen, until the cabin had been purged of nitrogen.

One irony of the Apollo accident was that Grissom, White, and Chaffee's workday was scheduled to end with a test of the new six-bolt escape hatch. This system had replaced the quick-release, explosively charged hatch that had been used on both Mercury and Gemini spacecraft, but which in 1961 had been blamed for prematurely blowing and almost sinking Grissom and his *Mercury Liberty Bell* 7 when he landed in the Atlantic. The new Apollo design returned to a quick-escape system; this improved hatch took only twelve seconds to release and opened outward so that internal pressure would not affect its functioning. Among the revelations of the full report on the Apollo accident was that even if the astronauts had managed to undo the six-bolt hatch, their efforts would have been futile. When the internal pressure exceeded the external pressure by more than 0.25 pounds per square inch, the hatch, which swung inward, became impossible to open.

If good can be said to come of such a disaster, it was that the American government and industry coalition became conscious of the need to raise design, workmanship, and safety standards. The fire both enabled and forced NASA to step back from its politically-controlled timetables and methodically build the high-performance spacecraft that ultimately proved capable of putting a man on the Moon.

LEONARD C. BRUNO

Where to Learn More

Bond, Peter. *Heroes in Space: From Gagarin to Challenger*. Basil Blackwell, 1987.

McAleer, Neil. *The OMNI Space Almanac: A Complete Guide to the Space Age*. World Almanac, 1987.

Murray, Charles and Catherine Bly Cox. *Apollo: The Race to the Moon*. Simon and Schuster, 1989.

"Electrical Malfunction Termed a Likely Source of Apollo Fire." *New York Times*, February 26, 1967, p. 39.

Wheeler, Keith. "Disaster—The Harsh Schoolmaster." *Life*, January 26, 1968, pp. 56-59.

Wilford, John Noble. "Apollo Fire Review Boards Finds 'Many Deficiencies'; Calls for Safety Moves." *New York Times*, April 10, 1967, p. 1.

Soyuz 1 spacecraft reentry crash

Orenburg, Russia

(1967)

Fatal crash due to spacecraft's landing parachutes becoming entangled after flight and reentry control failure sent the Soviet manned space program into decline at a crucial phase of the "space race."

Background

In the former Soviet Union's first space disaster, veteran cosmonaut Vladimir Komarov was killed when the *Soyuz 1* spacecraft's reentry into the Earth's atmosphere could not be controlled. The flight of this new class of Soviet spacecraft was apparently troubled from the onset, although Soviet officials maintained that the mission proceeded as planned until after reentry, when the sudden entanglement of the lines of the spacecraft's landing parachutes brought about the crash. Western experts, in attempting to reconstruct fully what happened, have concluded that solar panel and antenna malfunctions probably led to a major control failure during orbit, and that the spacecraft's "tumbling" through a forced reentry could have made the parachute foul-up and crash landing inevitable. Most observers also agree that political considerations played a major role in fatally rushing the *Soyuz 1* launch date.

When *Soyuz 1* was launched on April 23, 1967, it had been over two years since the last Soviet manned space flight. The Soviets had taken this time to develop the new Soyuz class of manned spacecraft. Soviet cosmonauts had flown in the older Vostok and Voskhod spacecrafts, but these spacecrafts had automatic controls and were essentially operated from the ground. "Soyuz" means "union" in Russian, and, as the name implied, the new spacecraft would be used for sophisticated rendezvous and docking operations that would rely on cosmonaut control of the navigations systems. For the first time, Soviet cosmonauts truly would function as pilots.

During the two-year hiatus in Soviet manned flights, the United States successfully executed ten consecutive two-man launches in the Gemini series. Therefore, the Americans were pulling significantly ahead in the race to put a man on the Moon until the fatal *Apollo 1* fire on January 27, 1967. The American disaster put the Soviet Union and their new Soyuz spacecraft back into the competition. The Soviets entrusted the first Soyuz mission to the veteran cosmonaut Vladimir M. Komarov, who would become the first Soviet to make a second trip into space.

The *Soyuz 1* spacecraft was 7.9 meters long, consisting of an orbital module, a descent module, and an equipment module. Weighing nearly 400 kilograms, it would be launched from the Baikonur space center by the A-2, a standard Soviet launch vehicle. Once in orbit, the launch vehicle would deploy a set of solar panels that unfolded like wings, and whose 14 square meters of cells powered the spacecraft's systems. Soyuz crews would spend their mission in the orbital module, but reenter the Earth's atmosphere in the descent module after jettisoning both the orbital and the equipment modules. To a degree, the American Apollo spacecraft relied on a similar system. In the American system, however, astronauts would "splashdown" into the sea and be retrieved by helicopters operating from an aircraft carrier. The Soviets opted to land their spacecraft, equipping it with two sets of parachutes and a special retro-rocket system to cushion the impact.

Although the *Soyuz 1* was a technologically groundbreaking mission, the Soviets were uncharacteristically open about their plans. In conferences granted to the Western press, Soviet officials impressively indicated that the mission was to involve two separate manned spacecraft, a rendezvous and docking, and perhaps a crew transfer.

Details of the Crash

Soyuz 1 was launched on April 23, 1967 at 3:35 A.M. Moscow time, the first Soviet manned spacecraft to be launched at night. Several hours after *Soyuz 1* entered into its nearly circular orbit 200 kilometers above the earth, Soviet television and radio suddenly became silent about its progress. Even so, Western had little definite indication that there was any trouble with the flight until nearly one day later, when the scheduled launch of the second Soyuz craft (which the orbiting spacecraft presumably was to link up with) did not occur. Subsequently, at 6:00 A.M. on April 24, Komarov was ordered to return to Earth. At this point, Soviet officials indicated that a modest one-day test flight had proceeded according to expectations.

After twelve more hours of ominous silence, Moscow announced that a freak accident had caused the spacecraft's parachute lines to tangle just prior to landing, and that a fatal crash had ensued. Komarov was the first person to be killed during a space mission, and was buried with full honors. But although the *Soyuz 1* crash was a global news event, the Soviets released no further details about the accident. Officials maintained that nothing untoward had happened during the mission, and that the freak tangling of the parachute lines was alone responsible for the disaster.

Since the *Soyuz 1* crash, nonetheless, Western experts have analyzed what information is available about the accident in an attempt to determine what happened. Most of these experts agree that the failure to launch the second Soyuz was due to a problem with the *Soyuz 1* spacecraft already in orbit. Additionally, there is evidence that one of the two solar panels of the original

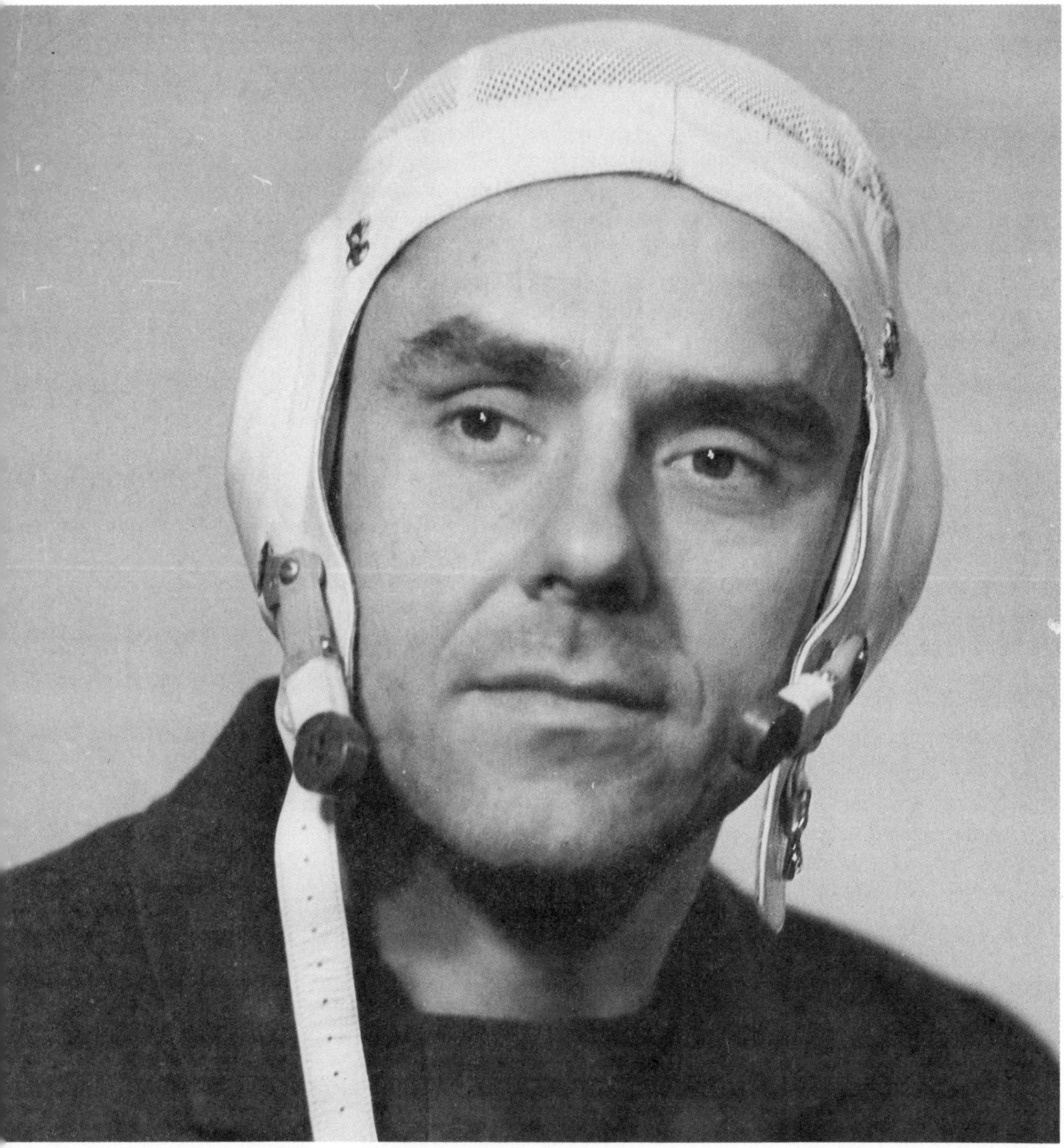

Veteran cosmonaut Vladimir M. Komarov was killed when the Soyuz 1 *spacecraft's reentry into the Earth's atmosphere could not be controlled.*

spacecraft did not deploy properly when the craft reached orbit. This meant that the panel's thermal radiator would be covered, reducing the spacecraft's electrical power supply. Most experts also hold that the spacecraft's television transmitter must have failed, and that other antennas also did not deploy properly. The most serious problem, however, was evidently a malfunction of the spacecraft's altitude control and guidance systems. This malfunction has been attributed to an overheating caused by the still-folded solar array, which could have prevented the heat generated by the spacecraft's electronic systems from dissipating.

The fact that the *Soyuz 1* spacecraft came down so far short of its planned landing site—600 miles off course, near the city of Orenburg on the steppes north of the Caspian Sea—supports the control failure hypothesis. Studies have shown that the 600 mile undershoot could have been caused by the spacecraft's spinning during reentry. If a spacecraft's guidance systems fail, moreover, a pilot might deliberately trigger such a spin or roll in a desperate maneuver to stabilize the craft. United States tracking facilities documented that the *Soyuz 1* spacecraft was tumbling or slowly rotating end-over-end during its fifteenth orbit.

By the seventeenth orbit, Komarov had been given permission to reenter. Evidently, however, he could not steady the spacecraft sufficiently to fire the reentry rockets accurately. Komarov apparently initiated the spin or controlled roll maneuver on the next orbit, and then fired the reentry rockets with the hope that the roll would balance all the aerodynamic forces set into play. The danger to this maneuver, however, is that the spinning descent that resulted could subject Komarov to up to 10 G's (10 times the force of gravity), double the normal reentry G's of 4 to 5. Thus, Komarov might have lost consciousness during reentry.

Even if he did not lose consciousness, Komarov probably was unable to stop the spacecraft's rotation before the parachute deployed at an altitude of four miles above the Earth. The spinning would have caused the shroud lines to become tangled around the spacecraft, preventing the parachute from unfolding or slowing down the spacecraft's rate of descent.

Even if he did not black out, Komarov probably was unable to stop the rotating of the *Soyuz 1* spacecraft before the parachute deployed at an altitude of four miles above the Earth. The spinning would have caused the shroud lines to become tangled around the spacecraft, preventing the parachute from unfolding or slowing down the spacecraft's rate of descent. The same factor would have rendered the emergency parachute useless. Whatever the case, there is little doubt that Komarov and the spacecraft hit the ground at a tremendous speed—between 300 and 500 miles per hour. No details about the wreckage have been released, but such a high-impact crash would have devastated the spacecraft.

The reluctance of Soviet officials to release details about what went wrong with the mission allowed all sorts of stories to flourish. One of these was the account a U. S. National Security Agency technician, who years later stated that he had overheard the *Soyuz 1* spacecraft-to-ground communications from his listening post in Turkey. He claimed that Komarov, fully aware of the dangers reentry would involve, protested to the Soviet ground control that "You've got to do something. I don't want to die." This technician also reported that Komarov was allowed to speak with his wife one last time. A separate account from Italy has Komarov screaming at the beginning of the fatal reentry maneuver: "You are guiding me wrongly ... can't you understand?"

Impact

Years after the crash, a Soviet engineer who emigrated to the United States discussed the *Soyuz 1* mission in his memoirs. He indicated that all the design bugs had not been eliminated when the Soyuz spacecraft was launched. Furthermore, the author charged that repeated failures had occurred in the pre-flight tests, especially in the crucial thermal control and parachute systems, and that everyone involved knew of these malfunctions. In fact, the author reports that the deputy chief designer (named Mishin) in charge of the space mission refused to sign the pre-flight papers endorsing the launch. Over the objections of the technical staff, however, the political arm of the Soviet government ordered the *Soyuz 1* mission to proceed. Soviet politicians feared that they would otherwise fall even further behind in their "space race" with the United States.

The subsequent loss of *Soyuz 1* signified more than the failure of an individual mission; the Soviet government and people were stunned and humiliated by their first space disaster. Given the international embarrassment surrounding the accident and the technological problems that obviously remained to be solved, it is not surprising that the Soviet manned space program entered a period of decline. The next test flight of a Soyuz spacecraft would not take place until 18 months after the disaster. With this test flight, the Soviets were just beginning to accomplish the manned orbital maneuvers that the Americans had achieved long before in the highly successful Gemini program. Twenty-seven months after Vladimir Komarov was killed, the United States became the first country to land men on the Moon and return them safely to Earth, and that facet of the "space race" had been won.

LEONARD C. BRUNO

Where to Learn More

Bond, Peter. *Heroes in Space: From Gagarin to Challenger*. Basil Blackwell, 1987.

Clark, Evert. "Soyuz Problems Discerned in U.S." *New York Times*, April 28, 1967, p. 20.

Newkirk, Dennis. *Almanac of Soviet Manned Space Flight*. Gulf Publishing, 1990.

"U. S. Aides Believe *Soyuz 1* Was Out of Control." *New York Times*, April 26, 1967, p. 3.

Oberg, James E. *Red Star in Orbit*. Random House, 1981.

———. *Uncovering Soviet Disasters: Exploring the Limits of Glasnost*. Random, 1988.

———. "Vladimir Komarov, First Space Casualty." *Space World*, June 1974, pp. 4-15.

Apollo 13 oxygen tank rupture

(1970)

Explosion in service module oxygen tank, forcing three-man crew to abandon their mission and use lunar module as an emergency "lifeboat," led to cutbacks in the Apollo program.

Background

On April 13, 1970, two days after the three-man crew of *Apollo 13* had been launched toward the Moon, an oxygen tank explosion disabled the spacecraft. This forced the astronauts to cancel the scheduled lunar landings and undertake a precarious completion of the voyage. With their command module steadily losing power, the crew transferred into the cramped lunar module and carefully rationed power, oxygen, and water supplies for the four-day trajectory around the Moon and back to Earth. The astronauts resorted to various measures to survive the trip back; in doing so, they were able to use the lunar module in ways never before attempted. Just prior to reentry, the crew transferred back into the command module, powered it for a last time, separated from their lunar module "lifeboat," and safely returned to the Earth.

Apollo 13 followed the highly successful manned lunar landing flights of *Apollo 11* and *Apollo 12*. Its mission was even more ambitious, since two moon-walks were scheduled. The spacecraft was made up of paired command and service modules (CSM); the former module housed the crew, the latter contained the engine and supply systems. The independent lunar module was attached to the command and service modules. This third module was designed to be used for the Moon landings while the CSM continued in a lunar orbit. The lunar module would then blast off from Moon, relink to the command and service modules, and eventually be discarded.

The crew was commanded by James A. Lovell, a veteran of two Gemini missions and the first circumlunar flight, *Apollo 8*. He was joined by two rookie astronauts: Fred W. Haise, who would pilot the lunar module, and John L. Swigert, the command module pilot. Swigert was a last-minute replacement for Thomas K. Mattingly, who had been exposed to German measles.

The *Apollo 13* flight seemed to both test and confirm the bad luck traditionally associated with the number thirteen. The 13th Apollo voyage, it was launched at the 13th minute of the 13th hour of the day, Houston time. On April 13, two days after its launch, the sudden crisis occurred.

Details of the Accident

The accident that handicapped the *Apollo 13* spacecraft took place in the initial part of the mission, approximately 56 hours and 205,000 miles into the flight. The astronauts heard a bang that was followed by a pressure drop in the service module's oxygen tanks and, correspondingly, in the command module. John Swigert matter-of-factly reported the malfunction in the service module oxygen tanks: "Houston, we've got a problem." No one in the spacecraft or at the NASA base in Houston understood why the oxygen pressure gauge for tank 2 soon read "zero" and the needle to the tank 1 gauge was dropping.

Thirteen minutes after the bang, the *Apollo 13* crew noticed a white, wispy cloud beginning to surround the service module. They correctly guessed that it was their oxygen. An oxygen supply was vital to the mission—both for the crew to breathe and for the fuel cells in the service module to produce electrical power and water. Its uncontrolled venting explained not only the loss in pressure, but the spacecraft's tendency to keep veering off course. The spacecraft would, in addition, gradually be encircled by debris from the explosion.

The controllers in Houston, however, were still struggling to interpret what their telemetry was telling them. The data seemed to indicate that the built-in redundancy of the Apollo systems was proving insufficient. But at this point no one could imagine that the spacecraft was actually undergoing the loss of a double redundancy system: two fuel cells and two oxygen tanks. Nor did anyone realize that a serious explosion had taken place. The problem was attributed instead to some sort of system breakdown or failure.

Despite this lack of understanding of what had transpired, it was evident to both Houston and the *Apollo 13* crew that the central command and service modules were becoming inoperative. About one hour and fifteen minutes after the accident, the only possible course of action became clear. The command and service modules had to be shut down; they had only about fifteen minutes' worth of power left, without which the spacecraft could not return to Earth. Thus, the crew was forced to deactivate the two modules and power the lunar module for use as a survival craft. This "lifeboat" option had been anticipated by the builders of the Apollo spacecraft, but no one had ever tested or fallen back on this strategy.

Once the crew transferred into the small lunar module, the NASA controllers in Houston had to improvise nearly every procedure that enabled the three astronauts to survive during the spacecraft's trajectory back to Earth. The lunar module was designed for the use of two men for 33 to 35 hours; this emergency required that it accommodate three men for up to 100 hours.

Houston also had to calculate how to use the lunar module's thrust engines to propel a complex spacecraft configuration.

All power, oxygen, and water supplies on the lunar module had to be strictly rationed. Power consumption was reduced to one-fifth of the normal level. Conserving power in part meant allowing the spacecraft's temperature to drop to a near-freezing 38 degrees Fahrenheit. As it got colder inside the modules, the humidity rose and frost formed on the inside of the windows. The cold became almost unbearable, and Fred Haise developed a kidney infection that was to plague him for the rest of the trip.

A buildup of carbon dioxide, moreover, was a serious threat. The lithium hydroxide canisters in the lunar module could not remove all the carbon dioxide that three men would produce. After the equivalent canisters from the command module were found useless because differently-shaped, the crew devised a makeshift system to neutralize the carbon dioxide. The astronauts also were restricted to six ounces of water each per day. While starvation was not at issue, the food packs could not be reconstituted, as there was no hot water. In addition, no waste was ejected into space for fear that any venting might affect the spacecraft's delicate trajectory back to earth.

The navigation of the crippled Apollo spacecraft required its own set of improvisations. Once the craft reached the far side of the Moon, the lunar module's engines had to be fired very precisely to free the spacecraft from the Moon's gravitational pull and send it back toward Earth. Once this maneuver was accomplished, three other engine firings were crucial to speeding the craft's return to Earth and correcting its course. The execution of these mid-course corrections was delicate and tense, since the debris from the original explosion now encircled the spacecraft like a cloud and gleamed in the Sun. This glittering debris made it impossible to find the familiar guide stars that the astronauts could have used for navigation and calculation of proper alignment. The NASA personnel in Houston, however, were able to conduct simulations and devise new strategies to navigate the craft.

With each day that the *Apollo 13* spacecraft came closer to Earth, the NASA controllers became more concerned about the physical state of the crew. In addition to the cold and the shortage of oxygen and water, there was no place to sit or recline in the lunar module, which had only standing room for two men. A third crew member, therefore, was forced to spend periods of time trying to sleep in the even colder command module. Houston later learned that none of the crew had more than 12 hours of sleep in the three and one half days it took to return to Earth. NASA was particularly worried about the effects of fatigue, cold, and dehydration on the crew's efficiency, since the astronauts would soon have to copy down correctly an emergency checklist of procedures and commands read to them from Houston. This task alone would take two hours.

Through these commands—which specified hundreds of switch, valve, and dial positions—the *Apollo 13* crew readied their craft for the final reentry and splashdown. When still several hours from Earth, they attempted the crucial

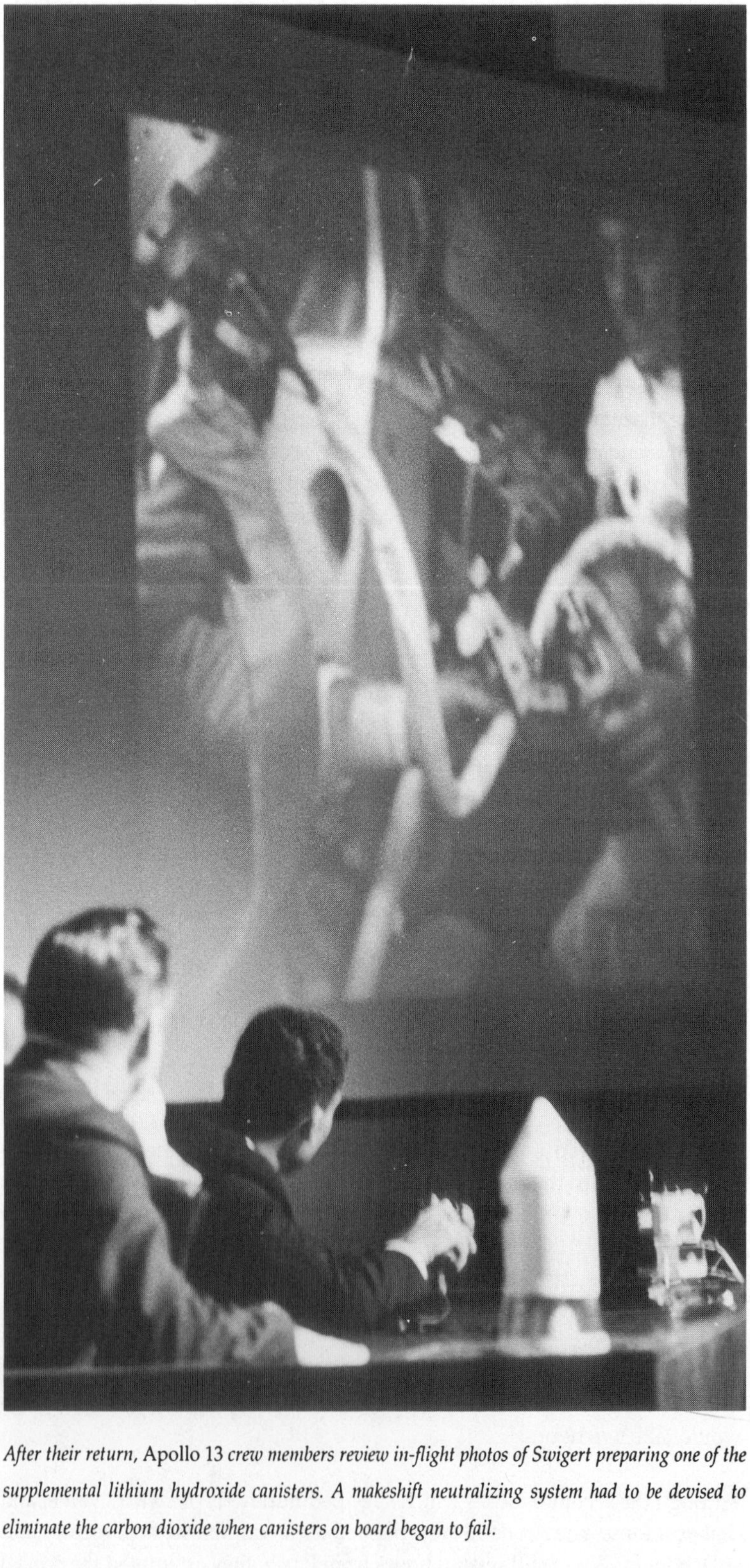

After their return, Apollo 13 *crew members review in-flight photos of Swigert preparing one of the supplemental lithium hydroxide canisters. A makeshift neutralizing system had to be devised to eliminate the carbon dioxide when canisters on board began to fail.*

power-up of the command module. It was this module that contained the heat-shielded capsule designed to protect them during the fiery reentry into Earth's atmosphere. If the crew could not activate this module, they would be left stranded in space. Fortunately, the power-up worked, and for the first time in days the astronauts began to feel warm.

On April 17, five hours before splashdown, the lunar module's thrust engines were fired a final time to initiate the reentry. A half an hour later, the crew used explosive charges to jettison the damaged service module. As it drifted away from them, the astronauts saw and photographed the extent of the damage the service module had suffered. Swigert exclaimed that: "There's one whole side of that spacecraft missing!" Subsequently, the three astronauts left the lunar module for the last time. Once inside the command module, 90 minutes before reentry, they cast off what had served as their lifeboat. "She was a good ship," Lovell said.

The remaining module soon was reentering the Earth's atmosphere. As it streaked through at some 25,000 mph, the heat shield (which ground controllers feared had been damaged in the explosion) was able to withstand a temperature of 4,000 degrees Fahrenheit. After an unusually long period out of contact, the capsule parachuted down within full view of the *Iwo Jima*, a U.S. carrier. Remarkably, *Apollo 13* benefitted from one of the most accurate splashdowns and the fastest crew-recovery operation in the history of the Apollo program. Once rescued, the exhausted astronauts were barely able to stand on the *Iwo Jima*'s deck. Lovell had lost 14 pounds, Haise had a severe kidney infection, and all three men were badly dehydrated. But, despite the odds, they had survived.

Impact

As one NASA official put it, *Apollo 13* reminded everyone "that flying to the Moon is not just a bus ride." A complete investigation released in June 1970 revealed that a heater in an oxygen tank in the service module's fuel cells had exploded. The explosion burst both of the service module's oxygen tanks, its main engine, and all of its fuel cells. The cause of the explosion was identified as defective thermostat switches that evidently had been damaged several years before the flight, when the oxygen tank was dropped a few inches. In subsequent pre-flight tests, ground crews had turned on heaters and fans when the tank would not empty. The resulting heat built up had welded the switches and burned the insulation off the wiring, creating the conditions that triggered an explosion when the *Apollo 13* crew turned on the stirring fan in the oxygen tank.

Additional fuel cells and oxygen tanks were designed into the service module, which had been built by North American Rockwell, in light of the post-flight technical review. The lunar module, manufactured by the Grumman Corporation, also underwent design modifications; these left it better equipped to serve as an emergency "lifeboat" in the future. In addition, NASA corrected quality control problems in the construction and prelaunch testing of equipment.

Taken by the Lunar Module/Command Module four days after the incident, this photograph shows the severe damage incurred after the explosion of one of the craft's oxygen tanks. The crew members inside were forced to evacuate, using the Lunar Module as a "lifeboat."

The *Apollo 13* crisis showed that thousands of NASA and space industry personnel could cooperate to solve an unprecedented technical dilemma of incredible complexity—with three lives hanging in the balance all the while. Nonetheless, the accident made many observers question the need for repeated manned lunar landings. In addition to causing a nearly ten month delay in NASA's timetable, the *Apollo 13* oxygen tank rupture led to the reduction of the Apollo program by three missions; *Apollo 18, 19,* and *20* were cancelled in the wake of budget cuts imposed by Congress.

LEONARD C. BRUNO

Where to Learn More

Cortright, Edgar M. *Apollo Expeditions to the Moon.* NASA, 1975.

Murray, Charles and Catherine Bly Cox. *Apollo: Race to the Moon.* New York: Simon and Schuster, 1989.

Apollo 13: "Houston, we've got a problem." NASA, 1970.

"*Apollo 13*: Three Who Came Back." *Newsweek,* April 27, 1970, pp. 21-27.

"What Really Happened to *Apollo 13.*" *Space World,* October 1970, pp. 20-37.

Strickland, Zack. "Pre-launch Electrical Overload Led to *Apollo 13* Tank Rupture." *Aviation Week and Space Technology,* June 8, 1970, pp. 18-19.

Soyuz 11 reentry disaster

(1971)

Death of three Soviet cosmonauts during reentry due to cabin seal malfunction represented a significant technical setback for the Soyuz manned spacecraft program.

Background

After occupying the first orbiting space station for over three weeks and achieving many scientific and technological feats, the three-man *Soyuz 11* crew were killed in the process of reentering the Earth's atmosphere. Upon separation from the service module just prior to reentry, the misfiring of explosive bolts caused a pressure equalization valve to open prematurely, exposing the command module to the vacuum of space. The Soviet crew were unable to close the valve manually before they lost consciousness and died due to the rapid decompression of the cabin. Their spacecraft was on automatic control, and its reentry and touchdown proceeded normally. The ground recovery crew, however, found the cosmonauts dead inside.

By 1971, the Soviets had given up attempts to compete with the United States in terms of manned lunar landings, and had redirected their efforts to establishing the first manned laboratory in space. To this effect, the Soviets launched the long-awaited *Salyut 1* on April 19, 1971. This 20-ton laboratory was 65.5 feet long and 13 feet wide, and was placed into a circular orbit about 200 miles above the Earth. *Salyut 1* was powered by two pairs of giant solar panels that looked like wings, and was divided into several different modules, three of which were pressurized and habitable.

Four days after the *Salyut 1* space laboratory was launched, the Soviets sent *Soyuz 10* into space. Most Western observers assumed that its three-man crew would occupy the orbiting space station, but the spacecraft remained docked to *Salyut 1* for only five and a half hours, and returned to Earth just two days after having been launched. Soviet officials characteristically claimed that the *Soyuz 1* mission had achieved its goals, but most Western experts assumed that some technical malfunction had prevented the cosmonauts from entering the space station as planned.

It was no surprise, then, when the three-man crew of *Soyuz 11* was launched at 7:25 A.M. Moscow time on June 6, 1971, and approximately 24 hours later docked beside and entered the *Salyut 1* space laboratory. The crew of Georgi Dobrovolsky, Vladislav Volkov, and Viktor Patsayev substituted for the better-trained *Soyuz 10* crew, who could not undertake the mission because one of its members was ill; of the *Soyuz 11* crew, only Dobrovolsky had previously flown in space. The *Soyuz 11* rendezvous and docking procedure was especially slow and exhausting, but this second crew either did not encounter or resolved whatever technical problems the *Soyuz 10* mission experienced, and the space station was occupied for the first time.

Once inside, the crew transferred the station to a higher, more stable orbit and spent the next three days preparing it for extended occupancy. The Soviets took great pride in deploying and manning the first successful space station, and publicized this accomplishment as more significant than the American manned lunar landings. The *Soyuz 11* crew performed a broad range of scientific experiments in space, among which were the cultivation of a small garden and conducting medical and biological experiments on themselves. They made observations of celestial targets using a special astrophysical observatory, and took thousands of photographs of the Earth.

In their stay of more than three weeks in space, the *Soyuz 11* crew had carried out a highly successful mission in terms of both scientific achievement and enhancing Soviet prestige, and were immediately transformed into national heroes. When the crew were instructed by ground control to leave the station they had occupied and worked in for a record span of nearly 24 days, they could look back on a mission that not only achieved all its scientific and technical goals, but had proceeded almost flawlessly. With complete confidence, the Soviet government organized official celebrations for the cosmonaut's return to Earth.

Details of the Disaster

On June 29, the *Soyuz 11* crew prepared the *Salyut 1* space station for automatic operation, and transferred back to their spacecraft. At 9:28 P.M. on June 29, they undocked from the station and changed from full-pressure suits into space-saving woolen flight suits and leather helmets. At 1:35 A.M. on June 30, the crew strapped themselves into their couches and fired their retro-rockets to deorbit. Dobrovolsky reported at this time that "Everything is satisfactory on board. Our condition is excellent. We are ready to land." Twelve minutes later, additional rockets were fired to separate the spacecraft's command module, in which the crew would reenter the Earth's atmosphere, from the service module.

Twelve pyrotechnic bolts were designed to fire sequentially just before reentry to complete the separation of the two modules. These devices, however, malfunctioned and fired simultaneously. The combined explosion generated severe vibrations that in turn caused a pressure equalization valve to release a seal and open prematurely. This valve normally would open at

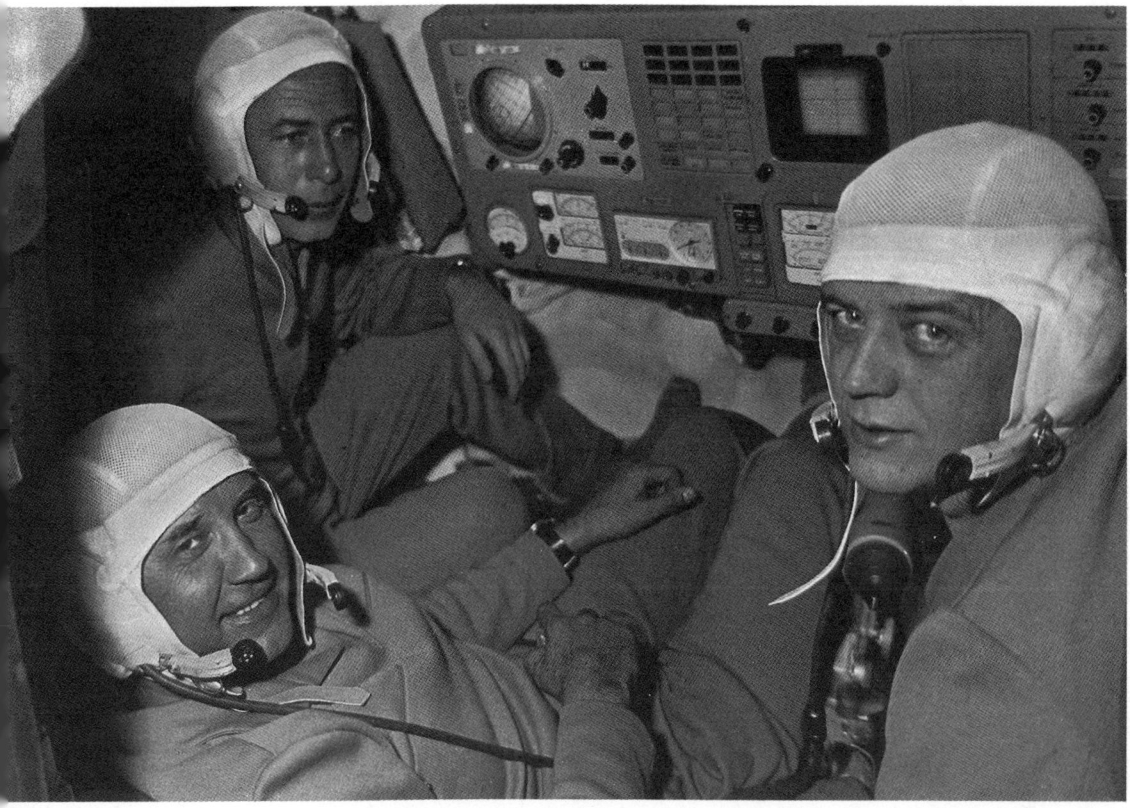

All three crew members of the Soyuz 11 *died during reentry when a pressure equalization valve opened early causing rapid decompression of the cabin.*

low altitudes within the Earth's atmosphere, just prior to touchdown, to equalize the command module and external air pressures.

The opening of this valve while the *Soyuz 11* spacecraft was still 120 miles above the Earth, in the vacuum of space, caused two things to happen. First, the air shooting out from the spacecraft's base skewed the vehicle off course. This was not in itself hazardous, since the spacecraft's stabilization rockets would fire automatically to return the vehicle to a proper reentry course. The second result of the escaping air, however, was a sudden decompression of the command module that immediately endangered the cosmonauts' lives.

The cosmonauts must have become aware of the leak almost instantly, and one crew member attempted to crank the valve shut manually. But he simply did not have enough time, and only halfway succeeded. The manual valve crank was difficult to reach because it was situated in the floor of the cabin below one of the crew member's couches. Once the crank was located, it would take well over 60 seconds to close the valve manually. Fifteen seconds after the valve had opened, half the cabin air had vented out. When only thirty seconds had passed, all the air was gone from the cabin. The three men at this point would succumb to anoxia (lack of air) and suffer the fatal effects of the "bends," which would cause their blood to boil as the nitrogen in their tissues turned into gas. Once these physical reactions set in, the *Soyuz 11* crew lost consciousness and died. The air pressure in the cabin remained at zero for eleven and a half minutes.

After hours of silence, the Soviet news agency announced the cosmonaut's deaths, the first human casualties in outer space, to the world. The effect of this disaster on the Soviet people has been compared to America's reaction to the Kennedy assassination.

During this time, the spacecraft's other systems operated automatically and took the cabin through reentry as planned. After the build-up of G's forces abated, a drogue parachute opened and the main canopy deployed. A recovery beacon sent signals to the ground crews. Braking rockets then slowed the spacecraft's descent even further, and *Soyuz 11* fell gently on the flat ground of Kazakhstan at 2:17 A.M. on June 30, 1971. The record-setting space flight had lasted 23 days, 18 hours, 21 minutes, and 48 seconds. Although air had by now seeped back through the open valve into the command module, it was much too late to save the three men inside. Helicopter recovery crews sped to the landed craft within minutes, aware that an ominous silence had been maintained since the command module separated from the service module. Despite this omen, the recovery crews were unprepared for what they discovered after they climbed aboard the eight-foot, bell-shaped capsule and opened its hatch. Expecting to greet three jubilant heroes, they found instead the dead men—two still in their couches, and another huddled in a corner. After hours of silence, the Soviet news agency announced the cosmonaut's deaths, the first human casualties in outer space, to the world. The effect of this disaster on the Soviet people has been compared to America's reaction to the Kennedy assassination.

Impact

Although the Soviets would not ordinarily release details about such an accident, a planned cooperative space venture with the United States compelled them to divulge what had happened to NASA (National Aeronautics and Space Administration). In this cooperative program, called the Apollo-Soyuz Test Project (ASTP), American and Soviet spacecraft were to link up in space and exchange crews. The Americans demanded to know what went wrong because they would be occupying a Soyuz spacecraft; the Soviets ultimately acquiesced and gave NASA a detailed report on the disaster. NASA then verified this information by analyzing the telemetry data provided by the Soviets.

As a result of the accident, the Soyuz spacecraft once again became a two-man craft. The use of spacesuits had been eliminated earlier in the Soyuz program to add room for a third cosmonaut. Making room for an extra man by dispensing with a redundant life-support system proved to be a fatal expediency; had the crew of *Soyuz 11* been wearing full spacesuits, they would have survived the loss of their cabin environment. With the Soyuz program returning to two-men crews, the cosmonauts would wear full-pressure spacesuits during ascent and reentry. This change implied a complete reconceptualization of the Soyuz spacecraft (including a redesigned pressure valve system) that delayed the Soviet manned space program for approximately one year. The *Soyuz 11* disaster also led to a top-echelon personnel change; General Nikolai Kamanin was forced to retire as head of the Soviet manned space program, and his position was filled by Vladimir A. Shatalov, a former cosmonaut.

The delay in the Soyuz space flight program eventually doomed the orbiting *Salyut 1* space station. Although the station had been boosted into a higher, longer-life orbit before the *Soyuz 11* accident, it could remain self-sufficient only for a limited period of time. Once the Soyuz program had been put on hold and there was no way to occupy the space station and replenish its propellant supply, the *Salyut 1*'s orbit inevitably decayed. The Soviets waited as long as they safely could, but then reluctantly intervened in the station's orbit so that it would reenter the atmosphere and burn up over the Pacific Ocean. This reentry occurred on October 11, 1971; *Salyut 1* had orbited the Earth for 175 days.

In 1972, the Soviets tested their redesigned Soyuz spacecraft in an unmanned flight. *Soyuz 12* finally flew in September 1973, with a two-man crew. It was not until November 1980 that the Soviets would again send a three-man crew into space.

LEONARD C. BRUNO

Where to Learn More

"Cause Sought in Soyuz Tragedy." *Aviation Week and Space Technology.* July 5, 1971, pp. 12-15.

Bond, Peter. *Heroes in Space: From Gagarin to Challenger.* Basil Blackwell, 1987.

Magill, Frank N., ed. *Magill's Survey of Science: Space Exploration Series.* Salem Press, 1989.

Newkirk, Dennis. *Almanac of Soviet Manned Space Flight.* Gulf, 1990.

Oberg, James. "Volkov: Space Engineer." *Space World.* January 1976, pp. 2-21.

"Triumph and Tragedy of *Soyuz 11.*" *Time.* July 12, 1971, pp. 38-39.

Skylab meteoroid shield failure

(1973)

Loss of the NASA orbiting space laboratory's meteoroid shield and one solar array was caused by failure to account for aerodynamic loads during launch.

Background

Approximately sixty seconds after the unmanned *Skylab Workshop* was launched on May 14, 1973, from the Kennedy Space Center, Florida, there was an indication that the workshop's meteoroid shield had inadvertently deployed. That mishap was the first in a series of events that made the Skylab program one of the most exciting and adventuresome in space exploration. Two hours after what appeared to be an otherwise successful launch, the *Skylab* was practically unusable. A rapidly planned and executed rescue saved the Skylab program from becoming a complete loss.

The Skylab Program was a logical continuation of manned space exploration that began with the Mercury, Gemini, and Apollo programs. The $2.5 billion program was planned to be a complex scientific laboratory, as well as a home-in-space where American astronauts could live and work in a shirtsleeve environment.

The project that became Skylab began in 1965 when NASA (National Aeronautics and Space Administration) first considered a "spent-stage experiment" which proposed that astronauts enter the empty upper-stage section of a Saturn rocket and perform experiments in weightlessness. By 1970, the project site had developed into an orbital workshop that would contain four modules, two of which were for laboratory experiments and living space.

Altogether, *Skylab* consisted of four major components: a multiple docking adapter for docking the manned spacecraft, an airlock module for providing astronaut access to the vacuum of space to perform extravehicular activity, a solar observatory (Apollo telescope mount) for conducting solar observations and providing power, and the orbital workshop to serve as crew quarters and general work areas. *Skylab* would be launched unmanned into earth orbit atop a two-stage version of the reliable *Apollo Saturn V* launch vehicle that was used to launch the Apollo astronauts to the moon.

When in earth orbit, the entire *Skylab* station, including the attached command module that would bring the crews to and from earth, was almost 120 feet (36.6 meters) long. The largest spacecraft ever launched by the United States, it was about the size of an average three-bedroom house. On earth it weighed almost one hundred tons. The orbital workshop area, which included living and workshop quarters, was the station's largest section at over 48 feet (14.6 meters) long and 21.6 feet (6.6 meters) in diameter. Once in orbit, *Skylab* was powered by two large arrays of solar cells that deployed like wings.

Details of the Failure

The two-stage *Saturn V* launched the *Skylab Workshop* precisely on time on May 14, 1973. As the great rocket began its climb into the atmosphere, all systems operated normally. Suddenly, a launch monitor reported, "Flight, we have some indication that the micrometeoroid shield has partially deployed," suggesting that *Skylab*'s shield had moved (although not scheduled to deploy until ninety-six minutes into the flight). Premature unlatching of one solar array was also observed. Normal separation of *Skylab* from the launch vehicle occurred ten minutes after launch, and eight seconds later, *Skylab* began its nearly circular orbit around the earth. A series of planned events then occurred to place *Skylab* in its desirable "solar inertial" orientation—the centerline of the solar observatory pointing toward the sun. As time passed, ground engineers monitoring the data from the orbiting *Skylab* suspected that the meteoroid shield had been lost, indicated by the presence of increasing heat on the exterior to two hundred degrees above its designed temperature. Motion sensors and loss of temperature and voltage measurements indicated loss of the solar array system. The first confirmed indication of a problem came forty-one minutes after launch when the planned deployments failed to occur. Two hours after what appeared to be a successful launch, the *Skylab* was in orbit, but overheated, underpowered, and uninhabitable.

The meteoroid shield was a deployable cover over *Skylab*'s cylindrical exterior that was designed to protect it from the heat of direct sunlight and high speed meteoroid penetration. The shield, when deployed, was a cylindrical structure 270 inches in diameter and 265 inches long that weighed 1,200 pounds. The shield's black and white paint patterns provided passive protection from the sun's heat. Meteoroid protection was provided by the 0.025-inch thickness of the shield—slightly less than the thickness of a credit card—which when deployed stood out five inches from *Skylab*'s exterior walls. After deployment in orbit, the shield required very little strength to serve its primary functions. Since the meteoroid shield was a very limber cover, it had to be wound tightly to withstand launch pressures.

Although the meteoroid shield was simple in concept, it was a very complex device. Straps, hinges, and torsion rods were provided to hold it in place and aid deployment when in orbit. A special spring-loaded fold-out panel provided the extra thirty inches in perimeter required when deployed. The torsion rods provided a twisting force to rotate the shield to its extended position.

Reconstruction of data determined the most probable sequence of events. When the rocket reached twenty-eight thousand feet, it approached the time of maximum aerodynamic forces (or loads) from the earth's atmosphere. At 60.12 seconds after launch, aerodynamic loads created internal pressures higher than the wrapped shield was designed to bear. This was caused by seals that were both deficient as well as missing. Consequently, the thin meteoroid shield was forced into the supersonic airstream as the launch vehicle tore through the atmosphere into space, and the hinges that held it in place failed as if they were unzipped. The shield then gave way like a label peeled from a can, tearing away one solar array completely and jamming the deployment of the other. As *Skylab* entered orbit, it found itself with no protection from the heat of direct sunlight and no solar panels to generate electricity.

The shield gave way like a label peeled from a can, tearing away one solar array completely and jamming the deployment of the other. As Skylab entered orbit, it found itself with no protection from the heat of direct sunlight and no solar panels to generate electricity.

Loss of the meteoroid shield allowed the sun to shine directly on the workshop's external gold foil coating, which absorbed most of the heat. This over-heating condition had the potential to reach over four hundred degrees Fahrenheit if not corrected. A temperature this high would damage on-board equipment and make *Skylab* uninhabitable. The immediate problem was to keep the project from further overheating. *Skylab*'s attitude was adjusted to a forty-five degree angle to the sun to keep its temperature as low as possible. Conditions then stabilized with an interior temperature of one-hundred-and-thirty degrees Fahrenheit but only enough electrical production—about a third of full capability—from the Apollo telescope mount solar arrays to operate essential equipment.

A decision was made to postpone, for ten days, the next day's scheduled first manned launch to allow an examination and analysis of the problem. National Aeronautics and Space Administration centers and their contractors executed emergency plans and worked around the clock to develop solutions to the problem. By May 16, it was decided to proceed with development of three concepts: a sunshade or parasol manually deployable through the *Skylab*'s scientific airlock; a deployable twin-pole window shade-type device to be deployed by booms; and an inflatable mattress-type shield. As a contingency, NASA was working on an improved version of a thermal shield rigged with ropes from the command service module (CSM), called the SEVA (standup extravehicular activity from the CSM hatch) sail.

The first manned visit to *Skylab* was eventually delayed to May 25. All design, development, manufacture, and qualification tests, including zero-gravity simulation in the neutral buoyancy facility, had to occur in the eleven days following the *Skylab* launch. At a May 23 review, NASA decided that use of the parasol would be primary, the window-shade device secondary, and the SEVA sail available as a contingency. Meanwhile, methods were being pursued to see if the crew could free the remaining solar array, believed to be stuck. Since its condition was unknown, the most versatile cutting and prying tools were stored in the CSM in preparation for the attempted release.

Soon after the first manned launch on May 25, astronauts Pete Conrad, Paul Weitz, and Joe Kerwin verified the loss of the shield and solar array. The next day, the crew entered the *Skylab* and deployed the parasol. The temperature immediately decreased and stabilized at about eighty degrees Fahrenheit.

On June 7, Conrad and Kerwin broke a restraining clamp and cut the embedded meteoroid shield joint that prevented the remaining solar array from opening. It gradually opened completely and remaining solar array electrical power capability doubled to almost 7,000 watts, less than the designed average of 10,500 watts—but sufficient to assure mission completion. The temperature and power capability remained satisfactory throughout the remainder of the program's missions. *Skylab* had been dramatically saved from the brink of disaster.

Impact

Bruce T. Lundin, director of the NASA Lewis Research Center, chaired the *Skylab* Investigation Board. The board held consultations and inquiries at various locations and reported to the NASA administrator, James C. Fletcher, on July 13, 1973. The board identified a major design omission—aerodynamic loads on the meteoroid shield during the launch phase had not been considered. Primary attention was devoted to achieving satisfactory deployment in orbit; and the question of whether the shield would stay in place during the dynamics of flight was not considered in any coordinated manner that the board could determine. Communications between aerodynamics, structures, manufacturing, and assembly personnel and systems engineering approval had broken down. Management procedures had not identified these problems.

The board directed its advice to strengthening the management processes of future complex endeavors, recommending that (a) design engineers have more firsthand experience in testing, operations, and failure analysis to better conceptualize how hardware behaves and fails; (b) project engineers be assigned oversight of all technical activity; and (c) "chief engineers," without administrative or managerial duties, be employed to integrate all elements of the system.

Despite this adversity, *Skylab* achieved and surpassed all expectations. The project was visited by nine astronauts in three different crews who stayed there for periods of twenty-eight, fifty-nine, and eighty-four days. During those manned missions, it orbited the earth some 3,900 times at an average altitude of 270 miles (434.5 kilometers). Its orbital path took it over seventy-five percent of the earth's surface before it finally reentered as planned, and burned up in the earth's atmosphere over the Indian Ocean on July 11, 1979. Until *Skylab*, NASA had little actual experience with prolonged weightless-

A U.S. Customs official in San Francisco examines the largest remaining piece of Skylab. *The chunk of debris was shipped to the U.S. from Australia after the craft fell through the Earth's atmosphere into the Indian Ocean (July 1979).*

ness, and the long-duration flights of *Skylab* provided the Administration with invaluable data for future space flights.

William B. Shapbell, Jr.

Where to Learn More

Belew, L. F., ed. *Skylab, Our First Space Station, 1977.* National Aeronautics and Space Administration, Scientific and Technical Information Office. Government Printing Office, 1977.

Bulban, E. J. "*Skylab* Array Deployed in Successful EVA." *Aviation Week and Space Technology,* June 11, 1973, pp. 26-8.

Fink, D. E. "*Skylab* Parasol: 6-month Job in 6 Days." *Aviation Week and Space Technology,* June 11, 1973, pp. 53-6.

George C. Marshall Space Flight Center. Skylab Program Office.

MSFC Skylab Mission Report-Saturn Workshop, 1974. NASA Technical Memorandum, NASA TM X-64814. National Technical Information Service, 1974.

National Aeronautics and Space Administration. *NASA Investigation Board report on the Initial Flight Anomalies of Skylab 1 on May 14, 1973.* NASA, July 13, 1973.

"Saving a Station in Space." *Science News,* May 26, 1973, pp. 336-37.

Schneider, W. C., and W. D. Green Jr. "Saving *Skylab.*" *Technology Review,* January 1974, pp. 42-53.

"Troubled *Skylab*: a Mission in Jeopardy." *Science News,* May 19, 1973, pp. 320-21.

U.S. House. Committee on Science and Astronautics. Subcommittee on Manned Space Flight. *Skylab 1 Investigation Report: Hearing, August 1,1973.* 93rd Cong., 1st session, 1973.

U.S. Senate. Committee on Aeronautical and Space Sciences. *Skylab. Hearing, May 23, 1973.* 93rd Cong., 1st sess., 1973.

———. *Skylab, Part 2: Report of Skylab 1 Investigation Board: Hearing, July 30, 1973.* 93rd Cong., 1st sess., 1973.

Wilford, J. N. "*Skylab,* Short of Power, Overheating; Further Delay for Astronauts Feared." *New York Times,* May 16, 1973, p.1.

Challenger explosion

(1986)

An explosion shortly after launch caused by faulty solid rocket booster seals destroyed the shuttle and killed its crew of seven, forcing a retrenchment of NASA's shuttle program.

Background

On January 28, 1986, NASA launched STS 51L, the twenty-fifth mission of its space shuttle program, from Cape Canaveral. The *Challenger* launch received considerable publicity because it carried the first Teacher-in-Space, Christa McAuliffe, who was to broadcast a series of lessons to schoolchildren throughout America. Seventy-three seconds after its launch, however, the shuttle's solid rocket O-rings failed, triggering an explosion that shattered the *Challenger* within full view of thousands of NASA personnel, relatives and friends of the crew, and spectators. The seven-person crew was inside a module that detached from the shuttle during the blow-up; the crew members evidently survived the explosion but died at impact after a nine-mile free-fall into the Atlantic Ocean.

In 1982, four years before the disastrous *Challenger* mission, a National Security Decision Directive indicated that NASA's highest priority would be to make its shuttle program "fully operational and cost-effective in providing routine access to space." Subsequent directives called for a schedule of up to twenty-four flights a year. But by the end of 1985, NASA had been unable to manage more than nine shuttle flights in any given year. Although the United States was dispensing with its expendable rockets and scheduling all of its satellites to be launched by shuttle, the nine-billion-dollar shuttle had not proven itself to be a cost-efficient, reliable space transportation system or achieved anything like the performance desired.

NASA intended 1986 to be the shuttle program's breakthrough year, reporting in January 1986 that it would launch fifteen missions, using all four of its shuttles, during the next twelve months. This total would be the largest to date, and in order to realize this goal, NASA would have to initiate the year's launches without delays. The year did not, however, get off to the desired start. After at least seven separate postponements, the first shuttle mission, 61-C, was launched on January 12. When the *Columbia* returned from space on January 18, with bad weather having prolonged the flight, NASA's tight 1986 schedule was already in jeopardy.

In the days that followed, everyone worked feverishly to get the *Challenger* (which had completed its last space mission on November 6, 1985) ready for its January mission. This ambitious mission involved not only broadcasting the much-publicized Teacher-in-Space programs, but launching a Data-Relay Satellite (TDRS) and the high priority Spartan-Halley comet research observatory. The mission was scheduled to last six days, during which time the Spartan observatory would be recovered from orbit. Because of launch window requirements, the Spartan could be orbited no later than January 31. This inflexible deadline, as well as the sequence of missions in line to follow, led NASA to prepare contingency plans to skip the *Challenger* mission if it could not fly by the end of the month.

Although an afternoon launch was originally planned for the *Challenger*, as recommended by the Spartan project scientists, a mid-morning launch was ultimately scheduled. Ironically, this shift was due to safety considerations. If the *Challenger* was to suffer an "engine-out" during launch, the shuttle would glide to its planned emergency landing site at Casablanca, on the west coast of Africa. NASA argued that an afternoon launch off Florida meant that the emergency landing in Africa would occur at night, a situation that should be avoided because the Casablanca runway was not equipped with lights.

On January 15, 1986, NASA had held via teleconference a Flight Readiness Review for the upcoming *Challenger* mission, linking up the various centers involved with the project. All systems were reviewed in detail, from the engineering of the spacecraft to the in-flight responsibilities of the Johnson Space Center in Houston and the Marshall Space Flight Center in Huntsville, Alabama. The conference concluded with a "Go for launch."

The seven-person crew chosen for the mission was commanded by Francis Scobee, who had piloted a 1984 shuttle mission. His pilot, Michael Smith, had not previously flown in space. Ellison Onizuka, Ronald McNair, and Judith Resnick, the mission specialists who ran the satellites and experiments, were all experienced space travelers. The payload specialist in charge of the TDRS satellite was Gregory Jarvis, who had no previous space-flight experience. The crew also included the 37-year-old Christa McAuliffe, the Teacher-in-Space from Concord High School in New Hampshire.

The launch of the *Challenger* was originally set for January 22, 1986. The first delays postponed the flight schedule to January 24 and then to January 25. A forecast of bad weather for the 26th held up the mission until Monday the 27th. On this date, a problem with a hatch bolt developed; by the time this problem was corrected, crosswinds had built up to a dangerous 30 knots. Liftoff had to be rescheduled for Tuesday, although the crew was ready to launch and the shuttle had been fueled.

Details of the Disaster

During the night of January 27, 1986, the temperature at Cape Canaveral dropped to well below freezing. This prompted a late-night review of the prospects for the *Challenger* launch by NASA managers and contractors, who were becoming increasingly concerned about the cold weather. In fact, no shuttle had ever been launched at a temperature lower than 53 degrees Fahrenheit. During these prelaunch discussions, engineers from Morton Thiokol, a NASA contractor, expressed concern about the O-rings on the shuttle's solid rocket boosters stiffening in the cold and thereby losing their ability to act as a seal. The engineers apparently were unable to make a fully convincing case, particularly since the O-rings had never been tested at low temperatures. With NASA managers pushing for an unequivocal go or no-go, Thiokol managers overruled their own engineers and signed a waiver stating that the solid rocket boosters were safe for launch at the colder temperatures.

During the final hours before launch on January 28, 1986, NASA turned its attention to the ice that had formed on the shuttle and launchpad overnight, when temperatures had ranged from 19 and 29 degrees Fahrenheit. The icicles formed could potentially break off during launch and damage the insulating tiles required to protect the shuttle's reentry into the Earth's atmosphere. The launch was therefore delayed from 9:38 A.M. to 10:38, and then to 11:38. During those hours, inspection teams surveyed the craft's condition and reported no anomalies due to the ice build-up. Finally, at precisely 11:38:00.010 Eastern Standard Time, the *Challenger* launched off from Complex 39B at Cape Canaveral to begin its tenth flight into space.

As the *Challenger* rose into a clear but cold blue sky, no one on the ground or in the shuttle realized that a tongue of flame was extending from the right-hand booster rocket to the giant fuel tank. The crowd of spectators, which included Christa McAuliffe's husband and two children as well as a group of her pupils, cheered at the shuttle's majestic ascent. The vehicle then rolled to align itself on the proper flight path and throttled back its engines. At about 59 seconds into the launch, the plume of flame became evident.

By 64 seconds in launch, a gaping hole had formed in the casing of the booster from which the flame was escaping. At 72 seconds into the flight, the flame loosened the strut that attached the booster to the external tank. At that moment, the cockpit's voice recorder captured the only indication that anyone onboard was ever warned of any serious trouble: pilot Michael Smith uttered "Uh oh." One second later, the loosened booster rocket slammed into and detached the *Challenger*'s right wing; it then crashed into the fuel tank and set off a massive explosion. The explosion occurred at an altitude of 48,000 feet, with the shuttle traveling at twice the speed of sound.

To many spectators unfamiliar with shuttle launches, the accident seemed like a spectacular staging or separation of the booster rockets. But the disaster became apparent when the fireball widened and debris began to scatter. The spectators fell silent, spellbound with disbelief. The shuttle itself was no longer visible. Although NASA began rescue operations immediately, the chances of finding survivors were very remote. During the segment of a flight

During its earliest moments in flight, the Challenger *exhibited no problems or complications.*

when the solid rocket boosters are thrusting, the crew has no survivable abort options. There was nothing anyone could do if something went wrong at that critical launch moment.

The *Challenger* exploded twenty miles off the coast of Florida, and the force of the explosion pushed its debris to an altitude of twenty miles. For the next hour, incendiary fragments of the shuttle continued to rain down on the recovery forces. Among the worst accidents in the twenty-five year history of manned spaceflight, the *Challenger* disaster marked the first time that American astronauts were lost during a mission. Footage of the explosion, replayed continuously on television, sent shock waves through the nation.

Impact

Within a few days of the *Challenger* disaster, then-President Ronald Reagan eulogized the crew during a nationally televised memorial ceremony at the Johnson Space Center in Houston. On February 3, 1986, Reagan established a Presidential Commission to investigate the accident, appointing former Secretary of State William B. Rogers as the commission chair.

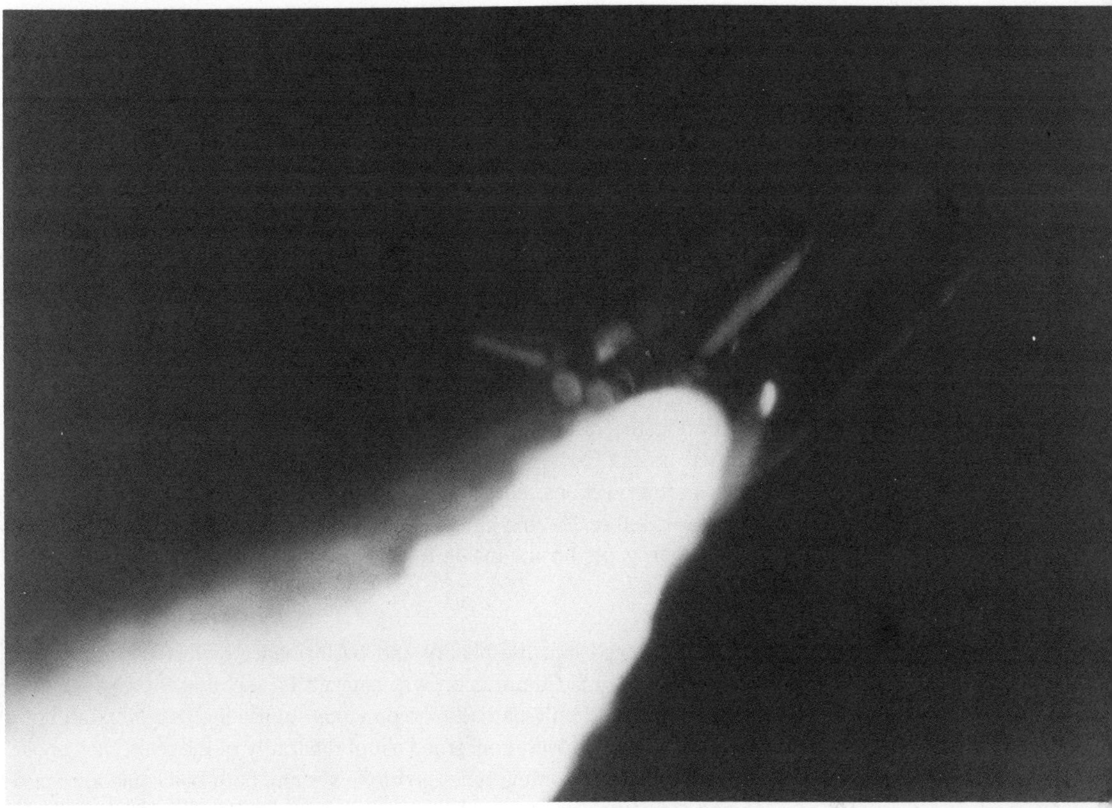

About one minute, 13.47 seconds after the launch, a tongue of flame extending from the right-hand booster rocket to the giant fuel tank became apparent.

Six weeks after the disaster, the shuttle's crew module was recovered from the Atlantic ocean floor; the crew members were subsequently buried with full honors. There was considerable speculation as to whether the crew had survived the initial explosion, and the evidence gathered by NASA indicates that the crew did survive the breakup and separation, and had begun to take emergency action inside the module. Whether the entire crew remained conscious throughout the two minutes, forty-five seconds free-fall into the ocean remains unknown, but at least two crew members were breathing from emergency air packs that they had activated. The crew module has never been exhibited publicly, but the cabin was essentially unrecognizable in the photographs that were eventually released. The force of the module's impact with the surface of the ocean has been estimated at a 200-G deceleration from 207 miles per hour. This impact, which compressed the sixteen and a half foot high cabin into a solid mass half the original size, certainly killed anyone still alive in the module. The module's thick windows were shattered, but there was no evidence of fire.

The story of the *Challenger* mission continued in the public investigations of NASA and the disaster. The Rogers Commission, which had been charged with assessing the accident and recommending preventative measures, conducted a three-month investigation that was to involve more than 6,000 people. The commission recorded 15,000 pages of testimony during public and closed hearings, collected 170,000 pages of documents as well as hundreds of photographs, and sponsored independent technical studies. In its methodical review of the events, the commission also evaluated flight records, film evidence, and recovered debris.

On June 6, 1986, the Rogers Commission released its determination that the immediate physical cause of the *Challenger* disaster was "a failure in the joint between the two lower segments of the right Solid Rocket Motor," and specifically "the destruction of the seals that are intended to prevent hot gases from leaking through the joint during the propellant burn . . . ." In attributing the disaster to the destruction of these seals, the commission focused on the O-rings. The shuttle's large strap-on booster rockets were built in four sections, and the rubber O-rings were required to seal the sections together. A zinc chromate putty was also used to keep the hot combustion gases on the inside from coming into contact with the rubber rings.

When it checked into the history and performance of this O-ring sealing system, the Rogers Commission was amazed to find that the O-rings had failed regularly, if only partially, on previous shuttle flights. Although both NASA and Thiokol were concerned about the frailty of the seals, they chose to forgo a time-consuming redesign of the system. Both had come to regard O-ring erosion as an acceptable risk because the seal had never completely failed. But when the *Challenger* flew on January 28, the frigid temperatures made the O-rings so brittle that they did not even provisionally seal the joint. Even before the shuttle had cleared the launch tower, hot gas was already "blowing by" the rings.

The Rogers Commission's 256-page report concluded that "the decision to launch the *Challenger* was flawed," and blamed the management structure of both NASA and Thiokol for not allowing critical information to reach the right people. This assessment was seconded by the U.S. House of Representatives' Committee on Science and Technology, which spent two months conducting its own hearings. The congressional committee determined that although the technical problem had been recognized early enough to prevent the disaster, "meeting flight schedules and cutting cost were given a higher priority than flight safety."

The consequences of these public indictments of NASA were grave. Not only was the nation's confidence in NASA shaken, but its own astronaut corps was extremely disturbed; they had never been consulted or even informed about the dangers that the current sealing-system exposed them to. Allowing astronauts and engineers a greater role in approving launches was among the nine recommendations the Rogers Commission made to NASA. The commission's other recommendations included a complete redesign of the solid rocket booster joints, a review of the astronaut escape systems towards achieving greater safety margins, regulation of the rate of shuttle flights to

This segment of the Challenger's *right wing was recovered, along with other debris, by Navy divers about 12 nautical miles northeast of Cape Canaveral and in 70 feet of water.*

maximize safety, and a sweeping reform of the shuttle program's management structure.

Following these revelations, several key people left NASA. The loss of personnel included a number of experienced astronauts who resigned due to disillusionment with NASA and frustration over their chances to fly, given the long redesign process now pending. Ultimately, it was not until September 29, 1988, that an American shuttle would again fly in space. NASA also built a shuttle to replace the *Challenger*; this vehicle, the *Endeavour*, first flew on May 7, 1992.

Since the *Challenger* disaster, NASA has been launching small American and foreign satellites with expendable rocket vehicles, and the shuttle's days as the sole deployer of unmanned satellites are past. The *Challenger* disaster also eroded the public's perception of NASA as a "can-do," high-achieving agency that justified the public expenditure involved. The funding problems NASA has encountered since, particularly with its new space station plans, indicate that NASA may face an uphill struggle to reassert its special status as a federal agency and scientific program.

LEONARD C. BRUNO

Where to Learn More

"NASA Identifies Failure Scenarios." *Aviation Week and Space Technology*. March 17, 1986, pp. 25-26.

Bell, Trudy E. and Karl Esch. "The Fatal Flaw in Flight 51-L," *IEEE Spectrum*. February 1987, pp. 36-51.

Boffey, Philip M. "NASA Had Warning of a Disaster Risk Posed by Booster." *New York Times*. February 9, 1986, p. 1.

Broad, William J. "Thousands Watch a Rain of Debris." *New York Times*. January 29, 1986, p. 1.

Lewis, Richard S. *Challenger: The Final Voyage*. Columbia, 1988.

McConnell, Malcolm. *Challenger: A Major Malfunction*. Doubleday, 1987.

"A Fatal 'Error of Judgment.'" *Newsweek*, March 3, 1986, pp. 14-19.

Report of the Presidential Commission on the Space Shuttle Challenger Accident, 5 volumes. The Commission, June 6, 1986.

Sanger, David E. "Shuttle Changing in Extensive Ways to Foster Safety." *New York Times*. December 28, 1986, p. 1.

Smith, Melvyn. Space Shuttle. Haynes, 1989, pp. 264-285.

Smith, R. Jeffrey. "Inquiry Faults Shuttle Management." *Science*. June 20, 1986, pp. 1,488-1,489.

593

Submarines

- *Squalus* (1939)
- *Thetis* (1939)
- *Thresher* (1963)
- *Scorpion* (1968)

Squalus sinking off New Hampshire (1939)

Mechanical failure in the operating gear of an air intake valve occurred during an underwater dive, allowing the aft part of the vessel to fill with water and sink the submarine.

Background

On May 23, 1939, the U.S. Navy submarine #192 known as the *Squalus* sank in 240 feet of water off Portsmouth, New Hampshire, with a loss of twenty-six men. A large, hydraulically operated induction valve had failed to shut, and the sea came pouring into the sub, filling its aft portion and driving the sub to the bottom. Despite the drowning deaths of twenty-six of the crew, thirty-three men were later rescued when the Navy used the newly invented McCann Rescue Chamber. Using this rescue bell, which attached itself to the sub's hatch, rescuers were able to extricate the trapped men who were still alive behind the sub's closed, watertight doors.

The *Squalus* was one of the U.S. Navy's newest patrol-type submarines—and one of the first of a new fleet of submarines that would go on to distinguish themselves in the coming war with Germany and Japan—when it was first launched from the Portsmouth Navy Yard (which is actually in Kittery, Maine) in September 1938. At 310 feet long, 27 feet wide, and weighing 450 tons, its streamlined design incorporated all the information the United States had garnered over a twenty-five-year period of submarine experience. Naval officials planned to test the innovative ship completely during the spring of 1939.

The *Squalus* ran daily trials beginning May 16, 1939. The ship made eighteen dives without incident and passed every test, meeting or exceeding all requirements. The required dives included the first by-the-numbers dive, which was slow and careful, followed later by running dives. Quick dives and torpedo-firing dives were also completed successfully by the *Squalus*. On May 23, 1939, its official trials (those necessary before its acceptance for service in the fleet) began, including the simulation of a real crash dive that very day.

The *Squalus* was to submerge fifty feet in sixty seconds. Before diving, the sub would be running on the surface on its diesel engines. For the *Squalus* to submerge, the crew was required to take certain preparatory measures in the correct sequence. The submarine's main indicator board, or "Christmas tree," would then confirm the proper execution of these actions with green, or "go," signals. Once all of the men were below and the watertight hatches securely shut, only the induction valves—large openings that provided the diesel engines with the necessary air and allowed exhaust fumes to escape—remained open to the sea. The final preparation for the dive involved closing these valves. On the indicator board, the green signal appeared only after the induction valves were closed. At this signal, the *Squalus* would dive to periscope depth, from where it would then plunge quickly to the prescribed depth.

This important test procedure was performed off the Maine and New Hampshire coasts. The submarine was manned by experienced and highly qualified personnel, and its skipper, Lt. Oliver F. Naquin, had already been commended by his admiral for his excellent performance. The ship proceeded from the Portsmouth River into the sea, with fifty-nine men aboard—the regular crew of fifty-one enlisted men and five officers, as well as three civilians from the Portsmouth Navy Yard.

Details of the Sinking

At 8:30 A.M. on May 23, 1939, Lieutenant Naquin ordered full speed for both diesel engines, and the ship reached sixteen knots (about eighteen miles per hour). This was the prescribed speed for its test dive, since it was to simulate a crash dive under attack. When the order to close the engine and hull induction valves was given, the second-class machinist's mate stationed at the hydraulic manifold on the forward bulkhead of the control room pushed the control handle upward through its thirty degrees of travel to the closed position. The control handle was connected via hydraulic and mechanical linkage to the large manhole-cover valve that measured thirty-one inches across. With the mate's action, the light on the indicator board showing the induction valve turned from red to green, and the Christmas tree board signalled green as well. With this apparent confirmation that the sub was watertight and ready to dive, its main ballast tanks were opened, allowing air to escape and water to rush in from the bottom. The sub dove, using the weight of the water it took on to make it heavy enough to submerge.

The *Squalus* started its crash dive at 8:40 A.M. The sub was running on its electric motors and plunged quickly, leveling off at a depth of sixty-three feet. Up to that point, everything seemed routine, and the captain and crew were pleased with the results of the initial test. Congratulatory remarks were interrupted, however, by a voice on the intercom yelling, "Take her up! The inductions are open!" Control room members only had time to look up when the voice was followed by another, more frenzied scream: "The engine rooms are flooding, Sir!"

The consequences of an open induction valve or ventilation duct were ominous. With these huge valves—two and a half feet across—open to the sea, the ocean would rush through with the weight and pressure of all the water above. The engine room would be smashed by columns of water, filling it and the rest of the sub until it sank. Knowing this, the captain gave immediate orders to blow ballast water in order to surface, and then ordered all watertight doors closed to prevent the rest of the sub from filling up. With the order to surface, the engines of the *Squalus* labored for about a minute, straining to overcome the seven and a half tons of water that had been added to its weight. It hung motionless for a short time, but then its stern started to drop and it slowly slipped downward until it came to rest in the cold mud, 240 feet below the surface.

The captain was fully aware that by ordering the watertight doors shut, he was probably condemning those on the wrong side of the door to death; but his command was not only by the book, it was imperative if he were to save any of the crew. After these doors were closed, the storage cells or batteries remained to be disconnected in order to save the lives of the crew. Unless the two disconnect switches were thrown immediately, the battery compartment would catch fire and explode. When the chief electrician's mate was finally able to accomplish this dangerous task, the ship was already sitting on the sea's bottom on an even keel with an eleven degree up-angle. The time was 8:45 A.M.

Captain Naquin then ordered a red smoke rocket fired, as well as the release of a forward marker buoy. After the rocket rose in a smoky arc over the water, an orange, tin-can-shaped buoy floated to the surface. It was connected to the sub below by a telephone wire, and on its top were printed the words, "Submarine sunk here. Telephone inside." The captain ordered the firing of smoke rockets at 10:07, 10:24, 11:40, and 12:40. Amidst inventories of food, fresh water, and supplies of air, the thirty-three men still alive in the sub's forward torpedo and control rooms had no choice but to minimize all activity and sit and wait.

The Navy had begun searching for the *Squalus* at 9:30 A.M. when its surfacing report should have been received but was not. At 12:40 P.M., the *Sculpin*, the sister sub of the *Squalus*, saw the red smoke rocket and headed toward it. At 12:55, four hours and fifteen minutes after submerging, the crew sitting on the bottom were able to discern the beat of their sister ship's propellers. They listened as the anchor was dropped and the engines cut, and finally, the buoy phone rang. The *Squalus* was only able to communicate the essential facts of its situation before the phone line snapped, but the link to the surface had been made.

As the men below waited silently, trying to keep warm and conserve oxygen, Navy officials above decided quickly that the only means of saving the crew was the new rescue chamber method. Invented by Commander Allen R. McCann—who was aboard the rescue vessel *Falcon* that had steamed to the Maine coast—the chamber would require much less time than trying to raise the sub. The diver and the chamber were not ready to be lowered until 10:00

The Squalus *being raised at Portsmouth Navy Yard in September 1939, nearly four months after the submarine sank off the coast of New Hampshire.*

A.M. the next day, however, and by then conditions in the stranded sub had worsened. The temperature in the sub was a bitter forty-five degrees and the air was increasingly toxic. Several members of the crew had begun to experience headaches, nausea, and vomiting—classic symptoms of carbon dioxide poisoning.

At 10:15 A.M., the diver's weighted boots clanged as they struck the sub, and the *Squalus* crew knew they were no longer alone. After twenty-five minutes, he had attached to the submarine the all-important cable down which the chamber would be lowered. When he had completed this job, the rescue chamber and its two-man crew was lowered. The rescue chamber was designed with a skirt to fit over the sub's escape hatch, out of which water could be blown. The McCann invention worked as planned, and at 12:56 P.M., it began to rise to the surface, carrying seven of the sickest crewmen. By 6:30 P.M., the rescuers had successfully brought up eighteen additional crew members in two more trips. The fourth and final trip left at 7:30 P.M., but after picking up the final eight men, the chamber became stuck 150 feet from the surface. To remove the obstacle—a tangle in the cables—a diver was forced to cut one of the cables that linked the chamber to the rescue ship at the surface. In order to gain buoyancy and allow the chamber to rise, therefore, ballast had to be carefully released. Maneuvering the chamber with ballast only was dangerous because it had no braking system and could have easily shot up through the bottom of the *Falcon*. With Comdr. McCann giving instructions by phone, the chamber slowly and safely rose, and the last eight of the *Squalus* survivors opened its hatch under the *Falcon*'s floodlights at 12:25 A.M.—nearly forty hours after their submarine went down.

Impact

Only a few months after the *Squalus* sank, the U.S. Navy assigned a three-man board of inquiry to determine the reasons for the tragedy. The inquiry revealed that the induction valve had failed to open properly some weeks before the accident, but that it had always closed correctly. It had been taken out and supposedly repaired before the mission on May 23. Besides the failure of the valve to close, the court ruled, the electrical signal system had also malfunctioned by falsely confirming the valve's closure with a green signal.

The inquiry revealed that the induction valve had failed to open properly some weeks before the accident, but that it had always closed correctly. It had been taken out and supposedly repaired before the mission on May 23. Besides the failure of the valve to close, the court ruled, the electrical signal system had also malfunctioned by falsely confirming the valve's closure with a green signal.

The inquiry also revealed that possible safety measures had not been taken by the Navy when the *Squalus* was designed. The Navy had considered interlocking a sub's air valve and ballast tanks to prevent the sub from taking on ballast water to dive if the valves were open. But the machinery necessary to do this was considered too bulky for a combat submarine like the *Squalus*. In addition, one member of the board questioned whether training the crew to close secondary valves as a habitual precaution might have lessened or prevented the disastrous sinking. The board of inquiry eventually concluded its investigation, blaming the sinking on the mechanical failure of the motor that should have closed the induction valve and recommending that certain technical improvements be installed in the nation's fleet of submarines to avert further disasters. The ruling absolved the crew of any blame whatsoever.

Another impact of the sinking was that the McCann chamber was proven to be a viable rescue method under certain conditions. The *Squalus* situation had been fortuitous: it sank in an area where there were no appreciable tides to affect the diver or the chamber; it landed on the bottom mostly upright and horizontal; it sank near the base, was missed immediately, and was located quickly; and it sank in generally good weather. Even under these exceptionally good conditions, everyone realized that the men of the *Squalus* were in very bad shape by the time they were rescued and could not have endured much longer. No submarine sinking since has been as lucky.

Nearly four months after sinking, the *Squalus* was raised, repaired, and given a complete overhaul. Recommissioned in May 1940 with a new name—*Sailfish*—it went on to fight in the Pacific in World War II and sank seven enemy ships.

LEONARD C. BRUNO

Where to Learn More

Barrows, Nathaniel A. *Blow All Ballast! The Story of the* Squalus. Dodd, Mead, 1940.

Lockwood, Charles A., and Hans Christian Adamson. *Hell at 50 Fathoms.* Chilton Company, 1962, pp. 203-27.

"Navy Affirms Finding *Squalus* Gear Failed." *New York Times,* February 2, 1940, p. 9.

Park, Edwards. "The Death Dive and Brave Rescue of the *Squalus.*" *Smithsonian,* January 1986, pp. 102-10.

Shelford, W. O. *Subsunk: The Story of Submarine Escape.* Doubleday, 1960, pp. 70-81.

Stafford, Edward Peary. *The Far and the Deep.* G. P. Putnam's Sons, 1967, pp. 107-34.

"Whole Truth: Inquiry into the Sinking of the *Squalus.*" *Time,* July 3, 1939, p. 12.

Thetis sinking

Liverpool Bay, England

(1939)

Sinking of British submarine, set off by human error, led to design improvements and recognition of need for sophisticated rescue procedures.

Background

On June 1, 1939, the third of Britain's new, T-class patrol submarines, the HM Submarine *Thetis*, sank during its first open-sea test dive. The test was being conducted in Liverpool Bay, at a water depth of 160 feet; the tugboat *Grebecock* escorted the new submarine. In the course of an unsuccessful trial dive, a torpedo officer opened the inner door of an uncapped torpedo tube; sea-water gushed in through the 21-inch gap. A series of mishaps and procedural failures prevented both the containment of the accident and the rescue of the crew and passengers. Only 4 of the 103 men aboard the submarine were to survive.

The *Thetis* was the first of the new British T-class submarines to be built by the Cammell Laird shipyard at Birkenhead, near Liverpool. The 275-foot long T-class was a relatively small submarine designed for coastal patrol; it displaced approximately 1,100 tons of water.

During preliminary trials, the submarine's "trim" (its buoyancy and balance fore and aft during a dive) had proven to be faulty. Most significantly, the *Thetis* was too buoyant. This over-buoyancy, however, was seen as erring on the safe side, given that a submarine can always valve sea-water into its tanks as ballast.

First dives were, of course, conducted very cautiously, and it seemed unthinkable that anything could seriously go wrong. Despite the operational problems revealed by the preliminary trials, the officials involved with the *Thetis* did not hesitate to attempt an open-sea dive. Their assurance was such that the submarine was actually overcrowded with crew and passengers for its test dive—caterers had even been hired, making for a holiday atmosphere.

Details of the Disaster

As the *Thetis* sailed out of Liverpool on June 1, 1939 and into the ocean off the north coast of Wales, the weather was clear and mild. Lieutenant Commander Guy Bolus was in charge of the new submarine, which was trailed by the *Grebecock*, a tugboat. By 1:30 P.M., the *Thetis* was 38 miles into the sea. At this point, before the submarine's first open-sea test dive was to be attempted, Bolus expected to transfer most of his extra passengers to the tugboat. Yet when Bolus announced that the *Grebecock* was prepared to take on any passengers not wishing to make the dive, not a single passenger spoke up. Even the caterers elected to remain on the submarine.

Besides the submarine's own crew of 53, the *Thetis* was carrying 50 extra passengers, including various naval officers and civilian construction engineers who were present as observers. To some extent, this was accepted practice; since many of these passengers were expert in some aspect of the submarine's systems, they could prove useful as trouble-shooters. Making a test dive with such crowded quarters, nonetheless, was unusual.

At 2:00 P.M., Bolus gave the orders to take the submarine down. The *Thetis* would attempt a slow, routine descent to check the diving machinery and adjust its underwater trim. As the submarine opened its ballast tanks, flooding them carefully in pairs, the crew of the *Grebecock* watched, prepared to see the *Thetis* disappear below the roiling surface. But even with all the ballast tanks filled—and the extra weight of 50 passengers—the submarine continued to float.

It was obvious to the *Thetis* crew that the trim was miscalculated, and that the submarine was indeed far too buoyant. The standard procedure to correct over-buoyancy was to fill up auxiliary water tanks. Even after this measure was taken, however, the *Thetis* remained afloat. Commander Bolus then tried to use the submarine's engines to drive it downward, but this resulted only in the slightest gain of depth. After almost one hour, the *Thetis* had been unable to perform the simplest, slowest dive.

Checks were then initiated throughout the submarine, one of which involved the torpedo tubes. For this first dive, the "trim statement" specified that two of the six torpedo tubes, numbers Five and Six, be filled with water. As each tube could hold 100 gallons (about 800 pounds) of water, the absence of this ballast could make the difference in the *Thetis*'s ability to dive.

The torpedo tubes were 21 inches wide and fitted with a door at either end. At the outer end of the torpedo tube, on the exterior of the submarine, was the "bow cap," which normally was lifted only prior to firing a torpedo. The inner ("rear") door of the tube was opened from within the torpedo room. Obviously, once a submarine is underwater, the bow-cap and inner door of a torpedo tube should never be open simultaneously. Nonetheless, the *Thetis* was not equipped with a safety mechanism to preclude this from occurring.

Instead, two independent procedures were in place to verify that a torpedo tube was not uncapped or filled with water, and the inner door of a torpedo tube was not to be opened before these two tests were completed. The first procedure involved a small test cock. When a lever was shifted, two miniscule openings would align, forming a small gap in the inner door. If the tube was full of water, a tiny jet of water would stream out. If no water appeared, a small, needle-like instrument was poked into the hole to ensure that the desired gap had been formed. The second procedure required that a drain valve be opened. If the torpedo tube was full, its water would begin to drain into the room. In addition, each torpedo tube had a bow-cap indicator that pointed to "Open" or "Shut."

Before the torpedo officer, Lieutenant Fred Woods, checked to see if any the tubes were full, he satisfied himself that all the bow-cap indicators indicated the "Shut" position. He then opened the small test cock, tube by tube, before unlatching the inner doors. In this process, however, Wood neglected the supplemental "needle" test and skipped the back-up procedure of testing the drain valve. When, after some difficulty, he opened the inner door to torpedo tube number Five, the door crashed out of his hands as sea-water poured into the *Thetis*.

As Woods screamed on the intercom to "Blow the main ballast!", the torpedo room crew scrambled into the next compartment and struggled to close its watertight door. This door, however, was fitted with 18 butterfly bolts, all of which had to be tightened individually. When a bolt jammed and the lights went out, the attempt had to be abandoned.

The *Thetis* at this point was twenty feet below surface. The resulting water pressure, combined with the force created by the forward-moving submarine, caused the sea-water to gush through the 21-inch opening at a tremendous rate. Shutting the door against this flood was impossible, and almost immediately water began filling the torpedo room and spilling into the next compartment.

A submarine can surface with one compartment flooded, but not with two. As Woods screamed on the intercom to "Blow the main ballast!" (force air into the main ballast tanks, expelling the heavy sea-water), the torpedo room crew scrambled into the next compartment and struggled to close its watertight door. This door, however, was fitted with 18 butterfly bolts, all of which had to be tightened individually. When a bolt jammed and the lights went out, the attempt had to be abandoned.

There was an additional pressure on the torpedo room crew. If any sea-water spilled into the next compartment, which contained the submarine's batteries, the water would react with the batteries' sulfuric acid and form clouds of chlorine gas. Any submariner would be aware of the horrors of choking on chlorine gas. Conscious of this danger, the men retreated into the third compartment and quickly shut its single-control, watertight door.

The *Thetis* all the while was plunging downward. If at one moment the crew could not force the *Thetis* to submerge, they were now powerless to stop its dive. The submarine soon hit the muddy bottom of Liverpool Bay, and stuck at an angle of between 30 and 45 degrees. It was now around 3:00 P.M. The submarine was angled in 160 feet of water, out of the normal sea lanes, and accompanied only by a tugboat whose crew had no reason to suspect that an accident had occurred. The *Thetis*, after all, was supposed to submerge.

There was, at most, 24 hours's worth of air in the overcrowded *Thetis*. Commander Bolus futilely tried to use engine power to free the submarine from the mud. Three individual escape attempts were then made using the DSEA (Davis Submarine Escape Apparatus), an oxygen bag worn on the chest and connected to a face mask by a breathing tube. Each volunteer who tried to ascend to the surface via the flooded escape chamber, however, was turned back by unbearable water pressure. At around 7:00 P.M., these attempts were abandoned.

Bolus then ordered the fuel and fresh water carried on the *Thetis* to be expelled. This measure lightened the stern and achieved its purpose; the stern rose toward the surface of the water, though the submarine's flooded bow remained fixed in the mud.

Because of a series of misunderstandings, the *Grebecock* did not initiate a search for the *Thetis* until 6:22 P.M. Further mistakes, and the arrival of darkness, prevented the *Grebecock*'s crew from locating the submarine. It was not until nearly 8:00 A.M. on the next day that a destroyer noticed Bolus's desperate signal—the pointed black hull of the *Thetis*, angling 18 feet out of the water.

By this time, conditions below had deteriorated. Even those in the best physical condition were suffering from severe nausea, sharp headaches, and blurred vision. Every breath and move became painful and difficult. With the submarine's two-man escape chamber now nearer the surface, Bolus resumed attempts to send men up using the DSEAs. At about 8:00 A.M., two men made it to the surface and were rescued. Two more men emerged about two hours later, but none followed. Conditions below were presumably so bad that no one was able to think or act clearly, and the escape-chamber hatch became jammed.

The surface vessels that had arrived at the scene of the accident were handicapped by their lack of rescue equipment and inexperience with submarines. Attempts were made, using a light wire cable, to pull the *Thetis* out of the mud. A single, inexperienced diver was sent down with an air hose, predictably to no avail. The surface crews even tried to cut through the stern with blow torches. When the tugging on the submarine finally dislodged it, around 3:00 P.M., the *Thetis* went under for the last time, eventually killing any of the crew or passengers who remained alive.

Impact

Four and a half months later, the *Thetis* was salvaged. With the aid of the few survivor's testimony, investigators were able to reconstruct exactly what had happened. The sinking was attributed to human error, since the torpedo officer had not fully followed the safety procedures.

The investigators also discovered, however, that it was in fact very difficult for the torpedo officer to do his job correctly. All six of the bow-cap indicators pointed in different directions to indicate "Shut," which must have initially misled Lieutenant Woods; they were also out of sequence in relation to the torpedo tubes. Tube number Five, the source of the disaster, had its test cock painted over on the inside, disabling the single safety test that the torpedo officer relied on. Finally, the bow-cap to this torpedo tube, against procedures, had been open from the beginning.

The sinking of the *Thetis* resulted in the British Admiralty introducing several design improvements. All submarine torpedo doors were fitted with "Thetis clips" that prevented the inner door from opening more than a fraction if the bow-cap was lifted. Quick-closing, single-action doors were installed between the submarine compartments, replacing the old-fashioned 18-bolt system.

The Admiralty also learned how quickly the air in a submarine can deteriorate, making the prompt deployment of trained divers with adequate equipment essential to rescue operations. The fact that 99 men suffered horribly and died in a submarine that was separated from help only by a few feet of water above the escape hatch and five-eighths inch of steel on the hull justifiably drew widespread criticism of the military bureaucracy responsible for the *Thetis* disaster.

LEONARD C. BRUNO

Where to Learn More

Ackerman, Paul. *Encyclopaedia of British Submarines: 1901-1955.* Anthony Rowe, 1989.

"Tragedy Below: Toll of 99 Britons on *Thetis* the Worst Submarine Disaster." *Newsweek,* June 12, 1939, pp. 22-23.

"*Thetis* Valve Failed, Survivor Testified." *New York Times,* July 6, 1939, p. 7.

Stafford, Edward Peary. *The Far and the Deep.* G. P. Putnam's, 1967.

Warren, C. E. T. and James Benson. *Only Four Escaped.* William Sloane Associates, 1959.

Thresher sinking

Atlantic Ocean

(1963)

A nuclear submarine plunged beneath its "crush depth" and sank, killing all 129 crew members.

Background

On April 10, 1963, an advanced American nuclear-powered submarine, the *Thresher*, attempted to dive to its maximum operating depth of 10,000 feet. At 9:13 A.M., communication from the *Thresher* to the accompanying surface vessel became erratic. At 9:17 A.M., the surface vessel heard the words "test depth" and then a muted thud. The entire crew of 129 men died in an accident whose cause remains undetermined. Many experts believe that carelessness and rushed construction resulted in a pipe failure and possibly a power loss, leaving the *Thresher* unable to prevent a descent to fatal depths.

The USS *Thresher*, with a price tag of $45 million, was the first of an advanced class of attack nuclear-powered submarines. These were to be true submarines, designed to spend prolonged periods of time in the depths of the sea. The *Thresher* was powered by a Westinghouse S5W nuclear reactor, which through a steam turbine drove the submarine's single propeller. The *Thresher* also had a small backup diesel-electric power plant. This class of submarine would eventually carry nuclear-tipped underwater-to-underwater Subroc missiles.

The *Thresher* featured an innovative tear-shaped hull which, though very efficient underwater, was primarily designed around the submarine's state-of-the-art sonar equipment, designated as BQQ-2. Setting this sonar equipment on the specially shaped hull left it as removed as possible from the interfering noise of the submarine's machinery and propeller.

Submerged, the *Thresher* could achieve speeds of 30 knots or better, and it could go deeper than any previous submarine. At 278 feet long and with a beam of 21 feet, its displacement was 3700 tons. Built at the Portsmouth Naval Shipyard in New Hampshire, the *Thresher*'s keel was laid on May 28, 1958, and the submarine was launched on July 9, 1960.

The most striking advance of the nuclear-powered *Thresher* over its diesel-engine predecessors was in underwater endurance. Diesel-driven submarines stored energy in batteries, which then propelled the submarine through electric motors. Since the energy storage capacity of batteries is limited, and since diesel engines required oxygen, a conventional submarine could stay submerged for only a moderate period of time. Nuclear power, however, not only required no oxygen, it could through electrolysis produce oxygen from water.

During its sea trials, nonetheless, the *Thresher* was subject to a number of problems. On April 31, 1961, after checking the submarine's trim (its buoyancy and ability to stay balanced) during fairly shallow dives, Commander Dean L. Axene ordered a deeper dive to be attempted. As the *Thresher* descended, its instruments suddenly began indicating that the submarine was near the pressure limit that the hull could withstand, although all other indicators confirmed that the submarine was not even approaching its maximum operating depth. Axene aborted the dive and returned the submarine to the surface. At the shipyard, the instruments were checked and found to have been faulty.

The submarine then resumed its sea trials. Throughout this trial period, the *Thresher* would often make brief stops at port to conduct routine maintenance and to rest its crew. In November 1961, at one such stop in San Juan, Puerto Rico, the *Thresher* underwent its routine of shutting down the nuclear reactor and switching to diesel power. After about eight hours, the diesel engines stalled, and the submarine switched to backup battery power. Once it was recognized that the diesel repairs would be time-consuming, and that the backup battery power would be inadequate, the nuclear power system was again activated, although this process would itself take several hours and risked overheating the reactor. During that time, the crew was forced to work at repairing the diesel engines in near-darkness. Having lost ventilation, the temperatures inside the submarine reached 140 degrees Fahrenheit before the diesel engines were repaired and the situation was brought under control.

The *Thresher* continued with the sea trials until it returned to the shipyard for Christmas. The trials resumed in February 1962, with the *Thresher* undergoing shock tests to determine its ability to withstand underwater explosions. On June 3, off Cape Canaveral, Florida, the submarine was the victim of an unscheduled blow: a tugboat rammed into it accidentally. A three-foot gash in the submarine's port side resulted. After being repaired at the shipyard, the *Thresher* was subjected to two more weeks of pounding by underwater blasts. Although it suffered some damage, the submarine passed all the shock tests. The *Thresher* then went into drydock for nine months for a complete overhaul.

On April 9, 1963, following this overhaul at Portsmouth, the *Thresher* went out to sea for a series of shallow test dives under its new skipper, Lieutenant Commander John W. Harvey. After these trials, the submarine crossed the Georges Bank area of the Atlantic to get beyond the continental shelf and into truly deep waters.

The nuclear-powered Thresher *before its 1963 sinking. The sub was 278 feet long with a beam of 21 feet.*

Details of the Disaster

On the morning of April 10, now in water between 7,800 and 8,500 feet deep, the *Thresher* began a depth dive. The *Thresher* was accompanied by the submarine rescue ship, the *Skylark*. In coastal waters, the *Skylark* could attempt a crew rescue. In these deeper waters off the continental shelf, however, it could provide only a communication link and navigation checks.

At 7:52 A.M., the *Thresher* was 400 feet down. By 8:53 A.M., it had approached its test depth of nearly 10,000 feet. Communication between the two vessels had been maintained up to this point. Starting at 9:13 A.M., however, communication from the *Thresher* became garbled and sporadic. Sometime between 9:14 and 9:17 A.M., disaster struck. During those minutes, the communications personnel on the *Skylark* heard the words "attempting to blow" and—just prior to a muted, dull thud—the words "test depth."

For one hour and forty minutes thereafter, the *Skylark* circled the area, continuously attempting to reestablish communications. It dropped hand grenades, a signal for a sub to surface, every ten minutes. There was no response from the *Thresher*.

Before dusk, another submarine rescue ship sent to the area spotted an oil slick and a few bits of cork and yellow plastic. Within a few hours, it became apparent that the *Thresher* must have exceeded its design depth and imploded. The Chief of Naval Operations notified the crew's families, scheduled a press conference to announce the loss, and prepared to mount the greatest search in United States naval history.

Impact

Among the scores of vessels used by the Navy to try to locate the lost submarine was the bathyscaphe *Trieste*. This submersible had recorded the deepest dive on record, and it was the only vessel capable of going down to the depths where the wreckage of the *Thresher* was thought to be located. The *Trieste* was a vertical-movement vehicle with minimal maneuverability.

The trailing cameras of the Navy's new research ship, the *Atlantis*, had photographed a mass of debris over an area of approximately 1,000 by 4,000 yards. The *Trieste* made its first dive to that spot on June 24, 1963. Several dives followed over the next two months, until on August 29 the *Trieste* spotted a large quantity of fairly large pieces of metal. Using its newly installed mechanical arm, it picked up a piece of twisted and mangled brass pipe. This piece of pipe was positively identified as part of the *Thresher*, since it carried its part number, job-order number, and the numerical designation of the *Thresher*, "593."

The only task remaining was for the Court of Inquiry to make its determination. Investigators quickly ruled out any enemy involvement, and then ruled it "inconceivable" that human error had caused the disaster. After accumulating 1,700 pages of testimony from 120 witnesses, the determination came that the "most likely" cause was a cracked fitting or faulty joint in a cooling pipe. Such a break would fill a small compartment with sea-water in seconds, possibly shorting out a key control panel or power circuit. This would have caused the nuclear reactor to shut down, resulting in the loss of propulsion power. As the ship filled with water, its auxiliary diesel power would be insufficient to counteract its plunge. Moreover, the system to expel sea-water ballast could not function at these depths due to the tremendous counterpressure. Under the increasing water pressure, the *Thresher*'s hull would have twisted and rippled like rubber before the submarine imploded.

Although most experts find this hypothesis the most reasonable, two additional explanations have been proposed. The first is that an open sea valve may have been to blame. Openings in the *Thresher*'s hull allowed water to be taken in for the nuclear power cooling system. If the submarine's electrical system had shut down because of a leak or any other reason, the valve could not have been closed electrically, leaving sea-water to flow in. A second alternative explanation is that the *Thresher* may have encountered a mammoth underwater wave. A large storm moved across the Gulf of Maine on April 8, 1963. This storm could have generated a sub-surface eddy that would raise an underwater wave of up to 300 feet in height. Such a subsurface wave may have driven the submarine far below its crush depth.

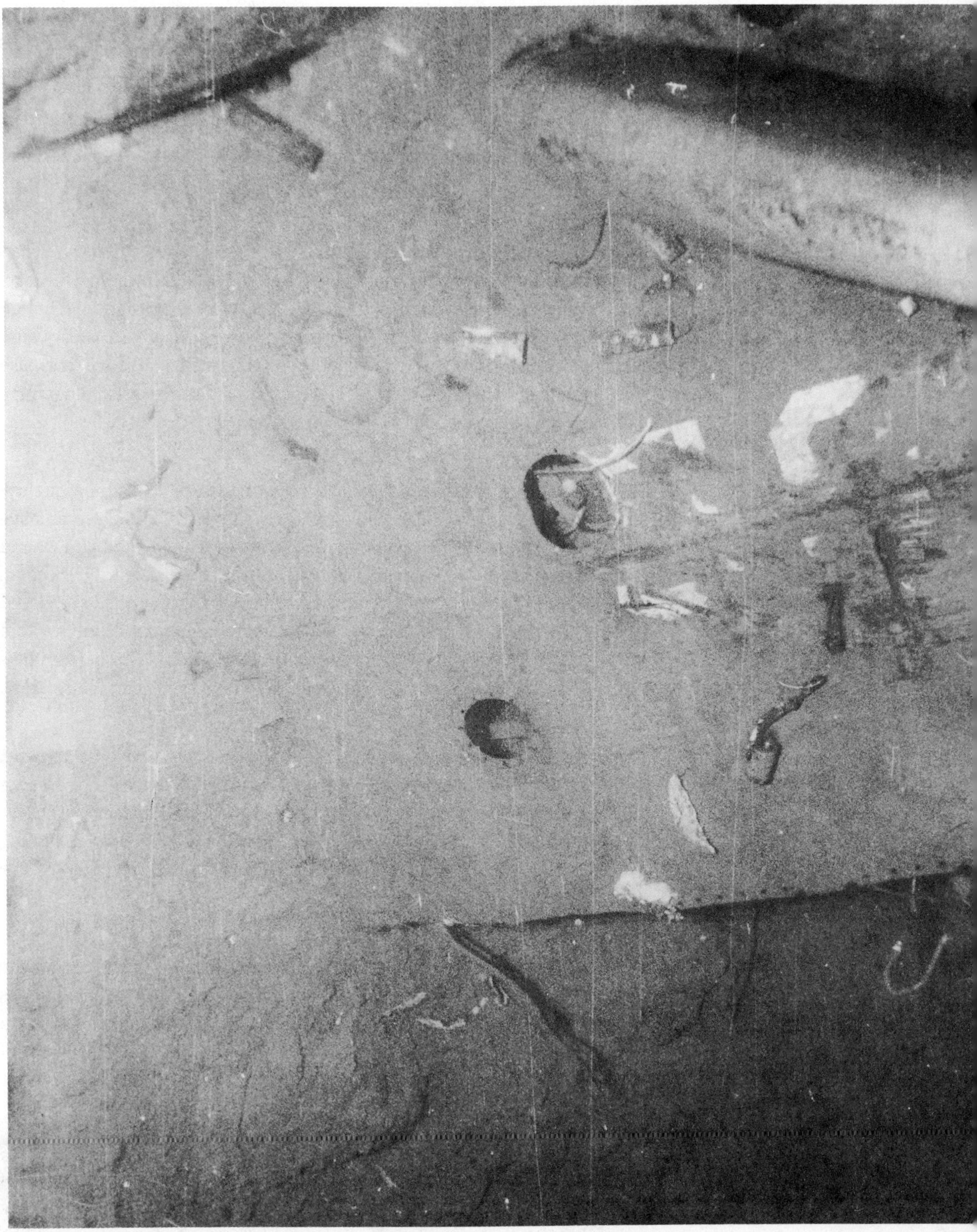

The wreckage of the Thresher *was photographed in 1964. The hull number, "593," is partially visible at right, slightly above center.*

During the Court of Inquiry's own investigation, a number of unsettling facts about the rushed construction history of the *Thresher* came to light. Its periscope control had been installed backwards. A spike was found in its air circulation system. For at least 20 percent of its hydraulic valves, the "open" and "close" indicators were reversed. Furthermore, tests on 145 of its pipe joints revealed that 14 percent were below standard. Yet the submarine was sent to sea with 2,855 other joints untested.

Although investigators ruled out human error as the source of the disaster, they did conclude, from this construction and drydock record, that "practices, conditions and standards existing at the time were short of those required to insure safe operation." The *Thresher* tragedy, although its exact cause remains undetermined, may trace back to this carelessness. The Court also acknowledged that the Navy's deployment of new design technology may have outpaced mastery of its practical application.

The *Thresher* disaster did lead to reform in Navy policies. First, a new quality-assurance program was implemented at naval shipyards. Fitters would have to fill out and sign a card verifying they had brazed each joint. Ultrasound replaced x-rays to inspect fittings and verify their integrity. Second, redundant hydraulic systems were installed that enabled valves to be closed without the use of electrical power. Finally, submarines were fitted with a new emergency "blow" system that introduced air directly into ballast tanks at a rate seven times faster than could the equivalent system on the *Thresher*.

The Navy also prioritized the development of highly maneuverable deep-sea submersibles. This resulted in the DSRV-1 (Deep Submergence Rescue Vehicle), which can rescue submarine crews at down to 3,500 feet; the DSRV-2, which can operate at depths to 5,000 feet; and the nuclear-powered NR-1, which operates at depths to 3,000 feet and can remain submerged for extended periods of time.

The Navy was subject to considerable political criticism for its handling of the *Thresher* disaster. The decision to classify all twelve volumes of its inquiry as secret meant that its determination could not be evaluated independently. In addition, Navy officials appeared uncooperative during the congressional hearings held by the Joint Committee on Atomic Energy.

The Joint Committee was justifiably concerned: a nuclear reactor now rests on the ocean floor 220 miles off the coast of Boston. Most experts believe that since the *Thresher* had undergone an extensive overhaul, possibly including renewal of the reactor core, the fission product content of its core would be low enough to render it harmless.

LEONARD C. BRUNO

Where to Learn More

Bentley, John. The *Thresher* Disaster. Doubleday, 1974.

Colley, David P. "The Lessons Learned from SSN 593." *Mechanical Engineering,* February 1987, pp. 54-59.

Gannon, Robert. "What Really Happened to the *Thresher.*" *Popular Science,* February 1964, pp. 102-109, 208-209.

Polmar, Norman. Death of the *Thresher.* Chilton, 1964.

"*Thresher* Post-Mortem." *Scientific American,* November 1963, pp. 66-68.

Scorpion sinking

Atlantic Ocean

(1968)

Unexplained sinking of submarine gave impetus to development of deep-sea rescue and search submersibles and raised concerns about safety of powering submarines with nuclear reactors.

Background

On May 27, 1968, the *Scorpion*, an American nuclear-powered submarine with a crew of 99, disappeared without a trace in the middle of the Atlantic Ocean. On October 28, 1968, following one of the most extensive searches in United States naval history, portions of the submarine were located about four hundred miles southwest of the Azores. While the naval Court of Inquiry failed to identify the cause of the *Scorpion* disaster, photographs of the wreckage rule out hostile action and indicate that the source of the accident was internal to the submarine.

The nuclear-powered *Scorpion* (SSN-589) was built in Groton, Connecticut, by the Electric Boat Division of the General Dynamics Corporation. The *Scorpion* was an attack submarine of the *Skipjack* class, designed primarily to hunt down and destroy enemy submarines. Accommodating 10 officers and 89 crewmen, it was over 251 feet long and 31 feet in beam, displacing 3,075 tons.

The *Scorpion* was armed with six torpedo tubes and powered by a water-cooled nuclear reactor. This reactor was a Westinghouse S5W with a thermal power of 70 megawatts. The S5W, combined with the submarine's Albacore-type hull shaped like an elongated teardrop, gave the *Scorpion* an exceptionally high submerged speed of 30 knots. The S5W propulsion plant was the workhorse of the Navy until the 1970s and featured extensive use of redundancy for its major components; if one component failed, another one could usually operate in its place. It had two steam generators, two pressurizers, two sets of turbines, and two turbo-generators. The sub also had conventional diesel engines and back-up batteries.

The *Scorpion* was launched on December 19, 1959, and joined the U.S. Atlantic Fleet during August 1960. It performed well during the next few years, and was overhauled from June 1963 to May 1964. Following another tour of duty, it drydocked at the Norfolk, Virginia, naval shipyard during February 1967 for another overhaul, which included refueling its nuclear reactor. During October of that year, the *Scorpion* began sea trials under a new skipper, Commander Francis A. Slattery. After a tour in the Mediterranean in early 1968, the sub left Gibraltar to return to the Norfolk naval shipyard.

Details of the Disaster

On May 21, 1968, during the voyage to Norfolk, the *Scorpion* reported its bearings south of the Azores, and noted that everything was operating normally.

The *Scorpion* did not subsequently communicate any sort of problem, but was never heard from again. On May 27, 1968, the submarine was reported overdue at Norfolk. On June 5, 1968, the Chief of Naval Operations announced it presumed lost. The Navy was never to learn any particulars about what actually transpired on the *Scorpion* prior to or during its demise.

Impact

Once the *Scorpion* was overdue, the Navy involved 55 ships and 23 aircraft in an extensive search. Eventually, experts identified wreckage of the submarine from photographs taken by the remote-controlled undersea cameras of the *Mizar*, a naval research ship.

Knowing where to search with these cameras was the result of sophisticated maritime detective work. If a *Skipjack*-class submarine went down in the deep Atlantic, it would descend past its crush depth of about 2,000 feet and implode. Navy antisubmarine warfare scientists knew that this event should have been picked up by the Navy's underwater listening system, called SOSUS. This consisted of an elaborate array of underwater devices that monitored the movements of foreign submarines.

A review of SOSUS tapes from the period that the *Scorpion* vanished revealed a peak noise event. Since the Navy had at least two listening devises picking up the noise, it was able to get rough bearings on the location of the event. Navy personnel then went to that area and detonated charges of TNT calibrated to generate acoustical energy comparable to what would have resulted from the *Scorpion*'s implosion. When the trace of these TNT noises on the SOSUS tapes matched that of the earlier event, the Navy dispatched the *Mizar*, with its trailing 35mm cameras, to photograph the area.

Several thousand undersea photographs provided views of the wreckage on the ocean floor. The bow appeared to have sunk partially into the sandy bottom. The submarine's "sail" or superstructure was lying intact more than 100 feet away; it had broken away from the body of the *Scorpion*. The nuclear reactor was not visible.

These photos and various forms of testimony were considered by the naval Court of Inquiry that had begun its work in June 1968. The seven-man court

The Scorpion, *shown in a photo taken about eight years before it sank in the Atlantic Ocean. The sub was over 251 feet long, with a beam of 31 feet.*

met intermittently for eleven weeks and heard ninety witnesses before issuing its final report on January 31, 1969. Nonetheless, the court submitted a mainly negative conclusion: "The certain cause of the loss of the *Scorpion* cannot be ascertained from evidence now available." The court discounted various potential causes, but it did not arrive at a hypothesis to explain the *Scorpion* disaster.

Before the court of inquiry received the *Mizar* photographs, it identified several possible causes for the *Scorpion*'s loss. One was that the submarine had collided with a seamount (an underwater mountain). Another was that the submarine had been the victim of hostile action by or collision with another vessel. A Soviet torpedo or depth charge, for example, could have sunk the *Scorpion*.

A popular rumor about the disaster was that the *Scorpion* was on a secret mission and had been lost in waters frequented by Soviet submarines and surface vessels. At this point in the Cold War, hardly a week seemed to pass without a clash or minor incident between Soviet and American ships. In fact, a crew member had written home two weeks before the sinking that the

Scorpion had been confronted by a Soviet destroyer with its guns trained on the American submarine.

The *Mizar* photographs, however, indicated to most observers that the origin of the *Scorpion* disaster was internal. After evaluating these photographs, the court eliminated both its seamount collision and hostile action/second vessel hypotheses. As one assessment of the disaster maintained, "had the *Scorpion* been hit by a torpedo or scraped by a surface ship while she was near the surface, this would have left identifiable damage."

Another possibility addressed by the court was that the *Scorpion*, prior to the accident, was unfit or damaged. The submarine, after all, was returning to the Norfolk shipyard. Testimony was given that the submarine had hairline cracks across its hull. Another witness stated that the submarine had a small oil leak around its propeller shaft. A crew member who was left ashore in Spain due to illness said that the *Scorpion* had been leaking hydraulic fluid around its periscope fitting, and that it was having trouble with its navigational equipment.

The *Scorpion* did report at least five mechanical problems before leaving Spain. Moreover, the submarine had collided with a barge during a storm in the Naples harbor one month before the *Scorpion*'s final disaster. This barge had been placed between the *Scorpion* and an American warship as a buffer, and the submarine's stern had taken the brunt of the impact. Divers made a partial inspection of the docked submarine but reported no significant damage. Finally, the Navy's own "Subsafe" program had not been completely implemented on the *Scorpion*. This program, initiated in response to the *Thresher* disaster in 1963, involved a series of safety modifications.

Even so, the court determined that the submarine's overall condition was "excellent" and that none of the pending repair work was of a nature to affect the *Scorpion*'s safety. It also concluded that the disaster could not be attributed to the delayed completion of the "Subsafe" program, and it ruled out the possibility that the *Scorpion*'s nuclear reactor had exploded. Such a catastrophe (like the relatively contained implosion that did occur) would have been detected.

The court also judged that the Scorpion's crew was well-trained and prepared for emergencies, and that it was improbable that any aberrant behavior on the part of the crew might have led to the disaster, noting that there was no indication that any officer or seaman was unstable or unreliable. The court therefore discounted the possibility that the loss of the *Scorpion* was caused by "the intent, fault, negligence or inefficiency of any person or persons in the naval service or connected therewith."

The court, however, to a degree refused to speculate on any such internal occurrences, and many outside observers hold that the court rejected this and other possible causes too readily. Among these potential causes is a control failure. The submarine, for example, could have been running very fast and very deep when its diving mechanism locked into the "dive" position. This would have sent it below its crush depth before correction was possible.

Once the political storm subsided, a logistical issue remained: the Navy had no submarine rescue capability for depths below 1,300 feet. After the 1963 loss of the Thresher, *the Navy had begun a research and development program aimed at producing deep-sea rescue vehicles. The loss of the* Scorpion *gave impetus to this program.*

Another possibility relates to the ever-present danger of leaks, particularly in light of the allegations about cracks in the *Scorpion*'s hull and a faulty propeller shaft. Under the severe pressure of great depths, such flaws could lead to a sudden, fatal breach. The malfunctioning of one of the submarine's own torpedos, perhaps involving a fire in the torpedo room, remains another possible cause of the disaster. The *Mizar* photographs do not rule this out, and some experts find this last hypothesis the most feasible. To date, nevertheless, no exact cause for the *Scorpion* disaster has been determined.

The loss of the *Scorpion* and the failure of the U.S. Navy to identify its cause generated a great deal of national controversy. Once the political storm subsided, a logistical issue remained: the Navy had no submarine rescue capability for depths below 1,300 feet. After the 1963 loss of the *Thresher*, the Navy had begun a research and development program aimed at producing deep-sea rescue vehicles. The loss of the *Scorpion* gave impetus to this program.

The Navy's first Deep Submergence Rescue Vehicle (DSRV), the DSRV-1, was launched in 1970. Shaped like a torpedo, with a fiberglass hull, this 50-foot vessel could operate at depths of up to 3,500 feet. The DSRV-1 could link to the submarine's escape hatch and take 24 men onboard to carry to the surface. The DSRV-2, built by Lockheed, was able to operate at up to 5,000 feet. Both submersibles could be transported by air or water, or carried on a nuclear submarine itself, which could launch and recover them without surfacing.

The Navy also built an experimental rescue craft called the NR-1. About 136 feet long, with an operating depth of up to 3,000 feet, the NR-1 looked like a small submarine. It had room for a crew of 7, and was powered by a small pressurized-water reactor. Its two propellers and tunnel thrusters made it highly maneuverable. Its special features included sonar for navigating and locating sea-floor objects, and a large recovery device.

Although none of these rescue submersibles have been required to aid sunken submarines, they have been used for comparable missions. In 1976, for example, the NR-1 helped in the rescue of an F-14 fighter that rolled off a Navy aircraft carrier and was sunken in 1,960 feet of water.

The possibility of nuclear contamination from the *Scorpion*'s reactor raises a final issue. The submarine's debris is scattered over a large area. As its hull was compressed from both ends during the disaster, it is not possible to assess the condition of the reactor compartment. However, the seabed around the wreckage of the *Scorpion* has been investigated three times to date. Samples show minor amounts of cobalt-60, which indicate that there is some leakage from the primary reactor system. In March 1975, Admiral Hyman G. Rickover reported to Congress that while the cause of the *Scorpion* disaster remained undetermined, there was still "no evidence that it was due to a problem with the nuclear reactors."

LEONARD C. BRUNO

Where to Learn More

Eriksen, Viking O. *Sunken Nuclear Submarines: A Threat to the Environment?* Norwegian University Press, 1990.

Middleton, Drew. *Submarine, the Ultimate Naval Weapon: Its Past, Present & Future.* Playboy Press, 1976.

"Presumed Lost." *Newsweek,* June 17, 1968, p. 74.

"Finding the *Scorpion.*" *Newsweek,* November 18, 1968, p. 104.

"Loss of *Scorpion* Baffles Inquiry." *New York Times,* February 1, 1969, p. 1.

"Sub Found—But Mystery Remains." *U. S. News & World Report,* November 11, 1968, p. 12.

Bibliography

Barclay, Stephen. *The Search for Air Safety: An International Documentary Report on the Investigation of Commercial Aviation Accidents*. William Morrow, 1970.

Bignell, Victor, Geoff Peters, and Christopher Pym. *Catastrophic Failures*. Open University Press, 1977.

Bishop, R. E. D. *Vibration*. The University Press, 1965.

Blockley, D. I. *The Nature of Structural Design and Safety*. Ellis Horwood, 1980.

Canning, John, ed. *Great Disasters*. Longmeadow Press, 1976.

Collins, J. A. *Failure of Materials in Mechanical Design: Analysis, Prediction, Prevention*. John Wiley & Sons, 1981.

Cornell, James. *The Great International Disaster Book*. Scribner's, 1976.

Davis, Lee. *Man-Made Catastrophes: From the Burning of Rome to the Lockerbie Crash*. Facts On File, 1993.

Ebert, Charles H. V. *Disasters: Violence of Nature, Threats by Man*. Kendall/Hunt, 1988.

Eddy, Paul, Elaine Potter, and Bruce Page. *Destination Disaster: From the Tri-Motor to the DC-10; The Risk of Flying*. Quadrangle, 1976.

Editors of Encyclopaedia Britannica. *Catastrophe! When Man Loses Control*. Bantam/Britannica Books, 1979.

Feld, Jacob. *Lessons from Failures of Concrete Structures*. American Concrete Institute, 1964.

———. *Construction Failures*. John Wiley & Sons, 1968.

Florman, Samuel C. *Blaming Technology: The Irrational Search for Scapegoats*. St. Martin's Press, 1981.

Ford, Daniel. *O-Rings and Nuclear Plant Safety: A Technical Evaluation*. Public Citizen, Critical Mass Energy Project, 1986.

Frank, Beryl. *Great Disasters of the World*. Galahad Books, 1981.

Godfrey, Edward. *Engineering Failures and Their Lessons.* [privately printed] 1924.

Godson, John. *Unsafe at Any Height.* Simon & Schuster, 1970.

Gordon, J. E. *The New Science of Strong Materials: Or Why You Don't Fall Through the Floor,* 2nd ed., Penguin Books, 1976.

———. *Structures: Or Why Things Don't Fall Down.* Da Capo Press, 1981.

Great Britain Navy Department Advisory Committee on Structural Steels. *Brittle Fracture in Steel Structures.* Butterworths, 1970.

Guide to Investigations of Structural Failures. Federal Highway Administration, 1980.

Hammond, Rolt. *Engineering Structural Failures: The Causes and Results of Failure in Modern Structures of Various Types.* Odhams Press, 1956.

Hertzberg, R. W. *Deformation and Fracture Mechanics of Engineering Materials.* John Wiley & Sons, 1976.

Janney, Jack R. *Guide to Investigation of Structural Failures.* American Society of Civil Engineers, 1979.

Keylin, Arleen and Gene Brown. *Disasters: From the Pages of the New York Times.* Ayer, 1976.

Kletz, Trevor A. *What Went Wrong? Case Histories of Process Plant Disasters.* Gulf Publishing, 1985.

Launay, A. J. *Historic Air Disasters.* Ian Allan, 1967.

LePatner, Barry B., and Sidney M. Johnson. *Structural and Foundation Failures: A Casebook for Architects, Engineers, and Lawyers.* McGraw-Hill, 1982.

Lewis, Elmer Eugene. *Nuclear Power Reactor Safety.* John Wiley & Sons, 1977.

Mair, George. *Bridge Down: A True Story.* Stein & Day, 1982.

McClement, Fred. *It Doesn't Matter Where You Sit.* Holt, Rinehart & Winston, 1969.

———. *Jet Roulette: Flying Is a Game of Chance.* Doubleday, 1978.

McKaig, Thomas K. *Building Failures: Case Studies in Construction and Design.* McGraw-Hill, 1962.

Nash, Jay Robert. *Darkest Hours.* Nelson-Hall, 1976.

Oberg, James E. *Uncovering Soviet Disasters.* Random House, 1988.

Osgood, Carl C. *Fatigue Design,* 2nd ed., Pergamon Press, 1982.

Perrow, Charles. *Normal Accidents: Living with High-Risk Technologies.* Basic Books, 1984.

Petroski, Henry. *To Engineer Is Human.* St. Martin's Press, 1985.

Ross, Steven S. *Construction Disasters: Design Failures, Causes, and Prevention.* McGraw-Hill, 1984.

Salvadori, Mario. *Why Buildings Stand Up: The Strength of Architecture.* McGraw-Hill, 1982.

Salvadori, Mario and Matthys Levy. *Why Buildings Fall Down: How Structures Fail.* Norton, 1992.

Serling, Robert J. *Loud and Clear: The Full Answer to Aviation's Vital Question, Are the Jets Really Safe*? Doubleday, 1969.

Stewart, Oliver. *Danger in the Air*. Philosophical Library, 1958.

Turner, Barry A. *Man-Made Disasters*. Crane, Russak & Co., 1978.

U.S. House of Representatives Committee on Science and Technology. *Structural Failures: Hearings Before the Subcommittee on Investigations and Oversight*. Government Printing Office, 1983.

———. *Structural Failures in Public Facilities*. Government Printing Office, 1984.

Webb, Richard E. *The Accident Hazards of Nuclear Power Plants*. University of Massachusetts Press, 1976.

Whyte, R. R., ed. *Engineering Progress Through Trouble*. Institution of Mechanical Engineers, 1975.

Index